AF698753

Perceptual Learning

Perceptual Learning

How Experience Shapes Visual Perception

Barbara Dosher and Zhong-Lin Lu

The MIT Press
Cambridge, Massachusetts
London, England

© 2020 Massachusetts Institute of Technology

All rights reserved. No part of this book may be reproduced in any form by any electronic or mechanical means (including photocopying, recording, or information storage and retrieval) without permission in writing from the publisher.

This book was set in Times New Roman by Westchester Publishing Services. Printed and bound in the United States of America.

Library of Congress Cataloging-in-Publication Data

Names: Dosher, Barbara, 1951– author. | Lu, Zhong-Lin, author.
Title: Perceptual learning : how experience shapes visual perception / Barbara Dosher and Zhong-Lin Lu.
Description: Cambridge, Massachusetts : The MIT Press, [2020] | Includes bibliographical references and index.
Identifiers: LCCN 2020000423 | ISBN 9780262044561 (hardcover)
Subjects: LCSH: Perceptual learning.
Classification: LCC LB1067 .D67 2020 | DDC 370.15/5—dc23
LC record available at https://lccn.loc.gov/2020000423

10 9 8 7 6 5 4 3 2 1

Science is beautiful when it makes simple explanations of phenomena or connections between different observations.

—Stephen Hawking

All models are wrong, but some are useful.

—George Box

Contents

Preface

We started doing research in perceptual learning in 1997. At the time, only a handful of researchers were focused primarily on this topic. The field has transformed since then, and this book tells the story of what we came to know about both the phenomena and the theories. This transformation occurred because of the fantastic contributions of many active researchers, from sophisticated investigations of the phenomenology by psychophysicists to insightful modeling and physiology.

In the late 1990s, we were working on a new model of the human observer, the perceptual template model (PTM). Our intention was to use this model to understand how visual perception depended on signal patterns and two kinds of noise—the noise in the external stimulus and the variability in the internal sensory response. We were also interested in using this model to tease apart the effects of visual attention on human perception ("the observer") by distinguishing improvements due to filtering out external noise in the stimulus from enhancing or amplifying the signal stimulus itself—previously elusive mechanisms easily distinguished with external noise methods.

At some point, we realized that the same analysis applied equally well to a major field of performance improvement—perceptual learning. Improvements with practice had been reported from the very beginning of experimental psychology in the late 1890s and were popularized in the 1950s by Eleanor Gibson as part of her interest in the early perceptual development of children. The role of experience in the performance of perceptual tasks in adults had been documented in numerous task domains, including acuity, motion, and stereopsis. Some of the best psychophysicists in the field of visual perception had studied learning, and sometimes the specificity of that learning to some aspect(s) of the task or the stimuli.

Then, in the late 1990s, prominent work by a number of scientists (Avi Karni, Dov Sagi, Merav Ahissar, Shaul Hochstein, Aniek Schoups, Robert Sekuler, and others) demonstrated a very curious form of specificity. Learned improvements in a task practiced in one location on the retina sometimes failed to transfer to new locations in the visual field. Specificity indeed! These observations led many researchers to attribute experience-dependent changes in performance to plasticity in the early visual cortex, a brain area long thought to be stable after the early years of development. Soon, the most prominent theory of perceptual learning involved plastic alterations of the sensory tuning in the early retinotopic visual cortex. Bolstered by similar reports in other modalities, a significant set of studies (by Rufin Vogel, Guy Orban, Geoff Ghose, John Maunsell, Charles Gilbert, Joshua Gold, Wu Li, and many others) began to investigate how learning affected the properties of cellular responses in the earliest levels of visual coding. How early in the visual cortex did learning reach? We have been avid followers of these physiological investigations.

Our first study of perceptual learning was a systematic analysis of the phenomenon using external noise methods and the PTM model. From the very beginning, we suspected that the dominant retuning theory of perceptual learning could only be one part of the picture; in order to influence behavior, sensory information must also be connected to decision. If the sensory system encoded the stimulus, this evidence also needed to be decoded. Even at this early stage, we developed an alternative reweighting theory in which changing how sensory information is weighted in a decision (changed readout) was perhaps the dominant mode of learning. If the early visual areas were the encoders of sensory information, then the brain also needed decoders to interpret the encoded information, and these decoders must also be key to learning. Based on this insight, we developed a reweighting (readout) theory in which evidence in many early visual channels determined how a decision was changed through reweighting. This was in 1998. Not until later did we realize that Mollon and Danilova had developed the same theoretical idea independently.

It wasn't until a few years later, with the help of a gifted postdoc, Alex Petrov, that we started work on a multichannel model of perceptual learning, the augmented Hebbian reweighting model (AHRM). This model built on network models of visual learning (by Tomaso Poggio, Shimon Edelman, Manfred Fahle, Michael Herzog, and others) from the 1990s and took advantage of significant recent developments in the field of neural networks. We joined this model to a physiologically inspired signal-processing front end. Experiments also became more complex in order to examine specificity in those situations where the two major learning theories (retuning and reweighting) made contrasting predictions. This pure reweighting model has subsequently been shown to be able to account for many of the major phenomena in visual perceptual learning. With another talented postdoc, Jiajuan Liu, and with insightful experimental work by graduate student Pam Jeter, the AHRM was extended in 2013 to form the integrated reweighting theory (IRT). This theory explains how certain forms of transfer occur when they do. This model has in turn been modified and taken further by other researchers (Aaron Seitz, Peggy Seriès, and others) in very clever ways. It is this story of models of perceptual learning that we tell in the Models section of this book.

Over the past 20 years, the field of perceptual learning has evolved significantly. There are now many studies that challenge the specificity of perceptual learning (by Cong Yu and others). Now, the ideas about the role of reweighting, or readout, in learning have become a prominent component of the integrated models by Takeo Watanabe and others that have positioned the field of perceptual learning within the broader considerations of human brain imaging. Models other than our own have either used or advanced the principle of multilevel reweighting in learning that we put forward. Meanwhile, the study of learning has increasingly made its way toward a substantial set of practical applications, from education to visual remediation, thanks to the laboratories of Michael Murzenich, Dennis Levi, Krystel Huxlin, John Anderson, Chang-Bing Huang, Uri Polat, Robert Hess, Ben Thompson, and many others.

By the mid-2010s, the field had reached a point where a systemic exploration of its recent development seemed to be called for. We began a detailed survey of the sometimes disparate literature on perceptual learning. Our goal was to evaluate the state of the various theories, understand the implications of the findings in physiology, and point toward possible fruitful directions for research. This book is the result of our efforts. It is meant for researchers of perceptual learning as well as scientists from other related fields. We have tried to discuss perceptual learning at several levels, hoping to be thorough yet concise, inclusive but not exhaustive.

Over the years, the many advances in the field, and our work in particular, were moved forward by key contributions from many prominent scientists in perceptual learning. In addition to reading papers and attending conference presentations, we have especially benefited from attendance at the International Workshop in Perceptual Learning, a smaller group that meets every two years to talk through exciting new ideas and methods. We certainly have personally and individually benefited from these open exchanges. One of the remarkable things about the study of perceptual learning—despite, or perhaps even because of, the highly divergent theoretical views of the participants—is that the field has remained collegial, cooperative, and committed to working together to advance the science.

These major contributions of many individuals and laboratories have now established perceptual learning as a major field, with a wide recognition that learning and plasticity play critical roles in human adaptation and must be accounted for in any investigation of the mechanisms of perception or, for that matter, learning.

Writing this book has taken us a number of years for a variety of reasons. During this period, we have had the support, both intellectual and personal, of many people, from colleagues and friends to family, who we wish to thank profusely.

Barbara Dosher comments, after 23 years of collaboration, it is still a pleasure to talk about science with my coauthor and friend Zhong-Lin Lu. The stimulating environment of the Cognitive Sciences Department at the University of California, Irvine, with smart and supportive colleagues, has contributed materially to work on this book. I am also thankful for the significant contributions of graduate students and postdocs in my laboratory (the Memory Attention Perception Lab), including Pam Jeter, Alex Petrov, Wilson Chu, Shiau-Hua Liu, Nate Blair, Richard Hetley, Emelien Tlapale, and many undergraduate students and assistants, including Anchit Roy, who assisted in the production of some figures for this book. I am still inspired by the theoretical vision of my graduate advisor, Wayne Wickelgren. I am also thankful to the colleagues and friends who first introduced me to vision science:

Norma Graham, George Sperling, and Eileen Kowler. This research was enriched by conversations with many colleagues in the perceptual learning and the attention communities. Friends and family have, in turn, offered fun, support, and accommodations. I am particularly grateful for the support of my friends Kristi, Liz and David, and Eileen, my sister Cathie, my retreat group, and many others. A special heartfelt acknowledgment goes out to my son Joshua Sperling, whose support, intellectual probing, and comments led to material improvements in the book in the closing stages. I am also thankful for the forbearance of all involved, including MIT Press, as the project extended longer than expected in part because of the calls of my aging and ill mother, Anne Dosher. Her lifelong intellectual curiosity, commitment to community, and gift for friendship has set an admirable example.

Zhong-Lin Lu comments, the publication of this book is a major milestone in our more than 20 years of collaboration. It is also the beginning of many new, exciting joint research projects. I really appreciate Barbara's friendship and the opportunity to work with her. Many thanks go to my wife, Wei Sun; son, James; and daughter, Mae. This book would not have been possible without their understanding and support. I have learned so much from my PhD advisors Samuel J. Williamson and Lloyd Kaufman, and my postdoc advisor, George Sperling. Their scientific spirit still inspires me today. For their contributions to the research program, I want to take this opportunity to thank members of the Laboratory of Brain ProcessES (LOBES), originally at the University of Southern California, then the Ohio State University, and now at NYU Shanghai and New York University, including Luis A. Lesmes, Wilson Chu, Simon Jeon, Debbie Dao, Jiajuan Liu, Chang-Bing Huang, Xiangrui Li, Gui Xue, Miao Wei, Fang Hou, Jongsoo Kim, Carlos Cabrera, Zhicheng Lin, Yukai Zhao, Pan Zhang, and Pengjing Xu. I would also like to thank C. Shawn Green, Daphne Bavelier, Alex Pouget, Tianmiao Hua, and Jinrong Li for collaborating with me on perceptual learning, and I thank organizers of the International Workshop on Perceptual Learning, Cong Yu, Dov Sagi, Takeo Watanabe, Merav Ahissar, Uri Polat, Shaul Hochstein, Yuka Sasaki, Mitsuo Kawato, Michael Herzog, Miguel Eckstein, Aaron Seitz, and Michael Silverman, for providing the opportunities for us to come together as a research community. I would also like to acknowledge my colleagues Irving Biederman, Adrian Raine, Antonio Damasio, Hanna Damasio, Michel Baudry, Judith Hirsh, James Todd, Roger Ratcliff, Brandon Turner, Deyue Yu, Alex Petrov, Tony Movshon, Paul Glimcher, Marisa Carrasco, Joanna Waley-Cohen, and Anqi Qian, as well as friends and colleagues I lost over the years: Patrick Suppes, Bosco Tjan, Stephen Madigan, and Richard F. Thompson.

Our experimental and theoretical work in perceptual learning has been supported over the years by grants from the National Institutes of Mental Health and the National Eye Institute. We appreciate this support. We are also grateful for the assistance of Robert Pryor, Anne Marie Bono, and Sherry Gerstein in the preparation of the book by MIT Press.

Barbara Dosher, Irvine, CA
Zhong-Lin Lu, Dublin, OH and New York, NY
August 2019

I

Overview

Figure 1.1
The human perceptual system uses all perceptual senses as the interface to a complex world. From www.freeimages.com (#1240544). (See plate 1.)

improve with training or practice. Major league baseball players,[1] expert billiard players,[2] and expert pilots[3, 4] are all better able to scan and process visual scenes than amateurs are. Although natural variance in ability is an important factor, expertise in sports or games will almost always require thousands of hours of practice and exposure to train specific skills. Baseball players are likely to be especially sensitive to motion cues;[5] expert video gamers are often able to rapidly detect elements in their visual periphery;[6] and avid bird-watchers are especially attuned to extracting texture from camouflage.[7] In all these domains, perceptual training is a primary avenue toward expert levels of performance.

This turns out to be true not only for advanced tasks but also for simple ones. By the time we are adults, we can see a few photons of light energy in the dark[8, 9] and hear a sound that displaces the cochlear membrane of the ear by the diameter of a hydrogen atom.[10, 11] We can smell a drop of perfume in a large room,[12, 13] taste a single teaspoon of sugar dissolved in several gallons of water,[14–16] and feel the lightest touch of a feather.[17] Our senses are surprisingly acute. Yet, in all these examples, our sensitivity can be improved through training. Extended practice can push these limits even further.

In the laboratory, most research has focused on perceptual tasks that fall somewhere between very precise judgments of minimal stimuli and the intertwined complexity of

1

Principles of Perceptual Learning

Experience plays a fundamental role in perception. The importance of domain-specific training to the development of perceptual expertise is immense, but while in principle the plasticity underlying such changes may alter the responses of the visual system at the earliest levels, overall system stability must also be taken into account. In this chapter, we present a synthetic framework of perceptual learning that balances plasticity and stability alongside a number of other dipoles: signal and noise, readout and encoding, and top-down as well as stimulus-driven factors. Our thesis is that reweighting, or changing the readout of sensory information, is the most likely candidate mechanism for optimizing learning across these many dimensions.

1.1 The Importance of Experience and Learning in Perception

Human perception is a necessary gateway to experience. It is integral to learning about our physical surroundings, discovering our place within a wider environment, and ultimately the planning and execution of purposeful behavior. Perception is also something we take for granted. Consider an everyday stroll through a street market. As you walked, you would encounter an active and varied environment, remarkable in its complexity, which nevertheless seems to cohere into a whole. The shoppers passing by, the rows of fruits and vegetables, and the sunlight coming from above the rafters—all this visual information would be accessible to you, along with input from other modalities: the sounds of people talking, the scents of the cut fruit, the feel of the breeze, perhaps even the taste of your cup of coffee. Registering any of these perceptions may seem to require little conscious effort, yet making sense of such a welter of stimuli in fact draws on a network of immensely sophisticated cognitive processes (see figure 1.1). Successful processing of perceptual input is critical to our lives, and *how we get better at* that processing will be the subject of this book.

We all know that some activities seem to rely more on advanced forms of perceptual analysis than others do. Anyone who has knitted a sweater or played a video game has made use of high-level visual and cognitive functions—functions that are tuned to our sensory machinery and that have developed over a lifetime of experience. In this sense, we are all perceptual experts, but we also know that many activities—not only knitting and gaming but also playing music or learning to detect letters in visual noise—almost always natural expertise. Even so, this covers a wide range of tasks. It includes judgments of everything from low-level visual features to high-level natural objects, from training that lasts a few minutes to many thousands of trials over many days.

In almost all these cases, experience and practice have been shown to greatly enhance the quality of visual perception, and although perceptual learning is not entirely ubiquitous—it fails to occur in some cases and can be of modest magnitude in others—the phenomenon is so widespread that perception itself cannot be fully understood without it. To understand perception also means to understand how it is modified by experience.

1.2 Perceptual Learning in the Laboratory

The first reports of improved perceptual judgments date back to the late nineteenth century, but it wasn't until the 1960s that perceptual learning was identified as an important subject for scientific inquiry. The birth of the field is most associated with the work of psychologist Eleanor Gibson, whose 1969 book *Principles of Perceptual Learning and Development* set out many of the fundamental phenomena still studied today. Gibson helped to put perceptual learning on the map. Still, the field remained relatively peripheral to mainstream cognitive scientific research until the 1990s, when bold claims about the role of neural plasticity in learning brought it to the fore. Since then, perceptual learning has experienced a resurgence of interest, research, and debate.[18]

In her seminal analysis, Gibson defined perceptual learning as a "relatively permanent and consistent change in the perception of a stimulus array, following practice with that array"[19] (p. 29). In the laboratory, this change has usually been measured by observing how practice improves the performance of a particular task. This almost always involves either detecting the presence of a stimulus or discriminating between two stimuli.

Performance, and thus learning, has been assessed in several ways (see figure 1.2). It can be measured as performance accuracy (and learning can be measured as an increase in accuracy), but it can also be measured as a threshold value of stimulus strength, or as the difference between stimuli at the threshold performance level (with learning in both cases seen as a decrease in threshold). In concrete terms, performance accuracy might be indexed by percentage correct or by a discriminability d'; by the stimulus contrast required to achieve a threshold accuracy in some other judgment, such as orientation; or as a difference in the judged dimension, such as the orientation difference required to achieve the threshold accuracy level. Reduced response times are also sometimes used as an index of learning.

In real-world cases of expertise, perceptual stimuli and judgments are almost always multidimensional, and the performance context is complex. The tasks studied in the laboratory, however, usually involve relatively simple stimuli and judgments, with controlled training or practice protocols. Likewise, many laboratory tasks use coarser judgments and only sometimes focus on acute judgments such as the minimum perceivable stimulus. This

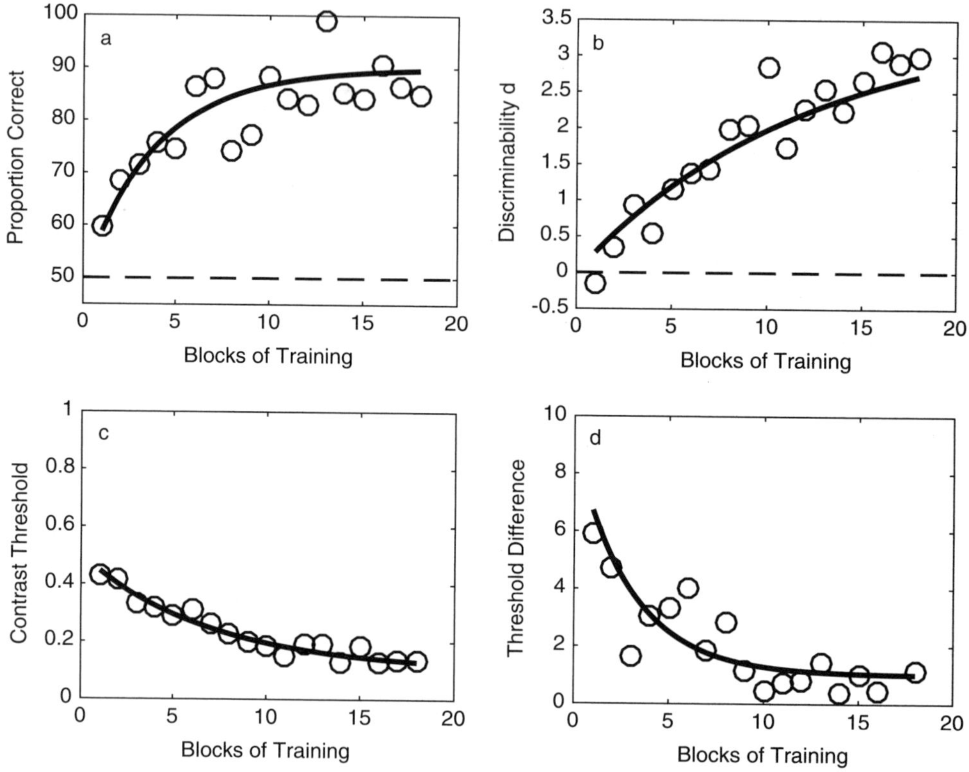

Figure 1.2
Simulated examples of perceptual learning measured as increases in (a) percentage correct, (b) discriminability, (c) decreases in contrast threshold, or (d) threshold differences over blocks of training or practice.

is not to say that the task domain in the laboratory is overly narrow, simply that it is more simplified and constrained than in natural contexts. As we will see, tasks are often grouped according to their complexity, as low-, mid-, and high-level.[20] Subjects may be asked to make judgments about basic visual features but also about natural objects. Most often, tasks involve judgments of mid-level visual features.

Although Gibson herself was interested in the role of perceptual learning in young children, whose visual systems are more labile, perceptual learning in the laboratory has most often been measured in adults, for whom the visual system is thought to be relatively stable (absent major injury). In fact, perceptual learning continues to occur throughout the individual's life span, from visual development to adulthood. It can even be used as a mitigating factor to stave off perceptual losses during aging and has also been studied as a route to remediation or rehabilitation in treatment regimens for clinical deficits.

One commonplace view that seemed to follow from observations of high plasticity during early development was that the visual system in adults was essentially stable (absent aging or damage). From this it was thought that perceptual learning, while measurable, would at best contribute marginally to performance. This turns out not to be the case. Perceptual learning in adults can have a significant impact on visual performance, even at the scale of laboratory practice. In some experiments, it has taken performance from slightly above chance to 90% correct or more.[21–23] Similar learning effects have been seen in many tasks, from spatial-pattern and texture discrimination to motion discrimination.[21, 22, 24] Although perceptual learning is more modest in some cases, the point remains that in adults it *can* make very significant contributions to perceptual functionality. (This has implications beyond the study of learning as such, as it is important to know the stage of practice even when the goal of perceptual testing is not to understand learning but to characterize the visual system and its functions.)

Perceptual learning is a very broad phenomenon. It occurs in many sensory modalities and task domains—although it is visual learning that is of special interest in this book. Over the many thousands of experiments in which it has been studied, it has occurred far more often than not. Learning has been found in the detection or discrimination of visual patterns of many kinds: spatial patterns, complex objects, textures, faces, motion, and stereo depth.[21–31] On the other hand, there are a few stimuli and tasks in which perceptual learning seems to have a relatively small effect or none at all. One such example is the discrimination between two patterns having different orientations around the horizontal or vertical in the fovea.[32] (It has been suggested that the stability of performance relative to the cardinal axes results from the frequency with which such judgments are performed in everyday life.[32]) Overall, however, initial task performance in almost any perceptual task is likely to be far from optimal and thus might be improved with practice or training.

Beyond the simple presence or absence of learning, however, a number of more specific questions emerge: How much and how fast can we learn? How might the extent and speed of learning depend on the nature of the perceptual task? To what degree does learning generalize to, or perhaps interfere with, new stimuli or related tasks? Are some training protocols better than others, and if so, how can we develop the best form of training?

In real-world cases, perceptual expertise is usually the result of an extensive amount of practice. Whether experts are judging the sex of baby chicks in commercial food production or processing rapid visual displays in first-person shooter games, thousands of hours will have been devoted to the activity.[33, 34] By contrast, learning in the laboratory is typically studied for periods as short as an hour, ranging upward from there. At one extreme, there have been some tasks in which very few exposures of easy stimuli altered the course of learning or where improvements in performance were seen over a few hundred trials within the first few minutes (e.g., stereo depth or illusory contours).[31, 35] At the other extreme, training protocols have sometimes extended over thousands of trials and several weeks[36, 37]—though no study has yet tracked performance over the multiyear duration sometimes found in real-world expertise. (This, however, may change with the rise of ubiquitous computing and massive data mining of individuals in situ.)

Another trademark characteristic of perceptual learning, especially as it has been studied in the laboratory, is its sometimes surprising specificity.[24] Specificity occurs when training in one task with a particular stimulus and judgment fails to transfer to improvements in seemingly related tasks and stimuli. Indeed, specificity has been reported for many aspects of both the tasks and stimuli involved in training or practice, including specificity to orientation, spatial frequency, motion direction, stimulus pattern, and even location in the visual field.[21, 22, 24, 28, 37] Furthermore, it was the seemingly odd specificity of trained improvements to a location in the visual field (e.g., training a visual task in the lower right quadrant can fail to transfer even to the same task and stimuli in the upper right quadrant), reported in the 1990s, that was especially provocative and attracted newfound interest in the field of perceptual learning as such.

Thinking about the places in the visual system that might represent such features led many investigators to infer that perceptual learning occurs by inducing plastic changes in the response of neurons in the early visual cortical areas. When placed against the dogma that saw plasticity as unique to early developmental stages, this was surprising indeed. For many researchers in the 1990s and later, perceptual learning became a gateway to understanding brain plasticity, overturning long-held assumptions about cognitive development in the process.[24, 38–44]

As we will see, however, the story is not so simple. The increased attention and enthusiasm in the field of perceptual learning was surely deserved, but the straightforward mapping of specificity onto physiology may have been overly simple. Specificity is a graded phenomenon in which there is some specificity of trained improvements but also some transfer to other stimuli and tasks.[24, 38, 39, 45] Understanding specificity, as we will explore further, turns out to be more complicated than one-to-one reasoning. What is undeniable, however, is that the debate over specificity helped attract justified attention to the field and push it forward. As with many fields, the need to explain provocative experimental findings has ultimately led to the development of new theories.

Specificity has been an incredibly useful tool for researchers as a means to investigate and localize plasticity in the brain network, yet it is almost always the *generalization* of training, rather than its specificity, that is important in practical applications. One of the fundamental tasks of the field, then, is to learn more about the two sides of the specificity versus generalization question. When is learning more specific and when is it more generalizable? And how might this ratio be influenced by the methods of training? In the chapters that follow, we consider both specificity and generalization, with an eye toward their possible applications. Indeed, the patterns of specificity in different tasks have much to tell us *theoretically* about the relevant stimulus representations and system architecture, just as generalization may point the way for *practical* applications of the theory.

Another significant feature involved in returning perceptual learning to the domain of real-world application is the *longevity* of training effects. Sometimes perceptual learning can be relatively ephemeral, but in many or even most cases it can persist for impressively long periods: it has been tested in some tasks after a delay of two to three years and been shown to be relatively strong.[39] It is not yet clear how common or fundamental a feature this kind of longevity is to all learning, with research currently investigating how it may depend on the task or the particular population. For example, if the goal of training is to improve the visual function in an amblyopic eye (an eye disease sometimes called "lazy eye," in which often one eye suffers from deficits in cortical processing), it is important to know not only the extent to which training improved the visual function but also how long such improvements can be expected to hold and with what frequency the functions should be retrained.[46–48] One of the issues we consider in subsequent chapters is the evidence for visual training used in these different practical contexts, from the development of perceptual expertise to remediation in specific cases.

From the extent and rate of learning to specificity, generalizability, and longevity, the study of perceptual learning moves in many directions. Despite dominant beliefs that the malleability of visual function was restricted primarily to childhood, it is now clear that adult visual processing remains highly plastic. Substantial improvements in performance can be achieved through training or practice. There have been many contributors to this extensive literature on the role of experience in diverse visual tasks, and this research is being continued today by an active set of researchers.[18]

Out of this basic observation, an exciting field of inquiry has evolved, with its many different approaches and clusters of research helping to structure this book. Part II (chapters 2 and 3) examines the basic *phenomena* of perceptual learning: learning and transfer. Part III (chapters 4 and 5) examines the *mechanisms* of learning: noise properties of the perceptual system, and evidence from physiology. Part IV (chapters 6–9) develops classic and new formal *models* of the phenomena and mechanisms of learning, with an emphasis on predictive, quantitative modeling. Part V (chapters 10–12) focuses on adjacent *modalities* and *applications* of learning, including possible real-world technologies, as well as the possibilities for optimizing the process. We hope to provide an overview of an exciting field as it has recently developed, while also suggesting promising avenues for future research.

1.3 Plasticity versus Stability

Perceptual learning is a consequence of brain plasticity. Plasticity allows the function of a system to change in response to changes in stimuli or the demands of a new environment. Understanding the biochemical and anatomical underpinnings of this process has been a dominant interest in several subfields of neuroscience, and perceptual learning is no exception.

However, a few researchers (including ourselves) have raised questions about the other side of the coin—stability, or the maintenance of (or return to) a stable or steady state in the face of plastic changes. We suggest that in addition to understanding the role of plasticity in learning, it is critical to recognize and understand this often-neglected counterforce.

Plasticity and stability intrinsically stand in a push-pull relationship: too much stability, and the system could not learn or adapt to new environments; too much plasticity, and it might no longer generate predictable outcomes or might suffer from a loss of prior experiences. Like Goldilocks in the story of the three bears, the system is looking for the porridge that is neither too hot nor too cold but "just right."

The *plasticity-stability dilemma* is one of the major issues that runs throughout this book. As a structural dialectic, it is of course central to any biological system that must operate in a dynamic environment. Analogous questions have arisen in the study of bone regeneration, the dynamic control of locomotion in animals, the responses of immune systems, and the interactions of complex biological systems such as beehives.[49] In all these systems, the advantages of change must be set against the requirements of homeostasis, with the importance of self-organization widely seen as one of the primary system challenges.

The tension between plasticity and stability plays out in biological systems at many different timescales, for the population and for the individual.

At one temporal extreme, the plasticity-stability dilemma is embodied in the *evolutionary* ideas of phenotypic plasticity and genetic robustness.[50, 51] Within this branch of evolutionary theory, a robust system is defined as one in which a stereotyped phenotypic characteristic emerges despite small genetic mutations and random environmental fluctuations.[52–54]

At the timescale of the individual (rather than the species), the development of perceptual ability in infancy and adolescence also embodies the plasticity-stability dilemma. In the field of developmental vision, for example, it has long been known that there are critical or sensitive periods where plasticity is more extensive, while the sensory and perceptual systems later shift to relative stability, which some call "putting the brakes" on plasticity.[55, 56]

At the shortest temporal extreme, perceptual response can change over a few seconds through the process of adaptation, or the altered sensitivity to stimuli as a function of recent stimulus experiences. In the case of adaptation, the perceptual system often returns to steady state over a relatively short time.[57–59]

Within any individual, then, perceptual learning operates within a system subject to lifetime developmental changes as well as immediate changes in the state of adaptation (and maybe even subject to epigenetic influences on gene expression). A vivid demonstration of some of the constraints and interactions between learning and development can be found in several recent case studies documenting the functional nature of recovered sight. The famous story of one individual, Mike May, who was blinded as a child and regained his sight as an adult is recounted in box 1.1. His experience was examined in extensive post-surgical testing by several groups.[60, 61] His story demonstrates the power of learning as well as its limitations and the important role of critical periods in early development. It also shows how the functional outcome of training depends on the nature of the visual skill involved.

Box 1.1
A Story of Recovered Sight

What would it mean to be blind and then recover your sight as an adult? For most of us, this is a thought experiment, but for Mike May, blinded at the age of three when a miner's lantern exploded, it was real. In his forties and already a record-breaking blind downhill skier, successful businessman, and passionate advocate for technical support for the blind, Mike took on his next challenge—the recovery of sight. The story of his postsurgical experiences highlights not only the power of learning but also the power of stability, and its interaction with development.

The explosion that damaged Mike's vision destroyed his left eye and scarred the cornea of his right eye. In the attempt to recover his sight, his right eye was surgically repaired (using stem cell therapy and transplanting a new cornea). "The light hit Mike May like a rush of air, moving through and around him in a burst of white that quickly turned to colors and shapes and movement"[62] (p. 128). The experience was immediate, yet recovering his sight was not only "going to be the greatest adventure of his life. … [I]t also would be one of the most difficult"[62] (p. 129).

Some visual impressions returned almost immediately (Mike was able to respond to pattern, form, color, and motion), yet compared to normally sighted individuals, all these visual functions were limited. Visual functions that are more complex, such as face recognition and three-dimensional depth perception, were especially poor, leading experts to conclude that these limitations reflected weak cortical responses.[61]

Although visual acuity and form processing improved in the first few years after surgery, mostly Mike seemed to have gotten better at *using* what he was seeing. He said, "The difference between today and over 2 years ago [just after surgery] is that I can better guess at what I am seeing. What is the same is that I am guessing." The effects of visual deprivation after age three were also profound, as other visual functions exhibited a slower developmental time course and were influenced by experience for many years afterward.

The story of Mike May vividly demonstrates that perception goes far beyond the projection of a picture on the back of the eye. Many aspects of visual perception reflect the physiological design of the system accrued through years of evolution; others are developed in the individual through experience during early development or through ongoing learning during the life span. Yet others may be affected by short-term adaptation. Perceptual learning operates in a complex system to optimize performance given the momentary constraints.

Mike May's story illustrates the power of perceptual learning but also the power of system constraints, requiring that some things be learned during development. The remaining ability to achieve the best performance will combine perceptual learning, cognition, and planning.

Alongside the study of plasticity in humans and other living organisms, concepts of learning and plasticity have also been studied in artificial computational systems, especially in the domain of artificial neural networks. In these networks, learned knowledge is coded in the strength of the connections between nodes (sometimes also called units), which are meant to be analogs for neurons or ensembles of neurons. The structure of the network consists of sets of input nodes and output nodes, and can also include many other nodes in one or more "hidden" layers (figure 1.3). Activating the input units with a stimulus input

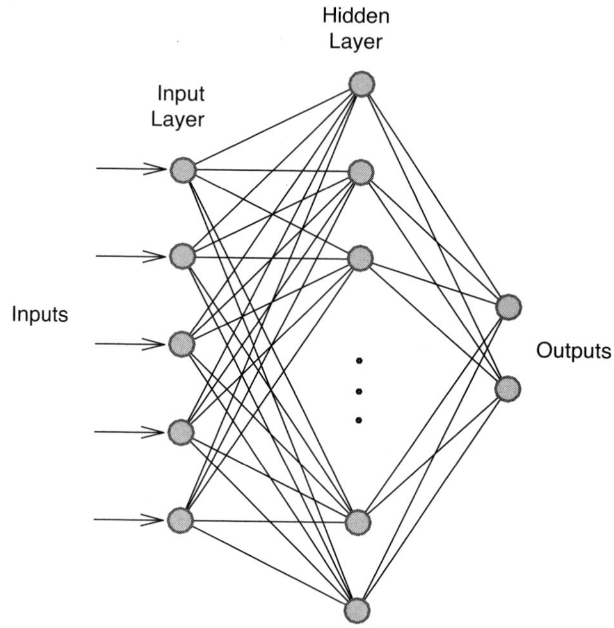

Figure 1.3
This diagram of an artificial neural network shows the nodes (units representing the stimuli or the output responses) and lines (connections), each with a different weight, that pass activation from the input layer to the hidden layer and then to the output layer. Such networks may include additional hidden layers.

drives the activation of units to which they are connected in proportion to the weight (strength) of the connection.

Neural networks have come to play a significant role in the theory and modeling of human cognition. It has been shown that even the simplest neural networks, consisting only of input and output layers, can learn to classify input stimuli into output categories by adjusting connection weights to achieve the desired activity pattern in the target or output unit(s), so long as the required classification is relatively simple. Hidden layers, if they are present, increase the capacity for learning classifications that are more complex. Changing the weights—increasing weak connections or reducing or eliminating strong ones—affects the behavior of the network. As each new stimulus-response classification is encountered, a learning rule or algorithm adjusts the weights in a way intended to improve the response to that item the next time it is encountered.

How these issues have played out in neural network theory has had a direct influence on the study of learning and plasticity in perceptual learning and cognitive science far more than in biological systems. Neural networks may also be a purer way of thinking through the theoretical costs and benefits involved in these domains. Indeed, essentially all the existing quantitative models of visual perceptual learning are neural network models, beginning with the groundbreaking work of Poggio, Fahle, and their colleagues.[23, 37, 63]

The costs and benefits of full plasticity play out fairly directly when considering how a network learns when it is first exposed to one set of classifications and later exposed to another. After a sequence of exposures to training experiences with an initial set of stimuli, the system will have learned the correct classifications in the first task (up to the capacity limit of the network). If the network is then trained in a sequence of exposures on another set of stimuli and classification responses, it will then have learned the recent associations—but unfortunately it may no longer correctly classify previously learned associations or tasks. In this case, the system is so plastic that the most recent learning will have altered the weights to improve classification of the new items, but in so doing it will have disrupted many previously learned weights.

When learning new information causes dramatic forgetting of earlier information, *catastrophic forgetting* or interference is said to have occurred. Stephen Grossberg, among the first neural network theorists to focus on the opposition between plasticity and stability in artificial neural networks, documented the impact of catastrophic forgetting in sequential phases of learning.[64–66] This was first done by showing that interleaved training on two stimulus sets tended to lower the performance for both, thus reflecting the system's limited capacity to learn the two sets simultaneously.[67] It was also done by showing that adding hidden units to the network did not by itself always eliminate catastrophic forgetting, although it did increase the system's capacity.

In the context of neural networks, training will always change the weights to improve them for the current set of stimuli, and since many or all the hidden units may participate in coding any learned response, new learning will corrupt weights that were learned previously. A number of different solutions have been proposed to overcome this problem (and, as we will see, similar approaches have been used in models of visual learning). Some solutions have involved segregating the most important learned connections for a given stimulus into a small number of weights,[68–71] while others used methods such as cyclic rehearsal to continuously refresh the memories of earlier stimuli, effectively converting sequential training into interleaved training through the assumption of implicit or hidden rehearsal processes.[72] Both seek to retain system stability in the face of ongoing plasticity.

Neural network theory provides an illuminating methodological foil to perceptual learning as currently investigated. Unlike the neural network-based approach, the study of perceptual learning in animals and humans has taken a pronounced turn away from system-level theory. In an independent line of inquiry, many theoretical claims have been heavily influenced by physiology. In the study of neural networks, researchers focus on the properties of learning and plasticity *for the system as a whole* and on the architectures and algorithms that support them. By contrast, in the field of perceptual learning, the ideas and claims have often focused on *where* plasticity occurs in the brain. If they were asked directly, of course, most researchers would surely recognize that learning must engage many parts of the complex and interconnected brain network,[73] yet the focus of the field has still overwhelmingly been on *localizing* learning and plasticity, often placing it at the earliest levels of the visual cortex.

This book takes a more synthetic approach. Our consideration of stability is not meant to challenge the reality of plasticity in the biological system, where synapses and neurons may be in constant flux. Instead, it recognizes the simultaneous imperative for maintaining stability at the functional or system level, even in the face of local plasticity. It also acknowledges that apparent plasticity at the earliest levels of the sensory cortex may itself be a transitory result of top-down influences. That is, while recognizing the essential role of plasticity in learning, we have also argued for a simultaneous consideration of the value—indeed requirement—for maintaining system-level stability. Likewise, though behavioral observations of perceptual learning are often thought of in terms of increasing signal sensitivity, our approach carefully grounds these improvements not only in the sensitivity and response to the signal but also in the stochastic and system-level properties of the internal and external noise, which together limit performance.[74] (Such analyses have also proven useful in characterizing aging or special populations.[75, 76])

While much research in the field has focused on which early cortical brain areas may be involved in representing important stimulus features and the plasticity of their responses, we have added a corresponding emphasis on the upstream use of this information. We also emphasize the ways by which the use of this information can change through an improved, reweighted readout of sensory evidence. Our hypothesis is that a change in *readout* (or the "reweighting of evidence," in neural network terminology), more than changes in the original representations themselves, best accounts for the bulk of learning. Our analysis is also interested in understanding the less frequently considered role of top-down factors such as task requirements, attention, and reward—all of which must, in principle, impact perceptual learning.

As we survey the current state of the field in the chapters to come, we appreciate ideas that have been proposed by active researchers but also question implicit assumptions. We hope to consider what has been left out of the picture and what the opposite side of the coin might be for any given explanation. This means valuing stability as well as plasticity, noise as well as signal, readout as well as encoding, and top-down as well as stimulus-driven processes. Only when we consider both sides of these dipoles can we arrive at a better, more synthetic understanding of the astounding balancing act that is perceptual learning.

1.4 Improving the Signal-to-Noise Ratio in Human Performance

Perceptual learning is an interdisciplinary science. It has benefited not only from biology and computational neural networks but also from mathematical methods developed in a range of related fields. Perhaps the most important of these is signal detection theory (SDT), which describes how signals are segregated from noise.[77, 78]

First introduced in psychology in the context of auditory perception,[79] signal detection approaches are now widely used in essentially all areas of psychology and cognitive science. In the study of perceptual learning, they are a methodological cornerstone, giving researchers both a conceptual framework and a quantitative toolkit for identifying the mechanisms by which learning occurs for a particular task and stimuli.

Whereas biological or artificial systems are characterized by the push and pull of plasticity and stability, the key duality in signal detection theory concerns the relative sizes of *signal* and *noise*. All brain systems are intrinsically noisy. Performance, and therefore learning, requires that any relevant signal be understood in the presence of external variability in the stimulus and variability in its internal representation. The tenets of this framework are implicit in the field's understanding of such a fundamental metric as percentage correct or the parallel discriminability measure d'. The underlying measures of discriminability in signal detection theory also relate directly to modern understanding of threshold measures.[80]

No matter how wonderfully sophisticated and sensitive human perception may be, the underlying neural responses and the corresponding performance of perceptual tasks will be variable and imperfect. Even in the absence of stimulus variability, neural responses are almost always variable or *noisy*.[27, 74, 81] This variability in the activity of neurons that form the earliest representations of the stimuli and in the corresponding signals is then transmitted to other areas in the brain.[82] Whatever the mechanism(s) of perceptual learning may be for any given task, if learning improves performance, it must improve the ratio of signal to noise.

Consider even the simplest human performance task—the detection of a stimulus. An observer is asked to determine whether a signal is present or absent in the stimulus array. Although humans are very sensitive, brain responses to stimuli are imperfect, resulting in different distributions in the representation, depending on whether the signal is present or absent (figure 1.4). In any given trial or response episode, variability in the neural representation—beginning with the first cortical responses to stimuli and moving upward through a hierarchy of processes—generates variability at the point of decision. This variability leads to *distributions* of the decision variable in two (or more) states of the world, such as a signal being present or absent, or two different stimuli to be discriminated. Such states are easier to distinguish if their distributions are quite different.[79, 83] Furthermore, human decision itself is not perfect, instead being subject to additional noise as well as variability in setting decision criteria.[84, 85]

Human performance, then, is based on the representations of signals and the decision process in the brain. Improving human behavior through training must reflect one or more of the mechanisms inherent in this schema (visualized in figure 1.4): it may improve the value of the signal (increasing the separation between the distributions), reduce the noise (reducing the variability or spread within the distributions), or both. To pursue a neural analogy, the improvements in the signal-to-noise ratio must in turn reflect either improved tuning of sensory representations, improved selection of sensory inputs for that decision, or improved connections to the areas controlling the choice or execution of behaviors.

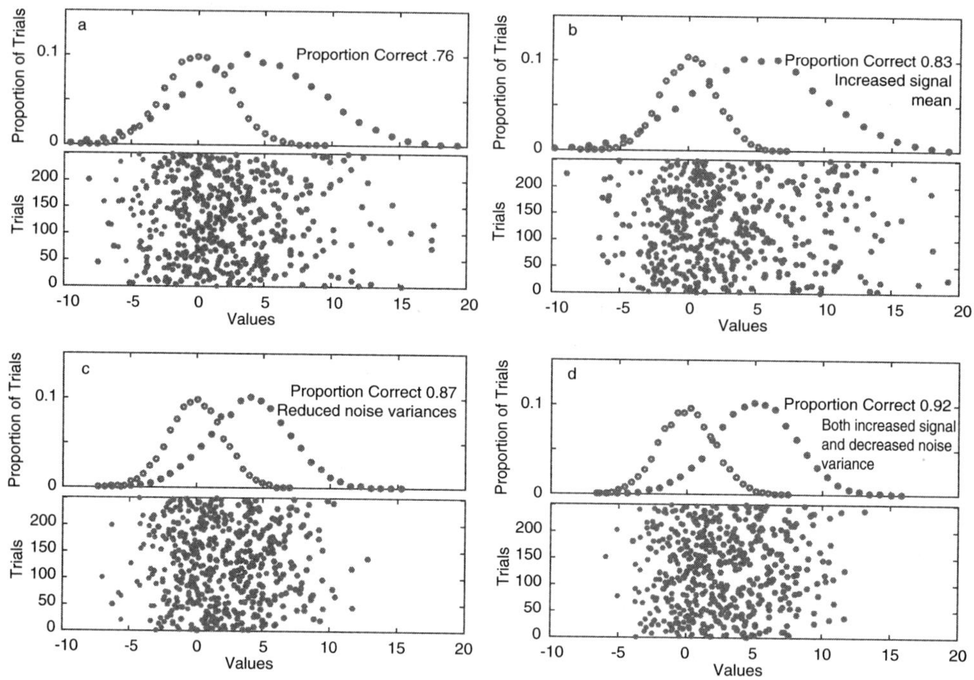

Figure 1.4
Performance depends on the signal-to-noise ratio—the responses to two different stimuli are noisy, which limits discrimination. (a) Histograms of responses from two distributions (n=10,000), with means and standard deviations of (0, 2.5) for light symbols and (4, 5) for dark symbols, and samples of each stimulus on each of 250 trials. The proportion correct is for a two-alternative forced-choice task, which depends critically on the noise. Histograms and samples show reduced noise variability and/or increased the signal mean, with (b) means and standard deviations of (0, 2.5) for light symbols and (5, 5) for dark symbols; (c) means and standard deviations of (0, 2) for light symbols and (4, 3) for dark symbols; and (d) for both increased signal and decreased noise variability, with means and standard deviations of (0, 2) for light symbols and (5, 3) for dark symbols. All three changes, in decreased variance, increased signal mean, or both, increased the signal mean and decreased the variance, improving performance.

Whether in physiology or in network models, a full account of learning will correspond with specific changes in signal and/or noise in the system.

Throughout this book, we connect this broad theoretical framework to particular experimental processes. We analyze the circumstances in which training or practice improves performance and how specific these improvements will be to the trained task and stimuli—the bread and butter for much of the empirical study of visual perceptual learning. In this approach, we are indebted to the work of a large group of researchers using various methodologies. As we will see, it turns out that there is an entire technology of models that can help us analyze the nature of learning. These so-called observer models, as well as their experimental testing methods, will be pursued in chapter 4 examining the different mechanisms by which learning can alter the signal-to-noise ratio. Chapter 5 will examine what we know from physiology about how learning is implemented in the brain and what this might suggest about the trade-offs between plasticity and stability at the system level.

1.5 Reweighting versus Representation Change

The need to balance plasticity and stability, together with how the signal-to-noise ratio might potentially be improved, led us to the idea of learning through *reweighting*, or changing the *readout* of evidence. The full implications of this idea, and the computational models that embody it, are fleshed out in the Models section of the book (chapters 6–9).[86–88]

Our underlying thesis is that changing the evidence used in making a given perceptual decision is almost surely a dominant component of perceptual learning. This change occurs by increasing the weights given to relevant sensory representations and reducing the weights given to irrelevant ones. If one also assumes, as we do, that there must be some level of constancy in sensory representation during normal learning (as distinct from systemic adjustments that might follow an injury), then reweighting becomes the logical means of negotiating the plastic demands of learning with the overall system virtue of stability.

This idea was especially controversial when initially proposed, set as it was against the backdrop of putative changes to neural tuning in the earliest levels of sensory representation in the cortex.[89–91] Though it is true that visual stimuli are processed through a complex network of brain areas, progressing from early sensory registration to decision in the prefrontal cortex (figure 1.5 illustrates the visual processing network from monkey physiology), and that learning could, in principle, occur across a number of brain regions (and the many connections between them), the observed specificity of perceptual learning to retinal location, orientation, and spatial frequency led many researchers to propose that learning occurs through changes in the receptive fields and the sensitivity of neurons at the lowest levels of the visual system. (Again, researchers who have focused on the earliest few areas in this complex diagram surely recognize that plasticity is unlikely to be *restricted* to them,[92–94] but until recently the emphasis nonetheless was on the early visual cortex.) If these changes to the neural responses in the early sensory cortex were persistent outside the context of the learned task, however, they would then affect performance in many other tasks that also rely on them.[89, 95] The result would be a systemwide vulnerability to some of the consequences of catastrophic forgetting.

The traditional theories of learning have tended to fall on either side of this divide. One class of theories assumes that plastic changes occur in the representations of the stimuli at the earliest levels of analysis, whereas the other class assumes that perceptual learning improves the selection of information used to make a decision or carry out a task, perhaps at several levels of representation and processing. These two classes of theories have been labeled respectively *representation enhancement* and *information reweighting*.[27, 45] (More recently, reviews of perceptual learning have acknowledged both as possible contributors.)[86, 96–98]

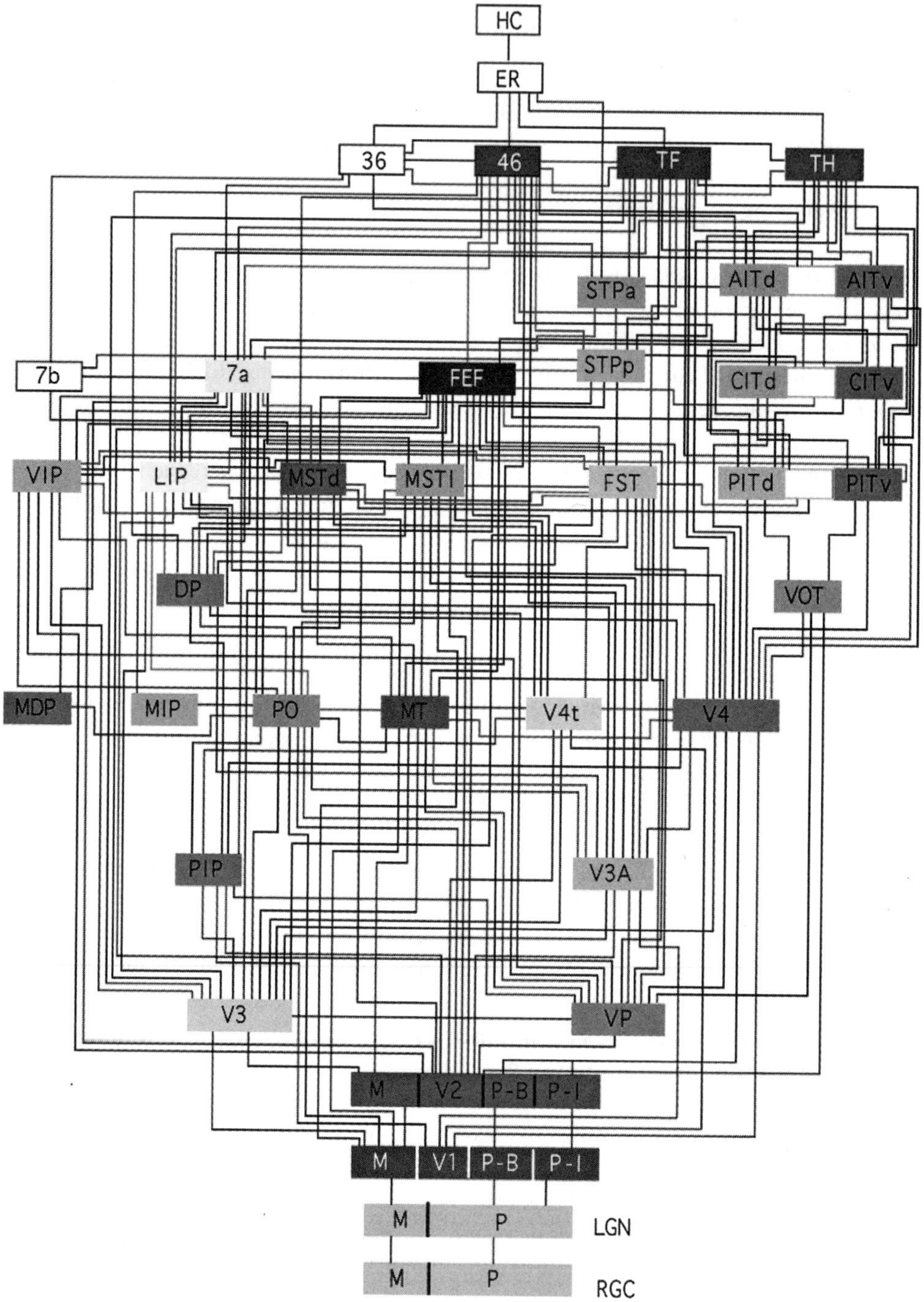

Figure 1.5
Diagram of the connected network of visual brain areas, based on monkey physiology. After Van Essen, Anderson, and Felleman,[99] figure 2, with permission. (See plate 2.)

Representation enhancement theories, sometimes also called representation change or retuning theories, focus on plastic changes of the responses and tuning of very early neural sensory representations before and after training or practice. Information reweighting theories focus instead on how information coded in the responses of sensory representations is selected and combined, perhaps through several levels of representation and decision, to carry out a specific task. There has been (and continues to be) vigorous debate in the field between these two views of plasticity in perceptual learning.[45, 100–104]

Following the reports that perceptual learning could be specific to stimulus characteristics coded early in the visual system, representation enhancement became the dominant theory of perceptual learning.[24] It remains a popularly held view today (garnering significant currency in the visual domain, perhaps because of related earlier claims made in the domains of tactile and auditory learning). As mentioned, the strong form of this theory claims that learning alters the field tuning properties of neurons as early as V1, or, as Fahle stated in an early study, "orientation specificity [of hyperacuity line offset judgments] ... requires that the *neurons that learn* are orientation specific ... [and] ... the position specificity ... suggest[s] Area V1 as the most probable candidate for learning of visual hyperacuity"[105] (p. 418). In this view, specificity occurs because changes to the sensory representations of one pool of neurons would not transfer to untrained neurons. For example, if neurons in V1 representing the lower right visual field were altered during learning, the effects would be specific to that location, since different neural populations represent the upper left visual field. Indeed, many researchers have interpreted the presence of specificity in learning to features coded at early levels of the cortex as literal proof of changed representations at those levels.

Opposed to this view, our proposal of reweighting states that learning changes the "read-out" connections from early visual representations through the hierarchy to decision in the task.[27, 106] Indeed, this view was first proposed as a principle by Mollon and Danilova.[107] Learning through reweighting can occur while leaving some or even most early representations unchanged and stable. From this explanatory hypothesis, we developed the reweighting theory, which, like other network theories, embodies learning in changes to the connections between layers in a multilayer system.

It is important to point out that, in principle, reweighting of evidence could occur at any level in the network or at several levels. In this view, specificity of learning occurs even if early sensory representations remain stable and unchanged. What changes instead is the reweighting (or readout) that may alter the responses of representations in other layers, ultimately connected to decision by weighted links.

Figure 1.6 illustrates one of our early proposals for learning through reweighting. In this framework, the early sensory representations (shown here as filters tuned for spatial frequency and orientation, analogous to neurons in the early visual cortex) remain unchanged, while the weights on the evidence from these channels to a perceptual decision are

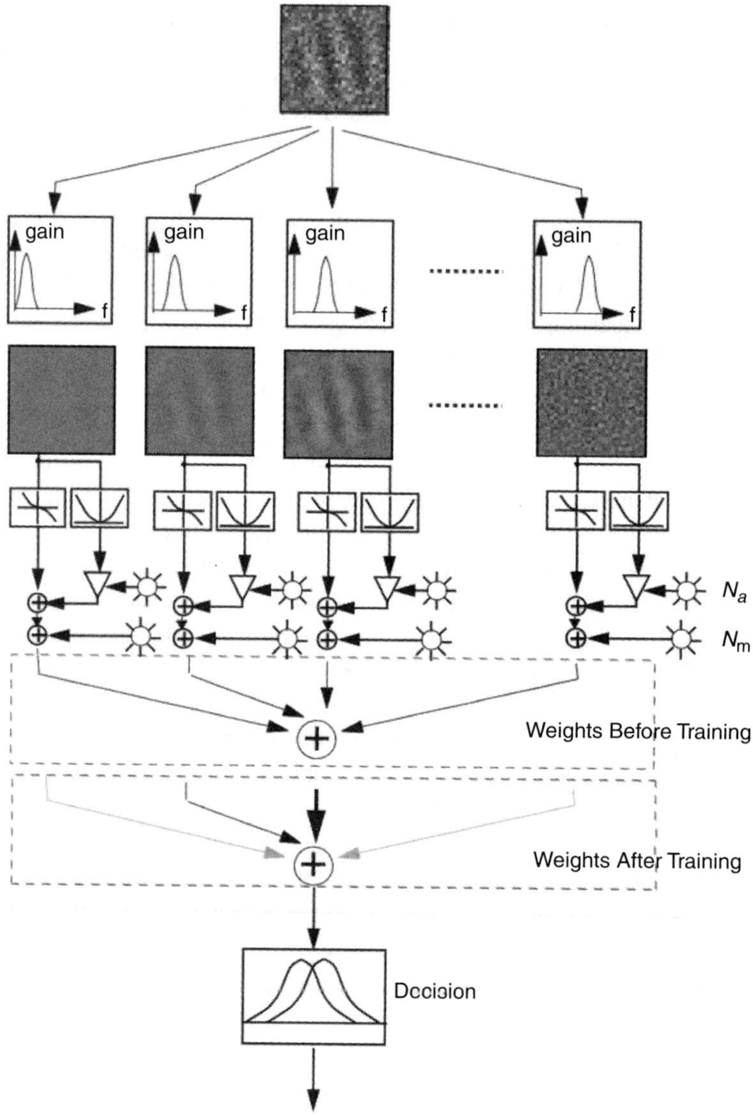

Figure 1.6
A schematic shows perceptual learning through reweighting of evidence from stable early representations to decision. Learning alters connections between stable early sensory representations to decision from an initial state (top) to a later state (bottom) in training. A stimulus image (top center) is processed and represented by units sensitive to spatial frequency and orientation, shown as filters and a filtered image, followed by nonlinearity and processing noises. After Dosher and Lu,[27] figure 3, and Dosher and Lu,[106] figure 11.

modified by experience.[27, 106] Separate weight structures almost surely would be needed to perform different perceptual tasks, even for the same stimuli. This early model also focused on gain control in the early visual response and noisiness in the responses, well-known properties of early visual system processing, visible in the flowchart boxes that show nonlinear transformations of the response and the sources of internal noise (displayed as circles with rays) associated with each channel of stimulus response.

One important conceptual difference between representation change and reweighting is functional: permanently changing neural representations in the early visual cortex could affect many different tasks and percepts, while reweighting of evidence from these neural representations to a task-relevant decision would help to restrict the effects of learning to the same or similar tasks (or, alternatively, change them, but only in the top-down context of a particular task). If perceptual learning actually does alter the more permanent coding in the early sensory layers in response to training in a particular task, then that training will affect the performance on any task that also uses those representations. For this reason, representation enhancement intrinsically has limited capacity for learning, in much the same way that a similarly structured neural network would. Such conditions provide a perfect opportunity for catastrophic forgetting. In the strong form of reweighting, by contrast, a certain amount of stability in visual performance is promoted via the relative stability of the lowest levels of visual coding (i.e., V1), holding stable the earliest stimulus representations that send information to many other areas in the visual hierarchy. Perceptual learning primarily takes place through updating connection weights from early sensory areas to intermediate representations and, ultimately, a decision structure that is specialized for a given task. Reweighting helps prevent catastrophic forgetting if distinct readout weights (weight structures) are used for different tasks, essentially multiplexing the information of early representations, coding these inputs in intermediate layers.

Specificity was the primary evidentiary foundation on which theories of representation enhancement were based.[39] In the reweighting framework, specificity still occurs, but it is explained differently. It occurs if two tasks (and their associated stimuli) rely on separate sensory representations, if the weight structures connecting the representations to decision in the two tasks are different, or both. In a related claim, Mollon and Danilova pointed out that the specificity of perceptual learning does not necessarily imply that the site of learning is distal (early) in the visual system.[107] Instead, it was argued that learning may be central, and the specificity may arise from the sensory codes incorporated in the learning. Given this explanation of learning, specificity seems the most likely, or default, outcome.

Especially in those cases where different tasks induce different decisions and weighting structures, learning through reweighting bears a certain relationship to the network approach taken by Grossberg.[108] In this system, a top-down process selects representations relevant to the perceptual task and initiates learning with a set of task-specific weights that are then further altered by subsequent learning. The capacity to respond to different task types or stimuli derives from segregating alternative tasks into distinct task weight networks—although these networks may rely on the identical sensory or perceptual inputs. The result is a robust system that still expresses plasticity.

In a highly multilayered network, reweighting is at least as flexible as representation change as a mode of plasticity. Reweighting could occur at multiple levels, early or late in the visual system, even as early as LGN to V1; it could change the lateral interactions within a layer; and it could furthermore introduce feedback from higher levels back to lower levels. In such a multilayer network, reweighting of information from one layer to the next will *look like* representation change when the new activation patterns become inputs for the next layer. The resemblance is superficial, however, as optimizing the performance of the entire system contributing to the decision is likely to require the reweighting of information at several, perhaps many, of these levels.

Finally, even if representation enhancement were to occur in the earliest visual representations, this would almost surely require subsequent reweighting, because upstream connection strengths might no longer be optimal and would therefore need to be changed. In other words, if representation enhancement or modification occurs at the early sensory levels, then reweighting is still going to be required to optimize performance. In this sense, the two theories are not wholly exclusive of one another, with representation enhancement now appearing as a particular, if relatively infrequent, subspecies of reweighting. In such cases, representation change modifies the *encoding* of the stimulus, and a different *decoder* will be required if the encoder changes sufficiently.[45, 109]

All these ideas will be fleshed out in further detail in later chapters of the book. The theoretical stakes are significant, yet the empirical details are often complex and nuanced. As we consider different computational models of visual perceptual learning, we hope to show that reweighting models often provide the best account of actual learning and transfer (or specificity) across a broad range of tasks and paradigms. Furthermore, perceptual learning phenomena are not restricted to vision but also occur in other modalities and bear a similarity to other forms of categorization tasks. Therefore, they may have implications for understanding broader learning principles.[110–112]

1.6 The Importance of Generative Models and Optimizing Perceptual Learning

Perception must be understood as a skill that is at least partly developed through experience. Whether oriented around behavior or complex brain systems, research in the field of perceptual learning has thus endeavored to explain how perception and learning work together in symbiosis. Given the impact that learning could have on nearly every perceptual task, it is important for researchers to characterize the state of learning, even in basic research on the fundamental properties of the human visual system. In this sense, much of what has driven research in perceptual learning has derived from a more fundamental interest in perception as such.

At the same time, the relationship can run in reverse. Another rationale for research into perceptual learning derives from the possibility that understanding learning in the perceptual domain may also contribute to understanding learning in other domains. If this were the case, the benefits could be decidedly practical, even commercial. Indeed, a burgeoning industry in cognitive and brain training, of which visual training is a subdomain,[113, 114] has grown up around the field. There are now proposals for visual training enterprises in a range of applications. These include mathematics education, training to overcome limitations in early readers, and training for (partial) remediation of eye conditions such as amblyopia or myopia.[115–119] Training protocols may also include top-down factors such as attention or reward.[120, 121]

A strong theoretical understanding of perceptual learning provides an opportunity to bolster the often intuitive or haphazard (at present) approaches used to design these protocols. Sound theory would allow optimal training protocols to be more efficiently determined through an *optimization framework*.[122, 123, 124] The general idea of this framework is to use models of perceptual learning to make predictions about different protocols *in advance*, harnessing computer simulations to identify which new combinations of training are likely to be the most promising. As we will explore in greater detail toward the end of this book, the optimization paradigm has the potential to replace exhaustive and expensive experimental testing, largely motivated by heuristic intuition, with computation. Successful use of the optimization framework requires a number of components: objective metrics for judging performance in the tasks of interest; a generative model that can predict behavioral outcomes for different training experiences; a similarly robust method for searching the potentially large set of possible training protocols; and, finally, selective use of experimental testing to validate the predictions as they emerge from simulated studies. These more pointed tests could then be used to evaluate and improve the generative model.

Though this approach is only beginning to be pursued systematically, the application of optimization methods to perceptual learning has the potential to accelerate effective protocol design. Approximate heuristic relationships could still be used to explore new training alternatives, but an optimization framework would allow a more systematic and efficient search process.

As we will see throughout this book, strong theory requires robust modeling and vice versa. In the context of optimization, maximizing a given objective function will only be as useful as the generating model behind that function is in predicting measurable behavioral outcomes. Models in this sense need to be theoretically informed and quantitatively defined. Good models will serve a key role not only in furthering our understanding of human perceptual performance but also in the design of practical applications meant to improve real-world deficits and conditions.

1.7 Summary and Overview

Human activity depends on successful perceptual contact with the world. This means not only the successful registration of sensory input but also the meaningful interpretation and analysis of sensory signals. For this reason, perceptual *learning*, or the study of experience-based *improvements* in sensory processing, is important. It is both a classic area of investigation and a new and exciting research enterprise. The last several decades, in particular, have seen an explosion of research. The rise of computational modeling methods, technologies for brain imaging, and the development of experimental protocols that are more sophisticated have all pushed the field toward new horizons.[27, 106]

In this chapter, we have briefly highlighted some of the most important conceptual dipoles that help explain this burgeoning field. We have highlighted and discussed six principles that have guided our consideration of both theory and experiment:

1. Learned plasticity must be balanced by stability in order to optimize overall system performance.
2. Perceptual learning improves the signal-to-noise ratio limiting human performance, by enhancing the signal and/or reducing the noise.
3. Perceptual learning occurs within a complex set of brain networks and may be the result of plasticity at multiple levels.
4. Reweighting evidence from one level of representation to another (or within a single level) is the dominant form of perceptual learning.
5. Learning is often mediated by top-down influences of task, attention, and reward.
6. Finally, formal models must be specified to the point that they make quantitative predictions about behavior in specific experiments that use specific training protocols. Such models are critical to codifying our understanding of the field.

The parts of this book are organized as follows.

Part I—Overview

Chapter 1 was intended to provide an introduction to perceptual learning alongside the key concepts and principles that will structure our discussion.

Part II—Phenomenology

Chapter 2 synthesizes the behavioral phenomena of perceptual learning and highlights some of its major findings. Chapter 3 examines specificity and transfer or generalization, how they might be measured, and their relationship to the concepts of reweighting and representation enhancement.

Part III—Mechanisms

Chapter 4 introduces the observer model and a range of tests to measure and understand the signal and noise properties of perceptual learning. Chapter 5 examines what is known about the physiological basis of perceptual learning, with an eye toward questioning some of the assumptions.

Part IV—Models

Chapter 6 reviews the classical computational models of perceptual learning, introduces the augmented Hebbian reweighting model (AHRM) as a theoretical framework, and applies the model to a number of major findings. Chapter 7 reviews the empirical literature on the role of feedback and the account of feedback phenomena within the AHRM model. Chapter 8 provides a theoretical account of specificity and transfer based on the integrated reweighting theory (IRT), a multilayer reweighting model in which transfer is based on higher-level invariant representations. Chapter 9 discusses the possible roles for task, attention, and reward in perceptual learning, the empirical evidence that supports them, and possible ways to integrate these effects into learning rules.

Part V—Comparisons, Applications, and Optimization

Chapter 10 positions perceptual learning within several temporal scales of plasticity from evolution to adaptation and compares the phenomena of visual learning to those of other modalities: auditory, tactile, olfactory, and multisensory learning; and to category learning. Chapter 11 examines some of the major existing applications of perceptual learning used in education as well as visual remediation and considers possible future directions in which to expand the useful applications of visual learning. Chapter 12 develops and discusses an optimization framework that could be applied to training protocols to improve the magnitude and generalization of perceptual learning.

This book is written for anyone who wants to understand the phenomena and theories of perceptual learning or to apply the technology of learning to the development of training methods and products. It is intended for a variety of readers at a range of levels. Parts of this book are meant to provide an introduction to students just entering this exciting new field, while other parts are meant more for active researchers. We have used a heading structure within each chapter to help readers navigate the material. Our overall goal is to provide an integrated treatment of the field to date, to describe the basic techniques and principles needed to successfully incorporate perceptual learning into applied developments, and to suggest new avenues for future research.

References

1. Mann DL, Spratford W, Abernethy B. The head tracks and gaze predicts: How the world's best batters hit a ball. *PloS One* 2013;8(3):1–11.

2. Crespi S, Robino C, Silva O, de'Sperati C. Spotting expertise in the eyes: Billiards knowledge as revealed by gaze shifts in a dynamic visual prediction task. *Journal of Vision* 2012;12(11):30,1–19.

3. Adamson MM, Taylor JL, Heraldez D, Khorasani A, Noda A, Hernandez B, Yesavage JA. Higher landing accuracy in expert pilots is associated with lower activity in the caudate nucleus. *PLoS One* 2014;9(11):e112607.

4. Yang JH, Kennedy Q, Sullivan J, Fricker RD. Pilot performance: Assessing how scan patterns & navigational assessments vary by flight expertise. *Aviation, Space, and Environmental Medicine* 2013;84(2):116–124.

5. Kellman PJ. Perceptual learning. In Pashler H, ed., *Stevens' handbook of experimental psychology*. Wiley;2002, 259–299.

6. Green CS, Bavelier D. Effect of action video games on the spatial distribution of visuospatial attention. *Journal of Experimental Psychology: Human Perception and Performance* 2006;32(6):1465–1478.

7. Tanaka JW, Taylor M. Object categories and expertise: Is the basic level in the eye of the beholder? *Cognitive Psychology* 1991;23(3):457–482.

8. Hecht S, Shlaer S, Pirenne MH. Energy, quanta, and vision. *Journal of General Physiology* 1942;25(6):819–840.

9. Baylor D, Lamb T, Yau K-W. Responses of retinal rods to single photons. *Journal of Physiology* 1979;288(1):613–634.

10. Von Békésy G, Wever EG. *Experiments in hearing*. Vol. 8. McGraw-Hill; 1960.

11. Allaire P, Billone M, Raynor S. Extremely small motions of the basilar membrane in the inner ear. *Nature* 1970;228(5272):678–679.

12. Lancet D. Vertebrate olfactory reception. *Annual Review of Neuroscience* 1986;9(1):329–355.

13. Reed RR. How does the nose know? *Cell* 1990;60(1):1–2.

14. Hänig D. Zur psychophysik des geschmackssinnes. *Philosophische Studien* 1901;17:576–623.

15. Stevens S. Sensory scales of taste intensity. *Perception & Psychophysics* 1969;6(5):302–308.

16. Moskowitz HR. Ratio scales of sugar sweetness. *Perception & Psychophysics* 1970;7(5):315–320.

17. Weinstein S. Intensive and extensive aspects of tactile sensitivity as a function of body part, sex, and laterality. In Kenshalo DR, ed., *The skin senses*. Charles C. Thomas;1968: 195–222.

18. Fahle M, Poggio T, Poggio TA, eds. *Perceptual learning*. MIT Press; 2002.

19. Gibson EJ. Perceptual learning. *Annual Review of Psychology* 1963;14(1):29–56.

20. Fine I, Jacobs RA. Comparing perceptual learning across tasks: A review. *Journal of Vision* 2002;2(2):5,190–203.

21. Fiorentini A, Berardi N. Perceptual learning specific for orientation and spatial frequency. *Nature* 1980;287(5557):43–44.

22. Ball K, Sekuler R. A specific and enduring improvement in visual motion discrimination. *Science* 1982;218(4573):697–698.

23. Poggio T, Fahle M, Edelman S. Fast perceptual learning in visual hyperacuity. *Science* 1992;256(5059):1018–1021.

24. Karni A, Sagi D. Where practice makes perfect in texture discrimination: Evidence for primary visual cortex plasticity. *Proceedings of the National Academy of Sciences* 1991;88(11):4966–4970.

25. Furmanski CS, Engel SA. Perceptual learning in object recognition: Object specificity and size invariance. *Vision Research* 2000;40(5):473–484.

26. Gold J, Bennett P, Sekuler A. Signal but not noise changes with perceptual learning. *Nature* 1999;402(6758):176–178.

27. Dosher BA, Lu Z-L. Perceptual learning reflects external noise filtering and internal noise reduction through channel reweighting. *Proceedings of the National Academy of Sciences* 1998;95(23):13988–13993.

28. Schoups AA, Vogels R, Orban GA. Human perceptual learning in identifying the oblique orientation: Retinotopy, orientation specificity and monocularity. *Journal of Physiology* 1995;483(3):797–810.

29. Yu C, Klein SA, Levi DM. Perceptual learning in contrast discrimination and the (minimal) role of context. *Journal of Vision* 2004;4(3):4,169–182.

30. Adini Y, Sagi D, Tsodyks M. Context-enabled learning in the human visual system. *Nature* 2002;415(6873):790–793.

31. Ramachandran V, Braddick O. Orientation-specific learning in stereopsis. *Perception* 1973;2(3):371–376.

32. Lu Z-L, Dosher BA. Perceptual learning retunes the perceptual template in foveal orientation identification. *Journal of Vision* 2004;4(1):5,44–56.

33. Biederman I, Shiffrar MM. Sexing day-old chicks: A case study and expert systems analysis of a difficult perceptual-learning task. *Journal of Experimental Psychology: Learning, Memory, and Cognition* 1987;13(4):640–645.

34. Green CS, Bavelier D. Action video game modifies visual selective attention. *Nature* 2003;423(6939):534–537.

35. Rubin N, Nakayama K, Shapley R. Abrupt learning and retinal size specificity in illusory-contour perception. *Current Biology* 1997;7(7):461–467.

36. Dosher BA, Lu Z-L. The functional form of performance improvements in perceptual learning: Learning rates and transfer. *Psychological Science* 2007;18(6):531–539.

37. Fahle M, Edelman S. Long-term learning in Vernier acuity: Effects of stimulus orientation, range and of feedback. *Vision Research* 1993;33(3):397–412.

38. Karni A, Bertini G. Learning perceptual skills: Behavioral probes into adult cortical plasticity. *Current Opinion in Neurobiology* 1997;7(4):530–535.

39. Karni A, Sagi D. The time course of learning a visual skill. *Nature* 1993;365(6443):250–252.

40. Ahissar M, Hochstein S. Eureka: One shot viewing enables perceptual learning. Paper presented at Investigative Ophthalmology & Visual Science; 1996.

41. Ahissar M, Hochstein S. Task difficulty and the specificity of perceptual learning. *Nature* 1997;387(6631):401–406.

42. Polley DB, Steinberg EE, Merzenich MM. Perceptual learning directs auditory cortical map reorganization through top-down influences. *Journal of Neuroscience* 2006;26(18):4970–4982.

43. Recanzone GH, Schreiner C, Merzenich MM. Plasticity in the frequency representation of primary auditory cortex following discrimination training in adult owl monkeys. *Journal of Neuroscience* 1993;13(1):87–103.

44. Recanzone GH, Merzenich MM, Jenkins WM, Grajski KA, Dinse HR. Topographic reorganization of the hand representation in cortical area 3b owl monkeys trained in a frequency-discrimination task. *Journal of Neurophysiology* 1992;67(5):1031–1056.

45. Dosher BA, Lu Z-L. Hebbian reweighting on stable representations in perceptual learning. *Learning & Perception* 2009;1(1):37–58.

46. Zhou Y, Huang C, Xu P, Tao L, Qiu Z, Li X, Lu Z-L. Perceptual learning improves contrast sensitivity and visual acuity in adults with anisometropic amblyopia. *Vision Research* 2006;46(5):739–750.

47. Lev M, Ludwig K, Gilaie-Dotan S, Voss S, Sterzer P, Hesselmann G, Polat U. Training improves visual processing speed and generalizes to untrained functions. *Scientific Reports* 2014;4:7251,1–10.

48. Levi DM, Li RW. Perceptual learning as a potential treatment for amblyopia: A mini-review. *Vision Research* 2009;49(21):2535–2549.

49. Brans RG, Kahn RS, Schnack HG, van Baal GCM, Posthuma D, van Haren NEM, Lepage C, Lerch JP, Collins DL, Evans AC, Boomsma DI, Pol HEH. Brain plasticity and intellectual ability are influenced by shared genes. *Journal of Neuroscience* 2010;30(16):5519–5524.

50. Debat V, David P. Mapping phenotypes: Canalization, plasticity and developmental stability. *Trends in Ecology & Evolution* 2001;16(10):555–561.

51. Kaneko K. Phenotypic plasticity and robustness: Evolutionary stability theory, gene expression dynamics model, and laboratory experiments. In Sawyer OS, ed., *Evolutionary systems biology*. Springer;2012:249–278.

52. Wagner A. *Robustness and evolvability in living systems*. Princeton University Press; 2013.

53. Waddington CH. Canalization of development and the inheritance of acquired characters. *Nature* 1942;150(3811):563–565.

54. Waddington CH. *The strategy of the genes: A discussion of some aspects of theoretical biology.* Allen & Unwin, 1957.

55. Lewis TL, Maurer D. Multiple sensitive periods in human visual development: Evidence from visually deprived children. *Developmental Psychobiology* 2005;46(3):163–183.

56. Bavelier D, Levi DM, Li RW, Dan Y, Hensch TK. Removing brakes on adult brain plasticity: From molecular to behavioral interventions. *Journal of Neuroscience* 2010;30(45):14964–14971.

57. Bao M, Engel SA. Distinct mechanism for long-term contrast adaptation. *Proceedings of the National Academy of Sciences* 2012;109(15):5898–5903.

58. Webster MA. Adaptation and visual coding. *Journal of Vision* 2011;11(5):3,1–23.

59. Webster MA, Mollon J. The influence of contrast adaptation on color appearance. *Vision Research* 1994;34(15):1993–2020.

60. Beyeler M, Rokem A, Boynton GM, Fine I. Learning to see again: Biological constraints on cortical plasticity and the implications for sight restoration technologies. *Journal of Neural Engineering* 2017;14(5):05103.

61. Fine I, Wade AR, Brewer AA, May MG, Goodman DF, Boynton GM, Wandell BA, MacLeod DIA. Long-term deprivation affects visual perception and cortex. *Nature Neuroscience* 2003;6(9):915–916.

62. Chapman E. Mike May. *Sacramento* June 2007:128–130.

63. Poggio T, Edelman S, Fahle M. Learning of visual modules from examples: A framework for understanding adaptive visual performance. *CVGIP: Image Understanding* 1992;56(1):22–30.

64. Grossberg S. Competitive learning: From interactive activation to adaptive resonance. *Cognitive Science* 1987;11(1):23–63.

65. Carpenter GA, Grossberg S. The ART of adaptive pattern recognition by a self-organizing neural network. *Computer* 1988;21(3):77–88.

66. McCloskey M, Cohen NJ. Catastrophic interference in connectionist networks: The sequential learning problem. *Psychology of Learning and Motivation* 1989;24:109–165.

67. Ratcliff R. Connectionist models of recognition memory: Constraints imposed by learning and forgetting functions. *Psychological Review* 1990;97(2):285–308.

68. Carpenter GA, Grossberg S, Rosen DB. Fuzzy ART: Fast stable learning and categorization of analog patterns by an adaptive resonance system. *Neural Networks* 1991;4(6):759–771.

69. French RM. Catastrophic forgetting in connectionist networks. *Neurocomputing* 2012;87:79–89.

70. French RM. Using semi-distributed representations to overcome catastrophic forgetting in connectionist networks. Paper presented at 13th Annual Cognitive Science Society Conference; 1991.

71. Lewandowsky S, Li S-C. Catastrophic interference in neural networks: Causes, solutions, and data. In Dempster FN, Brainerd C, eds., *Interference and inhibition in cognition.* Academic Press; 1995:329–361.

72. Abraham WC, Robins A. Memory retention—the synaptic stability versus plasticity dilemma. *Trends in Neurosciences* 2005;28(2):73–78.

73. Crossley NA, Mechelli A, Vértes PE, Winton-Brown TT, Patel AX, Ginestet CD, McGuire P, Bullmore ET. Cognitive relevance of the community structure of the human brain functional coactivation network. *Proceedings of the National Academy of Sciences* 2013;110(28):11583–11588.

74. Lu Z-L, Dosher BA. Characterizing observers using external noise and observer models: Assessing internal representations with external noise. *Psychological Review* 2008;115(1):44–82.

75. Bower JD, Andersen GJ. Aging, perceptual learning, and changes in efficiency of motion processing. *Vision Research* 2012;61:144–156.

76. DeLoss DJ, Watanabe T, Andersen GJ. Improving vision among older adults: Behavioral training to improve sight. *Psychological Science* 2015;26(4):456–466.

77. Peterson W, Birdsall T, Fox W. The theory of signal detectability. *Transactions of the IRE Professional Group on Information Theory* 1954;4(4):171–212.

78. Tanner WP Jr, Swets JA. A decision-making theory of visual detection. *Psychological Review* 1954;61(6):401–409.

79. Green D, Swets J. *Signal detection theory and psychophysics.* Wiley;1966.

80. Treisman M, Watts T. Relation between signal detectability theory and the traditional procedures for measuring sensory thresholds: Estimating d' from results given by the method of constant stimuli. *Psychological Bulletin* 1966;66(6):438–454.

81. Lu Z-L, Dosher BA. Characterizing human perceptual inefficiencies with equivalent internal noise. *Journal of the Optical Society of America A* 1999;16(3):764–778.

82. Faisal AA, Selen LP, Wolpert DM. Noise in the nervous system. *Nature Reviews Neuroscience* 2008;9(4):292–303.

83. Macmillan NA. Signal detection theory as data analysis method and psychological decision model. In Keren G, Lewis C, eds., *A handbook for data analysis in the behavioral sciences: Methodological issues.* Lawrence Erlbaum Associates;1993:21–57.

84. Cabrera CA, Lu Z-L, Dosher BA. Separating decision and encoding noise in signal detection tasks. *Psychological Review* 2015;122(3):429–460.

85. Wickelgren WA. Unidimensional strength theory and component analysis of noise in absolute and comparative judgments. *Journal of Mathematical Psychology* 1968;5(1):102–122.

86. Roelfsema PR, van Ooyen A, Watanabe T. Perceptual learning rules based on reinforcers and attention. *Trends in Cognitive Sciences* 2010;14(2):64–71.

87. Dosher BA, Jeter P, Liu J, Lu Z-L. An integrated reweighting theory of perceptual learning. *Proceedings of the National Academy of Sciences* 2013;110(33):13678–13683.

88. Herzog MH, Fahle M. Modeling perceptual learning: Difficulties and how they can be overcome. *Biological Cybernetics* 1998;78(2):107–117.

89. Ghose GM, Yang T, Maunsell JH. Physiological correlates of perceptual learning in monkey V1 and V2. *Journal of Neurophysiology* 2002;87(4):1867–1888.

90. Adab HZ, Vogels R. Practicing coarse orientation discrimination improves orientation signals in macaque cortical area v4. *Current Biology* 2011;21(19):1661–1666.

91. Raiguel S, Vogels R, Mysore SG, Orban GA. Learning to see the difference specifically alters the most informative V4 neurons. *Journal of Neuroscience* 2006;26(24):6589–6602.

92. Poort J, Khan AG, Pachitariu M, Nemri A, Orsolic I, Krupic J, Bauza M, Sahani M, Keller GB, Mrsic-Flogel TD, Hofer SB. Learning enhances sensory and multiple non-sensory representations in primary visual cortex. *Neuron* 2015;86(6):1478–1490.

93. Gold JI, Law C-T, Connolly P, Bennur S. The relative influences of priors and sensory evidence on an oculomotor decision variable during perceptual learning. *Journal of Neurophysiology* 2008;100(5):2653–2668.

94. Law C-T, Gold JI. Neural correlates of perceptual learning in a sensory-motor, but not a sensory, cortical area. *Nature Neuroscience* 2008;11(4):505–513.

95. Schoups A, Vogels R, Qian N, Orban G. Practising orientation identification improves orientation coding in V1 neurons. *Nature* 2001;412(6846):549–553.

96. Seitz A, Watanabe T. A unified model for perceptual learning. *Trends in Cognitive Sciences* 2005;9(7):329–334.

97. Kourtzi Z, Betts LR, Sarkheil P, Welchman AE. Distributed neural plasticity for shape learning in the human visual cortex. *PLoS Biology* 2005;3(7):1317–2327.

98. Zhang G-L, Li H, Song Y, Yu C. ERP C1 is top-down modulated by orientation perceptual learning. *Journal of Vision* 2015;15(10):8,1–11.

99. Van Essen DC, Anderson CH, Felleman DJ. Information processing in the primate visual system: An integrated systems perspective. *Science* 1992;255(5043):419–423.

100. Sotiropoulos G, Seitz AR, Seriès P. Perceptual learning in visual hyperacuity: A reweighting model. *Vision Research* 2011;51(6):585–599.

101. Talluri BC, Hung S-C, Seitz AR, Seriès P. Confidence-based integrated reweighting model of task-difficulty explains location-based specificity in perceptual learning. *Journal of Vision* 2015;15(10):17,1–12.

102. Seriès P, Seitz A. Learning what to expect (in visual perception). *Frontiers in Human Neuroscience* 2013;7:668,1–14.

103. Watanabe T, Náñez JE, Koyama S, Mukai I, Liederman J, Sasaki Y. Greater plasticity in lower-level than higher-level visual motion processing in a passive perceptual learning task. *Nature Neuroscience* 2002;5(10):1003–1009.

104. Zhang J-Y, Zhang G-L, Xiao L-Q, Klein SA, Levi DM, Yu C. Rule-based learning explains visual perceptual learning and its specificity and transfer. *Journal of Neuroscience* 2010;30(37):12323–12328.

105. Fahle M. Human pattern recognition: Parallel processing and perceptual learning. *Perception* 1994;23:411–427.

106. Dosher BA, Lu Z-L. Mechanisms of perceptual learning. *Vision Research* 1999;39(19):3197–3221.

107. Mollon JD, Danilova MV. Three remarks on perceptual learning. *Spatial Vision* 1996;10(1):51–58.

108. Carpenter GA, Grossberg S. *Adaptive resonance theory*. Springer; 2010.

109. Dosher B, Lu Z-L. Visual perceptual learning and models. *Annual Review of Vision Science* 2017;3:343–363.

110. Wright BA, Zhang Y. A review of the generalization of auditory learning. *Philosophical Transactions of the Royal Society B: Biological Sciences* 2009;364(1515):301–311.

111. Harrar V, Spence C, Makin TR. Topographic generalization of tactile perceptual learning. *Journal of Experimental Psychology: Human Perception and Performance* 2014;40(1):15–23.

112. Ashby FG, Alfonso-Reese LA. A neuropsychological theory of multiple systems in category learning. *Psychological Review* 1998;105(3):442–481.

113. Green CS, Bavelier D. Action video game training for cognitive enhancement. *Current Opinion in Behavioral Sciences* 2015;4:103–108.

114. Lu Z-L, Lin Z, Dosher BA. Translating perceptual learning from the laboratory to applications. *Trends in Cognitive Sciences* 2016:20(8):561–563.

115. Kellman PJ, Massey CM. Perceptual learning, cognition, and expertise. *Psychology of Learning and Motivation* 2013;58:117–165.

116. Tallal P, Miller S, Bedi G, Byma G, Wang X, Nagarajan SS, Schreiner C, Jenkins WM, Merzenich MM. Fast element enhanced speech improves language comprehension in language-learning impaired children. *Science* 1996;271(5245):81–84.

117. Tallal P, Merzenich MM, Miller S, Jenkins W. Language learning impairments: Integrating basic science, technology, and remediation. *Experimental Brain Research* 1998;123(1–2):210–219.

118. Campbell FW, Hess RF, Watson PG, Banks R. Preliminary results of a physiologically based treatment of amblyopia. *British Journal of Ophthalmology* 1978;62(11):748–755.

119. Huang C-B, Lu Z-L, Zhou Y. Mechanisms underlying perceptual learning of contrast detection in adults with anisometropic amblyopia. *Journal of Vision* 2009;9(11):24,1–14.

1201. Donovan I, Szpiro S, Carrasco M. Exogenous attention facilitates location transfer of perceptual learning. *Journal of Vision* 2015;15(10):11,1–16.

121. Yang X-S. Introduction to mathematical optimization from linear programming to metaheuristics. Cambridge University Press;2008.

122. Zhang P, Hou F, Yan F-F, Xi J, Lin J, Yang J, Chen G, Zhang MY, He Q, Dosher BA, Lu ZL, Huang CB. High reward enhances perceptual learning. *Journal of Vision* 2018;18(8):11,1–21.

123. Eckstein MP, Abbey CK, Pham BT, Shimozaki SS. Perceptual learning through optimization of attentional weighting: Human versus optimal Bayesian learner. *Journal of Vision* 2004;4(12):3,1006–1019.

124. Miettinen K, Ruiz F, Wierzbicki AP. Introduction to multiobjective optimization: Interactive approaches. In Branke J, Deb K, Mittenen K, Słowiński R, eds. *Multiobjective optimization*. Springer;2008:27–57.

II

Phenomenology

2

Perceptual Learning in Visual Tasks

Perceptual learning is a widely occurring phenomenon. It ranges from the effects of modest training in the laboratory to the specialized expertise resulting from extensive practice. In this chapter, we classify learning into tasks occurring at three levels of visual representation: low-level features, mid-level patterns, and objects or natural scenes that involve high-level visual coding. Learning at early levels, where the visual system represents many individual features, occurs through selection of the relevant representations, while higher-level learning tasks require the recruitment or creation of higher-level representations of natural objects that reflect unique combinations of features. Learning of low-level tasks can be slow and is often affected by external noise in the stimulus and the typicality of the stimuli, while learning of high-level tasks is often rapid and robust.

2.1 Perceptual Expertise and Perceptual Plasticity

Experts are not made overnight. As every guild or professional culture knows, seniority requires years of practice and training. Wine tasters spend years at the sommelier, musicians train extensively at conservatories, and radiologists have ongoing training in the reading of medical images.

Perceptual expertise is not only the province of vintners and musicians, however. Psychologists have also studied it, though the principles they focused on have tended to be more general. As the discipline of psychology developed in the late nineteenth and early twentieth centuries, experimenters tended to focus not on exceptional cases but rather on the early stages of more general improvement with training or practice. One of the founders of experimental psychology, William James,[1] devoted a section of his landmark work *The Principles of Psychology* to the role of practice in improving sensory discrimination, citing prior work by Volkmann, Fechner, and others. For James, the central example involved tactile two-point discrimination, a test developed in 1846 by Ernst Weber,[2] in which the two points of a drawing compass were placed on the skin of a subject. Weber discovered that the ability to distinguish the separation between the two points improved to less than half the initial value over an hour of practice, and that this learning transferred to varying degrees to other skin locations, but was only partially retained the following day.[2] Other nineteenth-century psychologists studied instances of perceptual plasticity that were more radical. George M. Stratton, founder of the first psychology laboratory at the University of California, placed prism glasses on subjects that reversed the visual world from left to right and turned it upside down. He published three reports between 1896 and 1898[3, 4] detailing his discovery that initial symptoms—nausea, disrupted motor interactions, and "out-of-body" experiences—first reported by subjects lessened over time. By the fourth day, a subjective sense of an upright world had returned, suggesting a fragile remapping of perception. During this famously long experiment, subjects were able to tentatively remap perceptual stimuli to support newly calibrated motor function, but after days in the new environment, Stratton took the experiment one step further. When he had subjects remove the prism glasses, they reported several hours of inverted percept before the visual world returned to normal.

Contemporary psychologists tend to use methods that are more sophisticated (if less intrepid). Although recent experiments have challenged certain details of Stratton's reports,[5] these early studies still make the now widely accepted point that the perceptual system is remarkably plastic, even in adults, and that plasticity is crucial to functional perception. This is true for short and long durations of training, for the subjects of an outlandish laboratory experiment, or for vintners training to be the best of their class. Whether in the early stages of perceptual learning or later in the achievement of perceptual expertise, to understand perception we must also understand plasticity.

2.2 Visual Perceptual Learning

As the preceding examples illustrate, the documentation of perceptual learning and expertise has a long and varied history. The last 30 years, however, have seen an animated surge in laboratory research, focusing especially on visual perception. Aided by a number of factors—the availability of display systems, technical instruments that are more complex, and computational modeling, as well as the funding structures to support basic research—scientists have been able to produce new knowledge at an accelerated rate. For a variety of reasons, the bulk of this new research focuses not on exceptional cases of expertise (an immensely subtle and multifactored phenomenon) but rather on the more general and testable phenomena of perceptual learning resulting from modest levels of practice in controlled laboratory environments. Just as the study of particle physics required complicated equipment and restricted conditions to test fundamental theories, modern research in visual perceptual learning uses better technologies to validate principles and theories that might, like the discoveries in physics, form the basis of understanding for additional naturally occurring phenomena.

Recent experimental and physiological research has produced exciting discoveries in the study of perception, with implications and applications far beyond what Stratton or James could have imagined. The initial suggestion in the 1990s, originating from frequent observations of specificity in visual perceptual learning, was that plasticity affects low levels of the visual system, long thought to be stable after childhood. This hypothesis inspired research into the nature of learning in many specific visual domains: contrast, color, texture, motion, and other features that dominate our perception of natural scenes.[6, 7]

Perceptual learning has now been documented in many visual functions and domains, with almost all examples showing some degree of learning, depending on how the observer has been trained. However, simply observing that a phenomenon exists—which was the focus of many laboratory demonstrations—is not the same thing as understanding or explaining it. In order to do that, it is necessary to discover fundamental principles and evaluate how these principles explain and predict outcomes.

To do this requires asking questions that are more precise, and researchers have only begun to ask them. These include: What are the factors that promote plasticity? Are there circumstances or domains where training does not improve performance? What changes in either the behavioral outcome or the brain process occur with training or practice? Where in the brain do these plastic changes occur or not occur, and does this depend on the nature of what is learned? How does training improve the extraction of the signal from the noise that limits performance and, relatedly, what are the functions and mechanisms of these changes? And can perceptual learning be modeled quantitatively?

The most exciting recent research in perceptual learning, especially in visual perceptual learning, aims to answer these questions. The answers promise to improve our theories of both perceptual learning and perception itself.

2.3 Learning through Representation Selection versus Creation

At the most basic theoretical level, perceptual learning in the visual domain (and perhaps in others) seems to reflect the *selection* of existing representations and the *creation* of new representations and associations, or some mixture of each. In some cases, the existing representations may also be retuned. Certain kinds of visual judgments are based on feature attributes already coded in the early visual cortex. In the case of discriminating the orientation of a pattern, for example, the observer can likely make the required decision based on already existing representations; here the problem solved in visual perceptual learning is to find the *right* representations to focus on. A higher-level visual task, by contrast, might require recognizing *combinations* of features—combinations not yet coded in the visual cortex—such that new representations need to be created (or recruited). To recognize a particular video game avatar, for example, will require that an observer discern the shape of the avatar's head, its body color, and the texture on its torso, among other features. While the basic features of shape, color, and texture are likely coded in the visual cortex, not every possible combination of every basic feature is likely to be coded in advance. It would thus seem probable that the observer must create new cortical representations (recruit new neural ensembles) to code the new combinations of those features defining these objects. A new cortical representation would be created for the avatar.

This interplay between *selection* and *creation* will be a central organizing principle in our survey and discussion of the current research. A given task focused on precoded low-level features may focus primarily on selection of low-level representations early in training, perhaps with some small retuning or increases in response emerging late in learning. In another task, however, training may strengthen associations connecting an *ensemble* of representations in the higher visual cortex to represent complex multifeatured stimuli. In this second task, learning would likely involve recruiting a high-level representation primarily sensitive to one low-level feature to become sensitive to a combination of features. Attention will often also have a role to play, especially early in the learning process.

The distinction between selection and creation has further implications for the degree of specificity of what is learned (the topic of chapter 3). In this instance, learning's

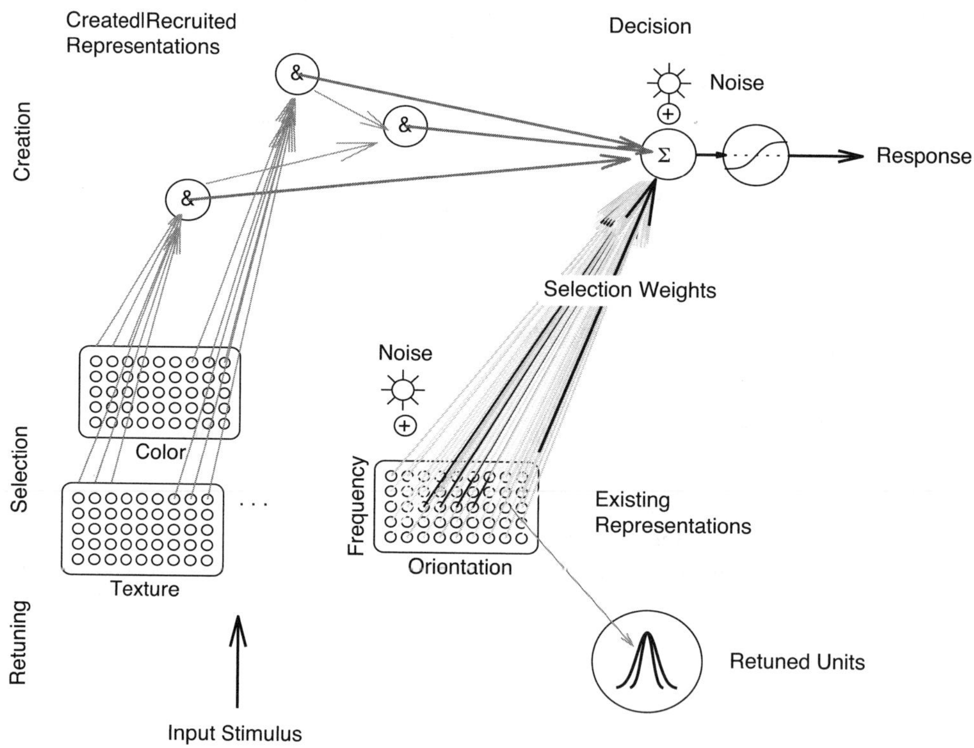

Figure 2.1
Perceptual learning can reflect learned retuning of low-level representations or reweighting of selected preexisting lower-level representations or creation of higher-level units that represent new combinations of features. Learning of low-level visual tasks generally reflects selection, while learning of higher-level visual tasks reflects learning by creating or recruiting new representation units. This schematic illustration includes preexisting representations of orientation, texture, and color, and created units representing combinations. Retuning early representations in any event will require selected weighting to decision. Reprinted with permission of the authors.

specificity to the stimuli or judgment may result either from selecting low-level sensory representations whose receptive fields are selective or from creating a representation that is sensitive to a unique combination of features.

Although a number of ideas related to the hierarchy of representations have been proposed, it is our belief that this dialectic between selection and creation is one of the fundamental principles—perhaps *the* fundamental principle—at work in visual perceptual learning. At a time when dozens of new observations of learning are reported yearly but governing principles remain elusive, this theoretical dialectic has the power to organize swathes of the field and bring separate branches of inquiry into dialogue with one another.

Alongside theory, of course, there are also important questions about practical applications. Can perceptual learning protocols significantly improve real-world function? It has been claimed, for example, that video game training may broadly improve visual perception and visual attention[8, 9] or that training may play an important role for special populations, such as anisometric amblyopes.[10, 11] These twin orientations of research—the theory of the phenomenon and how best to put new discoveries into practice—are of course closely related. As basic research develops new theories, this will not only aid our general understanding but may also become the means for building new technologies. The theoretical advances can be examined in the laboratory but also in the context of practical applications. Likewise, practical discoveries might in turn lead to new questions about theoretical interpretations. In what follows, we begin with a focus on the basic phenomenology of perceptual learning in laboratory research, but we return to practical applications in chapter 11.

2.4 Structure of a Typical Perceptual Learning Study

A typical perceptual learning experiment includes an observer (usually human but sometimes animal) and a perceptual task, which is defined by the perceptual judgment required and a set of stimuli being tested. The observer receives practice or training, and performance improvements are observed. Typically, in each trial, a stimulus is presented and the observer classifies the stimulus and makes a judgment. As in any scientific experiment, variables are manipulated—the amount or schedule of training, presence or absence of feedback or reward, complexity of the judgment, and so on.

At first glance, this framework seems straightforward enough, but many variations present themselves. Stimuli may be as simple as Gabor patterns or as complex as faces, textures, or compound patterns varying in motion, depth, or color. Stimuli must have a given contrast but can be presented briefly or longer and may include external noise or a mask.[12–17] The variations are essentially endless.

The observer's judgment, set by the experimenter, requires detection, discrimination, or identification of a *training feature*, while other *characteristic features* of the stimulus may be fixed or may vary. In the literature, if not the real world, the judgments usually involve a bipartite decision such as present/absent, left/right, or same/different, although recently the class of judgments has been expanded. Responses may be recorded by pressing a key, by a verbal response, or by neural responses measured using a range of devices.

Often, on each trial, the experimenter provides feedback about the accuracy of the response; however, sometimes only an average performance over a block of trials, or no feedback at all, may be provided. Rewards or other incentive structures, although rarely used in human research, have sometimes been introduced to motivate learning. The experimenter may manipulate the stimuli, the mixture of different stimuli or judgments, the number of training trials in a session, and/or the number and timing of sessions. In some cases, transfer to a new judgment or stimuli is assessed after training.

To make things concrete, figure 2.2 illustrates an example task in which the observer is trained to judge the orientation of a visual stimulus. An oriented Gabor (a windowed sine

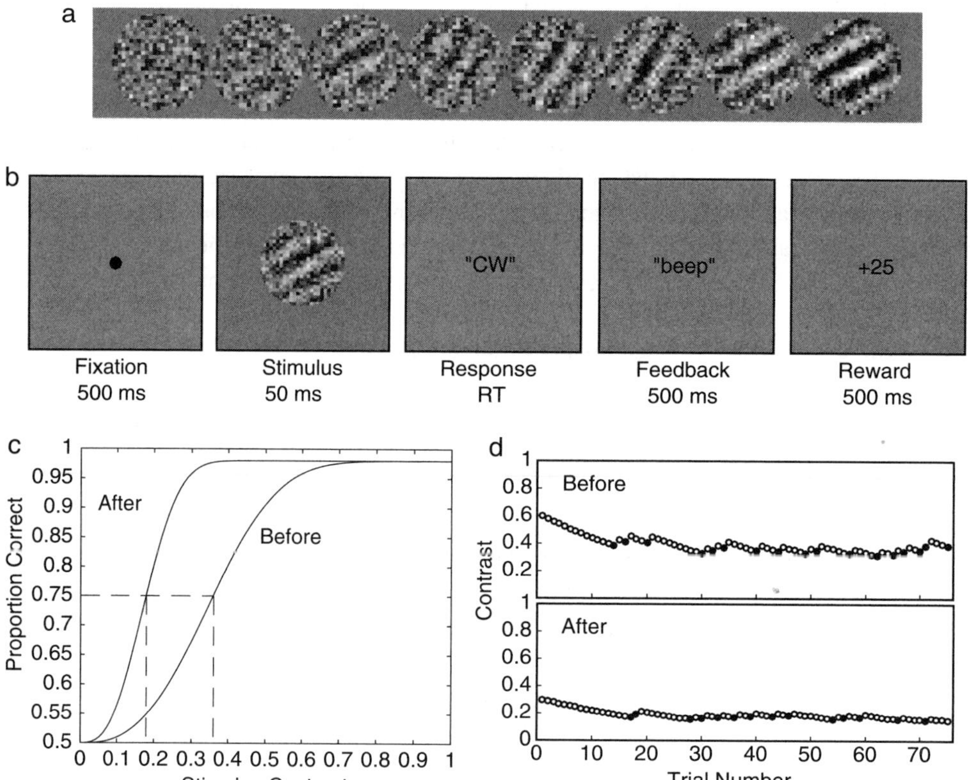

Figure 2.2
A sample experiment: (a) example stimuli (45°±12°) of different contrasts in an external-noise task; (b) a trial sequence with fixation, stimulus, response, (auditory) feedback, and reward; (c) contrast psychometric functions before and after learning, marked with 75% correct thresholds; (d) adaptive staircases estimating thresholds before and after training by increasing the contrast by 10% after errors (dark marks) or decreasing contrast by 10% after 3 correct responses (light marks).

wave) is presented briefly at the center of gaze, embedded in external noise. The observer then decides whether this pattern is rotated clockwise or counterclockwise relative to a reference angle. A tone ultimately sounds if the response was an error, and a point reward is indicated. In the case in figure 2.2, the angular difference is ±12° clockwise or counterclockwise of 45° off vertical, and the contrast of the briefly presented pattern is set from trial to trial to achieve 75% correct responses using adaptive methods. Hundreds of trials are performed in each of several daily sessions. The paradigm is a two-alternative forced choice, the training and assessment vary contrast to achieve the target accuracy, and trial-by-trial feedback is provided as well as a payoff or reward signal. The dependent measure is contrast threshold as a function of training.

Although the space of possible experimental protocols is clearly vast, the stimulus space *actually* tested tends to be quite simple. By far the most prevalent task is two-alternative identification. Feedback is usually trial-by-trial, and explicit rewards or payoffs are almost never included. Training schedules usually involve large numbers of trials per session and multiple sessions. When transfer of training *is* measured, this almost always is a test of immediate transfer, and only rarely is subsequent learning in the new condition measured. Finally, performance is almost always analyzed only at the level of either blocks or sessions.

This historically typical mode of studying perceptual learning may be expanded in several ways with the availability of new methods of measurement. Sometimes the nature of the performance assessment itself (percentage correct, estimated threshold) requires a modestly large number of trials, and this determines the temporal grain at which performance improvements can be measured. This requirement then cascades to determine the scale at which initial performance levels and estimates of the rate of learning are measured.[18] The recent developments of rapid adaptive testing methods for estimation may allow us to assess perceptual learning with many fewer trials and therefore at a finer temporal grain or even trial-by-trial.[19–22] Finer-grained or trial-by-trial measurements could then improve the estimates of both initial performance and the form and rate of learning.[24, 25] The sample size requirements of measurement in current experiments have also led to design limitations in which the training trials are the same as the assessment or measurement trials. As methods for quick estimation are developed, it should become possible to decouple training and assessment, allowing designs in which training is interspersed with quick assessments of performance on a target task (as discussed in several later chapters).

As research has grown more sophisticated, certain experimental paradigms have emerged, along with a functional terminology for describing them, lending researchers a shared methodological toolkit for building on each other's experimental approaches. At the same time, the recurrence of simple stereotyped experiments about perceptual learning may say more about the habits and limitations of current research than about the fundamental nature of the phenomena under study. As the field progresses, other theoretically relevant factors must be examined in a broader range of paradigms. Such an expansion would build on the existing literature of the field, which has already integrated many powerful insights and discoveries.

2.5 Training Features and Task Types

The existing literature on visual perceptual learning can be usefully parsed into three fundamental categories based on the complexity of the training feature required by the task. These include: (1) *basic visual features*, (2) *visual patterns*, and (3) *objects or natural stimuli*. Basic visual features require low- to mid-level analysis, visual patterns require mid-level to high-level analysis, and objects or natural stimuli require high-level analysis[25] (see table 2.1 for examples).

To invoke the distinction between *selection* and *creation*, it seems plausible to say that learning will move from selection to creation as the level of analysis required by the target task moves from low to high. Indeed, as we will see, learning in tasks requiring discrimination of basic single or compound features largely involves the dynamic selection of relevant sensory representations, including the level of representation, from among the many resulting from the parallel processing of the stimulus in different brain areas. In higher-level tasks, however, simply identifying any given feature may not be the limiting factor, and what must be learned is instead a unique combination of features defining an object, a category structure, an identifier, or a name.

Any task, regardless of complexity, necessarily engages a number of brain regions at different levels of analysis. Even if the trained perceptual judgment focuses on low-level features, the visual stimulus will of course be processed not only in the early visual pathway but also at higher levels of the visual cortex, and—because the observer takes an action—the judgment involves the expectation, reward, and decision systems of the brain. Nevertheless, the *essence* of the learning process will depend on the nature of the specific task. In some relatively simple tasks, this will involve the selection or weighting of the correct set of preexisting sensory representations—for example, winnowing down decision inputs to the appropriate set of neurons. In other more complex tasks, especially in

Table 2.1
Perceptual learning studied for judgments of features, patterns, and natural objects

Features	Patterns	Natural objects
Orientation	Compound stimuli	Contours
Spatial frequency	Texture	Shapes
Phase	Global patterns	Objects
Contrast	Search	Faces
Color	Depth	Avatars
Acuity	Motion	Biological motion
Hyperacuity		

domains with so many potential features (or feature levels) that particular combinations are unlikely to be precoded in the early cortex, learning almost surely requires the recruitment or creation of new representations. What changes from task to task is not so much the brain regions engaged but rather the degree to which each region plays a part and the mode of their interrelation.

Alongside the three classes of training features, there is a second axis by which experiments in the field can be categorized. This has to do with how performance is measured. Almost all the literature uses one of three primary task paradigms: (1) the *Type I*, or *feature-difference*, paradigm; (2) the *Type II*, or *visibility*, paradigm; and (3) the *Type III*, or *performance*, paradigm. These paradigms differ in their approach to measurement and training. In each, certain aspects (or all aspects) of the stimulus are held constant, while others may be modified to measure behavioral performance. In some, the stimulus is clear and highly visible but the observer is asked to make very fine judgments and the performance is measured by how fine a judgment is possible. In others, the judgments are coarse, the task is made difficult by controlling the visibility, and the performance is measured by how visible the stimuli need to be for the task to be completed successfully. In yet others, the fineness and visibility of the stimulus are unchanged and the performance is measured for accuracy.

Type I, or *feature-difference*, paradigms almost always use easily visible stimuli and vary discrimination along a feature-difference dimension (e.g., differences in orientation or direction of motion) until a particular criterion for a threshold is met. As the observer's performance improves with practice or training, the experimenter adjusts the feature difference to more precise discriminations in order to hold judgment accuracy constant. Observers performing orientation discrimination, for example, may initially need large orientation differences to achieve a set accuracy (e.g., 75%), but as the observer gains perceptual expertise, the orientation difference decreases (e.g., from 20° to 3°) while the threshold accuracy remains the same. The observer experiences a shifting set of stimuli—with these shifts occurring especially rapidly during early training, when performance improves the fastest.

Type II, or *visibility*, paradigms hold the judged feature difference constant while varying some other visibility variable, such as contrast, to achieve a given accuracy. Here, the stimulus patterns remain constant, while a lower stimulus contrast is needed to keep performance accuracy the same. The threshold contrast for detection or discrimination decreases with practice. The observer experiences changes in visibility (e.g., contrast or presentation time), but the stimulus patterns that they are looking for are fixed.

Type III, or *performance*, paradigms use an unchanging set of stimuli and measure improvements in behavioral performance with training, while the stimuli stay constant.

At first, it might seem that the choice of paradigm is unimportant, essentially unrelated to the fundamental phenomenon of learning. In reality, the two are often intermingled.

For many experiments, the method of assessment is also the method of training, which in turn will influence learning outcomes—a point recently argued on empirical grounds.[26] For such experiments, there is often a chicken-or-egg conundrum at work: what might first appear to be broad-based discoveries may in fact be influenced by the chosen experimental paradigm. Correspondingly, models of perceptual learning often make quite different predictions for different paradigms (see chapter 6).

The point is that the choice of paradigm clearly brings with it a number of corollary consequences. It may change the decision rule or determine the difficulty of the training task—in turn affecting the speed at which the observer learns and the extent to which learning generalizes. Other choices, such as whether to train several tasks at once, can also have profound consequences. Intermixing an easier version of the same task with a harder one can improve the learning (chapters 6 and 8), while training on task mixtures that vary along the training dimension has been shown to disrupt or eliminate learning entirely, even when learning is robust when the tasks are trained individually—a phenomenon called *roving*.[27–30] By the same token, variations in characteristic stimulus properties (e.g., those irrelevant to the defined judgment) need not disrupt learning (see chapter 7).

As the study of perceptual learning has grown over the past few decades, so has the number and size of available datasets. Many articles on perceptual learning are published each year. In what follows, we organize the discussion of perceptual learning in human adults into the three levels of training features and note the choice of experimental paradigm(s) in each. We summarize perceptual learning in each domain in the first paragraph or two and then include a number of examples with full experimental details for concreteness and to help clarify the discussion for nonspecialist readers. As we go along, we refrain from repeating the full details of every experiment as these become more familiar and patterns emerge.

As we progress, we hope to convey a sense of the many ways in which perceptual learning has been investigated, while focusing on the most classical or representative experiments. The field by now includes several highly populated clusters of research, each defined by task domain and paradigm, where scientists have developed a shared methodological toolkit and language on which to build. Of course, between these clusters, vast areas remain to be explored.

2.6 Perceptual Learning of Single Features

The most common tasks in the study of perceptual learning involve judgments about single basic features coded in early cortical areas. Perhaps the most prototypical stimulus for the primary visual cortex is a spatially windowed sine wave, called a Gabor (which approximates the receptive field of neurons in the early visual cortex), though other stimuli have also been tested. Single-feature judgments in the literature have included orientation, spatial frequency, phase, contrast, color, acuity, and hyperacuity.

2.6.1 Orientation

The orientation of contours is one of the most basic features in natural scenes. It is also one of the most studied judgments in classic perceptual learning. Orientation discrimination has been trained in the fovea and in the periphery, both of which are relevant for perception. It has been examined, usually with lines or sine-wave patterns of low spatial frequency, in cardinal (horizontal, vertical) and noncardinal (oblique) orientations,[31, 32] in the presence of varying amounts of external external noise, and in all three training paradigms (Type I,[31] Type II,[12, 13] and Type III[33] tasks).

Perceptual learning is more robust for orientation judgments in the periphery, for noncardinal stimuli, and in high external noise. On the other hand, training may have little or no impact on orientation judgments at the fovea, with cardinal reference angles, and in the absence of external noise—perhaps because such judgments are so common in natural viewing. Similar results have been found in monkeys,[34–36] which show larger learning effects for training in the periphery and for noncardinal orientations (see chapter 5 on physiology). (Indeed, the robustness of perceptual learning in the periphery led one researcher to argue for a longer plastic period in the peripheral cortex in a paper titled "Are Visual Peripheries Forever Young?"[37])

In one typical experiment, observers practiced orientation discrimination as the angular-difference threshold of a large 15° long, 0.25° wide bar centered at the fovea improved, requiring smaller angular differences (Type I paradigm) (figure 2.3). Thresholds for noncardinal orientations were reduced from about 2° to about 1° of orientation angle over nearly 5,000 trials of training, while training had little effect near cardinal orientations[31, 32] (learning at the fovea may only occur for longer line stimuli[38]). In another example, practicing orientation discrimination in the periphery by using fixed orientation stimuli with varying amounts of external noise (Type II paradigm) improved contrast thresholds at each level of external noise from high external noise (highest threshold curve) to zero external noise (lowest threshold curve) over nearly 13,000 trials.[13] Training the orientation task in the lower right quadrant transferred only partly to the lower left (T1) or upper right (T2) quadrants. Not all orientation tasks show improvements, however; training did not improve contrast thresholds in nearly vertical orientation judgments at the fovea or around an oblique angle in the absence of external noise.[39]

2.6.2 Spatial Frequency

Natural stimuli can be synthesized from patterns of different spatial frequencies and/or scales. For this reason, many classic vision tests use sine-wave or windowed sine-wave stimuli varying in spatial frequency and/or contrast. Despite this, perceptual learning of spatial-frequency judgments has seldom been studied. In these few cases, observers judged fixed stimuli (Type III tasks), and learning (usually measured as improvements in percentage correct) was either weak or occurred only in a minority of observers. Across the few

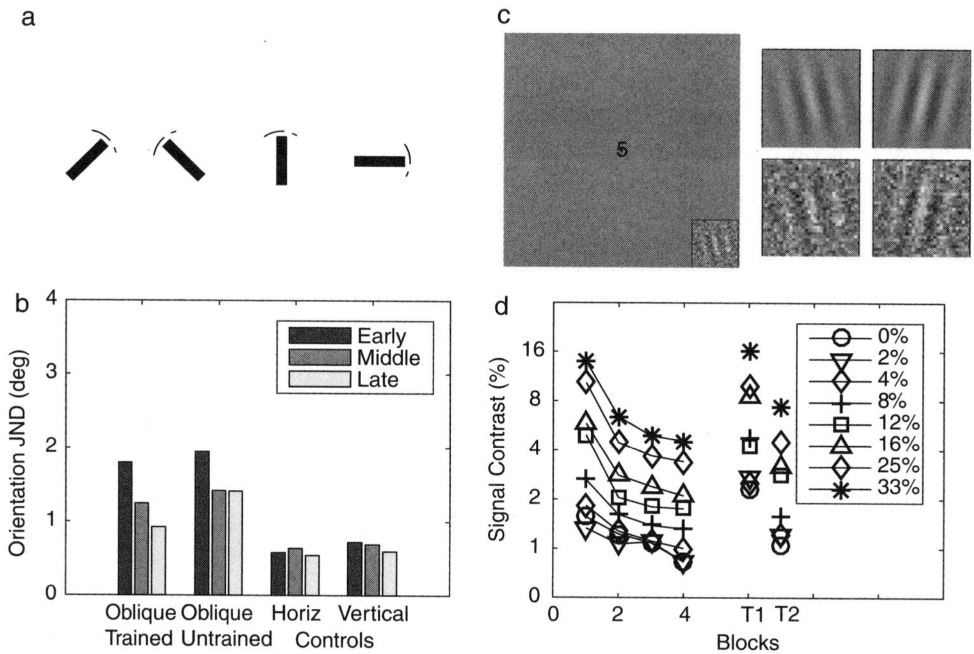

Figure 2.3
Perceptual learning in angular orientation difference thresholds and contrast thresholds, and some transfer tests. Stimuli (a) and (b) measured angular-difference threshold improvements. Stimuli (c) and (d) contrast threshold improvements in different levels of external noise (from high to zero, top to bottom, in curves) and transfer to two other quadrants. (b) Redrawn from selected data estimated from Vogels and Orban,[32] figure 2. (d) Redrawn from Dosher and Lu,[13] figure 6.

existing studies, the evidence for robust learning in discrimination or identification judgments of spatial frequencies is weak, but the scope of testing has been fairly narrow.

In one early study, researchers[40] reported no consistent improvements in discriminating two similar spatial frequency gratings at fovea over 200–500 practice trials, and weak and variable learning in the periphery even for discriminating very different stimuli,[41] although learning discrimination in the context of compound patterns (made of separate parts) has been reported (figure 2.4). However, learning in these spatial-frequency discrimination experiments may have been limited by stimulus roving; for example, discriminating f versus $2f$ while intermixing different base frequencies f. We know that roving can disrupt learning (see subsection 8.7.5). Recent work in our laboratory, however, showed improvements with practice in eight-alternative identification of different peripheral spatial-frequency stimuli (see section 7.5).

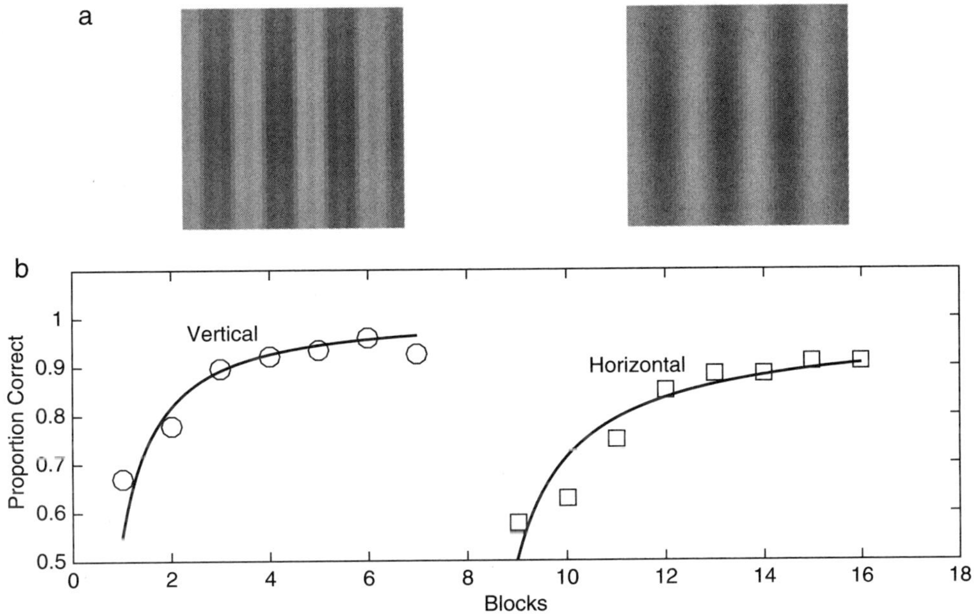

Figure 2.4
Learning to discriminate compound patterns differing in relative phase of 1*f* and 3*f* sine-wave components. (a) Sample vertical stimuli. (b) Practice improves performance independently for vertical and horizontal patterns with power-function learning curves (smooth curves). Data redrawn from Fiorentini and Berardi,[42] learning curves added.

2.6.3 Phase

Like spatial frequency, perceptual learning of phase has been studied rarely, with one study each trained in the fovea and the periphery, using Type III protocols with fixed stimuli. Learning has been reported for compound patterns differing in the relative phase of the components. In one case, discriminating the difference between two compound patterns formed from a high-contrast sine wave at frequency 1*f* (a pedestal) and a low-contrast sine wave at 3*f* differing in phase by 0° or 90° at the fovea improved over just a few hundred trials, and learning was specific to orientation.[42] The estimated learning curves are shown here (figure 2.4) as power functions of practice.[43] In another study, the ability to discriminate sine or cosine phase peripheral Gabors with or without external noise improved significantly over thousands of trials of practice in a study of object attention.[44]

2.6.4 Contrast

Luminance-contrast detection is "a fundamental and simple visual task"[45] (p. 1249). Yet sensitivity to contrast and contrast differences sometimes can still be improved by training or practice. Although some studies examine contrast discrimination in the periphery, it is usually studied at the fovea. Sometimes the effects of training one spatial frequency have been assessed on full contrast-sensitivity functions, which measure detection of patterns of different spatial frequencies.[46] Learning in contrast tasks is primarily studied in Type I or II protocols (equivalent in this case). Learning often occurs, but there are also some reports in which learning failed to occur; training is more likely to improve performance for patterns in noncardinal orientations or for testing in the presence of lateral or pattern masks.

One early report investigated the effect of practice on the detection of gratings of different orientations.[47] Practicing detection of oblique 10 cycle per degree gratings for 3,000 trials improved contrast detection, nearly eliminating the detection disadvantage relative to gratings in the cardinal directions (see also Sowden, Rose, and Davies[45]). In another early experiment, DeValois[48] reported substantial changes in the contrast-sensitivity function, especially at lower spatial frequencies, over the course of a year-long series of adaptation experiments. Other studies show improvements at higher spatial frequencies[48] using a method of adjustment for each spatial frequency. More recent studies[10] using two-interval forced choice (2IFC) detection revealed larger improvements in detecting high-spatial-frequency patterns after training with high-frequency stimuli (figure 2.5).

There are now many studies in contrast discrimination, and it appears that the presence of learning depends on the experimental details. Traditionally, these experiments have used 2IFC tasks in which observers choose the interval containing a contrast increment while the other interval contains a reference contrast (figure 2.6). One group[49, 50] reported no learning for isolated discriminating Gabors at the fovea but robust learning with Gabor flankers. Another group[51] reported learning even for isolated foveal patterns. Another study

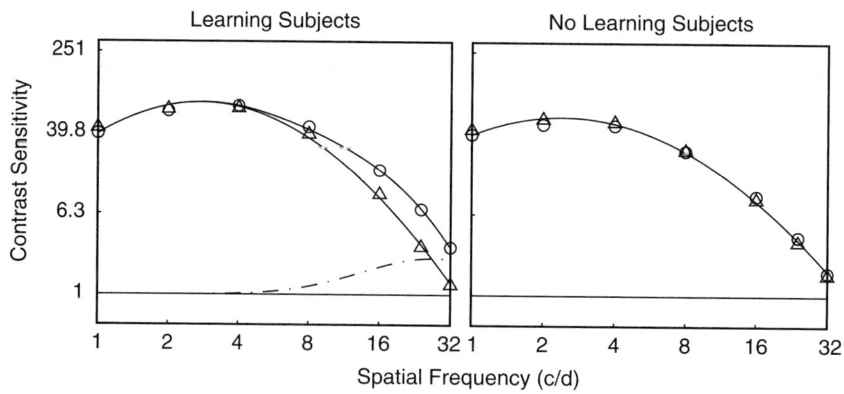

Figure 2.5
Training near the high-spatial-frequency cutoff (27 cycles per degree) improves the contrast-sensitivity function (CSF) in about half of observers, with improvements for learners primarily in high-spatial-frequency detection (broken dashed line, right). Based on Huang, Zhou, and Lu,[10] average data provided by Huang (personal communication).

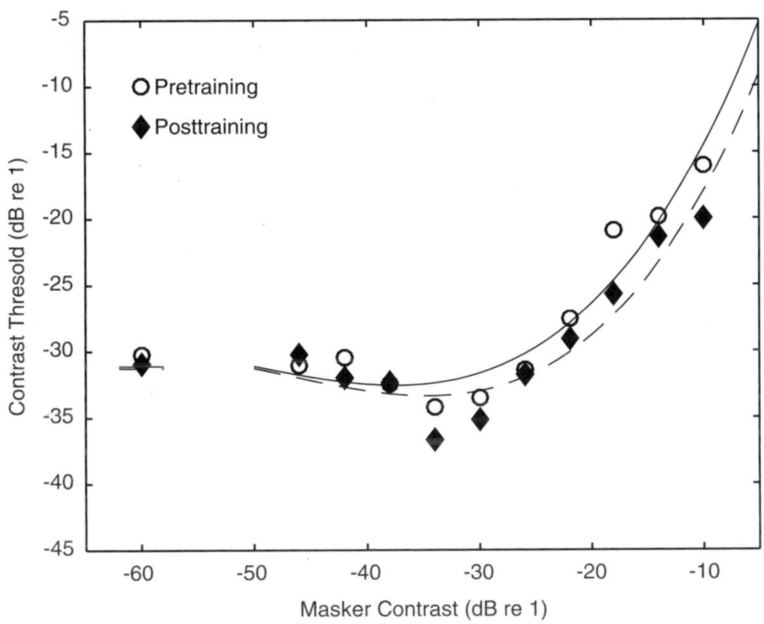

Figure 2.6
Perceptual learning occurs in contrast detection of a small foveal Gabor on a large (pedestal) masker, shown with a smooth masker function. Redrawn from the average of subject data read from Maehara and Goryo,[53] figure 3.

examined learning in the presence of fixed but not varying noise.[54] Learning sometimes depended on the contrast and presence of maskers, with substantial learning at high mask contrasts and no learning at low mask contrasts.[52] Overall, while learning may be more robust in the presence of flanking or masking stimuli, simple contrast discrimination at the fovea can show learning if the reference conditions are segregated during training, minimizing stimulus uncertainty,[54] while there may be no learning when stimuli are roved.[27]

2.6.5 Color

Color is an important feature of natural stimuli, yet there have been relatively few studies of perceptual learning involving color. This may reflect the experimental demands of calibrating truly isoluminant stimulus displays in the laboratory to isolate pure differences in color while eliminating differences in luminance. (Some experiments approach this by adding luminance noise to mask luminance contaminants.) The few existing studies focused on learning color discrimination (Type I or II) or color categorization for fixed color stimuli (Type III). From these, we know very little other than that training sometimes improves color judgments.

One study compared the detection of large luminance-defined and chrominance-defined gratings 9° in the periphery before and after training and found improvements in chromatic-contrast detection (in a 2IFC task) of about 40%.[55] Another study reported modestly improved isoluminant color discrimination 7° in the periphery.[56] Discriminating colors within compound gratings, or pattern learning, may also occur.[57] The accuracy of one-step hue or lightness discriminations in the constant-step Munsell color space (holding the other dimension constant) has been reported to improve with practice,[58] while training a color categorization can make color discrimination, equally difficult before training, easier for cross-category categorization and more difficult for within-category categorization.[58]

2.6.6 Acuity

Visual acuity is the sharpness or clarity of vision, measured by the ability to discriminate patterns, such as letters or numbers, at a particular distance. Acuity is limited by many factors, including the optics of the eye, the status of the retina, and neural processing. Of these, training can only improve neural processing or decision-making. Acuity measurements, such as the Snellen eye chart, are a standard clinical benchmark. The eye doctor usually measures acuity at the fovea, although the perception of patterns in the periphery can also affect visual functionality in the real world.

The effects of practice on acuity were among the earliest observations of perceptual learning. Most of the experiments varied the size of high-contrast patterns ("optotypes," or symbols), but other tests, such as dot or line resolution, have also been used. All three training paradigms (Types I, II, and III) have been used in different studies; they involve testing at the fovea or occasionally in the parafovea. The history of experimentation goes back over a hundred years.[41] Practice effects have been reported for the minimum display duration and size for letter recognition;[59] the threshold gap size for discriminating Landolt C orientation (a circle with a gap located up, down, left, or right)[60, 61] with transfer to similar size stimuli;[61] the threshold illumination for identifying the direction of a "tumbling E" (figure 2.7);[62, 63] the minimum separation to resolve a wider one-line stimulus from a two-line stimulus;[64, 65] and the highest-spatial-frequency sine wave that can be resolved.[66] There have also been reported failures of training even in the periphery. One experiment[67] failed to find improved Landolt C identification or peripheral two-line separation, and another failed to find improvements in line-gap discrimination in the parafovea.[68] The robust learning in many of these experiments occurred in paradigms that practiced one or a very small number of stimuli, while failures to improve acuity usually trained various sets of stimuli. So, failures of learning in the parafovea or periphery may reflect training stimulus mixtures.

Acuity is the most prominent end point in clinical vision. Even when other visual judgments are trained, the protocol sometimes evaluates pre- and posttraining visual acuity. Given the widespread importance of contrast sensitivity in clinical vision, it is perhaps surprising that we know little about visual acuity in Type II tasks that vary contrast, or how susceptible such tasks may be to learning.

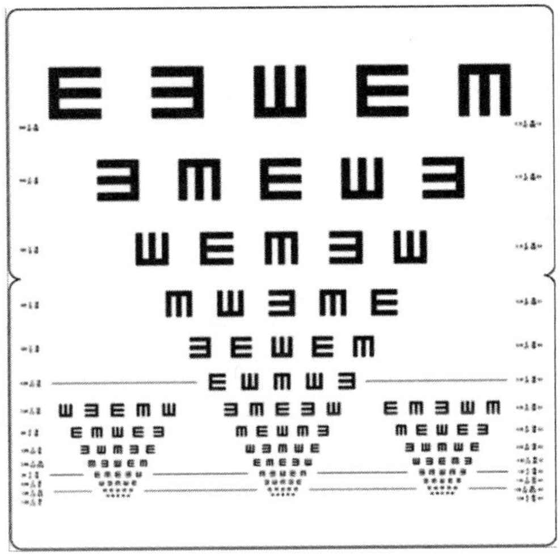

Figure 2.7
The tumbling E eye chart for measuring visual acuity.

2.6.7 Hyperacuity

Early researchers interpreted hyperacuity, or the ability to discriminate fine relative positions of pattern elements, as a form of perception that seemed at the sampling limits of the sensory receptors. For this reason, hyperacuity tasks are among the most extensively studied judgments in perceptual learning.[69, 70] Usually, identical stimuli are trained (Type III) and accuracy is the dependent measure, although some early studies[71] trained threshold offsets (Type I). Hyperacuity is usually studied in the fovea and only occasionally in the periphery or in the presence of visual masking. Type III tasks may allow observers to capitalize on the accidental properties of the retinal mosaic or of local noise in a specific retinal location, encouraging specificity in the later phases of training.[54]

In one standard task, named for seventeenth-century French mathematician Pierre Vernier (who invented a method for measuring distances between two marked lines in sextants and machine-tool devices), observers judge the offset of two lines abutting end to end. Vernier offset thresholds in humans are in the range of several arc seconds (1/3600 of a degree), far smaller than the one or two arc minute separations (1/60 of a degree) for resolving two lines from one line. A seminal paper[71] (figure 2.8) reported that practice reduced threshold offsets by half or more for some observers, that thresholds are higher for oblique (noncardinal) tests than for horizontal or vertical (cardinal) tests, but that learning occurs in all orientations. Other hyperacuity tasks include three-line bisection, in which an observer judges whether a middle line is closer to one or another flanking line, and three-dot Vernier and bisection tasks, which judge whether a middle dot is aligned left or right of two reference dots for Vernier or is in the middle for bisection.[72, 73]

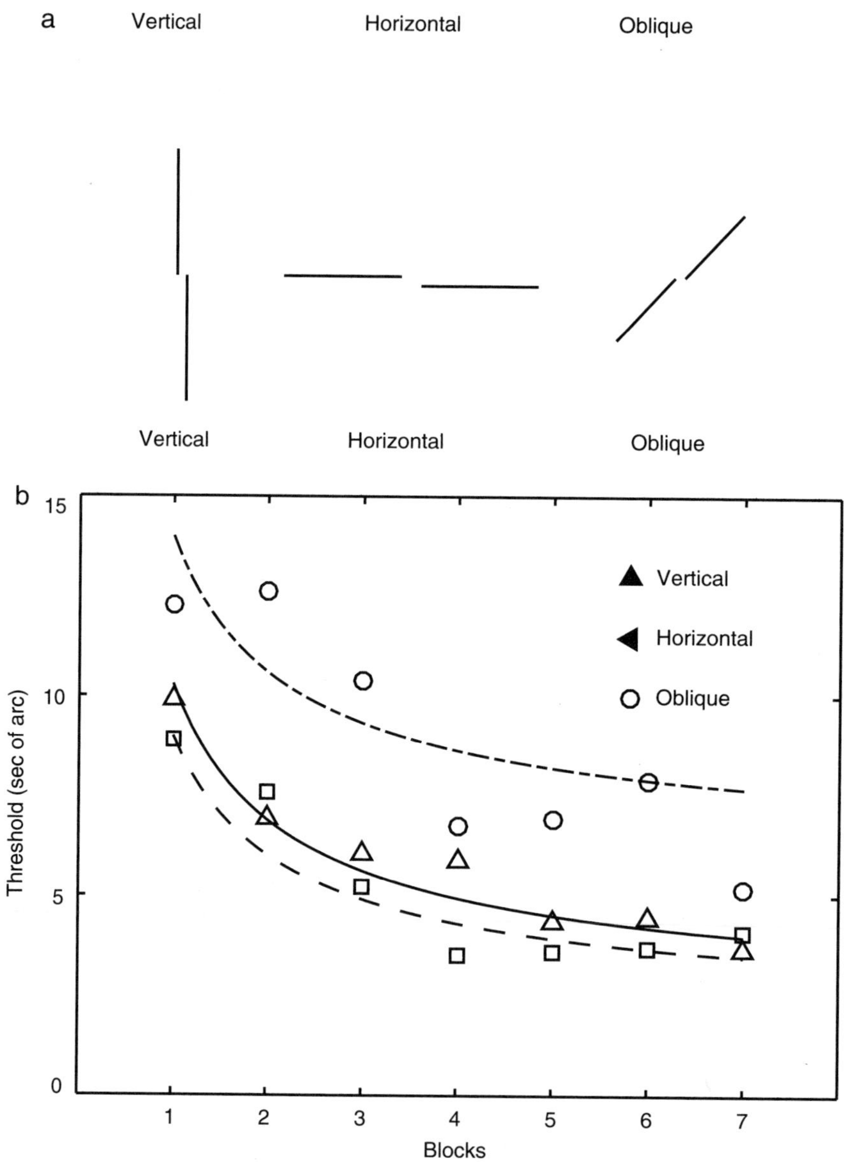

Figure 2.8
Perceptual learning of Vernier hyperacuity, showing (a) stimulus illustrations and (b) threshold learning data (one observer), fitted with exponential learning curves. Redrawn from data in McKee and Westheimer,[71] figure 1.

Learning in this domain requires trial-by-trial feedback[74–76] and can be specific to the retinal location of training and partially specific to the eye of training[73, 77–79] (chapter 3). Over all studies, there are individual differences in both initial performance and in how much is learned.[68, 72] The development of expertise in these high-precision hyperacuity tasks presents one fascinating example of perceptual learning and demonstrates many of the same results as in other perceptual tasks. These include individual differences in learning, differences between judgments in cardinal and noncardinal directions, the damaging effects of mixing stimuli during training, and differences between training regimens.

2.6.8 Summary

Research into the perceptual learning of single-feature judgments, generally associated with low- to low-middle-level vision, reported many vivid instances of substantial improvements. Learning has been shown to change performance by a factor of two or more, improvements that could potentially transform visual judgments from very low to quite respectable levels of accuracy. Aggregating over studies and tasks, a number of generalizable phenomena emerge. Special dominant forms of a feature, such as the cardinal versus noncardinal axes in orientation, often show less learning than the nondominant forms, which show worse initial performance and thus have room for learning that is more robust. Learning magnitude is likewise larger when the stimuli are masked or involve external noise, or when the task is carried out in the visual periphery. In short, training often has its largest effects for stimuli in the periphery, for nondominant feature values, and in the presence of external noise.

At the same time, learning these features—many of which are coded in the early visual cortical regions—almost surely requires *selection* (upweighting) of the most relevant representations among the many that preexist. The features of orientation, spatial frequency, phase, contrast, color, acuity, and hyperacuity (often related to the coding of orientation) are all densely represented in the cortex, because they all serve a general function in the perception and recognition of common visual patterns. From this, it follows that although including or weighting the relevant features for a particular judgment must be learned, the representations of those features are already present in the visual cortex.

Despite the important insights gleaned from years of concentrated research, a vast experimental terrain awaits exploration. Fundamental questions remain to be answered. Have observed improvements been fully optimized by the training protocols? Can improving basic visual feature judgments have a cascading effect on tasks relying on higher levels of analysis in the visual pathway? Although we know a great deal more now than we did only three decades ago, our theoretical understanding promises to expand as new experiments are designed to test domains and tasks that are less often studied. Likewise, the more we know about the physiology underpinning training effects (chapter 5) and the functional nature of optimization procedures (chapter 12), the more able we will be to build a thorough theoretical understanding of learning.

2.7 Perceptual Learning of Patterns

The second most commonly studied form of perceptual learning involves judgments about visual patterns. These patterns, often compositions of features or aggregations of features over space or time, are usually considered as being processed by mid-level vision. The pattern domain is also of special interest because learning might occur at multiple levels: at the early feature level and/or at the mid level. This section considers perceptual learning in compound stimuli, texture, depth, and several forms of motion.

2.7.1 Compound Stimuli

Combining several simpler patterns or elements creates compound stimuli that may better approximate images of real-world objects. Judgments about compound stimuli, in experiments that generally train on identical stimuli (Type III paradigms), often show very significant performance improvements. Most cases studied involve either "plaids" or spatial-frequency compounds combining several spatial-frequency patterns.

Relatively robust learning in pattern-based discrimination stands in contrast to the failures to learn to discriminate small differences in single spatial frequencies.[40] Early experiments[40, 42] used simple patterns combining sine-wave components of different frequencies and contrasts (e.g., 1 cycle per degree at 40% contrast and 3 cycles per degree at 13% contrast) and tested for changes in the contrast of one component, the presence or absence of one component, or the phase shift of one component relative to another (see figure 2.4). Learning was reported to be specific to orientation, spatial frequency, and visual field, but—consistent with reliance on mid- or high-level representations—not specific to the eye of training. Pattern-learning experiments have also used plaids, oppositely oriented sine-wave patterns of different spatial frequencies (figure 2.9).[80] Learning is especially robust when compound pattern masks require mid-level visual processes that aggregate information from low-level analyzers, implying that learning may be primarily localized in mid- to high-level mechanisms and not in low-level spatial-frequency or orientation analyzers.

2.7.2 Texture, Global Patterns, and Search

Learning in texture tasks has a unique role in the recent history of perceptual learning. Strong claims that learning is specific to retinal location—a finding that seemed to fly in the face of the idea that changes in early visual areas occurred only during a critical period—were first introduced with texture tasks. Texture patterns tile space with smaller patterns, and the observer identifies the position or orientation of elements differing from the background. In related visual search tasks, the observer identifies the presence or absence of one or more target element(s) or of an odd element in the display. Learning could occur at either the level of coding of individual elements and/or at the level of their aggregation in mid-level vision.

Training or practice can significantly improve discrimination accuracy in texture, global pattern, and visual search tasks. The studies generally use composite stimuli made of lines

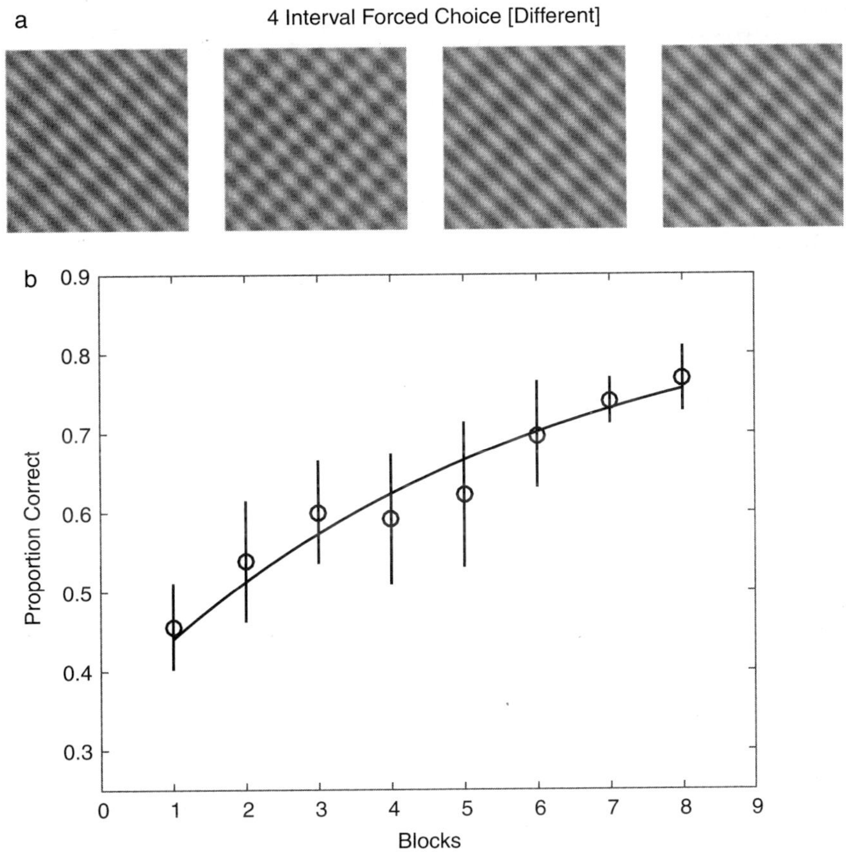

Figure 2.9
Learning to discriminate differences in compound plaid stimuli with orthogonal sine-wave stimuli. (a) Stimulus illustrations of the four-interval forced-choice task. (b) Improvement in proportion of correct detections with training, with fitted power-function learning curve. Redrawn from data in Fine and Jacobs,[80] figure 4a, with learning curve added.

or other features in the periphery and train with constant stimuli (Type III), measuring accuracy or response time. Some studies manipulate the stimulus asynchrony to a mask (Type III or Type II).

In perhaps the best-known example, Karni and Sagi[81] studied learning in a texture-discrimination task (TDT). They measured performance at different delays between a brief texture display and a pattern mask, and they found that threshold SOA (stimulus onset asynchrony, producing 80% correct) shortened by almost a factor of four over thousands of trials (figure 2.10). Learning was specific to the quadrant of the target orientation patch and the eye of training, leading to the widely influential theoretical conclusion that perceptual learning reflected plasticity in monocular cells of V1. Somewhat similar visual

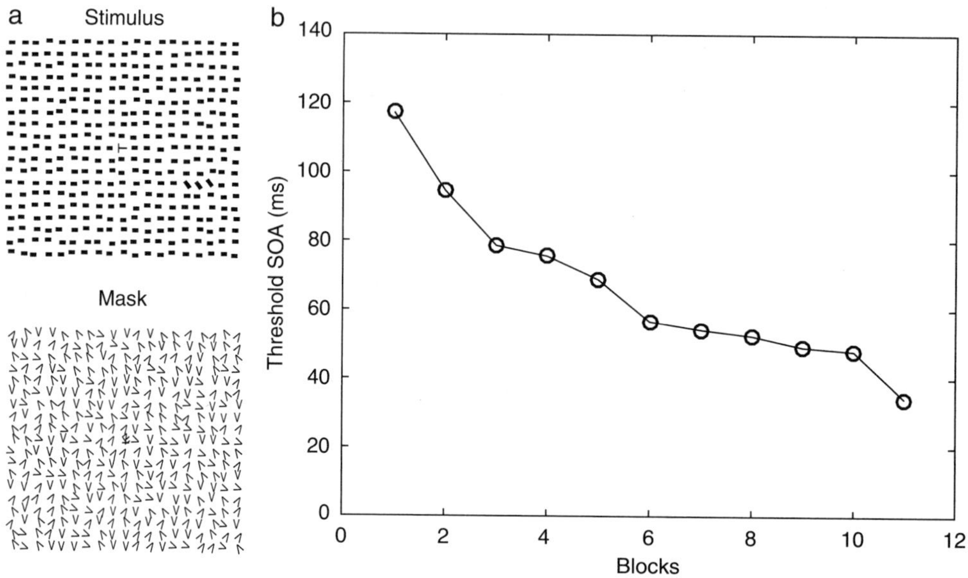

Figure 2.10
Learning in the texture-discrimination task. (a) Stimulus display with a horizontal texture patch (lower right) among distractors and a postmask. (b) Improvements in threshold SOA (stimulus onset asynchrony between the stimulus and the mask) at 80% correct with practice. Thresholds computed from data read from Karni and Sagi,[81] figure 1, upper right.

search tasks requiring the detection of a single odd element differing from the texture background, or pop out,[82] instead led to another influential proposal, the reverse hierarchy theory,[83] in which learning was hypothesized to begin high in the visual hierarchy and transition to low levels as required. An alternative claim is that improvements in these tasks reflects learning the temporal pattern of the stimulus sequence.[84]

Perceptual learning also occurs in standard visual search tasks. One study,[85] for example, examined perceptual learning in pop-out searches with a singleton of a different color, size, or orientation and visual searches for conjunctions of two features, such as color and orientation or color and size. Extensive practice improved the response time and/or accuracy for visual searches for individual features.[85–88] In contrast, some studies found little learning in conjunction searches,[89] others claim that perceptual learning can eliminate the disadvantage for conjunction searches compared to feature searches,[90] while still others report that perceptual learning improves conjunction searches but does not eliminate their characteristic dependence of response time on the number of elements.[85, 86, 88] The variation in evidence for learning in these domains, depending on the task and the training protocol, might be better understood in the context of specific computational models.

2.7.3 Depth

Another major feature dimension is perceived depth, derived from relative disparity. Stereo perception has been measured by something as simple as distinguishing the depths of short rods or as complicated as random-dot stereograms that depict objects by the disparity between dots in the left and right eyes. Not only can the accuracy of depth judgments improve, but researchers also sometimes report reduced response times for perceived depth to emerge (e.g., for random-dot stereograms), which can be quite slow for inexperienced observers. Most training studies have used Type III paradigms tested at the fovea, because stereo perception is limited in peripheral vision. Some reported performance improvements are quite large.

Two of the first reports of training of depth judgments[91, 92] studied how repeated viewing of a stereogram composed of short oriented line segments steadily reduced the perception time. This learning was specific to the orientation of the short line segments in the stereo images.[91] Similarly, training on somewhat more complex shapes in random-dot stereograms speeded up depth perception and was specific to particular regions of space.[92] Practice with random-dot stereograms can even show improvements in some observers in the perception of originally subthreshold patterns, measured by EEG responses.[93]

Improvements with practice can be quite large. One study[94] trained the ability to perceive relative depths of two outline squares at different distances from the fovea. Thresholds in the periphery improved by 60%–80% over the course of 3,000–4,000 trials of training, while improvements in the fovea were more modest.

2.7.4 Motion

Detecting motion is critical in a dynamic visual world. It helps highlight an object against camouflage and is required in judgments about moving objects. The perception of motion requires detecting the displacement of features of objects and integrating multiple motion signals over objects or regions to determine the direction and speed of motion. Motion is another domain where perceptual learning has been widely studied. Usually, this has involved Type III paradigms in which performance improves for the same stimuli, some have used Type I paradigms that track angular discrimination thresholds for detecting differences in motion direction, and a very few measured contrast thresholds in Type II studies. Learning often substantially improves motion perception, but the space is complicated—there are many possible forms of motion stimuli, and a dependence on the temporal frequency or speed of motion is possible—so even given the relatively large number of studies, the field has just started to evaluate this very large stimulus space.

Ball and Sekuler[97, 98] were the first to investigate perceptual learning for motion stimuli, documenting many important properties. They studied improvements over thousands of trials in discrimination of small differences in motion direction (3°) near a trained direction in random dot motion (figure 2.11). Improvements were larger for oblique directions, but there was some learning near cardinal directions as well; learning was moderately

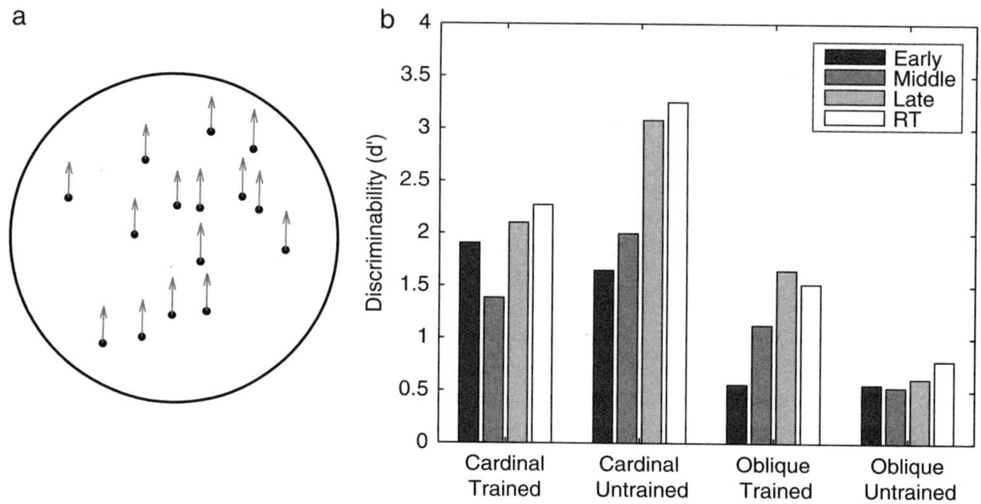

Figure 2.11
Direction-specific perceptual learning in random dot motion. (a) A random dot-motion display. (b) Improvement in same versus different judgments for 3° differences from a trained direction. Redrawn from selected data read from Ball and Sekuler,[95] figure 1.

specific to the trained direction (although surprisingly there was some generalization to directions at ±45° from the trained direction) and no transfer to motion detection. High interocular transfer suggested that learning likely involved representations above the primary visual cortex (V1), leading these researchers to propose that motion learning occurred in MT or other higher visual areas.[97]

Another study[38] showed learning, measured as motion thresholds, for a slowly moving single dot in both cardinal and oblique directions but only for oblique directions at higher motion speeds (see also Matthews et al.[97]). Yet other studies documented direction-specific perceptual learning in low-coherence displays (25% of dots moving coherently either left or right and all others in random directions)[98] even when any one dot occurred only on two successive image frames (two-frame dot lifetimes).

One influential study showed learning that reflected knowledge of motion at the local and global levels in succession, even when no individual dots moved in the global direction.[99] There were three conditions: (a) local dot motions drawn from ±5° relative to the global motion direction; (b) local dot motions drawn from ±30° relative to the global motion direction; and (c) local dot motions drawn from ±30° but not from ±5° relative to the global motion ("center missing"). Testing motion-direction discrimination around 11 directions before and at several points during training revealed learning in local motion discriminations first, followed later by learning in motion directions near the global motion direction—implying that global motion is learned after local motion coding (figure 2.12).

Learning has also been studied using other (not random dot) motion stimuli. Discrimination of motions of line stimuli designed to stimulate either just on-pathway or just

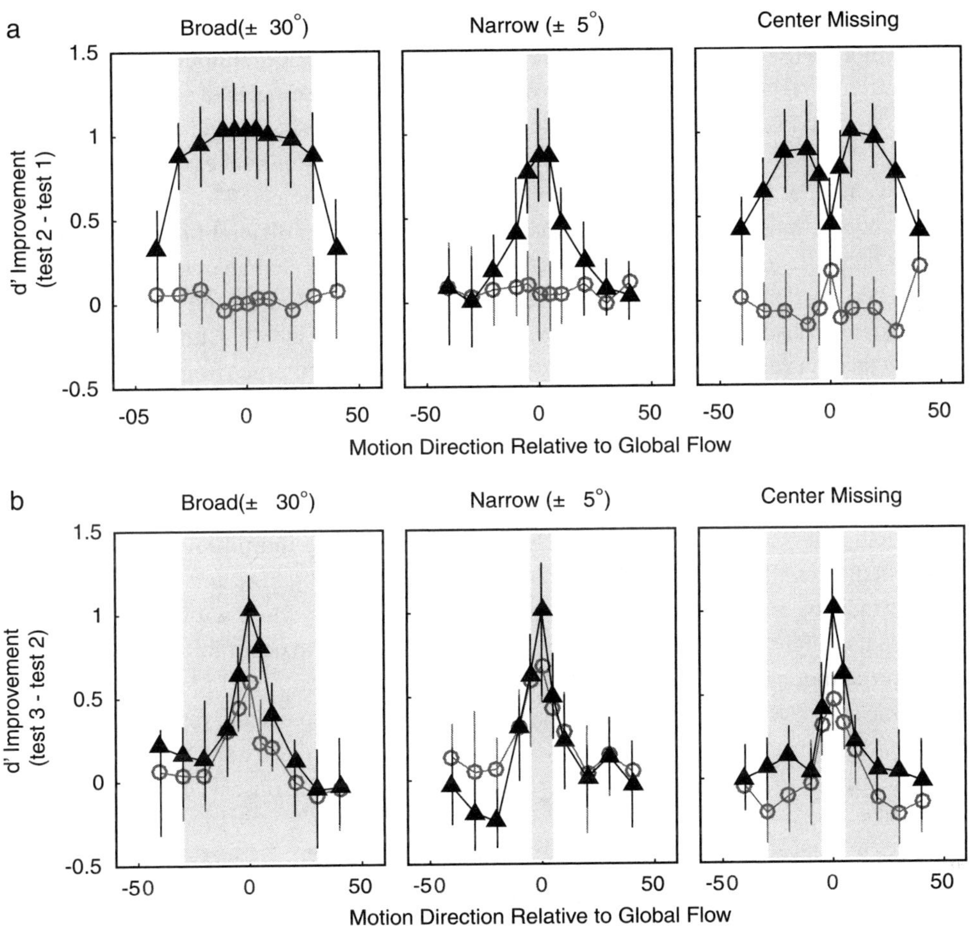

Figure 2.12
Learning to discriminate global motion direction in random dot-motion displays and changes in motion-direction discrimination (d' improvements) in the exposed (black triangles) and unexposed (gray circles) directions at two stages of learning, (a) and (b), and trained with broad, narrow, or center missing-dot motion distributions. After Watanabe et al.,[99] figure 3, with permission.

off-pathway motion[100] showed twofold improvements with practice in many observers, but this was not specific to on or off stimuli; other observers who started with good initial performance showed little learning. Learning also occurs for sine-wave motion embedded in varying levels of external noise;[101, 102] training improved performance in all levels of external noise, and training in low external noise transfers to tests in high external noise.[102]

Learning has also been reported in motion tasks that are more complex, such as when a moving object is defined by coherent motion against a dynamic noise background of new random dots on each frame. Moving objects defined by coherent motion in the same direction within an object region (Φ motion), by dots that are static (μ motion), and/or by dots moving coherently in a direction orthogonal to the object motion (θ motion) all showed improvements (reductions) of coherence thresholds with practice, although there is a complex pattern of asymmetric transfer between types.[103] The three types of stimuli correspond to the motion systems proposed by Lu and Sperling:[104] first-, second-, and third-order motions. The first-order motion system responds to moving luminance patterns (Φ)—here the movement of the luminance dots in the direction of the object. The second-order system responds to regions without luminance changes but with different properties of contrast or of flicker (μ). The third-order motion system tracks the change in location of regions of salience (θ). Perceptual learning in first- and second-order motion and different patterns of transfer has also been reported by several other laboratories.[105, 106] Although motion perception itself is very complex, learning of motion tasks can emerge from reweighting (see chapter 6).

2.7.5 Summary

Learning improves judgments about basic low-level visual features in some circumstances more than others (see section 2.3), while perceptual learning in mid-level tasks has been found in almost all cases—a ubiquity claimed in the review by Fine and Jacobs.[25] Certain caveats must be added, however. Some individuals, especially those with very good initial performance, may exhibit little learning. More learning tends to occur for noncardinal stimuli, such as motions or patterns in oblique directions, as in learning of low-level features. For some mid-level tasks, learning also seems to be limited to certain regions of the stimulus space, such as in very slow or very fast motion. Despite these qualifications, learning in these mid-level visual tasks is more robust than the learning observed in low-level feature tasks.

These mid-level tasks may also reflect a middle ground in the distinction between learning through the selection or creation of representations. Our interpretation is that perceptual learning in these mid-level tasks—compound pattern stimuli, texture, depth, and forms of motion—sometimes involves selection or winnowing of existing representations and sometimes involves the creation of new representations that participate in decisions. Although many simple forms of motion and texture may be precoded in visual areas (V1, MT, MST, etc.), the set of possible combinations of features in compound pattern stimuli or textures are unlikely to be precoded. It follows that the same would be true for all possible combinations of stimulus features supporting the perception of complex motions.[107] Learning in these cases likely involves not only selection (upweighting) of the feature representations most relevant to support the required judgments but also the creation or recruitment of new ensembles that represent new combinations.

The form of learning in pattern perception is especially interesting precisely because it could potentially occur at multiple levels—thus providing an opportunity to separate learning at a basic feature level, at the mid-vision level, or at even higher decision levels. To pursue these questions, techniques to measure learning at different levels must be devised on a case-by-case basis. One notable example is the study by Watanabe et al.[99] using random dot-motion stimuli that decoupled local motion cues from global motion cues, while other transfer paradigms address the same issue differently (see chapter 3).

Mid-level tasks present an opportunity to decipher whether learning occurs primarily at one level, at both levels simultaneously, or in sequence. The influential reverse hierarchical theory put forward by Ahissar and Hochstein posits that learning progresses from higher levels of the visual hierarchy and only later involves lower levels as required by task demands.[82, 83, 108, 109] Their hypothesis is intriguing, yet remains to be fully tested experimentally.

2.8 Perceptual Learning of Objects and Natural Stimuli

The stimuli that are closest to our everyday perceptual experience are also the most complex and difficult to study in the laboratory. Objects and other natural stimuli—shapes, faces, and complex motions—rely on configurations of features to define individual examples. They almost surely involve processes at multiple levels, starting with early representations of visual features, then mid-level feature aggregations or patterns, and ending with higher-level representations. In this section, we examine perceptual learning studies that have involved contours, shapes, and objects; faces and novel animal-like entities (e.g., Greebles); and animated biological motion. Whereas tasks requiring judgments at lower levels rely more heavily on the selection of existing representations, perceptual learning in many or most of these high-level tasks seems to involve creating new representations of specific objects.

2.8.1 Contours, Shapes, and Objects

Identifying contours, shapes, and objects in complex visual arrays is fundamental to visual perception. Seemingly effortless and autonomous in humans, these high-level visual functions actually involve several levels of analysis. They must represent local features, build up visual patterns and extended contours, and ultimately identify objects from different viewpoints.

The question of whether these processes can be enhanced by experience again presents itself. Essentially all reports in the literature suggest an affirmative answer, following certain standard patterns. The learning observed is often specific to the trained exemplars, suggesting that perceptual learning functions to develop new entries or new access to a shape lexicon. The vast majority of studies likewise use Type III training and assessment, demonstrating improvements in identification or classification of the same items with practice. That said, the few available studies are but a small sample from a vast but underexplored domain of natural object perception.

The identification of shape contours built from low-level pattern elements can be dramatically improved through learning when the object sets are small. In one study,[110] shape contours were outlines made up of small collinear oriented Gabor patches among a background of other oriented elements, similar to a shadow shape in a texture field. Shape contours took on high salience if the background elements were of a single orientation, or low salience if the background elements were randomly oriented. Training on a small set of contours improved detection and classification, affecting fMRI responses to the trained objects, while performance for untrained objects remained unchanged (figure 2.13). (Background element orientations were varied from trial to trial to eliminate learning to discount a specific background pattern).[82] In another study, learning a small set of arbitrary two-dimensional blob shapes was specific to retinal location,[111] while in yet another study, pattern recognition improved primarily when Gabor orientation was oblique or orthogonal to the induced contour lines, departing from the direction of the contour.[112]

Recognition of real three-dimensional objects from two-dimensional images is another classic visual function.[113] One of the long-standing theoretical questions in perception concerns what kinds of representations support three-dimensional object perception—whether it involves real three-dimensional shape representations, three-dimensional shape representations defined by component shapes,[114] or is instead inferred by interpolation from a set of two-dimensional view-dependent representations.[115–117] One robust finding is the preference for familiar views of objects,[118] often with some viewpoint dependence.[119, 120]

A few studies have looked at training on three-dimensional object recognition. In one of these,[121] the exposure duration for recognizing outline pictures of common objects at threshold durations showed long-lasting improvements that were specific to the trained objects. Unlike the previous examples using novel arbitrary shapes in which the object's identity or name was also learned, here all objects represented object categories known prior to the experiment.

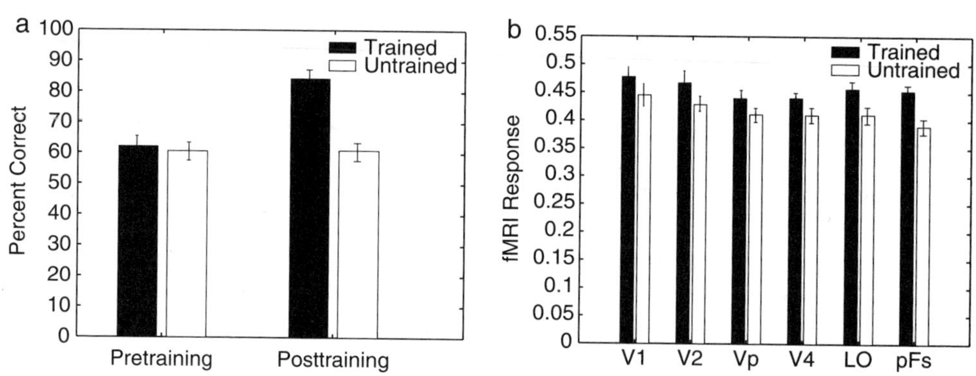

Figure 2.13
Perceptual learning improves detection of trained contours of oriented Gabor patches among orientation background elements. (a) Accuracy of symmetry judgments improved with training for a small set, (b) as did fMRI responses to the trained shapes, while performance for untrained stimuli was unchanged. After Kourtzi et al.,[110] figures 3a and 3c. Creative Commons, copyright 2005 Kourtzi et al.

2.8.2 Faces and Entities

Identifying a particular face or recognizing the emotion conveyed by a facial expression is a foundational aspect of human social interaction. Effective identification depends on configurations of features and their relationships to one another (e.g., space between eyes or distance from the nose to the mouth).[122–124] These functions involve a number of brain regions identified by dysfunctions associated with damage in specific regions.[125–127] Despite the wide interest in the perception of faces and other artificial entities, there is relatively little literature about learning to recognize, label, or name faces or entities. Perceptual learning or practice improves the identification of practiced faces or entities, just as the face and name of a person emerges with more familiarity. As with other studies of high-level perceptual learning, most used high-contrast stimuli in Type III paradigms.

In one study, observers learned to identify and label 10 previously familiarized faces in differing amounts of external noise.[128, 129] This study, one of the only Type II experiments in high-level vision, used adaptive methods to assess contrast threshold. Paralleling the results[12, 13] (figure 2.2) for Gabor orientation, perceptual learning occurred at all levels of external noise. The same data pattern occurred in identifying filtered texture patterns (which may or may not involve configural patterns like faces).[128, 129] When the experimenters arranged for faces to differ in just the eyes, the nose, or the mouth, observers can learn to differentially weight the most diagnostic regions of faces.[130]

Acquisition of expertise about novel stimuli has also been studied with artificial visual entities or avatars called *Greebles* (figure 2.14), each defined by a configuration of features. (In movie animation, the term *greeble* refers to adornment details added to basic shapes.) Expertise developed over 10 hours of mixed training that exposed individual Greebles with their gender labels, with name labels, or with family labels, while requiring that the observer produce or verify the gender, family, or name label of that Greeble.[131, 132] Not surprisingly, the ability to identify and name previously novel Greebles improved with training. What also emerged over time was a sensitivity to an upright orientation analogous to the differential processing of upright human faces.[133] These experiments broadly suggested that human expertise in face recognition might develop through analogous experiences.

2.8.3 Biological Motion

The ability to perceive and understand the motion of humans or animals, which is called biological motion, has been studied for decades. People can recognize the human figure—and other animate entities—even from very impoverished displays. First developed by Johannson,[134] early point-light displays attached lights to only a few points on major joints or limbs and posed the fascinating challenge of understanding how such displays are interpreted as human or animal and what kinds of information are extracted from them (see Blake and Shiffrar[135] for a review). The few studies with biological motion, all using Type III paradigms, show that training may be required to perceive challenging biological motion displays.

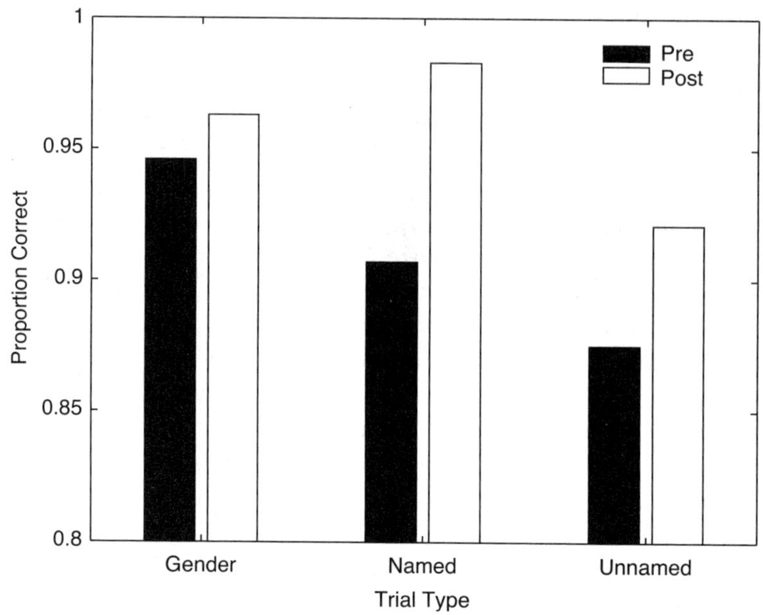

Figure 2.14
Practice creates Greebles experts as observers learn the names, families, and genders of avatars defined by particular configurations of visual features. Accuracy of performance for gender or individual names (trained with and without family names) before and after training is shown. Redrawn from selected data in Gauthier et al.,[131] figure 4.

In one study,[136] observers practiced discriminating biological motion displays from nonbiological ones with a varying number of noise dots masking the animations. The experimenters scrambled the starting locations of dots in biological motion animations to create controls for nonbiological motion animations. Following training, observers could tolerate a much larger number of noise dots and still distinguish biological motion displays from nonbiological ones. In another set of studies, observers learned to classify motion animation sequences generated from a skeletal structure—either biological or nonbiological—undergoing "feasible" motions very quickly, within just a few trials.[137–139] The observers were asked to decide whether two animation sequences were the same or different, with nonmatching animations selected from the same class category. The existence of a skeletal structure seems to allow observers to encode and remember animation sequences, at least for short periods.

2.8.4 Summary

Our perception of objects and other natural stimuli associated with high-level vision[25] can seemingly be trained quite rapidly, although learning may continue over extended periods. Such conclusions, however, are based on a very sparse set of studies and stimulus types. With few exceptions, learning has been assessed using small, fixed sets of easily visible items of high contrast (Type III). It remains to be determined whether training promotes the ability to distinguish increasingly similar exemplars (Type I). There are many open questions about learning in such high-level tasks.

In some ways, learning to identify these stimuli is the stuff of classical expertise. Humans are experts at recognizing faces and objects in the world, yet most existing studies train identification of novel objects, requiring observers to learn the response mapping or names of the items at the same time that they learn specialized perceptual features or configurations. Trained improvements in recognizing specific entities rarely seem to generalize to the processing of other examples of the same class.

Understanding the nature of learning in high-level tasks could very well play a critical role in the translation of perceptual learning from laboratory tasks to improved performance in the natural context. Even as current research remains confined to a limited experiment set, the existing studies set the stage for future work, including a more complete analysis of the development of exceptional perceptual expertise. Furthermore, the study of high-level perceptual learning opens up important theoretical questions. Future work is needed to determine whether learning occurs at multiple levels, as in some mid-level visual tasks, or exclusively at the higher level of creating new combinations.

In any visual task, the challenge is to create experiments that separate learning at different levels. Does training aimed at improving the processing of basic features or mid-level patterns impact the learning of objects, faces, and other natural stimuli? If so, how can this training be optimized? The answer to that question is likely to point the way for a host of real-world applications.

2.9 Conclusions

In this chapter, we organized our treatment of perceptual learning into three levels based on the nature of the training feature and argued that learning tasks at each of them involved a playoff between selection and creation of new representations, depending on the primary level of analysis required by the task. The three levels were *single-feature* tasks related to low-level vision, including orientation, spatial frequency, phase, contrast, color, acuity, and hyperacuity; *pattern* tasks related to mid-level vision, including compound stimuli, texture patterns, global orientation, search, depth, and motion; and *objects and natural stimuli* identification tasks that require more complex higher-level visual analysis, including contours, shapes or objects, faces or entities, and biological motion. At the same time, additional factors were consistently shown to affect the speed and extent of learning. These included whether the training occurred in the fovea, parafovea, or periphery; whether the stimuli involved noncardinal, or marked, feature values; and whether extraneous features or masks were present. Finally, we classified the testing paradigms used into three types of tasks (Types I, II, and III). These different paradigms (typically used for both training and assessment) create different learning experiences that may substantively affect not just learning but also transfer and generalization (for a related discussion, see chapters 3 and 12).

Based on this organization, we showed that the robustness of perceptual learning depended on the level of analysis that was the primary basis of the task judgment. Although perceptual learning occurs in most situations at every level, it is generally more robust for tasks requiring higher-level judgments. Still, the variation in learning rate and mechanism within each level can be substantial, and there is considerable overlap in the phenomenology in low-, mid-, and high-level tasks, corresponding to different levels of representations and processes in the visual hierarchy. Furthermore, we believe that underinvestigated factors, such as the type of task or the training experience, may prove to be almost equally important in determining learning (and transfer).

The relation between the behavioral and physiological phenomena of visual perceptual learning is complex. Even if a task focuses on representations or features coded early or late in the hierarchy of visual representations, plasticity may occur at multiple levels, including those beyond the visual cortex. Learning in tasks focused on low-level features may have cascading consequences for higher levels of visual analysis. Likewise, learning in tasks involving features coded in mid-level or high-level vision might also rely on learning features coded at low levels of visual representation. Plasticity certainly involves decision and, almost certainly, top-down processing. Future investigations will be required to fully identify how plasticity occurs across brain modules.

The distinction between learning through selection and learning through creation was critical for our theoretical analysis. The literature is broadly consistent with a trend from selection or winnowing of existing representations for tasks focused on features coded in early visual areas to the creation of new combinations of features and their configurations for tasks focused on high-level visual stimuli, as illustrated in figure 2.1. This in turn raises the question of how the difference between selection and creation relates to the difference between reweighting and representation change or to stability and plasticity.

In anticipation of a discussion threaded through subsequent chapters, it should be noted that reweighting processes are remarkably powerful. They can affect how evidence from stable early visual representations might be used to make decisions through a process of selection or winnowing. Alternatively, the reweighting of inputs from several representations at a lower level that feed into a higher representation can change the responses of higher-level representations, recruiting and then retuning new nodes (neural ensembles) to represent new objects and categories. We believe that the general class of reweighting models provides the strongest conceptual framework yet developed to model learning in most tasks. This point is further pursued in chapters 6, 7, and 8.

Despite the resurgence of research on perceptual learning since the 1990s, many issues remain largely unexplored. Existing experimental studies are densely clustered in small regions of the enormous territory of all perceptual learning tasks, involving different judgments, stimuli, and training paradigms. The functions of orientation discrimination,

Table 2.2
Potential new frontiers in perceptual learning

- Exploring the fine-grained temporal changes throughout perceptual learning, including rapid learning or within-session deterioration
- Evaluating the effectiveness of forms of training that are decoupled from the assessment of a target task or on a battery of target tasks
- Devising new tests to identify perceptual learning at multiple levels
- Testing theoretical predictions of quantitative models of learning

hyperacuity, and motion-direction kinomatograms (dot motions) have been studied fairly extensively. Other aspects of visual function, relying on sensitivity to spatial frequency, phase, contours, object identification, and so on, have been studied only sparsely. Often a single training paradigm dominates the studies in a given domain. Research aimed at investigating perceptual learning in understudied stimulus domains or using a wider range of training protocols may identify new and different forms of learning and plasticity. Existing work reveals the importance, magnitude, and potential complexity of visual plasticity, yet it leaves many regions essentially unexplored, and new techniques may open up entirely new approaches.

In addition to simply extending the visual feature domains and judgment tasks, new methods for performance assessment may expand how we can study perceptual learning in the future. As discussed earlier in the chapter, given the requirement for a large number of trials to assess performance, learning has almost always been measured at the scale of blocks or sessions of hundreds and sometimes many hundreds of trials. On the one hand, visual perceptual learning *can* continue to occur over many thousands of trials. On the other hand, learning sometimes includes a rapid component. Other temporal factors, such as deterioration or fatigue or consolidation, may occur within or between sessions. Recently developed fast-assessment methods have the potential to assess performance on a much finer timescale,[19, 21, 22, 140] sometimes trial-by-trial.[23, 24] Measurements on this timescale may reveal the fine temporal dynamics of learning that remain almost completely unexplored (see table 2.2).

In the existing literature, trials used to assess performance also serve as the method of training. The result is an unnecessary restraint on the form of training. Faster assessment methods would also permit decoupling of assessment on a target task from the methods of training used to improve performance on that task. Future experiments may also be able to assess performance changes in a battery of multiple tasks throughout the course of training with few trials, thus reducing the amount of learning during the assessments themselves.

References

1. James W. *The principles of psychology.* Vol. 1. Holt; 1890.
2. Weber EH. *E. H. Weber on the tactile senses.* 2nd ed. Ross HE and Murray DJ, eds. and trans. Erlbaum. 1996.
3. Stratton GM. Vision without inversion of the retinal image. *Psychological Review* 1897;4(4):341–360.

4. Stratton GM. Some preliminary experiments on vision without inversion of the retinal image. *Psychological Review* 1896;3(6):611–617.

5. Welch RB, Bridgeman B, Anand S, Browman KE. Alternating prism exposure causes dual adaptation and generalization to a novel displacement. *Attention, Perception, & Psychophysics* 1993;54(2):195–204.

6. Geisler WS. Visual perception and the statistical properties of natural scenes. *Annual Review of Psychology* 2008;59:167–192.

7. Simoncelli EP, Olshausen BA. Natural image statistics and neural representation. *Annual Review of Neuroscience* 2001;24(1):1193–1216.

8. Bavelier D, Levi DM, Li RW, Dan Y, Hensch TK. Removing brakes on adult brain plasticity: From molecular to behavioral interventions. *Journal of Neuroscience* 2010;30(45):14964–14971.

9. Bavelier D, Neville HJ. Cross-modal plasticity: Where and how? *Nature Reviews Neuroscience* 2002;3(6):443–452.

10. Huang C-B, Zhou Y, Lu Z-L. Broad bandwidth of perceptual learning in the visual system of adults with anisometropic amblyopia. *Proceedings of the National Academy of Sciences* 2008;105(10):4068–4073.

11. Zhou Y, Huang C, Xu P, Tao L, Qiu Z, Li X, and Lu, ZL. Perceptual learning improves contrast sensitivity and visual acuity in adults with anisometropic amblyopia. *Vision Research* 2006;46(5):739–750.

12. Dosher BA, Lu Z-L. Perceptual learning reflects external noise filtering and internal noise reduction through channel reweighting. *Proceedings of the National Academy of Sciences* 1998;95(23):13988–13993.

13. Dosher BA, Lu Z-L. Mechanisms of perceptual learning. *Vision Research* 1999;39(19):3197–3221.

14. Lu Z-L, Dosher BA. Characterizing human perceptual inefficiencies with equivalent internal noise. *Journal of the Optical Society of America A* 1999;16(3):764–778.

15. Lu Z-L, Dosher BA. Characterizing observers using external noise and observer models: Assessing internal representations with external noise. *Psychological Review* 2008;115(1):44–82.

16. Andersen GJ, Ni R, Bower JD, Watanabe T. Perceptual learning, aging, and improved visual performance in early stages of visual processing. *Journal of Vision* 2010;10(13):4,1–13.

17. Dobres J, Seitz AR. Perceptual learning of oriented gratings as revealed by classification images. *Journal of Vision* 2010;10(13):8,1–11.

18. Kattner F, Cochrane A, Green CS. Trial-dependent psychometric functions accounting for perceptual learning in 2-AFC discrimination tasks. *Journal of Vision* 2017;17(11):3,1–16.

19. Lesmes LA, Jeon S-T, Lu Z-L, Dosher BA. Bayesian adaptive estimation of threshold versus contrast external noise functions: The quick TvC method. *Vision Research* 2006;46(19):3160–3176.

20. Lesmes LA, Lu Z-L, Baek J, Albright TD. Bayesian adaptive estimation of the contrast sensitivity function: The quick CSF method. *Journal of Vision* 2010;10(3):17,1–21.

21. Lesmes LA, Lu Z-L, Baek J, Tran N, Dosher BA, Albright TD. Developing Bayesian adaptive methods for estimating sensitiizty thresholds (d′) in yes-no and forced-choice tasks. *Frontiers in Psychology* 2015;6,1–24.

22. Lesmes LA, Lu Z-L, Tran NT, Dosher BA, Albright TD. An adaptive method for estimating criterion sensitivity (d') levels in yes/no tasks. *Journal of Vision* 2006;6(6):1097.

23. Zhang P, Zhao Y, Dosher B, Lu Z-L. Evaluating the performance of the staircase and quick Change Detection methods in measuring perceptual learning. *Journal of Vision* 2019;19(7):14,1–25.

24. Zhang P, Zhao Y, Dosher B, Lu Z-L. Assessing the detailed time course of perceptual sensitivity change in perceptual learning. *Journal of Vision* 2019;19(5):9,1–19.

25. Fine I, Jacobs RA. Comparing perceptual learning across tasks: A review. *Journal of Vision* 2002;2(2):5, 190–203.

26. Hung S-C, Seitz AR. Prolonged training at threshold promotes robust retinotopic specificity in perceptual learning. *Journal of Neuroscience* 2014;34(25):8423–8431.

27. Herzog MH, Aberg KC, Frémaux N, Gerstner W, Sprekeler H. Perceptual learning, roving and the unsupervised bias. *Vision Research* 2012;61:95–99.

28. Parkosadze K, Otto TU, Malania M, Kezeli A, Herzog MH. Perceptual learning of bisection stimuli under roving: Slow and largely specific. *Journal of Vision* 2008;8(1):5,1–8.

29. Kuai S-G, Zhang J-Y, Klein SA, Levi DM, Yu C. The essential role of stimulus temporal patterning in enabling perceptual learning. *Nature Neuroscience* 2005;8(11):1497–1499.

30. Zhang J-Y, Kuai S-G, Xiao L-Q, Klein SA, Levi DM, Yu C. Stimulus coding rules for perceptual learning. *PLoS Biology* 2008;6(8):e197.

31. Schoups AA, Vogels R, Orban GA. Human perceptual learning in identifying the oblique orientation: Retinotopy, orientation specificity and monocularity. *Journal of Physiology* 1995;483(3):797–810.

32. Vogels R, Orban GA. The effect of practice on the oblique effect in line orientation judgments. *Vision Research* 1985;25(11):1679–1687.

33. Shiu L-P, Pashler H. Improvement in line orientation discrimination is retinally local but dependent on cognitive set. *Perception, & Psychophysics* 1992;52(5):582–588.

34. Ghose GM, Yang T, Maunsell JH. Physiological correlates of perceptual learning in monkey V1 and V2. *Journal of Neurophysiology* 2002;87(4):1867–1888.

35. Schoups A, Vogels R, Qian N, Orban G. Practising orientation identification improves orientation coding in V1 neurons. *Nature* 2001;412(6846):549–553.

36. Yang T, Maunsell JH. The effect of perceptual learning on neuronal responses in monkey visual area V4. *Journal of Neuroscience* 2004;24(7):1617–1626.

37. Burnat K. Are visual peripheries forever young? *Neural Plasticity* 2015;2015;307929.

38. Matthews N, Welch L. Velocity-dependent improvements in single-dot direction discrimination. *Perception & Psychophysics* 1997;59(1):60–72.

39. Lu Z-L, Dosher BA. Perceptual learning retunes the perceptual template in foveal orientation identification. *Journal of Vision* 2004;4(1):5,44–56.

40. Fiorentini A, Berardi N. Learning in grating waveform discrimination: Specificity for orientation and spatial frequency. *Vision Research* 1981;21(7):1149–1158.

41. Bennett RG, Westheimer G. The effect of training on visual alignment discrimination and grating resolution. *Perception & Psychophysics* 1991;49(6):541–546.

42. Fiorentini A, Berardi N. Perceptual learning specific for orientation and spatial frequency. 1980;287(5777): 43–44.

43. Dosher BA, Lu Z-L. The functional form of performance improvements in perceptual learning: Learning rates and transfer. *Psychological Science* 2007;18(6):531–539.

44. Dosher BA, Han S, Lu Z-L. Perceptual learning and attention: Reduction of object attention limitations with practice. *Vision Research* 2010;50(4):402–415.

45. Sowden PT, Rose D, Davies IR. Perceptual learning of luminance contrast detection: Specific for spatial frequency and retinal location but not orientation. *Vision Research* 2002;42(10):1249–1258.

46. Huang C-B, Lu Z-L, Zhou Y. Mechanisms underlying perceptual learning of contrast detection in adults with anisometropic amblyopia. *Journal of Vision* 2009;9(11):24,1–14.

47. Mayer MJ. Practice improves adults' sensitivity to diagonals. *Vision Research* 1983;23(5):547–550.

48. De Valois KK. Spatial frequency adaptation can enhance contrast sensitivity. *Vision Research* 1977;17(9):1057–1065.

49. Adini Y, Wilkonsky A, Haspel R, Tsodyks M, Sagi D. Perceptual learning in contrast discrimination: The effect of contrast uncertainty. *Journal of Vision* 2004;4(12):2,993–1005.

50. Dorais A, Sagi D. Contrast masking effects change with practice. *Vision Research* 1997;37(13):1725–1733.

51. Yu C, Klein SA, Levi DM. Perceptual learning in contrast discrimination and the (minimal) role of context. *Journal of Vision* 2004;4(3):4,169–182.

52. Swift DJ, Smith RA. Spatial frequency masking and Weber's Law. *Vision Research* 1983;23(5):495–505.

53. Maehara G, Goryo K. Perceptual learning in monocular pattern masking: Experiments and explanations by the twin summation gain control model of contrast processing. *Perception & Psychophysics* 2007;69(6):1009–1021.

54. Sagi D. Perceptual learning in vision research. *Vision Research* 2011;51(13):1552–1566.

55. Thurston C, Dobkins K. Stimulus-specific perceptual learning for chromatic, but not luminance, contrast detection. *Journal of Vision* 2007;7(9):469.

56. Sowden PT, Davies IR, Notman LA, Alexander I, Özgen E. Chromatic perceptual learning. In P G Biggam, C A Hough, C J Kay, and D. R. Simmons, eds., *New directions in colour studies.* John Benjamins Publishing Company; 2011:433–443.

57. Fiorentini A, Berardi N. Adaptation and learning in the visual perception of gratings. In Fahle M, Poggio T, eds., *Perceptual learning.* MIT Press; 2002:161–176.

58. Özgen E, Davies IR. Acquisition of categorical color perception: A perceptual learning approach to the linguistic relativity hypothesis. *Journal of Experimental Psychology: General* 2002;131(4):477–493.

59. Sanford EC. The relative legibility of the small letters. *American Journal of Psychology* 1888;1(3):402–435.

60. Bruce RH, Low FN. The effect of practice with brief-exposure techniques upon central and peripheral visual acuity and a search for a brief test of peripheral acuity. *Journal of Experimental Psychology* 1951;41(4):275–280.

61. Saugstad P, Lie I. Training of peripheral visual acuity. *Scandinavian Journal of Psychology* 1964;5(1):218–224.

62. Shvarts LA. Raising the sensitivity of the visual analyser. In Simon B, ed., *Psychology in the Soviet Union.* Stanford University Press; 1957:100–107.

63. Foley P, Lavery J, Abbey D. Scotopic acuity and knowledge of results. *Perceptual and Motor Skills* 1964;18(2):505–508.

64. McFadden HB. *The dependence of the resolving power of the eye upon the dynamics of the test field and upon practice effect.* PhD dissertation, Ohio State University:1940.

65. Wilcox WW. An interpretation of the relation between visual acuity and light intensity. *Journal of General Psychology* 1936;15(2):405–435.

66. Johnson CA, Leibowitz HW. Practice effects for visual resolution in the periphery. *Perception & Psychophysics* 1979;25(5):439–442.

67. Westheimer G. Is peripheral visual acuity susceptible to perceptual learning in the adult? *Vision Research* 2001;41(1):47–52.

68. Beard BL, Levi DM, Reich LN. Perceptual learning in parafoveal vision. *Vision Research* 1995;35(12):1679–1690.

69. Crist RE, Kapadia MK, Westheimer G, Gilbert CD. Perceptual learning of spatial localization: Specificity for orientation, position, and context. *Journal of Neurophysiology* 1997;78(6):2889–2894.

70. Poggio T, Fahle M, Edelman S. Fast perceptual learning in visual hyperacuity. *Science* 1992;256(5059):1018–1021.

71. McKee SP, Westhe G. Improvement in Vernier acuity with practice. *Perception & Psychophysics* 1978;24(3):258–262.

72. Fahle M, Edelman S. Long-term learning in Vernier acuity: Effects of stimulus orientation, range and of feedback. *Vision Research* 1993;33(3):397–412.

73. Fahle M, Edelman S, Poggio T. Fast perceptual learning in hyperacuity. *Vision Research* 1995;35(21):3003–3013.

74. Herzog MH, Fahle M. The role of feedback in learning a Vernier discrimination task. *Vision Research* 1997;37(15):2133–2141.

75. Herzog MH, Fahle M. Modeling perceptual learning: Difficulties and how they can be overcome. *Biological Cybernetics* 1998;78(2):107–117.

76. Herzog MH, Fahle M. Effects of biased feedback on learning and deciding in a Vernier discrimination task. *Vision Research* 1999;39(25):4232–4243.

77. Fahle M. Human pattern recognition: Parallel processing and perceptual learning. *Perception* 1994;23:411–427.

78. Fahle M. Specificity of learning curvature, orientation, and Vernier discriminations. *Vision Research* 1997;37(14):1885–1895.

79. Fahle M, Morgan M. No transfer of perceptual learning between similar stimuli in the same retinal position. *Current Biology* 1996;6(3):292–297.

80. Fine I, Jacobs RA. Perceptual learning for a pattern discrimination task. *Vision Research* 2000;40(23):3209–3230.

81. Karni A, Sagi D. Where practice makes perfect in texture discrimination: Evidence for primary visual cortex plasticity. *Proceedings of the National Academy of Sciences* 1991;88(11):4966–4970.

82. Ahissar M, Hochstein S. Attentional control of early perceptual learning. *Proceedings of the National Academy of Sciences* 1993;90(12):5718–5722.

83. Hochstein S, Ahissar M. View from the top: Hierarchies and reverse hierarchies in the visual system. *Neuron* 2002;36(5):791–804.

84. Wang R, Cong L-J, Yu C. The classical TDT perceptual learning is mostly temporal learning. *Journal of Vision* 2013;13(5):9,1–9.

85. Ellison A, Walsh V. Perceptual learning in visual search: Some evidence of specificities. *Vision Research* 1998;38(3):333–345.

86. Leonards U, Rettenbach R, Nase G, Sireteanu R. Perceptual learning of highly demanding visual search tasks. *Vision Research* 2002;42(18):2193–2204.

87. Sireteanu R, Rettenbach R. Perceptual learning in visual search: Fast, enduring, but non-specific. *Vision Research* 1995;35(14):2037–2043.

88. Sireteanu R, Rettenbach R. Perceptual learning in visual search generalizes over tasks, locations, and eyes. *Vision Research* 2000;40(21):2925–2949.

89. Treisman AM, Gelade G. A feature-integration theory of attention. *Cognitive Psychology* 1980;12(1):97–136.

90. Steinman SB. Serial and parallel search in pattern vision. *Perception* 1987;16(3):389–398.

91. Ramachandran V, Braddick O. Orientation-specific learning in stereopsis. *Perception* 1973;2(3):371–378.

92. Ramachandran V. Learning-like phenomena in stereopsis. *Nature* 1976;262(5567):382–384.

93. Skrandies W, Jedynak A. Learning to see 3-D: Psychophysics and brain electrical activity. *Neuroreport* 1999;10(2):249–253.

94. Fendick M, Westheimer G. Effects of practice and the separation of test targets on foveal and peripheral stereoacuity. *Vision Research* 1983;23(2):145–150.

95. Ball K, Sekuler R. A specific and enduring improvement in visual motion discrimination. *Science* 1982;218(4573):697–698.

96. Ball K, Sekuler R. Direction-specific improvement in motion discrimination. *Vision Research* 1987;27(6):953–965.

97. Matthews N, Liu Z, Geesaman BJ, Qian N. Perceptual learning on orientation and direction discrimination. *Vision Research* 1999;39(22):3692–3701.

98. Vaina LM, Sundareswaran V, Harris JG. Learning to ignore: Psychophysics and computational modeling of fast learning of direction in noisy motion stimuli. *Cognitive Brain Research* 1995;2(3):155–163.

99. Watanabe T, Náñez JE, Koyama S, Mukai I, Liederman J, Sasaki Y. Greater plasticity in lower-level than higher-level visual motion processing in a passive perceptual learning task. *Nature Neuroscience* 2002;5(10):1003–1009.

100. Wehrhahn C, Rapf D. ON- and OFF-pathways form separate neural substrates for motion perception: Psychophysical evidence. *Journal of Neuroscience* 1992;12(6):2247–2250.

101. Lu Z-L, Chu W, Dosher BA. Perceptual learning of motion direction discrimination in fovea: Separable mechanisms. *Vision Research* 2006;46(15):2315–2327.

102. Lu Z-L, Chu W, Dosher BA, Lee S. Perceptual learning of Gabor orientation identification in visual periphery: Complete inter-ocular transfer of learning mechanisms. *Vision Research* 2005;45(19):2500–2510.

103. Zanker JM. Perceptual learning in primary and secondary motion vision. *Vision Research* 1999;39(7):1293–1304.

104. Lu Z-L, Sperling G. The functional architecture of human visual motion perception. *Vision Research* 1995;35(19):2697–2722.

105. Chen R, Qiu Z-P, Zhang Y, Zhou Y-F. Perceptual learning and transfer study of first- and second-order motion direction discrimination. *Progress in Biochemistry and Biophysics* 2009;36:1442–1450.

106. Petrov AA, Hayes TR. Asymmetric transfer of perceptual learning of luminance- and contrast-modulated motion. *Journal of Vision* 2010;10(14):11,1–22.

107. Lu Z-L, Sperling G. Three-systems theory of human visual motion perception: Review and update. *Journal of the Optical Society of America A* 2001;18(9):2331–2370.

108. Ahissar M, Hochstein S. Task difficulty and the specificity of perceptual learning. *Nature* 1997;387(6631):401–406.

109. Ahissar M, Hochstein S. The spread of attention and learning in feature search: Effects of target distribution and task difficulty. *Vision Research* 2000;40(10):1349–1364.

110. Kourtzi Z, Betts LR, Sarkheil P, Welchman AE. Distributed neural plasticity for shape learning in the human visual cortex. *PLoS Biology* 2005;3(7):1317–1327.

111. Nazir TA, O'Regan JK. Some results on translation invariance in the human visual system. *Spatial Vision* 1990;5(2):81–100.

112. Schwarzkopf DS, Kourtzi Z. Experience shapes the utility of natural statistics for perceptual contour integration. *Current Biology* 2008;18(15):1162–1167.

113. Edelman S. Class similarity and viewpoint invariance in the recognition of 3D objects. *Biological Cybernetics* 1995;72(3):207–220.

114. Biederman I. Recognition-by-components: A theory of human image understanding. *Psychological Review* 1987;94(2):115–147.

115. Tarr MJ, Pinker S. When does human object recognition use a viewer-centered reference frame? *Psychological Science* 1990;1(4):253–256.

116. Bülthoff HH, Edelman SY, Tarr MJ. How are three-dimensional objects represented in the brain? *Cerebral Cortex*, 1995;5(3):247–260.

117. Bülthoff HH, Edelman S. Psychophysical support for a two-dimensional view interpolation theory of object recognition. *Proceedings of the National Academy of Sciences* 1992;89(1):60–64.

118. Edelman S, Bülthoff HH. Orientation dependence in the recognition of familiar and novel views of three-dimensional objects. *Vision Research* 1992;32(12):2385–2400.

119. Tarr MJ. Rotating objects to recognize them: A case study on the role of viewpoint dependency in the recognition of three-dimensional objects. *Psychonomic Bulletin & Review* 1995;2(1):55–82.

120. Liu Z. Viewpoint dependency in object representation and recognition. *Spatial Vision* 1996;9(4):491–521.

121. Furmanski CS, Engel SA. Perceptual learning in object recognition: Object specificity and size invariance. *Vision Research* 2000;40(5):473–484.

122. Tanaka JW, Sengco JA. Features and their configuration in face recognition. *Memory & Cognition* 1997;25(5):583–592.

123. Turk M, Pentland AP. Face recognition using eigenfaces. In *Proceedings of the IEEE Computer Society Conference on Computer Vision and Pattern Recognition, 1991*. IEEE;1991:586–587.

124. Valentin D, Abdi H, O'Toole AJ, Cottrell GW. Connectionist models of face processing: A survey. *Pattern Recognition* 1994;27(9):1209–1230.

125. Kanwisher N, McDermott J, Chun MM. The fusiform face area: A module in human extrastriate cortex specialized for face perception. *Journal of Neuroscience* 1997;17(11):4302–4311.

126. George N, Driver J, Dolan RJ. Seen gaze-direction modulates fusiform activity and its coupling with other brain areas during face processing. *Neuroimage* 2001;13(6):1102–1112.

127. Gauthier I, Tarr MJ, Moylan J, Skudlarski P, Gore JC, Anderson AW. The fusiform "face area" is part of a network that processes faces at the individual level. *Journal of Cognitive Neuroscience* 2000;12(3):495–504.

128. Gold J, Bennett P, Sekuler A. Signal but not noise changes with perceptual learning. *Nature* 1999;402(6758):176–178.

129. Gold JM, Sekuler AB, Bennett PJ. Characterizing perceptual learning with external noise. *Cognitive Science* 2004;28(2):167–207.

130. Peterson MF, Abbey CK, Eckstein MP. The surprisingly high human efficiency at learning to recognize faces. *Vision Research* 2009;49(3):301–314.

131. Gauthier I, Williams P, Tarr MJ, Tanaka J. Training "greeble" experts: A framework for studying expert object recognition processes. *Vision Research* 1998;38(15):2401–2428.

132. Gauthier I, Tarr MJ. Becoming a "Greeble" expert: Exploring mechanisms for face recognition. *Vision Research* 1997;37(12):1673–1682.

133. Valentine T. Upside-down faces: A review of the effect of inversion upon face recognition. *British Journal of Psychology* 1988;79(4):471–491.

134. Johansson G. Visual perception of biological motion and a model for its analysis. *Perception, & Psychophysics* 1973;14(2):201–211.

135. Blake R, Shiffrar M. Perception of human motion. *Annual Review of Psychology* 2007;58:47–73.

136. Grossman ED, Blake R, Kim C-Y. Learning to see biological motion: Brain activity parallels behavior. *Journal of Cognitive Neuroscience* 2004;16(9):1669–1679.

137. Jastorff J, Kourtzi Z, Giese M. Learning of the discrimination of artificial complex biological motion. *Dynamic Perception* 2002:133–138.

138. Jastorff J, Kourtzi Z, Giese MA. Learning to discriminate complex movements: Biological versus artificial trajectories. *Journal of Vision* 2006;6(8):3,791–804.

139. Jastorff J, Kourtzi Z, Giese MA. Visual learning shapes the processing of complex movement stimuli in the human brain. *Journal of Neuroscience* 2009;29(44):14026–14038.

140. Lesmes LA. A novel Bayesian approach to testing and analyzing visual acuity. *Investigative Ophthalmology & Visual Science* 2018;59(9):1073.

3

Specificity and Transfer

The specificity of learning to the task that was trained is a hallmark of perceptual learning. In principle, specificity could emerge from selecting or winnowing the most relevant early representations in low-level visual tasks, or it could emerge from the creation of new representations for high-level tasks. Peculiar forms of specificity to retinal location and other low-level aspects of the stimuli led to the proposal that perceptual learning reflects plasticity in the early visual cortex—changes in encoding. In this chapter, we argue that such learning may instead reflect reweighting—changes in readout—of sensory evidence, consistent with stable early representations. We show that the majority of the behavioral evidence is consistent with reweighting theories of learning, though a strong distinction between reweighting and representation change only occurs when representations and decisions are shared between training and transfer tasks. Finally, we present an analysis of the different paradigms and quantitative methods that provide accurate measures of specificity and transfer.

3.1 Specificity and Transfer in Perceptual Learning

One fundamental question to ask about perceptual learning is how specific the learning is to the task being trained. If an observer is trained for one task and then tested on a second related task, how much of the training will carry over?

The answer is that it depends. When performance on the second task shows little or no benefit, the learning is said to exhibit *specificity*. When, on the other hand, the training improves performance in the second task, the learning is said to *transfer*.

The phenomenon of specificity and transfer—but especially specificity—has been central to theories of visual perceptual learning because it might point to the physiological level(s) at which that learning occurs. When, in the 1990s, new studies reported specificity to trained visual locations, the finding seemed to overturn long-held assumptions about plasticity in the adult brain.[1] Early research on perception had assumed that the primary visual system, highly plastic early in life, remained relatively fixed in adulthood. But if learning in adults was shown to be specific to retinal location—as these reports claimed—then it stood to reason that V1, a region known for the small size of its receptive fields, was much more plastic than previously thought. (Several recent reviews[2–4] provide useful analyses of the literature.)

This hypothesis gave perceptual learning a newfound relevance. The long-standing questions regarding the plasticity of brain regions were now seen as being linked to the specificity of training tasks unique to those regions. If specificity were to be observed for a given task, researchers could then conclude (or so their line of reasoning went) that the neurons in the brain areas associated with that task had been *retuned.* It has since become a commonplace in the field to describe specificity as one of the hallmark properties of perceptual learning.[5–9] In effect, specificity became a window into the regions of brain plasticity.

In many ways, it is natural to associate perceptual learning with plasticity in early visual cortical representations. The association seems to explain the surprising specificity of features coded in the earliest levels of the visual cortex, but this hypothesis faces a serious challenge. If plasticity were to occur throughout the brain, even in low-level stimulus representations, then every new experience could conceivably change the response of the system to subsequent stimuli. If a few training sessions actually retuned early cortical neurons—as many researchers claimed—this retuning might produce a disruptive cascade of changed responses through the higher regions of the brain, thus impacting many other tasks. The costs of plasticity in this case could very easily outweigh the benefits.

The conundrum of cascading suggests that the one-to-one claims concerning specificity and retuning in the early visual cortex were likely overstated. Specificity is certainly an important phenomenon, but it is also a complex one, requiring careful interpretation. As discussed in chapters 1 and 2, the reweighting of sensory evidence from relatively stable early representations provides an alternative explanation for learning.[10–12] In this reweighting framework, perceptual learning is understood to derive from changes to the "readout" (weights) given to sensory information in early cortical areas in the course of making a behavioral decision. Specificity would then occur because the decision gives weight to evidence in representations tuned to specific features of the stimulus and/or its location. Mollon and Danilova made the point this way: "The *site* of the learning may … be central and what is specific may be *what* is learnt"[12] (p. 52).

This chapter asks two interrelated questions about how to interpret specificity. First, under what conditions does observed specificity inform the distinction between retuning and reweighting? Second, how much of the literature can be explained by reweighting alone?

In what follows, we argue that reweighting is one way for the system to balance stability and plasticity during learning. Early visual representations could then remain stable (or largely so) for use in multiple tasks, while plasticity may predominantly occur upstream, possibly at several levels or through top-down influences, allowing the push and pull between plasticity and stability to be successfully balanced. To better classify and interpret the literature, we also propose a taxonomy of training-transfer task pairs, noting their consequences for interpreting specificity and transfer.

Furthermore, specificity may arise whether learning involves reweighting that selects the most relevant representations for the task among many possible preexisting visual representations or through the creation of new representations that code for unique combinations of features likely not previously represented through recruitment and reweighting. The former likely describes learning in low-level and mid-level visual tasks, while the latter may be more relevant to high-level visual tasks that involve learning unique visual objects. Both forms of learning can produce specificity.

This chapter also examines what specificity has to tell us about the levels of representations used in different tasks. Even if plasticity does not alter early representations, specificity could still point to which physiological representations are central to the task. Specificity to retinal location or to the eye, for example, would indicate the task's reliance on early cortical representations, while transfer across location, eye, and/or scale would indicate the task's emphasis on higher-level representations. At the end of the chapter, we consider a final question: can certain kinds of training increase the likelihood of specificity—or, conversely, might other kinds of training enhance generalizability?

3.2 Example Paradigms for Assessing Specificity and Transfer

Scientists have studied learning and specificity for a range of tasks. (A *task*, by definition, consists of a judgment and the set of stimuli that are judged.) Specificity and transfer might be assessed for combinations of tasks—usually pairs—that may be very similar or may differ in many ways. Figure 3.1 provides several examples to give a sense of the many kinds of transfer that have been studied. The pairs of tasks may differ in the judgment required or the stimuli used, or both judgment and stimuli.

In one example (figure 3.1b), the initial training task T is orientation discrimination around a reference angle (+45°±10°), followed by a switch to transfer task X of orientation discrimination around the opposite reference angle (−45°±10°).[13] The extent of immediate specificity and transfer is inferred by comparing the performance in the transfer task to the learning in the training task (see figure 3.2). If learning is fully specific to the trained orientation, then the curves in the training task T and the transfer task X are identical ("full specificity"); once the reference angle is switched, performance returns to the initial level of the original training task (which here is equivalent up to the reference angle) and is then learned independently. If, on the other hand, learning transfers completely to the other reference angle, then performance continues in the transfer task where it left off in the training task ("full transfer"). Often, empirical results lie somewhere in between—what is learned is partially specific to the training task and partly transferable or generalizable to the transfer task ("mixture"). This simple analysis assumes that the two tasks are basically equivalent (otherwise performance in X could not be compared directly to performance in T). Often, X is assessed only once after the switch, despite the fact that continued training on X can be very informative. Section 3.8 (appendix A) provides a discussion of several ways to handle nonequivalent cases.

The literature typically plots data as learning curves or as bar graphs. One kind of display shows the full learning curves for T and either the single point (if there is only one

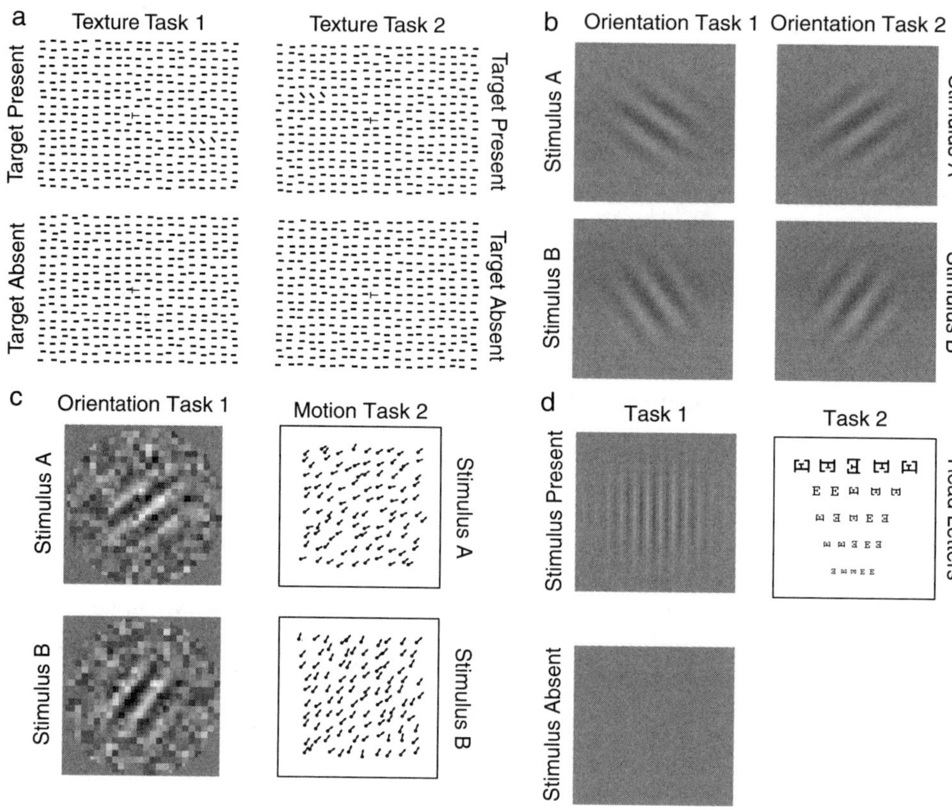

Figure 3.1
A sample of transfer tests: (a) texture-detection task (TDT) to another with a change in location (lower right to upper left); (b) orientation-identification task to orientation-identification task with different orientations; (c) orientation-identification task (in high external noise) to motion-orientation identification for the same angles; (d) grating-detection task with high spatial frequency to a rotated-E visual-acuity task.

postswitch assessment) or the learning curves for X. That said, many studies also report data as bar graphs showing performance before and after training and at the first point of transfer and (sometimes) at the end of further training on the transfer task, depending on the experimental design (see the bar graphs in figure 3.2). For the simple case, full specificity is seen when the "pre" to "post" (or "beginning" to "end" of training) bars are the same for the training and transfer tasks (given task equivalence) and are learned independently. In full transfer, performance in the transfer task simply continues learning from "pre" to "post" for the training task. Mixed cases are intermediate.

The degree of specificity is sometimes further quantified by an index. The index compares the amount of learning that does *not* carry over to the first assessment of the transfer task, expressed as the percentage of overall improvement in the training task (specificity

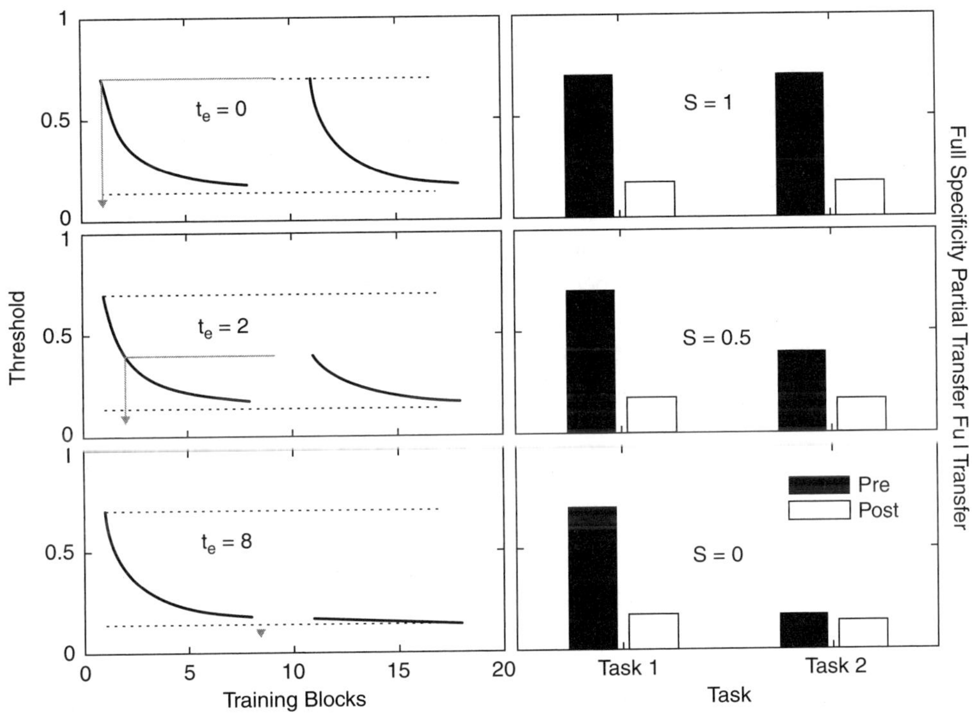

Figure 3.2
Illustrations of perceptual learning in a training task and a transfer task using a threshold measure, shown as learning curves (left) and corresponding bar graphs showing performance at the beginning ("pre") and end ("post") of training in the two tasks. There are three scenarios: full specificity (top), full transfer (bottom), and partial specificity and partial transfer (middle). The light gray lines and arrow measure transfer in training-block equivalents, labeled as t_e in sessions, and the specificity indices for the initial transfer performance are labeled as S on the right. Redrawn from Jeter et al.,[14] figure 1.

indices[14–16] of $S=1$, 0.5, or 0 and $t_e=0$, 2, or 8 in figure 3.2) or, if learning curves for the training task are available and the tasks are equivalent, an alternative measure of transfer to task X can be expressed as being equivalent to a given number of sessions in the training task T[14, 16] (gray lines in figure 3.2, and see section 3.8).

In reality, the situation is generally more complicated than the index might suggest. Although the two tasks can sometimes reasonably be assumed to be equivalent, more often than not they are different (e.g., training on orientation judgments with ±10° discriminations and transferring to orientation judgments with ±20° discriminations or to motion directions in the same ±10° directions). In such cases, assessing transfer performance requires either baseline measures or control groups. As discussed in chapter 2, the choice of paradigm has implications for data interpretation, and some experimental designs require sophisticated models in order for the data to be interpreted correctly. Section 3.8 explores the benefits and costs of five classic paradigms, and the corresponding ways of quantifying specificity and transfer in each. These include transfer without baseline, transfer with baseline, transfer with control groups, and mixture or alternation.

Another important conceptual point is that performance (in both the training and the transfer tasks) has so far almost always been measured at a fairly coarse grain, partly because of the relatively large sample sizes required for measurement and partly because of the typically extended time course of perceptual learning. Virtually all existing studies measure performance at the grain of scores of trials or use adaptively estimated thresholds at the end of relatively long blocks or sessions. Estimates of specificity and transfer of the rate of learning are then computed on performance at this coarse scale of measurement. Yet learning may still occur on a trial-by-trial basis, and most learning models make trial-by-trial predictions. We return to this point at the end of the chapter.

3.3 Task Structure Analysis

Researchers have tended to interpret almost all instances of specificity as though they were manifestations of the same phenomenon. This is an oversimplification. In fact, certain observations of specificity are far more powerfully diagnostic than others. What specificity implies will depend on the relationship between the training and the transfer tasks. When studying this relationship more closely, four *classes* of task relationships emerge, each with deep implications for whether the data can be accounted for by reweighting or readout, or by retuning or representation change. As we will see, the correct inference will depend on whether the perceptual tasks share sensory representations (stimuli), decision structures (judgments), both, or neither.

Figure 3.3 illustrates the four classes of relationships between training and transfer tasks in terms of simplified neural networks.[17] The small nodes are sensory features or representation units; the larger ones stand for the decision unit that selects a response. Light nodes and dark nodes represent the training and transfer tasks, respectively. The lines are connections that weight the representation information (activation) to make the decision. Although we illustrate simple two-layer networks, a similar analysis can be extended to more complex networks with hidden layers. In more experiments and data than originally understood, learning by changing the representations and learning by reweighting evidence make similar predictions. In a few important cases, the two forms of plasticity make different predictions and may be distinguished empirically. It should be emphasized that these simplified structures are idealizations.

In *class A* situations, the stimulus representations, the task decision or judgment, and the connections between them are all completely different. Performances in the training and transfer tasks T and X are independent regardless of whether perceptual learning alters the representations or reweights the connections between representations and decision. An example study in the literature pairs a line orientation task and a two-dot motion-direction task.[18]

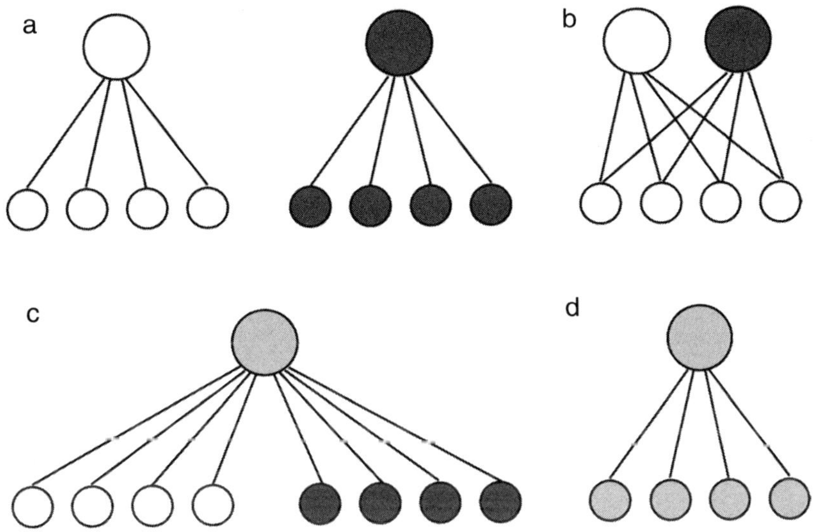

Figure 3.3
Training and transfer tasks are related in four ways, depending on whether they share sensory representations (stimuli) and/or decision structures (judgments). Sensory representation nodes (small circles) are connected to decision units (large circles) by weights (lines) (lighter and darker units represent training and transfer tasks, respectively): (a) class A, separate representations and decision structures; (b) class B, distinct representations but shared decision structure; (c) class C, shared representations but separate decision structures; (d) and class D, shared representations and decision structures. Class C and D task relationships may distinguish reweighting and representation change. Based on Petrov, Dosher, and Lu,[17] figure 1.

In *class B* situations, the two tasks share the judgment or decision unit (rules) but rely on separate stimulus representations. Again, both representation change and reweighting predict independent learning. An excellent example study examined three-line bisection at two different peripheral locations.[19]

In *class C* situations, the two tasks share stimulus representations but use different judgments, requiring distinct weighted connections from the (shared) sensory units to separate decision units. Here, representation change and reweighting can be distinguished. One excellent example study examined two hyperacuity judgments (left/right and up/down bisection) with nearly identical sets of dots as inputs.[7] If perceptual learning in the training task retuned the input representations, this must affect performance on the transfer task. Alternatively, if learning reweights or changes "readout" connections, performances in the two tasks may be independent—which was the case in the relevant experiments.

In *class D* situations, the stimuli and the decision are the same but the two tasks differ in some other way, such as external-noise context or luminance of the stimuli. With overlap in the representations, weights, and decision units, it is necessary to use a model to make predictions in order to distinguish representation enhancement and reweighting. In one example study, orientation discrimination was trained without external noise and then transferred to high-external-noise testing or vice versa, revealing asymmetric transfer between the two external-noise contexts that could be explained entirely by reweighting.[20]

The perceptual learning literature in the 1990s interpreted a wide range of observed specificities as evidence for representation enhancement (retuning), and for early sensory retuning in particular. Yet specificity can only distinguish forms of plasticity for class C or D task pairs. High specificity in class C experiments, which use the same representations (stimuli) in the two tasks, implies independent cortical processing on the same representations, or learning through reweighting. High specificity for class D task pairs, which share representations and tasks but differ in some other way, implies that learning is based on reweighting to decision, or that the different contexts in some way result on different low-level neural representations. The overwhelming majority of the literature has examined class A or class B task pairs and therefore cannot distinguish the two forms of plasticity. Only a very few cases are diagnostic. (Again, these classes are simplifications, as hierarchical representation structures might lead to task pairs that are mixtures of these classes, which might be used to explain systematically graded forms of specificity.)

High levels of specificity or transfer, or partial transfer and partial specificity, can often be interpreted within a modeling framework, with a simplified two-layer model potentially generalizable to forms that are more complex or hierarchical (see chapter 8, where, in principle, class mixtures might occur). In these hierarchies, reweighting at one level may look like changed representations higher in the hierarchy but the question remains: is it possible to account for learning in the presence of stable representations at some early level in the system?

In the following sections, we consider this question with reference to the behavioral evidence for specificity in different kinds of tasks and the corresponding implications for brain plasticity. Fortunately for researchers, the *kind* of specificity—whether to location, orientation, eye, or anything else—can point to the cortical area that encodes or preserves the representation property that must be used in learning.

3.4 Behavioral Evidence

So far, we have examined a few simple ways of measuring specificity and transfer, several existing examples of training and transfer tasks, and the importance of considering the relationship between the tasks when drawing inferences about plasticity, but what larger conclusions can be drawn from these principles, and how might we apply our schema to better understand the existing literature?

In this section, we review and analyze some of the most representative demonstrations of specificity and transfer. As with any survey, what follows is necessarily selective. We have organized our treatment according to the *kind* of specificity demonstrated: retinal location, eye of training, stimulus feature or object, nature of the judgment, or testing context. Each of these suggests the sites or cortical levels involved. The taxonomy of task relationships (figure 3.3) helps to frame interpretations concerning the nature of plasticity.

Our aim is to organize and classify the growing literature on learning and on specificity and transfer, and, ultimately, draw a number of conclusions. As we will see, although observations of specificity are widespread, the case of partial specificity and partial transfer is also a common pattern, and full transfer sometimes occurs, too. Furthermore, many observations originally thought to imply change in early sensory representations are in fact equally consistent with reweighting theories. Our overall argument is that reweighting provides a powerful and wide-ranging basis for perceptual learning. This hypothesis is bolstered by the few cases in which the two theories make conflicting predictions.

3.4.1 Retinal Location Specificity

Early demonstrations of specificity to retinal location have been among the most iconic findings in perceptual learning. They implicated early retinotopic visual areas with small receptive fields as the relevant sensory information used in the task. In some examples, specificity seemed mostly complete, while in others location specificity was partial. Most demonstrations, however, were of class B, in which the training and transfer tests used distinct representations but the same task. Specificity could thus have reflected either representation enhancement or reweighting.

One of the most influential early demonstrations of specificity—one that went on to inspire the heated interest in perceptual learning over the last few decades—showed retinal specificity in texture discrimination.[1] In this task (see chapter 2), observers identified the direction (horizontal or vertical) of a patch of line elements oriented differently from the background elements. The threshold stimulus onset asynchrony (SOA) between the texture display and a mask improved (decreased) with practice. Learning from extensive practice with the texture pattern in one quadrant did not generally transfer substantially to other quadrants. Instead, performance returned to near the initial baseline and had to be learned independently (figure 3.4). The specificity indices *S*, which measure the extent of return to the baseline, and the equivalent training measure of transfer t_e, were estimated from the learning curves, which are not shown in the figure.

Another extraordinarily influential early example did very precise assessments of retinal location specificity in orientation discrimination.[21] Threshold orientation judgments were practiced first at the fovea and then in a succession of locations around an annulus at 5° in the periphery (figure 3.5). An oriented pattern of black bars that masked white random-noise dots was rotated slightly clockwise or counterclockwise of the negative diagonal to measure the just noticeable difference (JND) threshold. Untrained performance in different peripheral locations at the same eccentricity should be approximately equal but worse than untrained performance at the fovea, however. Our interpretation is that there is significant transfer from the initial training at the fovea to the peripheral locations, because the initial performance at a peripheral location has a lower threshold than the initial performance at the fovea. Perceptual learning then occurred in each successively trained location, indicating some specificity to locations even within a visual quadrant. Locations that

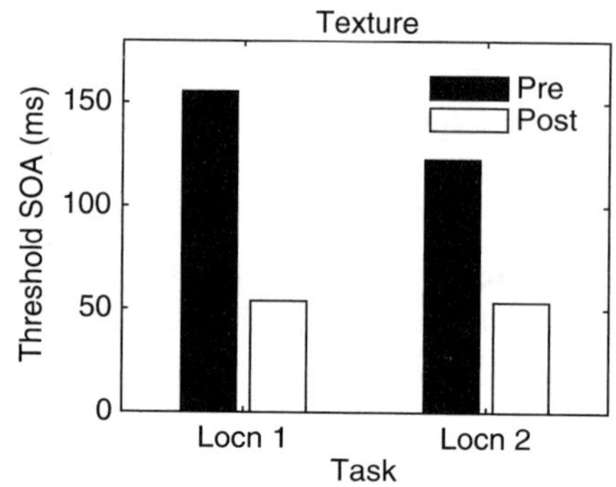

Figure 3.4
Perceptual learning in a texture-discrimination task (TDT) is mostly specific to retinal location. Performance is measured as the threshold SOA (stimulus onset asynchrony to a mask) in different retinal locations (with training equivalent and return to baseline measures of specificity and transfer: $t_e \approx 2$ of 10, $S \approx 0.75$; see the text for definitions). Redrawn from selected data in Karni and Sagi,[1] figure 2.

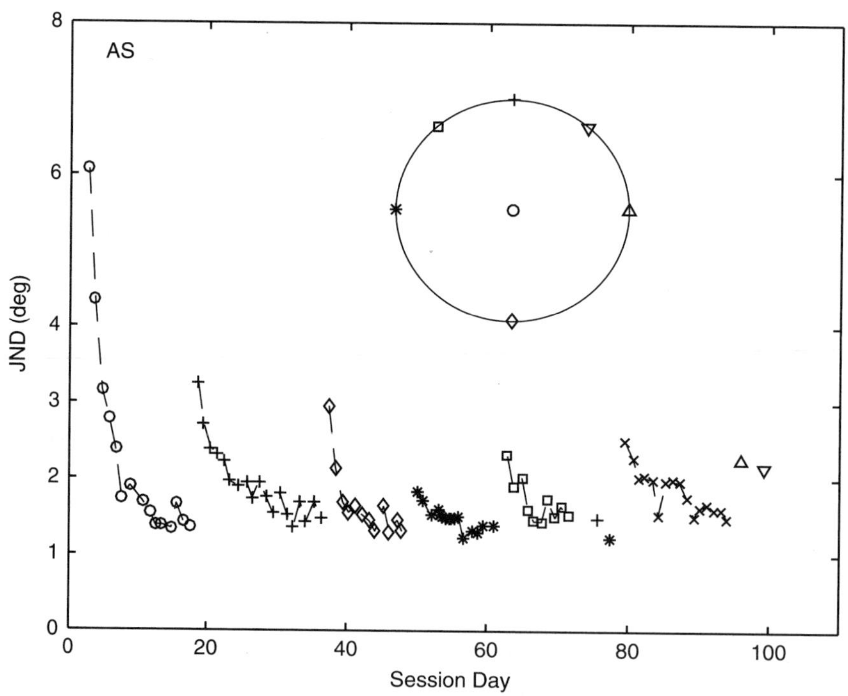

Figure 3.5
Perceptual learning in orientation discrimination is to varying degrees specific to retinal location. Data show the just noticeable difference (JND) in the orientation-discrimination task (deg) as a function of testing session or day for one observer. The first training function is at the fovea, followed by locations 1–5. AS is a subject identifier. After Schoups, Vogels, and Orban,[21] figure 4, with permission.

are symmetric across the midline from a previously trained location (i.e., 3, 4, and 5) may show less specificity (see also Shiu and Pashler [9]).

The specificity of learning to retinal location has also been shown to occur in motion-direction perception,[22] in depth perception from random-dot stereograms,[23] in localization,[19] and in object recognition.[24] In many or most of these cases, specificity to retinal location predominates, although some transfer can occur; the task pairs are of class B.

3.4.2 Eye Specificity

Tests of eye specificity assess whether monocular and/or binocular representations are relevant in learning. At least three representations are involved: two monocular representations, one per eye, and a binocular representation from both eyes together. This is especially interesting because learning could involve any combination of these representations. Complete or nearly complete transfer between eyes after monocular training involves learning at or above the level of the binocular representations. If there is significant eye specificity, then monocular representations must be involved. Researchers have associated monocular specificity of learning with retuning of monocular representations: "Absence of interocular transfer would imply that the changes accompanying learning remain restricted to monocular cells"[21] (p. 804). However, essentially all the tasks demonstrating eye specificity are of class B. So, while specificity of monocular training does imply reliance on monocular representations, eye specificity may occur either through representation enhancement or reweighting, and the reweighting could occur upstream of the low-level representations.

Different tasks show different degrees of specificity to the trained eye (figure 3.6). Texture-discrimination tasks and some motion tasks sometimes show specificity of perceptual learning to the trained eye, while orientation discrimination often does not, although

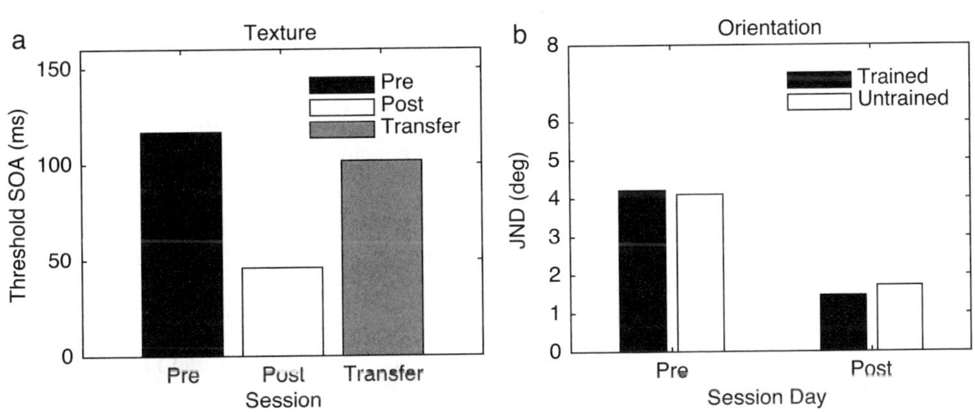

Figure 3.6
Perceptual learning is specific to the *eye of training* for a texture (a) but not an orientation task (b). Redrawn, respectively, from selected data in Karni and Sagi,[1] figure 1, and Schoups, Vogel, and Orban,[21] figure 7.

the results can depend on the paradigm. Paradigms starting with baseline measures in both eyes (transfer plus baseline; see section 3.8) seem more likely to transfer over the eye.[25] The amount of transfer seems also to depend on other details, such as whether the second eye receives a black or a mean-luminance image during initial training of the other eye (see our discussion of Sowden, Rose, and Davies[26]). One of the influential findings of specificity to the eye was in the texture-discrimination task.[1] Initial performance in the second eye returns almost to the initial performance level of the first eye (figure 3.6), with related results in fMRI.[27] Surprisingly, when a pretraining baseline was measured in the transfer eye, there was nearly full eye transfer in many texture conditions.[28, 29]

Full or nearly full interocular transfer occurs for threshold orientation discrimination (orientation JND task) in a pretraining baseline paradigm[21] and for the orientation-discrimination contrast threshold in varying levels of external noise, which generalized from the trained to the untrained eye in a no-baseline paradigm.[30] Learning to identify small differences in the motion direction of random dot stimuli substantially transferred between the two eyes, again in a design using pretraining baselines.[22] Learning to discriminate the motion direction (left or right) for sine-wave motion in different levels of external noise transferred completely to the untrained eye when tested in high external noise and expressed about half specificity when tested in zero external noise. This led to the conclusion that perceptual learning "is mostly binocular" in high external noise but "is largely monocular" in low external noise.[30]

In short, perceptual learning may rely on either monocular or binocular representations, depending on the feature domain, testing conditions, and—most interestingly—training protocol. Even relatively short evaluations of baseline performance in both eyes are sufficient to trigger learning at or above the level of binocular combination. Learning can occur at multiple levels of the visual hierarchy, depending on the circumstances, and, as class B task pairs, the specificity observed in these cases may be consistent with either form of learning.

3.4.3 Feature and Object Specificity

The specificity of learning to the trained stimulus features can point to the level of representation engaged. Feature specificity has been assessed in Vernier hyperacuity, orientation discrimination, motion-direction discrimination, and object identification. The results depend on the judgment. Some cases show very high specificity, some show a mixture of specificity and some transfer, and, in some cases, transfer is almost complete. Almost all this literature used class A or class B studies, in which specificity points to the neural areas involved but provides no information about whether the perceptual learning is mediated by representation change, reweighting or readout, or both. (Nevertheless, feature specificity may still play a role in helping to identify plausible visual regions that code the relevant visual representations. Learning that is specific to spatial frequency, for example, may suggest that the early visual cortex is most involved in learning.)

Orientation specificity is one of the most frequently cited forms in the literature (see figure 3.7). A number of studies have shown learning is highly specific to the orientation of the

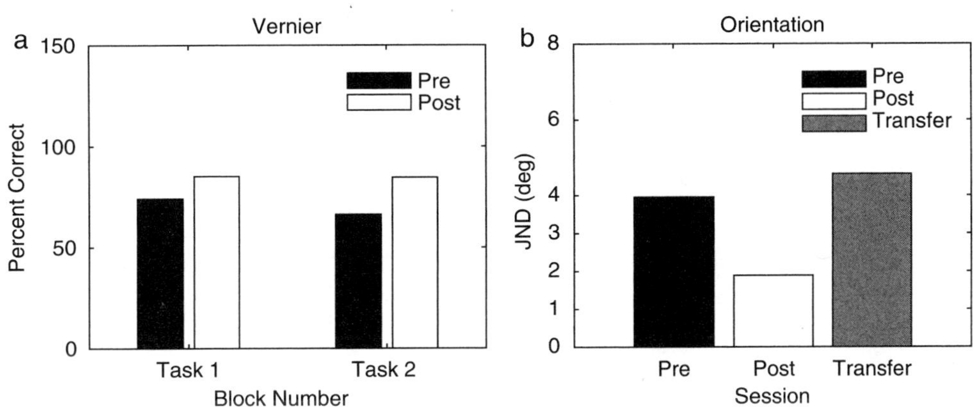

Figure 3.7
Perceptual learning that is specific to the *orientation* of trained stimuli. (a) High specificity in perceptual learning for vertical and horizontal of Vernier line-offset judgments, measured as percentage correct. (b) Specificity of orientation-difference thresholds (JND). Redrawn from data selected from Poggio, Fahle, and Edelman,[31] figure 3, and from Schoups, Vogels, and Orban,[21] figure 6.

trained stimuli. In training Vernier line judgments, for example, high specificity to horizontal or vertical stimuli has been observed.[31, 32] From these and other studies, several tentative generalizations can be drawn: orientation-difference learning is often specific to the trained orientations;[21] complex pattern discrimination is substantially specific when switched to a perpendicular orientation (though less so when switched to more similar orientations within 30° of the trained orientation);[33, 34] training in motion-direction discrimination is relatively specific to the trained directions;[22, 35] and, similarly, the time to settle on a stereo percept from random-line stereograms improves with practice and transfers to other images with the same orientation of the carrier lines (but not to those differing by 90°).[36]

Another less commonly studied feature, *spatial frequency*, has also shown specificity. Training contrast detection of stimuli of one spatial frequency in the periphery usually shows a moderate breadth of transfer along the contrast-sensitivity function (CSF), which expresses performance as 1/threshold (figure 3.8).[26] The bandwidth of perceptual learning, as seen in the difference between pre- and posttraining CSF measurements, is less than one octave of spatial frequency. This bandwidth has also been studied in amblyopes, for whom transfer seems to be broader (chapter 11).[37] In complex-pattern discrimination, perceptual learning seems to partially transfer to near spatial frequencies but not to very different ones.[33, 34]

By comparison, *spacial scale* shows relatively little specificity to the size or scale of the trained stimuli in identification tasks. Improvements in an object-naming task, as measured by shortened temporal thresholds to identify objects, were specific to the trained objects but transferred over moderate changes in size (figure 3.9).[38] Furthermore, reducing the viewing distance by one half, thus increasing its effective retinal size, showed complete

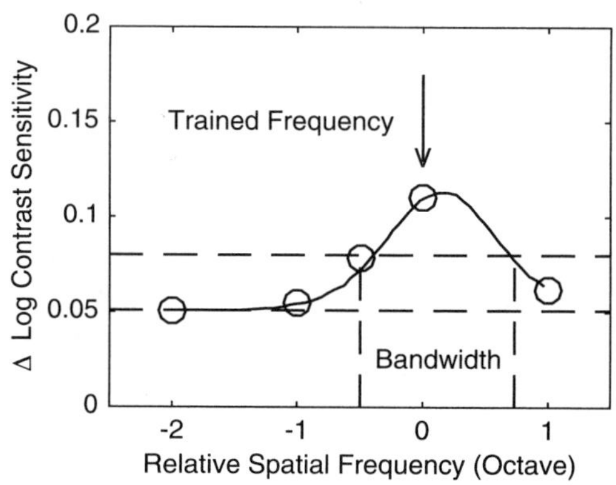

Figure 3.8
Spatial-frequency specificity of perceptual learning to detect a peripheral contrast defined sine wave. Post-training increases in contrast sensitivity show a spatial-frequency bandwidth of the perceptual learning effect of about half an octave. After Sowden, Rose, and Davies,[26] figure 5, with permission.

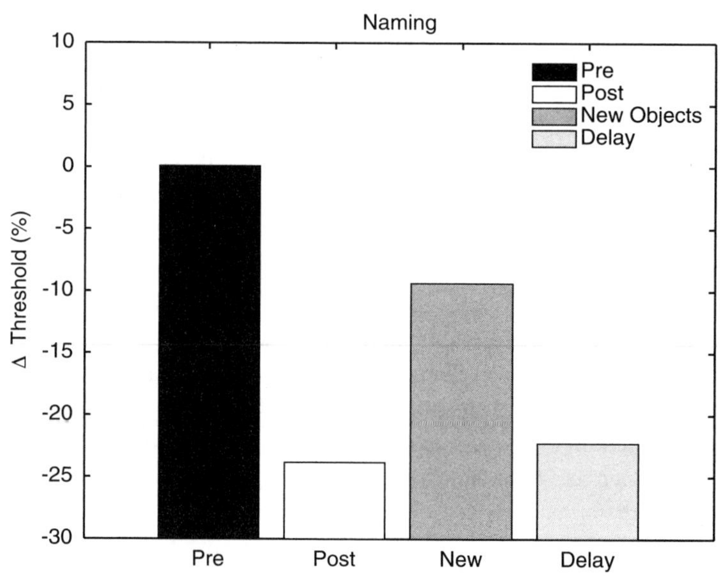

Figure 3.9
Perceptual learning of object naming with practice improves threshold mask delays. Improvements extend to a different context at a delay but do not extend to new objects. Redrawn based on selected data from Furmanski and Engel,[38] figure 6.

transfer of training in an orientation-discrimination task tested in external noise at the fovea.[39] The specificity of perceptual learning to the scale of the stimuli has also been reported in other situations, such as the size of the arrays in a texture-discrimination task[40] and in perception of illusory contours.[41] Whether perceptual learning is specific to spatial scale seems to depend on whether the tasks involve low-level visual features (although here, too, learning may involve higher-level representations, as in chapter 8) or higher-level natural objects, which may show invariance to low-level features such as location or scale but great specificity to the trained objects.

3.4.4 First- and Second-Order Specificity

Another interesting case in which learning operates in a multistage system involves first-order and second-order visual processing. Most visual pattern stimuli are defined by variations in luminance and processed through a *first-order system* of spatiotemporal frequency channels in the primary visual cortex.[42, 43] Other visual stimuli are defined by variations in other features, such as contrast,[44, 45] texture,[46–49] and orientation modulations[50] that are processed through a *second-order system*.[51, 52] This is thought to involve several stages of processing: a first stage of linear filtering, a (nonlinear) rectification, and a second stage of linear filtering.[51, 53–55] The initial linear filter stage is usually associated with cortical processing in V1, and the second linear filter stage is usually associated with cortical processing in higher visual cortical areas.[56] There is substantial behavioral and physiological evidence for distinct first-order and second-order processing systems,[52, 57–61] although there are some exceptions.[62, 63]

Most perceptual learning studies use first-order luminance-modulated stimuli. Other studies focused on second-order stimuli, defined by variation in features but not in luminance, and compared learning in both. The cases studied in the literature are in either class A or class B, which involve different representations and tasks that, although they may seem to be the same, may or may not be carried out differently. Training either the first-order or the second-order systems could lead to complete specificity for first- and second-order tasks. Transfer asymmetry—transfer in one direction but not the other—is often observed, requiring explanations in which some processing stages are separate and some are shared between training and transfer tasks.

One of the first investigations to compare learning with stimuli designed to predominantly stimulate the first- and second-order systems was in the domain of motion.[64] In first-order stimuli, a subset of luminance-defined random dots move in one direction; in second-order stimuli, an object that is created by dots moving differently from background dots is what moves. Learning in the second-order task largely transferred to the first-order task but not the reverse (figure 3.10). Some other studies reported this same transfer asymmetry,[65, 66] one showed no transfer (suggesting independence),[67] and one showed the opposite direction of asymmetry when training amblyopes to identify letters.[68] In another

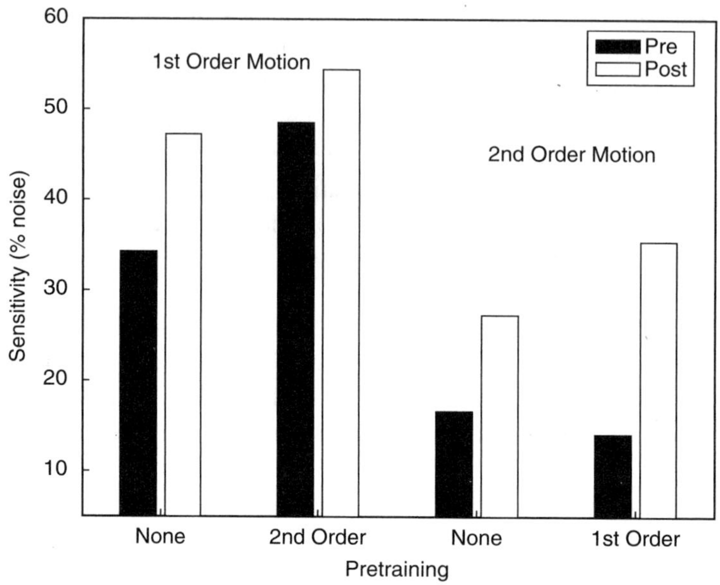

Figure 3.10
Perceptual learning of first- or second-order motion shows asymmetric transfer, as measured by sensitivity (coherence thresholds in percent) without pretraining (none), or after pretraining in the opposite motion type. Training second-order motion transfers to improved first-order motion judgments but not the reverse, suggesting a shared first-order stage is trained. Redrawn from selected data in Zanker,[64] figure 2.

study, training detection of second-order gratings generalized to some other spatial frequencies but not to the detection of first-order gratings.[69]

Studies using several possible stimulus features have yielded similar results. In one, orientation discrimination was trained for bars defined by color, luminance, motion, or a combination of these.[70] Training on bars defined by any single feature, or all three together, transferred to performance in all forms in the same retinal locations compared to pretraining baselines. Perceptual learning seemingly occurred at the level of the extracted shapes, regardless of how they were coded. In another example, perceptual learning of second-order orientation, curvature, or global pattern tasks all showed some mutual transfer.[71]

A hierarchy of first-order and second-order processes that combines first- and second-order stimulus information can produce either asymmetric transfer or none. This is because learning could involve either representation separately or learning about a shared representation or stage. First-order luminance stimuli are encoded directly in spatial-frequency- and orientation-selective channels, while contrast, texture, or other second-order stimuli first feed into a set of processors (filters and rectifiers) that extract their features, which are then integrated and fed into a final decision stage. The signals from the two paths are integrated at decision. It is believed that the first-order and second-order representations are integrated at decision because generally only one texture or motion is apparent in any location.

In a corresponding network model, there are multiple sets of weights: first-order channels to second-order channels, first-order channels to an integration stage, second-order channels to an integration stage, and an integration stage to decision. Training could change any or all these weights and thus could account for variations in transfer that would need to be explored with explicit modeling. Since these experiments involve separate first-order and second-order representations, they might either be of class A or class B—so learning and specificity may result from either improved reweighting or readout, or a combination of the two, at different levels of the system, similar to the multilevel systems modeled within an integrated reweighting theory (IRT) (see chapter 8).

3.4.5 Judgment Specificity

Learning often seems to be specific to the judgment required by the training task. Such specificity may seem to be a truism, especially if the training task and transfer-task judgments emphasize different evidence. If practice on the training task retunes shared representations (class C), then these changes must affect performance in the second task. Generally, the evidence is that they do not. Instead, the tasks almost always seem to be learned independently. For this reason, task specificity provides evidence *against* changes in the early representations, not for them. Fahle and Morgan, for example, concluded that "the neuronal mechanisms underlying the ... tasks are at least partially non-identical and ... learning does not take place on the first common level of analysis." [32]

Two classic studies briefly reported specificity to the task judgment. In one, training motion-direction discrimination left thresholds for two-interval same-versus-different judgments in the trained and untrained directions equivalent,[22] perhaps because discrimination by definition relies on different sensory evidence for one direction over another, while detection requires only evidence of motion. Similarly, orientation discrimination failed to improve subsequent line-luminance discrimination.[9] This implies that training did not alter the stimulus representations in either case.

Another classic experiment similarly showed independent learning of spatial bisection or Vernier judgments (i.e., whether a middle dot was aligned up or down or left or right relative to two reference dots in almost identical three-dot displays, thus a class C experiment) (figure 3.11).[7] These results were further challenged in a more powerful task-alternation design (figure 3.12)[72] (i.e., using the same four central dots for bisection and Vernier judgments). The two tasks were learned independently over several successive alternated phases of learning—indicating complete specificity (independence). In another often-cited study, observers performed either a global judgment or local judgment about a texture array (i.e., array shape or detecting an odd element). Here, too, learning in the two tasks showed full task specificity.[15] (Because the relevant stimulus features could depend on the task even though the stimuli were the same, this could be either class A or class C.)

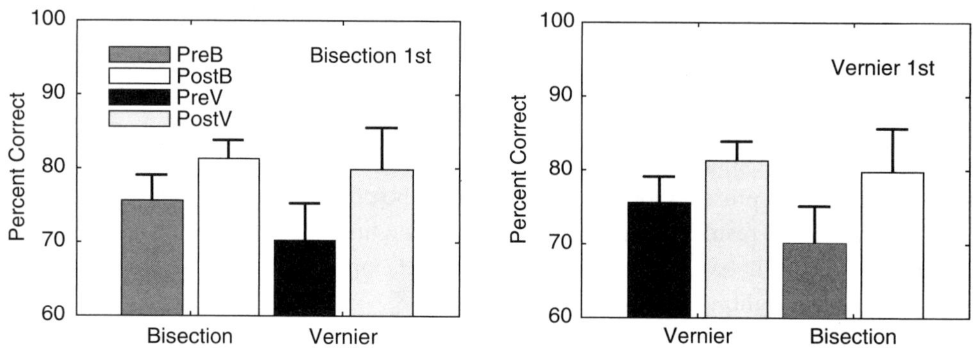

Figure 3.11
Specificity of perceptual learning with (nearly) the same stimuli for bisection and Vernier tasks, for a subject trained on bisection first (left) and a subject trained on Vernier first (right). Redrawn from selected data in Fahle and Morgan,[7] figure 3.

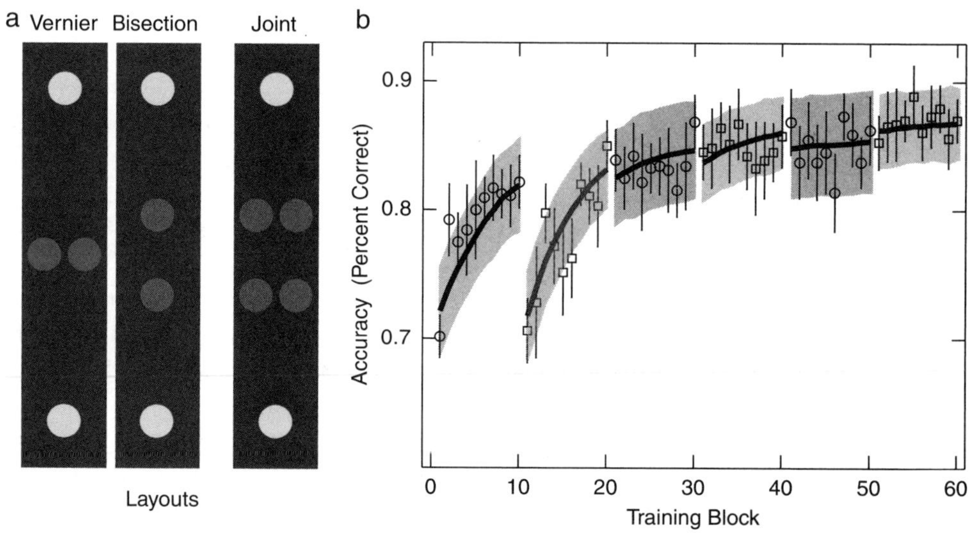

Figure 3.12
Specificity of perceptual learning in alternated bisection and Vernier tasks using identical stimuli. (a) Classic Vernier and bisection stimuli and the new joint stimulus layouts. (b) The data show independent training effects in each task over multiple cycles of training. After Huang, Lu, and Dosher,[72] figures 1 and 5.

3.4.6 Context Specificity

Is perceptual learning context specific? Does training in the laboratory or in a special testing environment extend to performance in other situations? Such questions are of both theoretical and practical importance. They have been investigated only rarely, however, usually by examining the specificity of learning to the visual noise or surrounding masks within which the task-relevant stimuli were embedded. In this section, we focus on the degree to which learning has been found to be specific to some aspect of the task context with otherwise identical stimuli and judgments (class D).

Training and transfer contexts certainly might differ in many ways. They could vary in surface characteristics (such as luminance level) or some other characteristics of the stimuli, or even in other aspects of the tasks (such as the levels of risk and reward and so on). Several early studies reported specificity of learning to stimulus context, in particular learning context detection and its specificity to the trained pattern mask.[73] Similarly, improvements in detecting a Gabor in one compound mask consisting of two diagonal Gabors failed to transfer to different compound masks or even to a mask with one of the two masking Gabors (although there was some transfer to a mirror symmetric mask).

Petrov, Dosher, and Lu[17] measured perceptual learning in training phases that alternated the external noise context. Observers discriminated the orientation of a Gabor embedded in either right- or left-oriented external noise alternated in long multiblock phases (i.e., Gabors that tilted either top right or left embedded in white external noise filtered to tilt right or left) (figure 3.13). The learning, indexed by discriminability (d'), showed an unusually interesting pattern of results. Training improved performance, yet there were also switch costs reflected in reduced performance whenever the external noise context alternated, and these costs persisted over many alternations. The improvements in accuracy with learning, as well as the higher accuracy for higher-contrast (more visible) Gabors, almost entirely reflected improved performance for incongruent stimuli (i.e., when the Gabor orientation was *opposite* to the orientation of the external noise). Observers were in effect "looking for" the stimulus evidence that differed the most from the current external noise context. A computational model that learns by reweighting can readily explain this complex data pattern (see chapter 6), while an explanation based on neuronal recruitment or sharpened tuning curves is more difficult to devise.

3.4.7 Summary

Specificity is widely cited as a landmark feature of perceptual learning. It takes many forms, including specificity to retinal location, eye, feature, processing systems, tasks, and contexts. Although specificity has been investigated using a variety of experimental designs, the simple transfer with and without baseline paradigms predominate. Our survey reflected this widespread practice of the field but also featured several examples of the more powerful but less often used task-alternation paradigm.

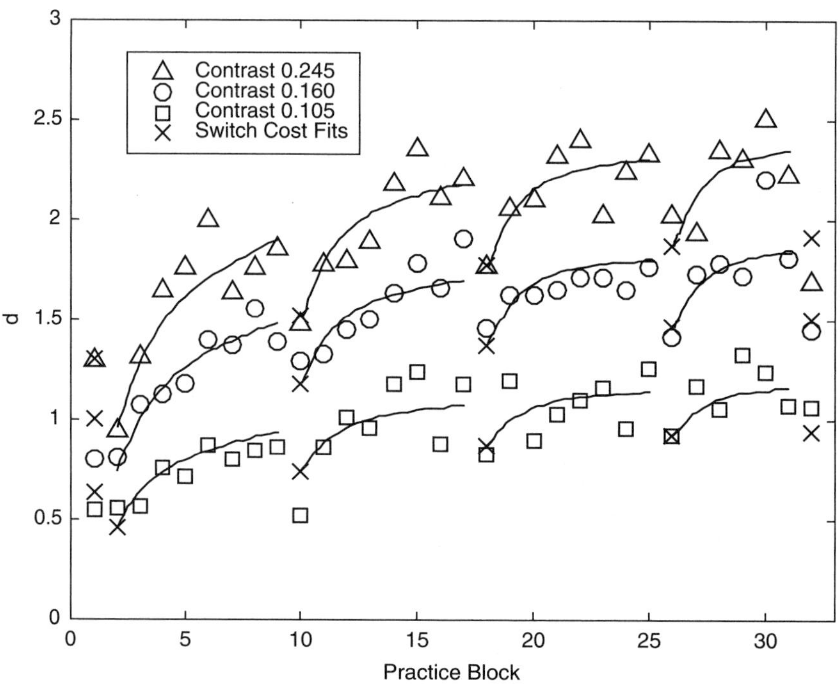

Figure 3.13
Perceptual learning shows ongoing switch costs for alternations of background noise. Performance (d') for three Gabor contrasts improved with practice, with switching costs at switches of external noise context. After Petrov, Dosher, and Lu,[17] figure 4.

In many of the most famously reported instances of specificity—such as specificity to retinal location—the study also showed some partial transfer as well. In other forms of specificity, such as changes of eye or scale, complete transfer has sometimes been reported. Specificity and transfer have often been evaluated purely qualitatively, but when both specificity and transfer occur in tandem, qualitative statements prove insufficient and quantitative evaluations that are paradigm-specific become necessary. (Subsections 3.8.2–3.8.6 describe five different evaluation paradigms along with their benefits and drawbacks.)

It has been our argument throughout that the visual processes central to the training and transfer tasks likely also determine the kind of specificity that occurs (in a way directly related to the analysis of learning described in chapter 2). Tasks involving low-level features coded in early visual analyses may express specificity (e.g., to the eye of training or retinal location or the first- or second-order sensory representations) and that learning occurs through improved readout of evidence from these representations. Learning that concerns natural objects coded in high-level visual representations, however, while highly specific to the trained objects, tends to transfer over scale or location—observations that are consistent with the idea that such learning recruits or creates new object representations found in higher (scale-invariant) cortical regions. It should also be added that regardless of the kinds of representations most centrally involved in learning, other task factors might also modulate the titration of specificity versus transfer. Several candidate factors are discussed in section 3.5.

Our analysis of the relationships between the training and transfer tasks had a further theoretical component. The sorting of pairs of training and transfer tasks into different classes helps to guide the interpretation of the empirical results and, importantly, the resulting implications for the relative roles of retuning and reweighting in plasticity. By definition, class A and class B task pairs rely on separate input representations and therefore separate weights to the same (class A) or different (class B) decision structures. Therefore, results from these experiments must be agnostic to the underlying form of plasticity. It is telling, then, that almost all experiments on specificity and transfer in the literature fall in the categories that fail to distinguish retuning from reweighting. Class C task pairs, by contrast, use the same input representations but distinct decision structures (independent connection weights), and in these cases plasticity based in retuning almost always predicts some form of transfer, either positive or negative. The few cases of class C reported in the literature instead show independence, bolstering the case for reweighting (and unchanged representations). Class D task pairs are unique in that they share everything—the input representations, the connections, and the decision units—differing only in some other aspect of task context. This class of tasks is also unique in that a full modeling analysis is required to adjudicate between learning through reweighting, representation enhancement, or both. These principles remain fundamentally the same (although the specifics are made somewhat more complex) if the simple diagrams of figure 3.3 are replaced with hierarchical networks.

We believe that the reweighting model provides a more compelling account of the vast majority of behavioral phenomena observed in the current literature. The strong early conclusions favoring representation enhancement should be reevaluated; in the vast majority of these reports, either reweighting or creation of new high-level representations through recruitment and reweighting can provide equally good or better explanations of the data. Future studies should be designed to better reveal the forms of plasticity that can occur simultaneously at multiple levels of the visual hierarchy. In the meantime, more convincing evidence for representation change at early visual levels during learning is required (Indeed, the physiology does include some reports of slight retuning as early as V1; see chapter 5.) Even if significant changes in early representations *do* occur during learning (i.e., changes in the *encoder*), however, learning would still in most cases also require changes in readout (i.e., changes in the *decoder*).

Despite all these interpretive complications, the field is not wrong to think of specificity as a powerful behavioral indicator. Even when representation enhancement or reweighting cannot be definitively ruled in or out, observed specificity can still tell us something about the cortical level(s) involved in learning. If there were specificity to retinal location, to the eye, or to certain other features, such observations would indicate that the representations involved were at a level that preserves those properties (thus identifying early levels of analysis in the visual cortex as the site of task-relevant representations). By contrast, if there were transfer or generalization over location or scale, this would identify a greater reliance on higher cortical levels of analysis. Future studies that investigate the connections between function and physiology would almost surely shed light on these questions and on the relationship of specificity and transfer to plasticity in the brain.

3.5 Factors Affecting Specificity and Transfer

Given that many experiments demonstrate partial specificity and partial transfer, two questions emerge: What factors influence this balance? And how can we encourage transfer (or generalization), which is almost always of more use in the real world? These questions, still largely unresolved, hold the key to further work in both the theories and practical applications of learning.

In principle, many things can influence transferability. So far, however, the field has accumulated sufficient evidence to support only a limited number of hypotheses about what might drive specificity and transfer. Although there may be others, four factors have emerged: the difficulty of the task, the state of adaptation, the amount of training, and the presence of cross-task training. Each hypothesis or factor has found support in specific experimental contexts. Since the empirical outcomes are likely to depend on variations in experimental implementation, further work will be necessary in order to more definitively identify them as causal factors (and preferably work that uses parametric manipulations and quantitative measures). A generative model—one that can make specific predictions about the relative success of different training protocols in supporting transfer—would also lead to new ideas for training protocols. In what follows, we explore the evidence supporting some of these factors.

3.5.1 Task Difficulty and Stimulus Precision

One of the earliest and most influential ideas about the effect of training on transfer was that specificity depended on task difficulty. The experiments designed to test this hypothesis generally showed a mix of specificity and transfer that seemed to depend on task difficulty. In these studies, "task difficulty" actually referred to manipulations of judgment precision: the size of the stimulus difference to be discriminated. These early experiments did not independently vary (i.e., they confounded the difficulty of) the training and the transfer tasks, which were either both easy (low precision) or both difficult (high precision). The initial conclusions of these studies claimed that the nature of the training task controlled the degree of specificity. Subsequent studies that cross-manipulated the precision of the training and transfer tasks instead found that specificity primarily reflects the precision of the transfer task. This makes sense because a highly precise judgment in the transfer task requires especially close evaluation of the sensory evidence. That said, the applicability of this principle to other judgments and stimulus domains awaits further investigation.

In these early papers, the task was texture discrimination, and difficulty was manipulated by the size of the orientation difference between odd element(s) and the background (we would call it the required judgment precision).[15] The transfer task swapped the orientations of the target and background lines and the target locations, either two or many locations, and the dependent measure was the threshold stimulus onset asynchrony (SOA). The more "difficult" tasks showed more specificity (figure 3.14), leading to the claim that training in more difficult tasks led to more specificity. (Notice, however, that even the highest-specificity indices were about 34% and 62%.) Similar effects occurred in motion discrimination: training to discriminate small (4°) angular differences in random dot-motion direction near one reference angle showed only about 13% transfer to discriminating 4° differences near another reference angle, while another study found that despite little immediate transfer to the new reference angle in the 4° task, postswitch learning of the transfer task was almost twice as fast.[8] (This speedup in learning has not been observed in several similar studies in other task domains.)

These studies did not independently vary the precision of the training and transfer tasks.[15, 74] In our subsequent but analogous study involving an orientation-discrimination task, they were decoupled.[14] The precisions of the training and transfer tasks (±5° versus ±12°) were crossed in four conditions of training and transfer; observers in each group

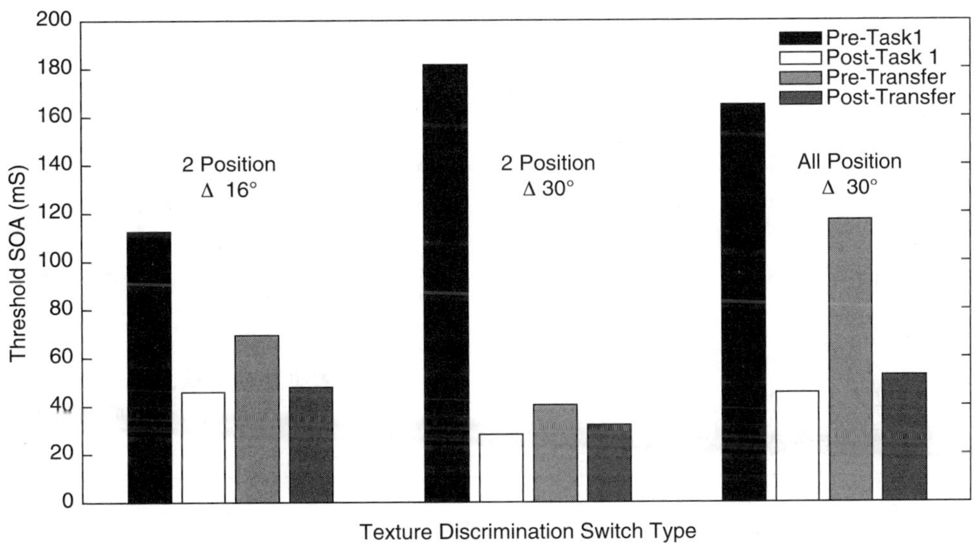

Figure 3.14
The degree of transfer between a texture-discrimination task (task 1) and a transfer task depends on whether the difference between target and background elements ("easy," Δ30°; or "hard," Δ16°) and the number of relevant locations (two or all). Derived from data in Ahissar and Hochstein,[15] figure 2.

were trained in different diagonal locations in the periphery, and adaptive methods were used to track contrast thresholds separately in trials with zero and high external noise, all intermixed during practice (figure 3.15).

In the study that decoupled the training and transfer tasks, specificity was found to depend on the precision of the *transfer task*, which exhibited more specificity if it required higher-precision judgments. The relevant pattern is easily visible in the data: threshold learning curves in the transfer phase overlapped almost exactly, regardless of the level of precision of the training task. This remarkably exact overlay of the data in the transfer tests occurred in this study because adaptive methods targeted the same percentage correct. Subsequent studies replicated these findings in an odd element texture task,[75] motion of filtered visual noise textures,[76] and dot motion direction.[77] The last of these, which did not control threshold accuracy but instead measured percentage correct, found a small interactive effect of the training task in addition to the substantial effect of the precision of the transfer task. A quantitative model (see chapter 8) illuminates these results and the limits of their generality.

3.5.2 Adaptation and Specificity

Another proposal—and a provocative one—has suggested that specificity is sometimes a by-product of adaptation. The idea is that specificity to stimuli will be more pronounced if the stimuli are shown repeatedly in the same location;[29, 78] for example, that observers learn to read out from adapted sensory responses resulting from continuous exposure in

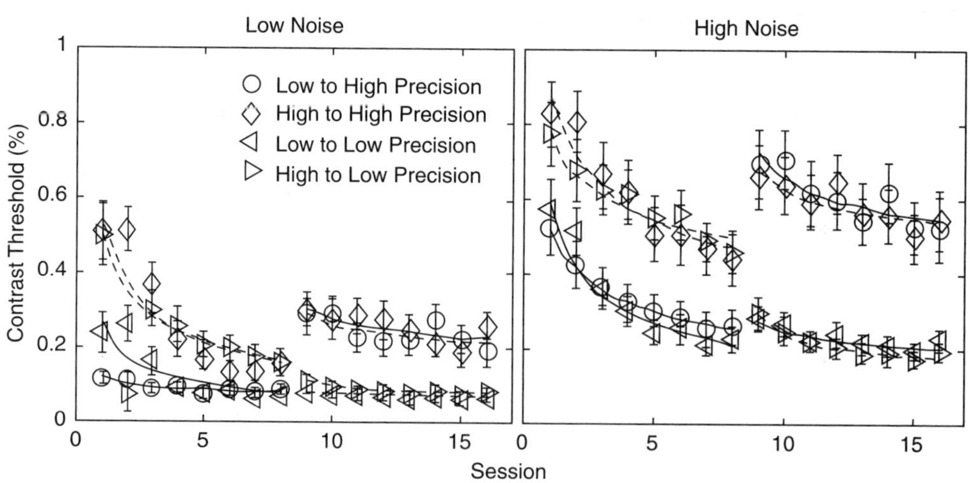

Figure 3.15
Transfer depends on the precision of the transfer task, with high-precision tasks showing more specificity. High (±5°) or low (±12°) precision orientation discrimination is trained and then switched to high- or low-precision judgments near an opposite reference angle and in different positions, trials with zero or high external noise being intermixed. Performance after the switch (session 9) shows more specificity for the high-precision task: nearly identical regardless of the precision of the training task. After Jeter et al.,[14] figure 2.

the same location. When tested in a new location, the learned readout would then be set neither for the location nor for unadapted sensory responses.

The adaptation hypothesis has been explored in several studies using the texture-discrimination task. These studies examined specificity in conditions that either did or did not include stimuli designed to create a release from adaptation during training.[78] In conditions showing the most specificity,[1] the target occurred in only one location and the target, background, and mask line orientations remained unchanged, thus creating a dense repetition of the same stimuli in the same locations. In other conditions, interspersed frames with differently oriented lines (but no target) reduced adaptation.[79] Initial learning was observed to be somewhat more rapid in the interspersed condition, and there was a marked difference in which group showed significant threshold elevation (specificity). At the first postswitch block, the group that received interspersed frames to reduce adaptation showed much more generalization. Interleaving standard displays with blank displays also showed specificity (see figure 3.16).

Sagi[29] has suggested that the interaction between readout and adaptation might be a general principle that explains a certain class of specificity. Paradigms that repeat stimuli

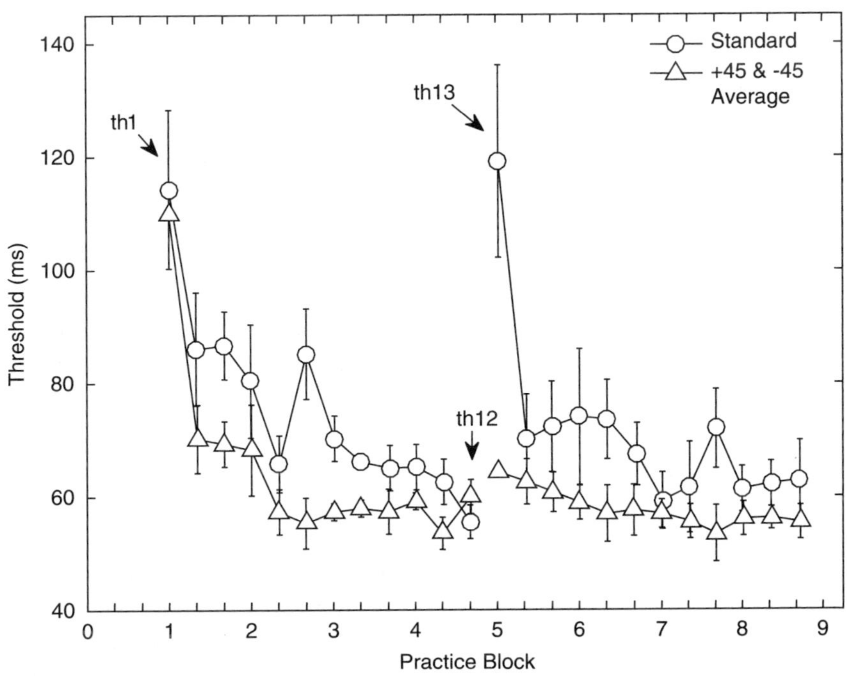

Figure 3.16
Specificity in a texture-discrimination task with and without interleaved trials, including lines rotated 45° to reduce adaptation, as measured by threshold SOA in initial training and transfer tasks. After Harris, Glicksberg, and Sagi,[78] figure 2, with permission.

may create states of local adaptation leading to "overfitting" features coded by different networks of units in the visual cortex (i.e., location, orientation, eye). This theory is based almost entirely on data obtained from the texture-discrimination task and must be tested in other domains and tasks before its generality can be assessed.

3.5.3 The Extent of Training and Specificity

Another possible factor controlling specificity, one motivated by intuition but also by hierarchical reweighting models (which are introduced in chapter 8), is the *duration* of training: the more an observer is trained in a given task, the hypothesis goes, the more specificity there will be. The current evidence is mixed on this point. The number and distribution of training trials differs widely from experiment to experiment, and there have been very few systematic studies of the consequences of these choices. Nevertheless, it has been observed that the distribution of training trials over different sessions or days, even when the total is held constant, may substantially influence not only the amount of learning but also the degree of specificity. Only a few studies have explicitly manipulated the extent or pattern of training and measured transfer, however. The existing data support the principle that training that is more extensive leads to more specificity, although additional research may reveal interactions with task precision or other factors. (We return to these ideas in the context of reweighting models for transfer in chapter 8.)

One of the earliest reports to pursue this idea briefly stated that very short initial training was less specific to the eye.[80] Although eye specificity appears later in training in texture discrimination,[1] the first few blocks of training with above-threshold times to the mask led to rapid improvements that transferred immediately to the other eye. Another more recent study compared several distributions of 1,600 training trials over sessions and days in a hyperacuity (Chevron) task.[81] Too few trials of training in a session (fewer than 200 trials per day or per week) failed to produce learning, while a sufficient number (either 800 trials per session for two sessions in two days or 400 trials per session over four weeks) led to robust learning. More intense two-day training failed to transfer to a new implicit reference, while half the training per session over four weeks led to about 25% transfer. This led the authors to conclude that specificity increases with dense practice over a few sessions of training, while the learning in a less dense (but adequate) training schedule over a longer period of time was more transferable.

Our laboratory explicitly tested the idea that more training leads to more specificity in a high-precision orientation-discrimination task[16] that was transferred to different orientation angles and retinal locations (figure 3.17).[14] The amount of training before a task switch was manipulated (one, two, four, or six sessions, with 1,248 trials per session), measured by contrast thresholds in zero and high external noise at two retinal locations. Unsurprisingly, groups that trained longer learned more. At the point of transfer, however, conditions with the least initial training led to the best performance on the transfer task, while those with the most initial training showed the worst performance (most specificity). Specificity indices (see section 3.8) ranged from about 10% for the least to almost 80% for the most training in

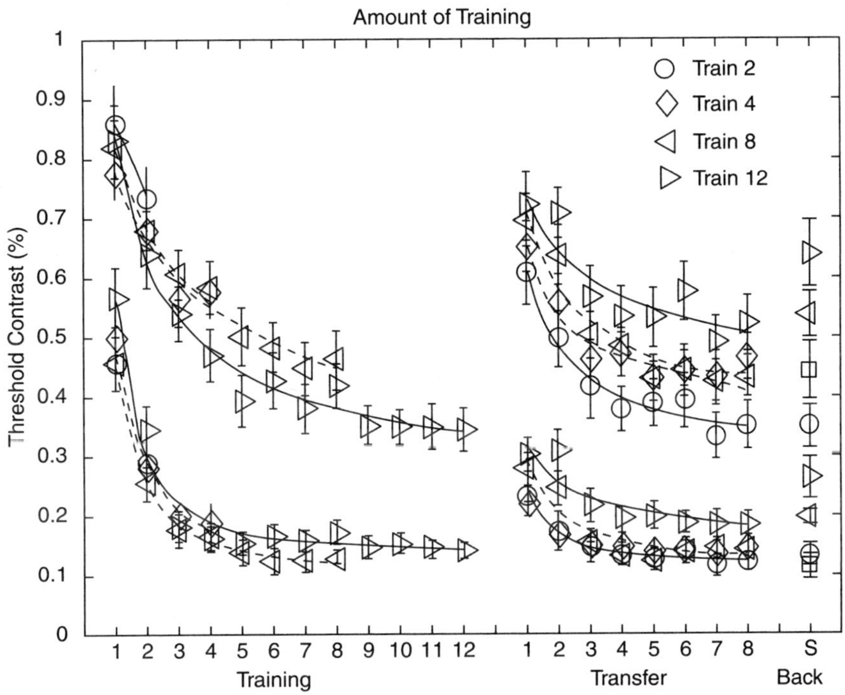

Figure 3.17
More initial training before the task switch increases the specificity of perceptual learning. Contrast-threshold performance improvements in groups of observers that trained for two, four, eight, or twelve blocks for no-external-noise (lower curves) and high-external-noise (higher curves) trials are shown. See the text for specificity indices for different groups. After Jeter et al.,[16] figure 4.

high-external-noise tests and from about 10% to about 35%–40% in zero-external-noise tests. More training led to more specificity. Subsequent practice on the transfer task, however, resulted in no or only small disruptions in performance on the initial training task (in chapter 8, we show that this is predicted by a reweighting model).[16] Such studies are experimentally demanding (requiring large numbers of observers and trials) but also promising. The effect the amount of training has on the immediate specificity and subsequent rate of perceptual learning should be evaluated for other stimuli, judgments, and protocols.

3.5.4 Enabling Transfer with Cross Training

Another more widely studied hypothesis in the field is that *cross training* improves transfer over retinal locations.[23, 82–84] A cross-training protocol trains a secondary or promoter task in a transfer location to increase transfer to that location of an otherwise specific primary task initially trained in another location. A number of clever examples have been investigated experimentally. The extent of transfer in these designs can depend on the details, as shown in several recent studies, with slightly different training or assessment protocols producing relatively significant differences in either initial learning or specificity and transfer.[85, 86] Overall, however, the claims associated with these investigations have typically been strong: that cross training eliminates specificity and releases full generalizability. Experimentally, however, the outcomes can be more graded and may be better understood with more fine-grained performance measurements.[87]

So far, three cross-training protocols have been studied: double training, piggyback training, and training plus exposure, each training a secondary as well as a primary task. *Double training* was proposed first.[82] The primary task in this protocol is one that normally shows retinal specificity, while a secondary task, trained at another location, is designed to *promote* transfer of the primary task to that new location. Either training on the two tasks is intermixed or training in the secondary enabling task occurs after initial training of the target task. In one study,[82] contrast increment detection of a vertical Gabor (the primary task) was trained at one location, and Gabor orientation discrimination (the secondary, promoter task) was trained at a second location, both using two-interval difference-threshold paradigms. The design begins with baseline measures of contrast discrimination at both locations. Although relatively little transfer was exhibited following primary-task training alone (about 78% specificity), almost full transfer was observed following intermixed double training (figure 3.18). Successive rather than interleaved training on the primary and secondary tasks was shown to improve transfer in another experiment.

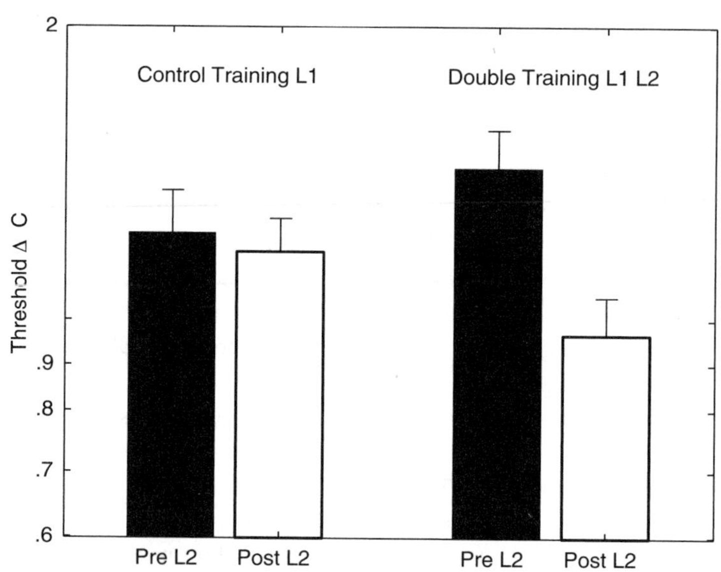

Figure 3.18
A double-training paradigm uses a different task in a second retinal location (L2) to improve transfer of a primary task to that location. Alternating training on contrast increment detection in location 1 and orientation discrimination in location 2 (double training) increases the transfer of increment detection to location 2. Redrawn from data in Xiao et al.,[82] figure 1.

Piggyback protocols mix the training of two tasks in a different way.[83] Here, the secondary task naturally tends to transfer over location. Intermixing training on the primary task and the more transferable secondary task in location 1 may promote transfer of the primary task to location 2. In one experiment,[85] training on a Gabor Vernier task was intermixed with blocks of Gabor orientation judgments in the upper left quadrant. Vernier performance improvements, usually specific to location, transferred about 75% (25% specificity) to a location in the lower right quadrant.

The *training-plus-exposure* protocol uses a secondary task with the same stimuli but a different judgment to promote transfer.[25] In one case, the primary task was orientation discrimination near the positive diagonal, and the transfer task was orientation discrimination near the negative diagonal, which showed little transfer. The secondary task was contrast discrimination, called "passive exposure." The orientation task was measured at a pretraining baseline, after primary-task training (which showed about 23% transfer), and then again after secondary-task contrast discrimination training (which showed about 87% transfer). An interleaved training protocol led to similar results.

These demonstrations of cross training have attracted significant interest, corresponding with the claims put forward in the papers, such as "additional location training enabled a complete transfer of feature learning (e.g., contrast) to the second location";[82] "with a training-plus-exposure procedure ... perceptual learning can completely transfer to the second orientation in tasks known to be orientation-specific";[25] or, finally, "This finding challenges location specificity and its inferred cortical retinotopy as central concepts to many perceptual-learning models and suggests that perceptual learning involves higher non-retinotopic brain areas that enable location transfer."[82]

As seen in this collection of studies, cross-training protocols often yield quite substantial benefits, although the exact magnitude may be influenced by experimental details[85] and the tasks.[85, 88] Based on findings in other paradigms, observed specificity may also depend on several other factors: precision and variability of the tests during training and transfer,[85, 89] the presence of a pretraining baseline transfer assessment,[25] and/or the length of training of the primary task.[90, 91] There may also be individual differences in learning and transfer,[82, 88, 90, 91] so there may be individual differences in the magnitude of cross-training effects as well. The sequential training designs often used in these studies typically rely on three data points to assess specificity (baseline, intermediate, and final assessments, and how much learning would naturally have occurred between these assessments should perhaps be considered.[13] Also, cross-training designs should compare net gains to the sum of the individual benefits from each training part alone. (See the discussion in section 3.8 regarding methods that could be used to discount this source of learning in different protocols.)

As research on this influential idea continues, the influences of cross-training manipulations are likely to be shown to be more complex and graded than originally thought. Understanding their potential for promoting transfer would benefit from future parametric investigation and a generative model capable of predicting task interactions. Several such models have been used to predict some of the relevant transfer phenomena (see chapter 8).[87, 92]

3.5.5 Summary

In our overview of the existing literature, we reviewed four factors that have been proposed to influence the balance between specificity and transfer, sometimes motivated by intuition and sometimes suggested by prior data. These factors included task or judgment precision, adaptation, intensity of training, and cross training in multiple tasks. Consideration of this literature led to the set of provisional principles in table 3.1. Each of these has been shown in some cases to affect the degree of specificity and, conversely, transfer—generally within a given task and paradigm. Further empirical investigation in different task domains (and using a variety of experimental paradigms) promises to move the field closer to generative models that can predict the outcomes of different training experiences. Only with such models can we clarify and test these and other principles. Indeed, model accounts for a number of these phenomena of transfer have been developed within integrated reweighting theory (IRT), which explains transfer by learned reweighting of higher-level invariant (i.e., location-invariant) representations (see chapter 8).

3.6 Measurement Scale, Adaptive Estimation, and Decoupling Training and Transfer Assessment—Directions for Future Research

Perceptual learning and transfer have historically been measured in paradigms focused on block- or session-level measures of performance. The same trials have often been used both to train and to assess performance. Transfer has also generally been measured for only a single transfer task. Such scientific habits arise from a need to derive a single behavioral performance measurement (e.g., contrast threshold, difference threshold, or percentage correct) from a sufficient (fairly large) number of trials. The consequence, as mentioned in chapter 2, is that perceptual learning has almost always been measured at a relatively coarse grain, perhaps about every 80–150 trials, or sometimes at the end of blocks of hundreds of trials or sessions of thousands of trials.

Although rarely discussed, such practices can have important consequences for the research enterprise. The grain of measurement can affect estimates of initial performance, the rate of learning, and/or the initial estimate of transfer, especially for domains in which there is a phase of rapid learning. New techniques promise to yield performance measurements that are more efficient and therefore allow finer-grained analysis. With such new techniques, scientists would be better able to estimate initial performance in both the

Table 3.1
Potential principles influencing specificity and transfer

- Switching *to* high-precision tasks generally increases specificity
- More extensive training on a task with a set of stimuli increases specificity
- Cross training may increase transfer over locations and features
- Measuring baseline performance in a transfer task may improve later transfer

training and transfer tasks, both of which are key to the assessment of specificity. These more efficient measures could also reveal rapid early learning if and when it occurs. Adaptive procedures, currently under development, that estimate performance levels in relatively few trials may prove very valuable for this purpose.[93–96] These new adaptive methods can also be used to validate (and estimate the parameters of) the functional forms of learning, such as the exponential or power function,[13] on a trial-by-trial basis.[97]

It is our belief that these new rapid estimation methods represent one of the most promising avenues for future research in the study of specificity and transfer, as well as other properties of learning. Such rapid estimation methods could allow for more complex and precise measurements of aspects of performance. These would include the ability to measure transfer of training on one task to multiple transfer tasks and to measure the changing rate of training and transfer for more complex measures of performance throughout the course of learning. Such methods would yield quick estimates of a number of functions, including the functional form of more detailed measures of performance throughout the course of learning or the functional form of detailed metrics such as the contrast-sensitivity function.[94, 98, 99] Growing out of similar principles behind adaptive methods used for the rapid assessment of the contrast-sensitivity function or the threshold versus contrast function,[93] these methods could also be extended to other measures, such as acuity or temporal-frequency sensitivity. Even more importantly, adaptive estimation would enable the decoupling of training and performance assessment on an important target measure. If performance on a target task could be assessed very rapidly, this would then allow the possibility of using combinations of different kinds of training and assessment trials (e.g., using easy detections or discriminations to promote perceptual learning while observing changes in more difficult tasks in rapid assessments throughout the course of learning). Some examples of rapid, single-trial assessment methods are shown in section 3.9 (appendix B). We believe that these technical innovations will soon extend the range of what can be measured and improve how transfer is assessed.

3.7 Conclusions

The landmark observations of specificity in the early 1990s sparked a resurgence of interest in perceptual learning and its relationship to plasticity in the adult human brain. Most researchers proceeded under the assumption that specificity was a direct indicator of the site of plasticity. This led to strong claims that specificity of training to small regions in the visual field was caused by plastic changes in the early visual cortical regions, as early as V1. Given how determinant such early areas are to higher levels of cortical processing and how widely visual functions are affected by experience, we have argued that a deeper, more nuanced understanding of plasticity and learning is necessary.

As we have tried to show, the original claims about specificity tended to overinterpret their results. Specificity alone is insufficient to infer that experience changes the tuning of the earliest levels of the visual system. A more sophisticated explanation allows specificity to result either from changes in early representations (representation enhancement) or changes in the readout of evidence from relatively stable early representations to decision (reweighting), possibly through a multilayer system. We proposed a classification of the relationships between training and transfer tasks to guide this interpretation, from which we concluded that the bulk of the existing literature does not provide a genuine test between the two explanations. So far, in those cases where the observed specificity *is* diagnostic, the behavioral evidence seems to favor reweighting explanations of perceptual learning. Nevertheless, future models are likely to incorporate potential roles for both modes of plasticity, perhaps deployed in different situations or used synergistically.[100, 101]

Specificity may in fact be a consequence of selecting those existing visual representations (among many) that are most relevant for the task. We believe that such *selection* of evidence that influences decision is likely to be the basis for judgments of stimulus features coded in early visual areas as much as or more than plasticity within these areas. In other cases, specificity may reside in the *creation* of new representations for complex objects, with their own forms of invariance, coded in higher visual areas (see chapter 2).

Some tasks in the literature show full specificity, a few show complete transfer, but many exhibit something in between—partial specificity and partial transfer. Many factors in the training and test paradigms likely modulate the quantitative balance between the two. In these intermediate cases, correct interpretation of the data will depend on quantitative treatments that are specific to the paradigm. A number of such cases are discussed in detail in section 3.8.

Despite the necessary caveats, empirical investigations of specificity have nevertheless generated many exciting discoveries. Forms of specificity have helped to identify the relationships between visual functions, operations, and representations. Furthermore, even though certain examples of specificity fail to persuasively distinguish between representation enhancement and reweighting, they nonetheless can help to identify candidate regions of the visual cortex for the relevant stimulus representations. Observations of specificity might in turn influence the selection of model architectures and the implementation of representations in models. The creation of nuanced computational models that make predictions about transfer as well as learning will promote more powerful evaluation of the models themselves, thus improving our ability to optimize learning and transfer in both theory and practice (see chapter 12).

3.8 Appendix A: Experimental Paradigms, Methods of Analysis, and Indices of Specificity and Transfer

Scientists studying perceptual learning must choose an experimental paradigm. As we know from other fields (most famously in the Heisenberg uncertainty principle), how we choose to measure a phenomenon could very well influence the result. It thus follows that the different paradigms used in the assessment of perceptual learning and of specificity and transfer may themselves influence the results and the conclusions that can be drawn. This appendix considers the pros and cons of several major experimental paradigms used to study perceptual learning and how best to analyze and interpret the results. Our purpose is to provide a pointer to the appropriate theoretical tools.

In the following discussion, T refers to the training task and X refers to the transfer task. Although there are three broad types of tasks in perceptual learning (subsection 2.2.3), each with associated performance measure(s), the following examples are mostly presented for contrast threshold, a Type II task. In most but not all cases, parallel analyses can be developed for Type I and Type III tasks.

The traditional and proposed analyses developed here are shown for the typical, fairly coarse-grained measures of performance over block or session measures. As alternative adaptive or estimation methods are developed and validated, we suggest that it will become possible in many cases to estimate performance at the grain of a small number of trials or even at the level of trial-by-trial performance. The coarse grain of analysis leads inevitably to estimates of quantities such as learning rates or initial performance in training or transfer sessions that, while appropriate for the grain at which the measurement is carried out, may be biased estimates of the learning and transfer measured at fine grain (see section 3.9).

3.8.1 Power Function or Exponential Learning and Specificity Measures

Exponential and power-function curves are the two functional forms most often used to characterize learned improvements resulting from practice.[13, 14, 16] (They also can be used to estimate and quantify transfer between the training and transfer tasks T and X, although this has been done infrequently.) Power laws provide the most common descriptive function for the effects of practice.[102] Even if learning functions for individual observers are better described as exponential, averaged learning curves of several individuals in a group will approximate power-function forms because of variability across individuals.[13, 103] If learning takes a more complex form with several stages or components of learning, composite functions can be constructed by averaging or joining several functional forms.[13]

In this discussion, we focus on contrast-threshold measures of performance over the course of learning. The equation for exponential improvements in contrast threshold with practice is

$$C(t) = \lambda e^{-\beta t} + \alpha, \tag{3.1}$$

where α is the lower (minimum threshold) asymptote after extended practice, λ is the initial incremental threshold (i.e., the initial performance is $\lambda + \alpha$), β is the exponential rate parameter for the improvement, and t is the number of practice blocks (or trials or sessions, depending on the grain of measurement). Transfer improvements from T to X may be estimated as equivalents of transfer practice t_e in an expanded exponential:

$$C_X(t) = \lambda e^{-\beta(t+t_e)} + \alpha. \tag{3.2}$$

The corresponding equation for power-function improvements is

$$C(t) = \lambda t^{-\rho} + \alpha, \tag{3.3}$$

and the generalized power function explicitly incorporating transfer from prior experience is

$$C_X(t) = \lambda(t+t_e)^{-\rho} + \alpha. \tag{3.4}$$

(For experimental applications, see Dosher and Lu,[13] Jeter et al.,[14] and Jeter et al.[16]) In either form, the transfer task X has received a transfer benefit of t_e practice units. The parameter t_e is a measure of training-equivalent transfer in which performance in the transfer task behaves as though it had already benefited from that number of training units, called the training-equivalent transfer index. Estimated values range from $t_e = 0$ for no transfer to $t_e = k$ for full transfer, where k is the number of units (trials, blocks, or sessions) of practice in the training task T.

The numerical value of the training-equivalent transfer index t_e is estimated using fits of a model to data, which requires an explicit or implicit comparison to some data for the transfer task X without pretraining. This is accomplished differently in different paradigms: by assuming equivalence of the training and transfer task, by comparison to a control group, or by testing for discontinuities from a subsequent learning rate.

This quantitative analysis of learning functions is carried out within a model comparison framework. Nested tests can compare a model in which t_e is free to vary or is set to zero using either an F-ratio test or related tests that discount model complexity, such as the Akaike information criterion (AIC), Bayesian information criterion (BIC), or Bayes factor to compare the fuller model to the restricted model (see Lu and Dosher [104] for a treatment of model comparison). Alternatively, estimates of transfer may be carried out wholly in the context of hierarchical Bayesian modeling.[105]

Functional model fitting could be used to quantify the specificity or transfer from the training task T in order to model the initial performance after the task switch; the rate of subsequent learning; and the final level, or total magnitude of learning in the transfer task X. The latter two are only defined in experimental designs in which the transfer task itself is trained after the switch, which occurs infrequently in the literature. Otherwise, specificity or transfer is assessed only in a snapshot, at the point of initial transfer to X.

The following sections take up five basic paradigms for the assessment of learning and transfer and when each may be appropriate.

3.8.2 Transfer-without-Baseline Paradigm

One of the two most commonly used paradigms for studying specificity and transfer is *transfer without baseline*. This implicitly makes an *equivalence assumption*, T ≈ X—that the initial performance and rate of learning would be about the same for T and X if assessed independently. There is a real benefit to including further training on X, as this allows estimation not just of the initial level of X, which is compared to the initial level of T, but also the rate of learning and final level of performance in X.

One example in which the equivalence assumption was approximately correct[13] used contrast-limited (Type II) orientation discrimination (±10°) around a reference angle of either +45° or −45° (see section 3.2). The training and transfer phases were rotational equivalents of one another, so the equivalence assumption is plausible (orientation discrimination around the positive diagonal should be equivalent to orientation discrimination around the negative diagonal). The equivalence assumption might fail in many other situations, and the initial accuracy or the rate of learning could differ (e.g., if the switch was from a diagonal to a cardinal reference angle, 0°).

Specificity at the point of initial transfer has often been evaluated by eye. Given the task equivalence of T and X, the transfer-without-baseline paradigm is relatively simple to interpret. If learning is completely specific, the training curves for T and X are essentially identical. If transfer is complete, then continued training on X takes off where training on T ends. For intermediate cases, researchers have used specificity indices to quantify the amount of specificity (and its inverse, transfer). Transfer is the improvement in performance at the point of the task switch (e.g., the first performance measurement for X) relative to an untrained baseline, while specificity is the return toward untrained baseline performance.[106] Ahissar and Hochstein[15] defined a specificity index as the proportion of improvement during initial training that does *not* transfer. For contrast-threshold measurements (with input values in units of performance), the specificity score is

$$S=\frac{\left(C_{X_1}-C_{T_{end}}\right)}{\left(C_{T_1}-C_{T_{end}}\right)}. \tag{3.5}$$

The contrasts for the first and last blocks of practice in the training task T are C_{T_1} and $C_{T_{end}}$, and C_{X_1} is the contrast threshold for the first block in the transfer task X. Performance in task X is directly compared to that of task T, which requires task equivalence (otherwise a paradigm using control measures of learning in X should be selected; see subsection 3.8.4). This index also works best if $C_{T_{end}}$ is measured at nearly asymptotic levels (figure 3.19). Otherwise, $C_{T_{end}}$ may require correction for where it *would have been* at the next time point, or $\hat{C}_{T_{end+1}}$, to substitute for $C_{T_{end}}$; see Jeter et al.[16] If full training functions of both T and X are available, fitted functions may provide better estimates of values entered into *S*. However, the literature has almost always used the empirically measured values.

In the context of testing model forms, assuming the power function for group data for example, a system of two equations is jointly fit to the data:

$$C_T(t_T)=\lambda t_T^{-\rho_T}+\alpha \tag{3.6}$$

and

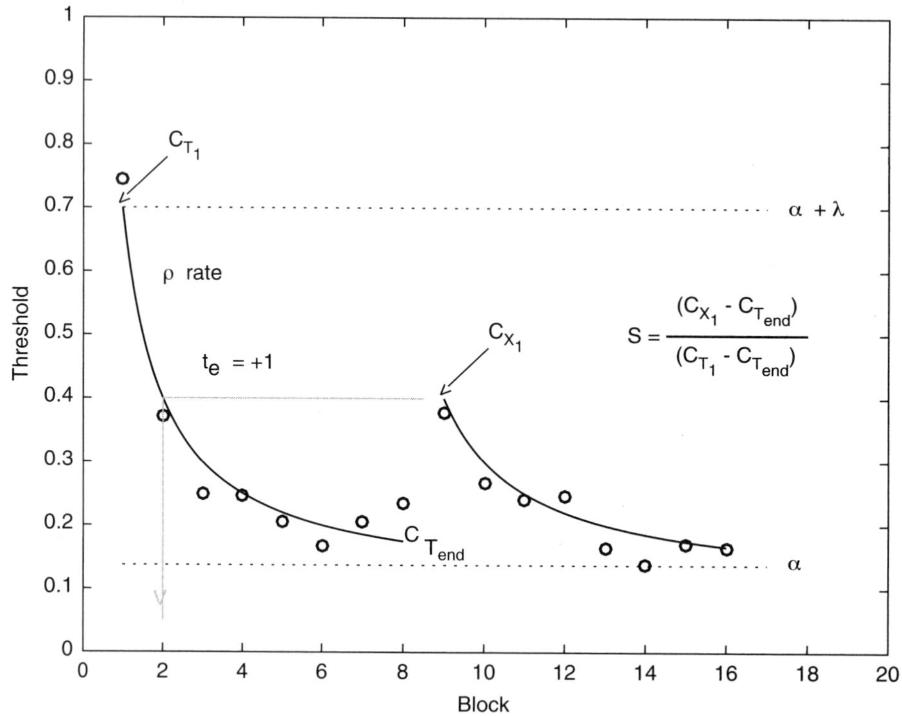

Figure 3.19
Contrast-threshold learning in a *transfer-without-baseline paradigm*, which assumes equivalence of the training and transfer tasks, $T \approx X$, including illustrations of specificity index S and the training-equivalent transfer index t_e=+1 ($\lambda_{T=X}$=0.6, $\alpha_{T=X}$=0.1, $\rho_{T=X}$= 1). Definitions of the values of contrast C are explained in the text.

$$C_X(t_X) = \lambda(t_X + t_e)^{-\rho_X} + \alpha. \tag{3.7}$$

As a result of the equivalence assumptions, these equations constrain λ and α to be the same for X and T. The immediate transfer from T to X, t_e, can be estimated from the data by assuming a single learning rate, $\rho = \rho_T = \rho_X$. This learning-rate equivalence can also be further tested in model comparisons. If only a single data point C_{X_1} is available for the transfer task, a single-point estimate of t_e is estimated by interpolation on the function T. See figure 3.19 for an illustration.

3.8.3 Transfer-with-Baseline Paradigm

In most situations, T and X are *not* equivalent, and this demands another paradigm. There are many such examples: assessing whether orientation discrimination transfers between the fovea and a peripheral position, between orientation discrimination near cardinal and oblique directions, or between two distinct judgments on the same stimuli. The most frequently used approach in such cases involves baseline measures, or *transfer-with-baseline* paradigms. The transfer task X is assessed (sometimes briefly) to yield a measure X_{pre}, or sometimes both X and T are assessed in baseline measures, then T is trained, and then X is measured to generate a posttraining measure X_{post}, sometimes followed by more training on X. Generally, researchers simply compare X_{post} to X_{pre} and only occasionally see if more is learned in X with more training.

Although very commonly used, interpretation of the transfer-with-baseline paradigms is actually very complicated because baseline assessments themselves also provide practice. Often, the fastest change in performance occurs early in training, so even in the absence of the training task, one would expect learning between the baseline $X_{pre}=X_1$ and the first postswitch block $X_{post}=X_2$. Removing these contaminants in assessments of transfer and specificity therefore requires estimation of normative performance at X_2 and modified indices of specificity. Unfortunately, this also limits inferences about whether the learning rate for X is the same as it would have been without training on T. In short, interpretation of transfer-with-baseline studies is often challenging.

In some cases, however, the expected effects of baseline training can be estimated by extrapolation from the subsequent learning curves back to predicted initial baseline levels[20] using a functional equation for the learning curves of the transfer task:

$$C_X(t_X) = \lambda(t_X + t_*)^{-\rho_X} + \alpha\,, \tag{3.8}$$

in which t_* is set to 0 for the baseline measures prior to training on T, and is set to t_e, to incorporate transfer from training for blocks following T. A schematic of transfer-with-baseline paradigms is illustrated in figure 3.20. Fitting discontinuities in the learning curve between the pretraining baseline and the first posttraining block of practice with estimates of transfer t_e requires a significant assessment of the learning curve for the transfer task, something that rarely occurs. Sometimes, researchers try to mitigate learning during baseline assessment by keeping the baseline assessment brief or by eliminating feedback. However, without explicit comparisons to controls, we cannot know with full certainty whether these approaches are successful.

3.8.4 Transfer-of-Training Paradigm

A more direct but infrequently used approach to measuring specificity and transfer is the *transfer-of-training* paradigm, which compares a trained group and a control group. T and X need not be equivalent but can differ in arbitrary ways. Training in X after training in T is compared with training a separate control group X_0 not previously trained by T. A more complete option goes on to train the X_0 group on T (e.g., training without baseline with two groups that differ in the order of training).

The advantage of this paradigm is that all aspects of the learning curves X and X_0 can be compared—the impact of training T on the initial level and the rate of learning of X. The disadvantage is that transfer and specificity cannot be assessed for individual observers but only between groups of observers. Since perceptual learning often shows

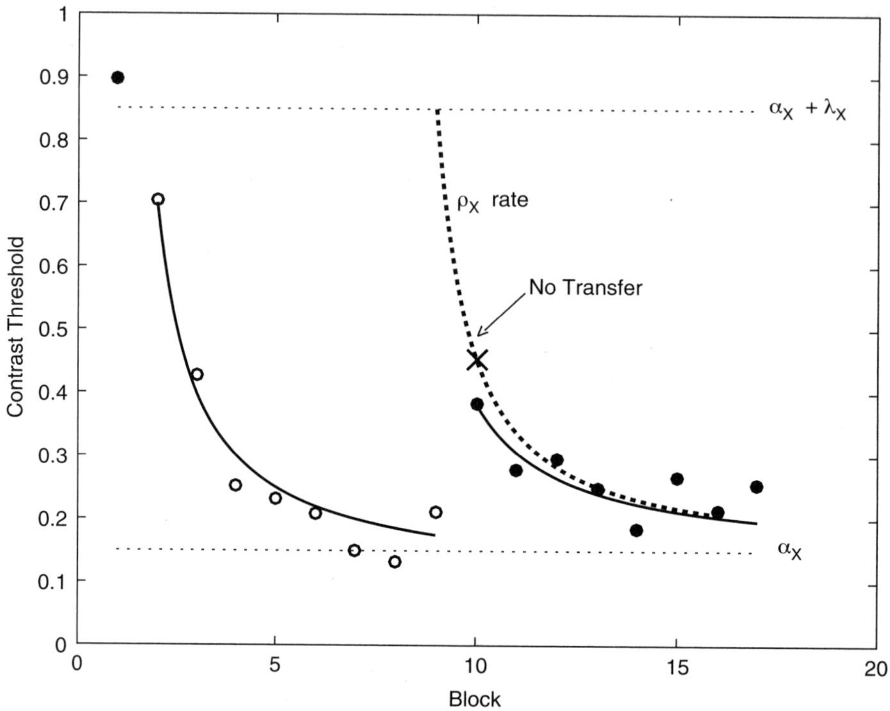

Figure 3.20
Contrast-threshold learning curves in a *transfer-with-baseline paradigm* with nonequivalent T and X tasks; learning from the baseline assessment of X (block 1) should be incorporated in the analysis. The dashed line shows (hypothetical) training on X without transfer, with x marking expected improvement from the pretraining baseline to the posttraining baseline. The solid curve shows some transfer, with t_e=+1.2 (λ_T=0.6, α_T=0.1, ρ_T=1; λ_X=0.7, α_X=0.15, ρ_X=1.2). This approach requires fitting functions, equivalent to using function-derived inputs to specificity indices.

substantial individual variation, large groups of subjects may be required. In these designs using a control group, the same analyses can be carried out as in transfer-without-baseline design (subsection 3.8.2), substituting the measures from the control group on the transfer task for the measures in the training task. Some of the issues are illustrated in figure 3.21.

3.8.5 Alternation-Training Paradigm

Another rarely used design is the *alternation-training* paradigm. In this paradigm, stimuli or tasks T and X are trained in alternation for several cycles of training; this requires a choice of the rate of alternation (i.e., every 20 trials or every 2,000 trials). This paradigm is especially powerful for assessing cases in which T and X are learned completely independently or may interact in push-pull competition in which changes that improve one task

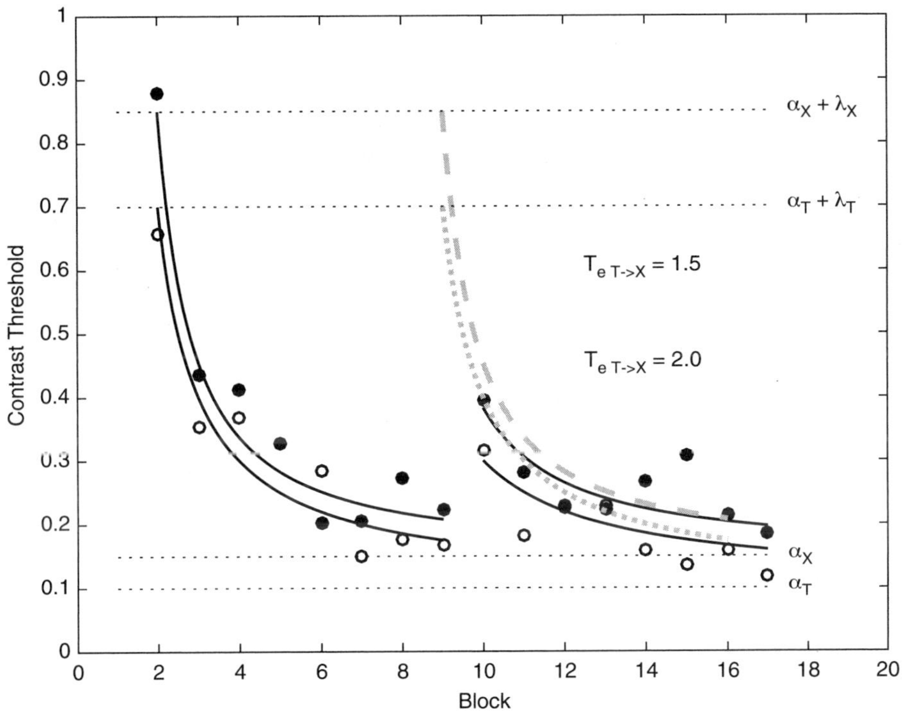

Figure 3.21
Contrast-threshold learning in a *transfer-of-training paradigm* comparing performance after training on one task (right) to (control) performance without training (left) (λ_T=0.6, α_T=0.1, ρ_T=1; λ_X=0.7, α_X=0.15, ρ_X=1.2), with $t_{eT\to X}$=+1.2 and $t_{eX\to T}$=+2.0 (light and dark circles show data, and the dashed and dotted curves project control curves from *T* and *X*). Transfer indices t_e are estimated from model fits to data; specificity indices, *S*, are estimated by reading C_{T_1} and $C_{T_{end}}$ from control curves (left) and C_{X_1} from the posttraining curves (right) (or vice versa for pretraining *X* on *T*) or from fitted estimates (see the text).

may damage the other. If T and X are learned independently—are completely specific—then alternation measures slices from the respective independent learning curves. The learning curves for each task independently can be reassembled by graphing performance as a function of the blocks of training on each task alone. The left panel of figure 3.22 shows independent learning curves for tasks T and X, with one (the thin line) shifted right; the horizontal dotted lines indicate the maximum ($\lambda + \alpha$) and minimum (α) thresholds, which are assumed to be the same in the two tasks for this illustration. The right panel shows the result for independently learned tasks, where training segments of the original curves for T and X are simply shifted into the alternated training windows.

This paradigm has the power to reveal either independence, cross transfer, or switching costs that would be ambiguous or difficult to document in shorter nonalternation designs. Sometimes, the first two cycles of training are ambiguous with respect to push-pull

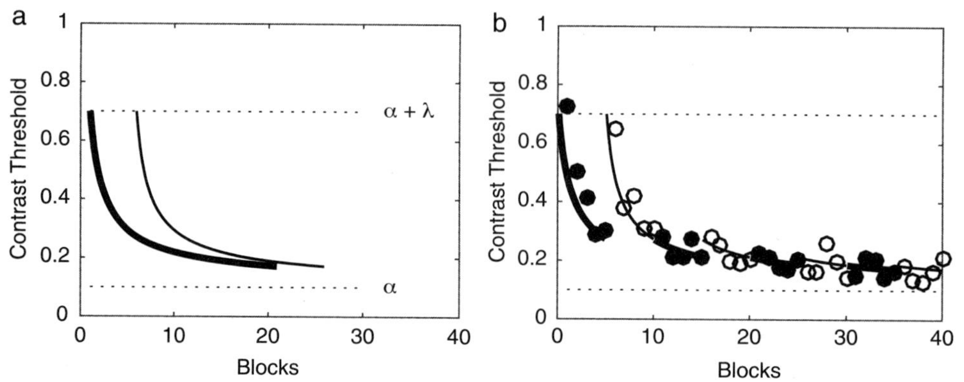

Figure 3.22
Schematic illustrations of the *alternation-training paradigm* to test for task-training interactions resulting from specificity or transfer. Independent training of two stimulus/task conditions X and Y with one shown offset to the other in training trials (a); alternation blocks of training X and training Y with tasks that are colearned independently with full specificity (b) (see the text for explanation).

competition or switching costs between tasks. Explicit consideration of push-pull competition leading to transfer or switching costs also emphasizes the obvious fact that transfer may be either negative or positive, or a mixture of both. Such situations require model-based treatment. See Petrov, Dosher, and Lu,[17] Huang, Lu, and Dosher,[72] and Petrov, Dosher, and Lu[107] for experimental examples.

3.8.6 Unequal Trial Mixture Paradigm

The *unequal trial mixture* paradigm, created by Liu and Vaina[108] as another way to test transfer, alternates T and X on a regular trial-by-trial sequence of T-T-X. They argue that (we have substituted our notation of T and X in the quotation) "if, however, the learning is not stimulus specific, but transfers between the two attributes T and X, we would expect that more improvement occurs in condition X than in condition T, because X is lagging behind T in the learning sequence.... After 3n trials (2n T and n X), if learning transfers from T to X, then the amount of improvement for the n X trials should be greater than for the first n T trials" [108] (p. 347). This paradigm was used to assess transfer within a short single session of training for motion in two directions.[108, 109] If learning of T and X is completely specific and independent (and both show learning), then performance should be ordered T1 = X < T2 (T1 refers to the first half and T2 to the second half of all T trials; the order refers to positive performance such as percentage correct); for example, T1 is the same as X, and T2 is better. If there is some positive transfer between T and X, then performance should be ordered T1 < X < T2. This paradigm, as applied, makes the same strong equivalence assumptions as the training-without-baseline paradigm. T and X must have the same initial performance level and the same learning rate if trained separately. In some circumstances, the paradigm may also require special tests for alternation costs (not examined in the initial study); for example, if T and X are different judgments, then in a T-T-X… trial sequence, X trials always involve a judgment change that is only true for the first of the pair of T trials. Whether this matters can be tested by comparing performance estimated from only the first or only the second of a pair of Ts.

3.8.7 Summary

Obviously, each of these experimental paradigms for assessing specificity and transfer has advantages and disadvantages. The theoretical question, the nature of stimulus constraints, and practical considerations may all influence the researcher's selection of an experimental paradigm. Often, researchers have elected to use the abbreviated versions of the most frequent designs. For example, the transfer-without-baseline and the transfer-with-baseline paradigms generally use only a single assessment of X rather than including subsequent training in X. Such abbreviated assessments cannot assess whether the training T affects the subsequent rate of perceptual learning in X. Observationally, it seems that a particular choice of paradigm often becomes habitual in a particular domain. This appendix sought to provide information that will guide the selection of the paradigm and analysis best suited to the question being pursued.

3.9 Appendix B: Effects of Measurement Grain

Any empirical assessment of learning necessarily chooses a grain of measurement. The primary measures of performance used in the literature (whether percentage correct, discriminability d', contrast threshold, or difference threshold) each require significant sample sizes to yield reasonable estimates and therefore typically use 60–500 trials per point. As with any other measurement, any resulting estimates of learning depend on the scale. For example, weather temperature can be listed as variations within a day or as monthly averages, and while tied together, these measures reveal different properties. One reason that measuring threshold at the end of blocks (or sessions) has been seen as reasonable is that learning in perceptual tasks in many cases seemed to extend over thousands of trials and many days of practice. However, the observation that learning continues for a long time does not rule out a more rapid early component of learning. Even if there is no separate rapid early phase, block-level estimates of the initial levels of performance in the training task or in the first block in the transfer task can be both biased and highly variable.[110, 111] The traditional measures of specificity and transfer are unusually sensitive to these initial levels.

The ability to carry out rapid and efficient performance measurements, typically with sophisticated adaptive methods that allow changing state,[110, 111] may open the field to very flexible methods of training and assessment. Such adaptive methods permit more accurate assessments of performance at the very beginning of practice. They permit estimates of rapid early phases of learning (if they exist), and fine-grained assessments of the functional forms of change. They also provide a window into perceptual learning that is closer to the trial-by-trial, or experience-by-experience, learning implemented in most quantitative models. This appendix illustrates some of the issues with such methods.

Learned improvements in performance with training or practice may be either rapid or slow, and the challenge is to provide accurate and precise measures. Existing methods of measurement (e.g., percentage correct or d') that aggregate performance over a certain number of trials tend to use a coarse scale of measurement. Other commonly used adaptive methods (e.g., adaptive "n-down/m-up" staircases,[112, 113] Quest,[114, 115] the stochastic approximation method,[116] and the accelerated stochastic approximation method[117]) have other issues. They were developed to estimate an unchanging performance level over the series of trials used to produce a threshold measurement and are therefore not statistically optimized for estimation when performance is in flux. Either of these kinds of measures may suffice if the learning is slow (although there are consequences for the variability of the estimates). On the other hand, derived measures of specificity and transfer may require measures that are more accurate and less variable.

Although the detailed implications require the simulation of underlying processes, with different learning rates, as they interact with different assessment measures, the general point is that since it may take 80 or 100 trials, for example, to estimate contrast threshold with a staircase, this limits the grain of measurement. Also, some methods, such as staircases, tend to eliminate trials early in the measurement block, so rapid learning in the first 50 trials may be missed, the initial level is more variable, and fits to the learning curve are consequently less constrained.

An alternative method assumes a learning curve and uses each new data point to update the parameter estimates of that curve on a trial-by-trial basis.[110, 111] One of these, the quick-change-detection (or qCD),[111, 118] was designed to estimate the threshold on a trial-by-trial basis. It assumes the threshold learning curve is an exponential function $T(\vec{\theta}, n) = \lambda \exp\left(\frac{-n}{\gamma}\right) + \alpha$, where $\vec{\theta} = (\lambda, \gamma, \alpha)$ are the exponential parameters (called the generating parameters), α is the asymptotic performance after extensive practice, λ is the amount by which the initial performance exceeds this, γ is the learning rate, and n is the training trial. The method sets priors over $\vec{\theta}$ and computes an optimal value (e.g., stimulus contrast) used for testing on the first trial. Then, Bayesian methods are used to update the probability distributions for these parameters based on the accuracy of the response and then select a most informative stimulus value to test on the next trial.

The qCD provides an estimate of the threshold on each trial, as well as measures of the credible interval for each parameter (see figure 3.23). Then, at the end of training, the estimates of the threshold on each trial can be revised using all the trial information to give the best estimates of the generating parameters. Several simulation studies have shown that these trial-by-trial estimates are close to unbiased and significantly less variable (by

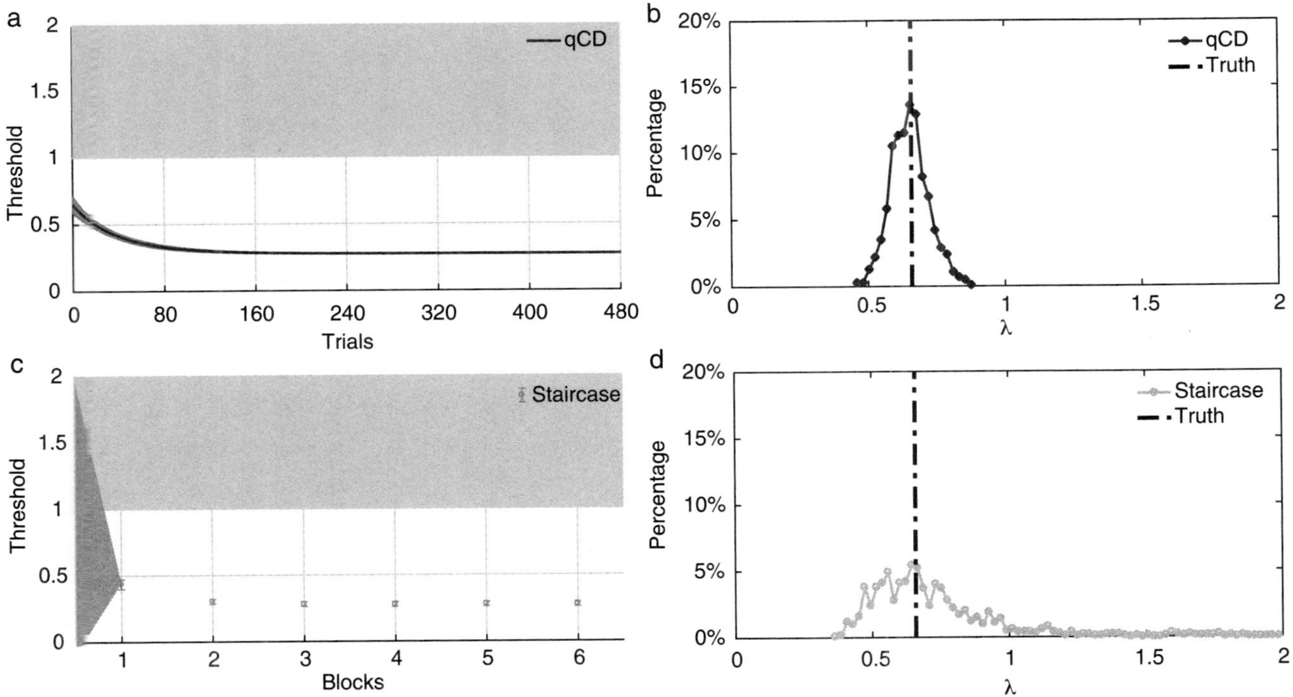

Figure 3.23

Comparisons of assessments of exponential perceptual learning $T(\vec{\theta}, n) = \lambda \exp\left(\frac{-n}{\gamma}\right) + \alpha$ using (a) a quick-change-detection (qCD) method with trial-by-trial estimates of threshold and (b) using a standard 3:1 staircase once a block. Fitting the two forms of data with the exponential form with simulated distributions of the parameter estimates for λ (the initial level above asymptote) are shown for (c) the qCD and (d) the staircase methods. The qCD estimates are less biased and have significantly smaller standard deviations.

a factor of more than four in many cases) than measures acquired through the typical three-down, one-up staircase methods with 80- or 160-trial blocks. These more precise estimates of initial thresholds in the training and transfer tasks in turn lead to more accurate estimates of transfer and more accurate specificity indices.[110, 111]

Using these improved trial-by-trial estimates of performance, especially of the performance early in learning, then flows through to better estimates of specificity—because specificity has often focused on the comparisons of initial performance in the first training task and in the transfer task. The specificity indices take initial training points as key inputs (together with the easier-to-estimate asymptotic levels late in learning). It then follows directly that improving the estimates of initial performance will also improve the estimates of specificity and transfer and their corresponding indices.[110]

References

1. Karni A, Sagi D. Where practice makes perfect in texture discrimination: Evidence for primary visual cortex plasticity. *Proceedings of the National Academy of Sciences* 1991;88(11):4966–4970.

2. Ahissar M, Laiwand R, Kozminsky G, Hochstein S. Learning pop-out detection: Building representations for conflicting target-distractor relationships. *Vision Research* 1998;38(20):3095–3107.

3. Gilbert CD. Early perceptual learning. *Proceedings of the National Academy of Sciences* 1994;91(4):1195–1197.

4. Karni A, Bertini G. Learning perceptual skills: Behavioral probes into adult cortical plasticity. *Current Opinion in Neurobiology* 1997;7(4):530–535.

5. Ahissar M, Hochstein S. Eureka: One shot viewing enables perceptual learning. Paper presented at Investigative Ophthalmology & Visual Science; 1996.

6. Crist RE, Li W, Gilbert CD. Learning to see: Experience and attention in primary visual cortex. *Nature Neuroscience* 2001;4(5):519–525.

7. Fahle M, Morgan M. No transfer of perceptual learning between similar stimuli in the same retinal position. *Current Biology* 1996;6(3):292–297.

8. Liu Z, Weinshall D. Mechanisms of generalization in perceptual learning. *Vision Research* 2000;40(1):97–109.

9. Shiu L-P, Pashler H. Improvement in line orientation discrimination is retinally local but dependent on cognitive set. *Attention, Perception, & Psychophysics* 1992;52(5):582–588.

10. Dosher BA, Lu Z-L. Perceptual learning reflects external noise filtering and internal noise reduction through channel reweighting. *Proceedings of the National Academy of Sciences* 1998;95(23):13988–13993.

11. Dosher BA, Lu Z-L. Mechanisms of perceptual learning. *Vision Research* 1999;39(19):3197–3221.

12. Mollon JD, Danilova MV. Three remarks on perceptual learning. *Spatial Vision* 1996;10(1):51–58.

13. Dosher BA, Lu Z-L. The functional form of performance improvements in perceptual learning: Learning rates and transfer. *Psychological Science* 2007;18(6):531–539.

14. Jeter PE, Dosher BA, Petrov A, Lu Z-L. Task precision at transfer determines specificity of perceptual learning. *Journal of Vision* 2009;9(3):1,1–13.

15. Ahissar M, Hochstein S. Task difficulty and the specificity of perceptual learning. *Nature* 1997;387(6631):401–406.

16. Jeter PE, Dosher BA, Liu S-H, Lu Z-L. Specificity of perceptual learning increases with increased training. *Vision Research* 2010;50(19):1928–1940.

17. Petrov AA, Dosher BA, Lu Z-L. The dynamics of perceptual learning: An incremental reweighting model. *Psychological Review* 2005;112(4):715–743.

18. Matthews N, Liu Z, Geesaman BJ, Qian N. Perceptual learning on orientation and direction discrimination. *Vision Research* 1999;39(22):3692–3701.

19. Crist RE, Kapadia MK, Westheimer G, Gilbert CD. Perceptual learning of spatial localization: Specificity for orientation, position, and context. *Journal of Neurophysiology* 1997;78(6):2889–2894.

20. Dosher BA, Lu Z-L. Perceptual learning in clear displays optimizes perceptual expertise: Learning the limiting process. *Proceedings of the National Academy of Sciences* 2005;102(14):5286–5290.

21. Schoups AA, Vogels R, Orban GA. Human perceptual learning in identifying the oblique orientation: Retinotopy, orientation specificity and monocularity. *Journal of Physiology* 1995;483(3):797–810.

22. Ball K, Sekuler R. Direction-specific improvement in motion discrimination. *Vision Research* 1987;27(6):953–965.

23. O'Toole AJ, Kersten DJ. Learning to see random-dot stereograms. *Perception* 1992;21(2):227–243.

24. Nazir TA, O'Regan JK. Some results on translation invariance in the human visual system. *Spatial Vision* 1990;5(2):81–100.

25. Zhang J-Y, Zhang G-L, Xiao L-Q, Klein SA, Levi DM, Yu C. Rule-based learning explains visual perceptual learning and its specificity and transfer. *Journal of Neuroscience* 2010;30(37):12323–12328.

26. Sowden PT, Rose D, Davies IR. Perceptual learning of luminance contrast detection: Specific for spatial frequency and retinal location but not orientation. *Vision Research* 2002;42(10):1249–1258.

27. Schwartz S, Maquet P, Frith C. Neural correlates of perceptual learning: A functional MRI study of visual texture discrimination. *Proceedings of the National Academy of Sciences* 2002;99(26):17137–17142.

28. Schoups AA, Orban GA. Interocular transfer in perceptual learning of a pop-out discrimination task. *Proceedings of the National Academy of Sciences* 1996;93(14):7358–7362.

29. Sagi D. Perceptual learning in vision research. *Vision Research* 2011;51(13):1552–1566.

30. Lu Z-L, Chu W, Dosher BA, Lee S. Independent perceptual learning in monocular and binocular motion systems. *Proceedings of the National Academy of Sciences* 2005;102(15):5624–5629.

31. Poggio T, Fahle M, Edelman S. Fast perceptual learning in visual hyperacuity. *Science* 1992;256(5059):1018–1021.

32. Fahle M. Specificity of learning curvature, orientation, and Vernier discriminations. *Vision Research* 1997;37(14):1885–1895.

33. Fiorentini A, Berardi N. Perceptual learning specific for orientation and spatial frequency. *Nature*;1980:287(5777):43–44.

34. Fiorentini A, Berardi N. Learning in grating waveform discrimination: Specificity for orientation and spatial frequency. *Vision Research* 1981;21(7):1149–1158.

35. Ball K, Sekuler R. A specific and enduring improvement in visual motion discrimination. *Science* 1982;218(4573):697–698.

36. Ramachandran V, Braddick O. Orientation-specific learning in stereopsis. *Perception* 1973;2(3):371–376.

37. Huang C-B, Zhou Y, Lu Z-L. Broad bandwidth of perceptual learning in the visual system of adults with anisometropic amblyopia. *Proceedings of the National Academy of Sciences* 2008;105(10):4068–4073.

38. Furmanski CS, Engel SA. Perceptual learning in object recognition: Object specificity and size invariance. *Vision Research* 2000;40(5):473–484.

39. Lu Z-L, Dosher BA. Perceptual learning retunes the perceptual template in foveal orientation identification. *Journal of Vision* 2004;4(1):5,44–56.

40. Ahissar M, Hochstein S. Attentional control of early perceptual learning. *Proceedings of the National Academy of Sciences* 1993;90(12):5718–5722.

41. Rubin N, Nakayama K, Shapley R. Abrupt learning and retinal size specificity in illusory-contour perception. *Current Biology* 1997;7(7):461–467.

42. Campbell FW, Robson J. Application of Fourier analysis to the visibility of gratings. *Journal of Physiology* 1968;197(3):551–566.

43. Shapley R, Lennie P. Spatial frequency analysis in the visual system. *Annual Review of Neuroscience* 1985;8(1):547–581.

44. Dakin SC, Mareschal I. Sensitivity to contrast modulation depends on carrier spatial frequency and orientation. *Vision Research* 2000;40(3):311–329.

45. Schofield AJ, Georgeson MA. Sensitivity to contrast modulation: The spatial frequency dependence of second-order vision. *Vision Research* 2003;43(3):243–259.

46. Cavanagh P, Mather G. Motion: The long and short of it. *Spatial Vision* 1989;4(2):103–129.

47. Sutter A, Graham N. Investigating simple and complex mechanisms in texture segregation using the speed-accuracy tradeoff method. *Vision Research* 1995;35(20):2825–2843.

48. Werkhoven P, Sperling G, Chubb C. The dimensionality of texture-defined motion: A single channel theory. *Vision Research* 1993;33(4):463–485.

49. Regan D. *Human perception of objects*. Sinauer Associates; 1999.

50. Larsson J, Landy MS, Heeger DJ. Orientation-selective adaptation to first- and second-order patterns in human visual cortex. *Journal of Neurophysiology* 2006;95(2):862–881.

51. Chubb C, Sperling G, Solomon JA. Texture interactions determine perceived contrast. *Proceedings of the National Academy of Sciences* 1989;86(23):9631–9635.

52. Lu Z-L, Sperling G. The functional architecture of human visual motion perception. *Vision Research* 1995;35(19):2697–2722.

53. Baker CL. Central neural mechanisms for detecting second-order motion. *Current Opinion in Neurobiology* 1999;9(4):461–466.

54. Chubb C, Sperling G. Drift-balanced random stimuli: A general basis for studying non-Fourier motion perception. *Journal of the Optical Society of America A* 1988;5(11):1986–2007.

55. Wilson HR. Non-Fourier cortical processes in texture, form, and motion perception. *Cerebral Cortex* 1999;13:445–477.

56. Lin L-M, Wilson HR. Fourier and non-Fourier pattern discrimination compared. *Vision Research* 1996;36(13):1907–1918.

57. Lu Z-L, Sperling G. Three-systems theory of human visual motion perception: Review and update. *Journal of the Optical Society of America A* 2001;18(9):2331–2370.

58. McGraw PV, Levi DM, Whitaker D. Spatial characteristics of the second-order visual pathway revealed by positional adaptation. *Nature Neuroscience* 1999;2(5):479–484.

59. Nishida SY, Ledgeway T, Edwards M. Dual multiple-scale processing for motion in the human visual system. *Vision Research* 1997;37(19):2685–2698.

60. Schofield AJ, Georgeson MA. Sensitivity to modulations of luminance and contrast in visual white noise: Separate mechanisms with similar behaviour. *Vision Research* 1999;39(16):2697–2716.

61. Vaina LM, Cowey A, Kennedy D. Perception of first- and second-order motion: Separable neurological mechanisms? *Human Brain Mapping* 1999;7(1):67–77.

62. Allard R, Faubert J. Zero-dimensional noise is not suitable for characterizing processing properties of detection mechanisms. *Journal of Vision* 2013;13(10):25,1–3.

63. Johnston A, McOwan P, Buxton H. A computational model of the analysis of some first-order and second-order motion patterns by simple and complex cells. *Proceedings of the Royal Society of London B: Biological Sciences* 1992;250(1329):297–306.

64. Zanker JM. Perceptual learning in primary and secondary motion vision. *Vision Research* 1999;39(7):1293–1304.

65. Chen R, Qiu Z-P, Zhang Y, Zhou Y-F. Perceptual learning and transfer study of first- and second-order motion direction discrimination. *Progress in Biochemistry and Biophysics* 2009;36:1442–1450.

66. Petrov AA, Hayes TR. Asymmetric transfer of perceptual learning of luminance- and contrast-modulated motion. *Journal of Vision* 2010;10(14):11,1–22.

67. Vaina LM, Chubb C. Dissociation of first- and second-order motion systems by perceptual learning. *Attention, Perception, & Psychophysics* 2012;74(5):1009–1019.

68. Chung ST, Li RW, Levi DM. Learning to identify near-threshold luminance-defined and contrast-defined letters in observers with amblyopia. *Vision Research* 2008;48(27):2739–2750.

69. Zhou J, Yan F, Lu Z-L, Zhou Y, Xi J, Huang C-B. Broad bandwidth of perceptual learning in second-order contrast modulation detection. *Journal of Vision* 2015;15(2):20,1–10.

70. Rivest J, Boutet I, Intriligator J. Perceptual learning of orientation discrimination by more than one attribute. *Vision Research* 1997;37(3):273–281.

71. McGovern DP, Webb BS, Peirce JW. Transfer of perceptual learning between different visual tasks. *Journal of Vision* 2012;12(11):4,1–11.

72. Huang C-B, Lu Z-L, Dosher BA. Co-learning analysis of two perceptual learning tasks with identical input stimuli supports the reweighting hypothesis. *Vision Research* 2012;61:25–32.

73. Dorais A, Sagi D. Contrast masking effects change with practice. *Vision Research* 1997;37(13):1725–1733.

74. Liu Z. Perceptual learning in motion discrimination that generalizes across motion directions. *Proceedings of the National Academy of Sciences* 1999;96(24):14085–14087.

75. Meyer J, Petrov A. The specificity of perceptual learning of pop-out detection depends on the difficulty during post-test rather than training. *Journal of Vision* 2011;11(11):1025.

76. Petrov A. The stimulus specificity of motion perceptual learning depends on the difficulty during post-test rather than training. *Journal of Vision* 2009;9(8):885.

77. Wang X, Zhou Y, Liu Z. Transfer in motion perceptual learning depends on the difficulty of the training task. *Journal of Vision* 2013;13(7):5,1–9.

78. Harris H, Gliksberg M, Sagi D. Generalized perceptual learning in the absence of sensory adaptation. *Current Biology* 2012;22(19):1813–1817.

79. Greenlee MW, Magnussen S. Interactions among spatial frequency and orientation channels adapted concurrently. *Vision Research* 1988;28(12):1303–1310.

80. Karni A, Sagi D. The time course of learning a visual skill. *Nature* 1993;365(6443):250–252.

81. Aberg KC, Tartaglia EM, Herzog MH. Perceptual learning with Chevrons requires a minimal number of trials, transfers to untrained directions, but does not require sleep. *Vision Research* 2009;49(16):2087–2094.

82. Xiao L-Q, Zhang J-Y, Wang R, Klein SA, Levi DM, Yu C. Complete transfer of perceptual learning across retinal locations enabled by double training. *Current Biology* 2008;18(24):1922–1926.

83. Zhang J-Y, Wang R, Klein S, Levi D, Yu C. Perceptual learning transfers to untrained retinal locations after double training: A piggyback effect. *Journal of Vision* 2011;11(11):1026.

84. Zhang J-Y, Kuai S-G, Xiao L-Q, Klein SA, Levi DM, Yu C. Stimulus coding rules for perceptual learning. *PLoS Biology* 2008;6(8):e197.

85. Hung S-C, Seitz AR. Prolonged training at threshold promotes robust retinotopic specificity in perceptual learning. *Journal of Neuroscience* 2014;34(25):8423–8431.

86. Talluri BC, Hung S-C, Seitz AR, Seriès P. Confidence-based integrated reweighting model of task-difficulty explains location-based specificity in perceptual learning. *Journal of Vision* 2015;15(10):17,1–12.

87. Sotiropoulos G, Seitz AR, Seriès P. Performance-monitoring integrated reweighting model of perceptual learning. *Vision Research* 2018;152:17–39.

88. Zhang J-Y, Yang Y-X. Perceptual learning of motion direction discrimination transfers to an opposite direction with TPE training. *Vision Research* 2014;99:93–98.

89. Xiong Y-Z, Xie X-Y, Yu C. Location and direction specificity in motion direction learning associated with a single-level method of constant stimuli. *Vision Research* 2016;119:9–15.

90. Liang J, Zhou Y, Fahle M, Liu Z. Specificity of motion discrimination learning even with double training and staircase. *Journal of Vision* 2015;15(10):3,1–10.

91. Liang J, Zhou Y, Fahle M, Liu Z. Limited transfer of long-term motion perceptual learning with double training. *Journal of Vision* 2015;15(10):1,1–9.

92. Liu J, Lu Z-L, Dosher B. Multi-location augmented Hebbian re-weighting accounts for transfer of perceptual learning following double training. *Journal of Vision* 2011;11(11):992.

93. Lesmes LA, Jeon S-T, Lu Z-L, Dosher BA. Bayesian adaptive estimation of threshold versus contrast external noise functions: The quick TvC method. *Vision Research* 2006;46(19):3160–3176.

94. Lesmes LA, Lu Z-L, Baek J, Albright TD. Bayesian adaptive estimation of the contrast sensitivity function: The quick CSF method. *Journal of Vision* 2010;10(3):17,1–21.

95. Lesmes LA, Lu Z-L, Baek J, Tran N, Dosher BA, Albright TD. Developing Bayesian adaptive methods for estimating sensitivity thresholds (d′) in yes-no and forced-choice tasks. *Frontiers in Psychology* 2015;6:1070,1–24.

96. Lesmes LA, Lu Z-L, Tran NT, Dosher BA, Albright TD. An adaptive method for estimating criterion sensitivity (d') levels in yes/no tasks. *Journal of Vision* 2006;6(6):1097

97. Liu J, Dosher B, Lu Z-L. Perceptual learning trial-by-trial in a task roving paradigm. *Journal of Vision* 2018;18(10), 755.

98. Hou F, Huang C-B, Lesmes L, Feng L-X, Tao L, Zhou Y-F, Lu Z-L. qCSF in clinical application: Efficient characterization and classification of contrast sensitivity functions in amblyopia. *Investigative Ophthalmology & Visual Science* 2010;51(10):5365–5377.

99. Hou F, Lesmes L, Bex P, Dorr M, Lu Z-L. Using 10AFC to further improve the efficiency of the quick CSF method. *Journal of Vision* 2015;15(9):2,1–18.

100. Dosher BA, Lu Z-L. Hebbian reweighting on stable representations in perceptual learning. *Learning & Perception* 2009;1(1):37–58.

101. Seitz A, Watanabe T. A unified model for perceptual learning. *Trends in Cognitive Sciences* 2005;9(7):329–334.

102. Anderson JR. Acquisition of cognitive skill. *Psychological Review* 1982;89(4):369–406.

103. Heathcote A, Brown S, Mewhort D. The power law repealed: The case for an exponential law of practice. *Psychonomic Bulletin & Review* 2000;7(2):185–207.

104. Lu Z-L, Dosher B. *Visual psychophysics: From laboratory to theory.* MIT Press; 2013.

105. Gu H, Kim W, Hou F, Lesmes, LA, Pitt, MA, Lu, ZL. A hierarchical Bayesian approach to adaptive vision testing: A case study with the contrast sensitivity function. *Journal of Vision* 2016;16(6):15,1–17.

106. Dill M. Specificity versus invariance of perceptual learning: The example of position. In Fahle, M. & Poggio, T. *Perceptual Learning.* MIT Press;2002:19–231.

107. Petrov AA, Dosher BA, Lu Z-L. Perceptual learning without feedback in non-stationary contexts: Data and model. *Vision Research* 2006;46(19):3177–3197.

108. Liu Z, Vaina LM. Simultaneous learning of motion discrimination in two directions. *Cognitive Brain Research* 1998;6(4):347–349.

109. Jeter P, Dosher B, Lu Z-L. Simultaneous training of two high precision tasks is largely independent even when orientation or position is shared. *Journal of Vision* 2008;8(6):978.

110. Zhang P, Zhao Y, Dosher B, Lu Z-L. Evaluating the performance of the staircase and quick Change Detection methods in measuring perceptual learning. *Journal of Vision* 2019;19(7):14,1–25.

111. Zhang P, Zhao Y, Dosher B, Lu Z-L. Assessing the detailed time course of perceptual sensitivity change in perceptual learning. *Journal of Vision* 2019;19(5):9,1–19.

112. Levitt H. Transformed up-down methods in psychoacoustics. *Journal of the Acoustical Society of America* 1971;49(2B):467–477.

113. Wetherill G, Levitt H. Sequential estimation of points on a psychometric function. *British Journal of Mathematical and Statistical Psychology* 1965;18(1):1–10.

114. Watson AB, Pelli DG. QUEST: A Bayesian adaptive psychometric method. *Perception & Psychophysics* 1983;33(2):113–120.

115. King-Smith PE, Grigsby SS, Vingrys AJ, Benes SC, Supowit A. Efficient and unbiased modifications of the QUEST threshold method: Theory, simulations, experimental evaluation and practical implementation. *Vision Research* 1994;34(7):885–912.

116. Robbins H, Monro S. A stochastic approximation method. *Annals of Mathematical Statistics* 1951;22(3):400–407.

117. Kesten H. Accelerated stochastic approximation. *Annals of Mathematical Statistics* 1958;29(1):41–59.

118. Zhao Y, Lesmes LA, Lu Z-L. The quick Change Detection method: Bayesian adaptive assessment of the time course of perceptual sensitivity change. *Investigative Ophthalmology & Visual Science* 2017;58(8):5633.

III

Mechanisms

4

Mechanisms

Any perceptual judgment requires that an observer respond to meaningful signal information in the face of noise. This noise can be located in the stimulus or in the many sensory representation(s) of the brain. Finding the signal in the noise leads to successful perception. Perceptual learning must then improve performance by improving the signal-to-noise ratio, which requires either improving the signal or reducing the noise. This can occur through several mechanisms, which we outline in this chapter. It can occur by excluding or filtering external noise in the physical stimulus, by amplifying the stimulus input relative to internal noise, or by changing the response or gain properties of the system. The first two of these play an important role in perceptual learning and all three can be assessed by combining external-noise tests with a model of the observer. The observer model incorporates known properties of the visual system to form a "front end" for signal detection analysis that will play an important role in subsequent computational models of learning.

4.1 A Signal-and-Noise Analysis of the Mechanisms of Perceptual Learning

So far, we have looked at the major phenomena of perceptual learning and its specificity and transfer. In this chapter, we ask the next logical question: what are the *mechanisms* through which perceptual learning improves performance? The answers to this question use analyses that derive from signal detection theory and related models of the perceptual observer. These analyses examine an observer's ability to discriminate a target signal from two kinds of noise: noise intrinsic to neural processing and noise deriving from competing stimuli (i.e., external noise). How does the observer detect the presence of a signal and/or discriminate one signal from another? In either case, the relevant signal(s) must be extracted from the surrounding noise in the stimulus input (external noise) and/or the noise generated by variability in processing (internal noise). The process of extraction, as we will see, is central to the mechanisms of perceptual learning.

An everyday example may help illustrate the possible mechanisms involved. Imagine that you are speaking on your cell phone to a friend and the volume is at a middle setting. If your friend is speaking from a noisy party with many competing conversations, your ability to hear the signal message will be limited by the extraneous sounds. In this case, it is said to be limited by the *external* noise—noise in the physical stimulus. Turning up the volume on your phone will not solve the problem; it simply turns up the volume of the other voices (the external noise) at the same time that it increases the volume of your friend's voice (the signal). Instead, you might ask your friend to put her hand around her phone or to ask the background speakers to be quieter. Either action excludes or filters out the external noise in the background. On the other hand, if your friend is speaking from a very quiet environment and you are having trouble hearing her, then the limitations are intrinsic to your auditory system. In this case, boosting the signal either by increasing the volume setting on the phone or by asking your friend to speak more loudly would help make the message interpretable. (This amplification would be relative to the internal noise of your auditory system.) A third possible mechanism of learning is more intuitively grasped by using the example of your phone's camera. If a section of the image seems too dark or too light, touching it adjusts the overall range for dark and light in the image before you take a photograph. If the camera is facing the light so a face in the image is too dark, touching the face will lighten it and change the response to the very bright light. Processes that change the response to inputs are called changes in *gain*. Unlike changes in sensitivity to internal or external noise, this changes the response of the system to inputs.

Each of these three mechanisms—*filtering* of external noise, *amplification* of the stimulus relative to internal noise or limits, and changing the *gain* of the system—represents a distinct way by which you can improve the signal-to-noise ratio in perception. Regardless of the perceptual task, learning must use one or another of these mechanisms or perhaps more than one.

In this chapter, we put forward a framework of quantitative tests that identify which of the three mechanisms sketched here underlie learning in any given task. We go on to show what certain specially designed experiments have revealed about the mechanisms actually used in a number of classic perceptual learning paradigms. Although much of what follows will be technical, the essential predictions of the models reveal signature patterns of performance associated with each mechanism. The empirical results paint a picture of the observer as a very intelligent agent: even when initially focusing on relevant evidence used to perform a task, the observer may yet learn to do an even better job of finding the signal in the noise.

4.2 Signal Detection Theory (SDT)

Signal detection theory (SDT)[1, 2] is one of the most important and widely used theoretical paradigms to estimate and analyze the discriminability of different signals, and the decision factors related to setting criteria for responses. As will be familiar to anyone well versed in SDT, the distributions of evidence that derive from a signal in noise or noise without a signal differ in their mean values, and a "signal present" response occurs when the sampled value of evidence exceeds a decision criterion. (The two categories may also be stimulus type A or type B rather than stimulus present or absent.) The SDT framework is primarily designed to distinguish discriminability from the decision criteria that separate the evidence distribution into responses.[1, 2] By assuming that a given form, such as the

Gaussian distribution, describes the evidence from different stimuli, the discriminability (i.e., the difference between the means of the evidence distributions relative to their variability) and the criterion can be estimated.

The SDT framework is one of the most powerful and widespread approaches to the behavioral analysis of human performance and has a long history in a number of fields, from cellular physiology to human memory, perception, and certain forms of abstract decision-making.[3–7] And yet, as a descriptive framework rather than a predictive one, there are important processes that it does not address. It does not reveal how the signal is processed into an internal representation, for instance, nor does it tell us anything about the sources and characteristics of the noise. In this sense, the SDT is so widely applicable precisely because it does not impose a domain-specific structure.

The technology that we use in this chapter, combining external-noise paradigms with an observer model, seeks to specify how the distributions of evidence arise from the external stimulus and internal processes and noises. Such an analysis, though used far less frequently in the field, could theoretically be applicable to different levels within a sensory domain: at the cellular level of individual neurons or groups of neurons, at the level of a functional module, or at the level of the observer as a whole. This model thus ties the physical properties of the external stimulus to the internal representations that are created from them, which in turn become the distributions assumed by signal detection theory.

In what follows, we begin with a discussion of a simple *perceptual template model* (PTM)[8, 9] of the observer and sample experiments that can be used to specify the model and its parameters. The PTM provides a model of the whole observer that characterizes the input-output relationships based on behavioral data. Each component of the PTM is essential to a quantitative model of human behavior. This framework can also be used to quantify the mechanisms by which perceptual learning occurs.[9, 10] Each of the three different mechanisms can be detected from measured changes in specific parameter values, each of which leads to signature patterns of the behavioral data.[10] As we will see, these mechanisms are analogous to the concepts of *amplification*, *filtering*, and *gain control* in the signal-processing domain. Related concepts, under slightly different names (see table 4.1), are discussed again in chapter 5.

The appendix at the end of the chapter (section 4.7) develops further technical details about how to specify models and mechanisms. It also discusses other methodological or theoretical elaborations, including band-pass masking to specify the spatial, temporal, and spatial-frequency and orientation sensitivity of the observer; related methods of reverse correlation; and how to constrain properties of the model using double-pass experiments.

4.3 A Systems Analysis of Performance in the Observer

Behavioral psychologists and neuroscientists are often interested in characterizing the functional relationships between visual input stimuli and behavior. One of the most common frameworks to do this uses psychophysical methods. Researchers can use observer models, combined with manipulated variations in the amount and kind of external noise, to powerfully investigate not only what limits accuracy in performance but also how perceptual learning changes these functional relationships. These tools can also be used to specify the signal and limiting noise sources and which of the three mechanisms are at work in the perceptual learning of any given task. Any learning model must in turn include the components of the PTM and address these mechanisms.

4.3.1 Observer Models of Human Performance

An observer model quantifies the input-output function of the observer, or the psychophysical relationship that associates behavioral decisions with different stimulus manipulations. In particular, the model specifies the transformation from the stimulus to the internal representation and then to the response—with this last stage typically using standard SDT to specify the response for any given value on an evidence axis. Modeling the noises (internal variability) in the internal representation is key to predicting performance. Noise makes the perceptual process stochastic, generating random internal variables and the corresponding trial-to-trial variability in responses to the same physical stimulus, and it is the noise that typically limits performance accuracy.[11]

Every observer model has a *template* (or templates) for the target pattern(s) of the task. A template is essentially a detector tuned to a task-relevant stimulus. The output of the template is then processed through nonlinear gain stages, and several sources of noise are added at different stages. The SDT decision module then converts the noisy internal response into an overt behavioral response. Performance is better when the stimulus is a good match to the template, the contrast of the stimulus is high, and external and internal noises are low.

Good observer models possess robust predictive power. They make it possible to forecast an observer's behavioral performance for many varied stimuli after measuring performance in a few conditions. In that sense, they are *generative* models of human performance. By measuring a few observer characteristics (estimating parameters), we can predict the performance over a wide range of possible stimuli that vary in contrast or in external noise (visual noise masking). Absent the observer model framework, we would need to exhaustively tile a stimulus space that crosses several levels of contrast and several levels of external noise, perform a relatively large number of experimental tests, and then interpolate to make predictions about performance in untested stimuli. Ultimately, a good observer model can use a small experiment to estimate parameters of the observer in order to make many predictions about that observer's performance in a wide range of situations with different stimuli, different task demands, and different decisions.

Observer models also provide a theoretical framework to quantify the possible changes in performance caused by manipulations of perceptual learning. The model thus allows us to understand how training (or other manipulations, such as attention) affects the state of the observer by seeing how some estimated parameters of the observer model (such as signal responses, noises, or nonlinearity) change. By examining how these characteristics of the observer change as a result of a given manipulation, this framework allows us to identify the mechanisms by which perceptual learning alters performance—with one or more of the three mechanisms—and to make predictions over a wide range of stimuli.

In the simplest observer models, a single template (or sometimes the difference between two templates) generates a noisy decision variable that in turn leads to the response. More advanced or elaborated observer models can be more complicated—involving a network of templates and processes. These so-called *multichannel models* usually include many templates designed to detect different possible features or patterns that may operate at an array of different retinal locations. In this class of models, a decision module must transform (e.g., selectively weight) a pattern of activities across the responses of many templates into the final response variable and then a behavioral response. These more complicated observer models (which are described in the Model section in chapters 6–8) may include multiple channels or multiple templates, physiological response nonlinearities and interactions, multiple sources of internal noise, and even more complex decision rules.

4.3.2 The Perceptual Template Model (PTM)

A number of prominent observer models have been developed over the last few decades: the linear amplifier model,[12] the induced-noise model,[13] the linear amplifier model with decision uncertainty,[14] the induced-noise and uncertainty model,[15] and the perceptual template model.[8, 16] The *perceptual template model* (PTM) is the most powerful of these, as it incorporates the major components of prior observer models, incomplete in themselves, to account for the fullest possible range of experimental results. By estimating a few key parameters, the PTM can provide a quantitative functional form that specifies the distributions of evidence and ultimately predict the signal-to-noise ratio (d') of the behavioral performance. Many studies have shown that the PTM provides an excellent account of a wide range of psychophysical data.[8–10, 16–21]

As figure 4.1 illustrates, the PTM contains a number of components. For a discrimination task, these must include perceptual templates tuned to two stimuli in the task. For example, if the task requires discriminating the orientation of a Gabor as horizontal or vertical, one template would be tuned to each. (Alternatively, there could be a single template in the case of detecting a specific stimulus.) The PTM also incorporates a nonlinear transducer function (akin to known nonlinearities in neural response), which describes the relationship between the stimulus contrast and the internal response. It also includes two different internal noise sources; multiplicative and additive internal noises. These describe the stochastic nature of the internal representations in response to the same stimulus. The final component is a decision module. It is important to note that multiplicative internal noise increases as a function of the contrast energy in the stimulus and is related to Weber law behavior (in which larger differences are required for discrimination at higher

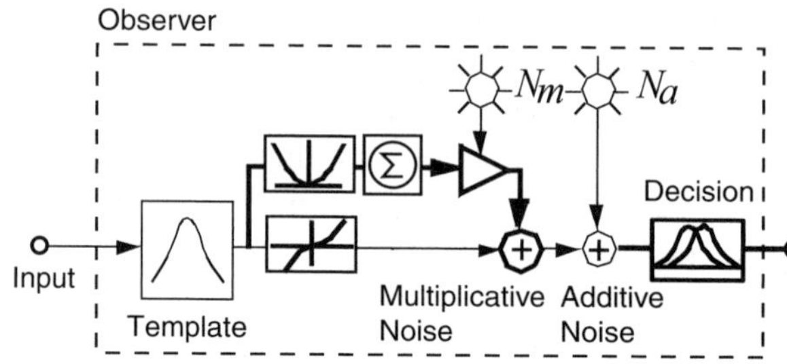

Figure 4.1
The perceptual template model (PTM) includes a perceptual template tuned to the signal stimulus, a nonlinear transducer function, multiplicative and additive internal noises, and a decision module. Modified from Lu and Dosher,[16] figure 15a.

contrasts), while additive internal noise limits the absolute threshold for very low-contrast stimuli. Taken together, the templates, nonlinearity, and external and internal noise determine the mean and the variability of a decision variable (e.g., internal evidence) that is passed to an appropriate signal detection module. This overall system is a step forward from simple SDT. General signal detection theory simply assumes that underlying distributions (with certain means and variances) exist, but does not tie these distributions to specific transformations of the physical properties of the stimulus. The PTM, by contrast, quantitatively describes precisely the translation of the stimulus input to noisy internal representations and then to behavioral decision.

The fully stochastic PTM model is somewhat complicated; fortunately, its predictions can be approximated by a simple set of analytic equations. Here we present equations for a simple two-alternative identification task that involves discrimination between orthogonal (or nearly orthogonal) stimuli. The input stimulus on any trial is a signal (target) stimulus with contrast c and possibly external (masking) noise with variance σ^2_{ext}. This stimulus is processed through two pathways in each of two templates. In the *signal pathway*, one perceptual template selectively tuned for the signal stimulus (e.g., +45° Gabor relative to vertical) responds to the input stimulus with gain β. An orthogonal (e.g., −45° Gabor) template's mean response to the signal stimulus is 0. The gain of the templates to white Gaussian noise at maximum contrast is normalized or set to 1.0. The signal pathway includes a nonlinear transducer function $Output = sign\ (Input)|Input|^{\gamma}$ with nonlinearity parameter γ. The variance of multiplicative noise is proportional (N_{mult}) to the output of the *multiplicative-noise pathway* which also has a nonlinear transducer γ' (which could equal γ). This form of nonlinearity is generally consistent with that observed in neural systems and the results from the pattern vision literature.[22, 23]

The outputs of the signal and noise pathways are combined with another internal additive noise (with variance of σ^2_{add}) and become the evidence for decision. If observers must discriminate between two stimuli, the response reflects the difference in the responses in the two templates. In the preceding example, if the observer is discriminating whether a stimulus is a +45° or a −45° Gabor, one of the templates provides a good match to one stimulus, and the orthogonal template provides a good match to the other stimulus.

In the template detector that matches the signal, the mean internal response is $\beta^{\gamma} c^{\gamma}$ and the total variance of the internal response is

$$\sigma^2_{total1} = \sigma^{2\gamma}_{ext} + N^2_{mult}[\sigma^{2\gamma}_{ext} + (\beta c)^{2\gamma}] + \sigma^2_{add}. \tag{4.1}$$

In the template that mismatches the signal (but matches the alternative), the mean of the internal response to the mismatched stimulus is 0 and the total variance of the internal response is

$$\sigma^2_{total2} = \sigma^{2\gamma}_{ext} + N^2_{mult}[\sigma^{2\gamma}_{ext}] + \sigma^2_{add}. \tag{4.2}$$

For such a two-alternative forced identification task, the PTM predicts that the average signal-to-noise ratio (d') is

$$d' = \frac{(\beta c)^{\gamma}}{\sqrt{\sigma^2_{ext} + N^2_{mult}\left[\sigma^2_{ext} + \frac{(\beta c)^{2\gamma}}{2}\right] + \sigma^2_{add}}}. \tag{4.3}$$

Correspondingly, the probability that the response is correct is computed as the probability that the response of the matching template exceeds the response of the mismatching template, assuming the distributions are Gaussian:

$$\begin{aligned} P_c = \int_{-\infty}^{+\infty} & g\left(x - \beta^{\gamma} c^{\gamma}, 0, \sqrt{\sigma^{2\gamma}_{ext} + N^2_{mult}[\sigma^{2\gamma}_{ext} + (\beta c)^{2\gamma}] + \sigma^2_{add}}\right) \\ & \times G\left(x, 0, \sqrt{\sigma^{2\gamma}_{ext} + N^2_{mult}\, \sigma^{2\gamma}_{ext} + \sigma^2_{add}}\right) dx. \end{aligned} \tag{4.4}$$

The PTM equations account for performance in a wide range of stimulus conditions with as few as four parameters: gain of the template to a matched signal β, nonlinearity γ, multiplicative internal noise N_{mult}, and additive internal noise σ_{add}. By estimating these four parameters, it predicts the d' or corresponding percentage correct for stimuli of different target contrasts c and different levels of external noise in the stimulus σ_{ext}.

Most observer models, including the PTM, were developed to account for detection or discrimination of *orthogonal* or nearly orthogonal stimuli. The PTM has been extended to handle the discrimination of very similar (*nonorthogonal*) stimuli, which involves overlapping templates, both of which may respond to the same stimulus, leading to an elaborated PTM[21] (see section 4.7).

4.3.3 Specifying the PTM with External-Noise Methods

External-noise experiments have been developed to constrain the parameter estimates of the PTM, especially the internal-noise and nonlinearity parameters.[24, 12] This is the method by which the observer can be efficiently described and within which the mechanisms of learning can be characterized. In such experiments, titrated amounts of external noise are added to the signal stimulus; the detection of a stimulus or discrimination between stimuli is measured for a number of levels (or types) of external noise (see figures 4.2 and 4.3). By adding external noise that the experimenter controls, the other sources of internal noise can be compared to an externally controlled and measurable quantity; they can be benchmarked against the external noise in the stimulus.

Experimental applications of the PTM model that use this method usually test detection or discrimination of a signal in noise at each of five to nine logarithmically spaced external-noise levels. These so-called equivalent internal-noise methods traditionally measure signal contrast thresholds as a function of external noise at a single performance level (e.g., 75% correct). This is insufficient to fully characterize the observer; full specification of the PTM requires either measuring contrast psychometric functions at each level of external noise or estimating threshold contrast while keeping performance accuracy constant at each of three performance accuracy levels in each external-noise condition (see section 4.7 for a technical elaboration of these points).

Graphs of contrast thresholds measured as a function of external noise contrast in the stimulus have a characteristic shape (figure 4.2). These so-called *contrast threshold versus external-noise contrast* (TvC) functions reveal that at high levels of external noise, performance is limited by the external noise (such that contrast threshold increases directly

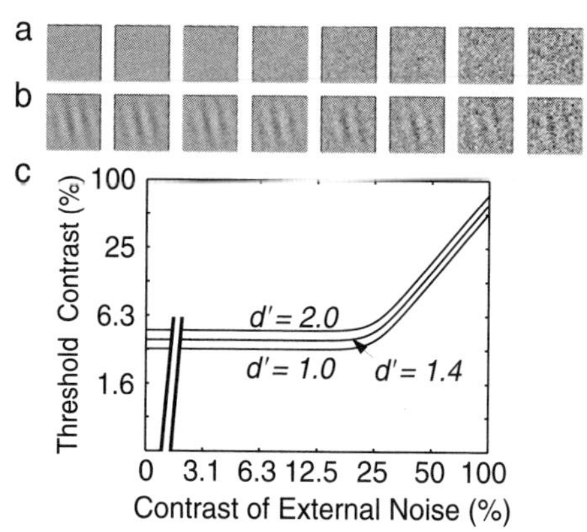

Figure 4.2
Examples of different external-noise levels without (a) and with (b) a Gabor signal stimulus, and (c) contrast threshold versus external-noise contrast (TvC) functions at three accuracy levels (*d′*). Internal noise limits performance at low external noise, while external noise limits performance at high external noise. After Dosher and Lu,[10] figure 3.

with external noise contrast, usually with a slope of 1), while at low levels of external noise, performance is limited by internal noise (such that contrast threshold does not depend on small variations in external-noise contrast).

This functional relationship between contrast threshold and the contrast of external noise can be modeled quantitatively. A formula for the threshold signal contrast c_τ as a function of external-noise contrast σ_{ext} and the contrast threshold corresponding to a d' performance level can be derived for the PTM by rewriting the fundamental signal-to-noise equation (4.3) to yield

$$c_\tau = \left\{ \frac{d'^2[(1+N_{mult}^2)\sigma_{ext}^{2\gamma} + \sigma_{add}^2]}{\beta^{2\gamma} - N_{mult}^2 \beta^{2\gamma} d^2 / 2} \right\}^{\frac{1}{2\gamma}}. \tag{4.5}$$

This equation characterizes the performance of the observer at a given signal-to-noise ratio (d') across multiple external-noise conditions. To get the best estimates of PTM parameters and good tests of the model, this equation is fit to experimental TvC measurements at several signal-to-noise ratios; measurement at three d' levels is generally sufficient. Other external-noise methods, such as band-pass masking and reverse-correlation regression methods, can further specify the perceptual template. (These methods are described more fully in section 4.7.)

4.4 Using External Noise to Study Perceptual Learning

Taken together, external-noise methods and the PTM observer models form a powerful tool. When the two are used in conjunction, it becomes possible to discriminate between the distinct *mechanisms* of perceptual learning, as each mechanism will make signature predictions about changes in the TvC curves with practice.

As described at the start of the chapter, there are three major mechanisms by which perceptual learning could modify behavior that are intuitively analogous to the concepts of *amplification*, *filtering*, and *gain control* in the signal-processing domain. To recapitulate, amplification increases the response to the signal stimulus, which improves the ratio of the signal to the internal additive noise; filtering refers to changing which aspects of the stimulus are weighted in the response, usually by filtering out external noise; and gain control refers to changes in the response of the system that allow it to shift sensitivity to different magnitudes of input stimuli, which often involves normalization or another nonlinearity. These three concepts are implemented within the PTM observer model as *stimulus enhancement*, *external-noise exclusion*, and *multiplicative internal noise reduction or nonlinearity change*. When training (or some other intervention, such as attention) modifies the response to a given stimulus, this change must in turn reflect a change in one or more of these mechanisms.

Table 4.1
Three analogous mechanisms of state change (perceptual learning) in the language of signal processing, the PTM observer model, and physiology

Signal processing	PTM observer model	Physiology
Amplification	Stimulus enhancement	Gain increase
Filtering	External-noise exclusion	Retuning
Gain control	Multiplicative noise reduction and/or nonlinearity change	Normalization

The same three mechanisms also have analogs in the physiological literature, especially in cellular recording. *Contrast gain* change, analogous to amplification, increases the gain of the system to stimulus contrast without changes in the tuning sensitivity of a neuron.[25, 26] *Retuning* is a change in the sensitivity profile of the neuron with respect to some stimulus feature, which may occur without changing the maximum gain or maximum response.[27, 28] *Normalization* refers to setting the maximum response of the cell based on the total energy in its input.[29]

The point here is not to equate these three analytic domains but rather to suggest that analogous mechanisms of perceptual learning are at work in each. The functional properties of each mechanism can thus be studied at different levels. In this chapter, we focus on a behavioral approach based on external noise manipulations and observer models; however, in principle, similar questions could be studied at the level of physiology, as changed behavior in individual neurons or populations of neurons. Indeed, the concepts, models, and technology of the PTM model could well be applied at the level of either the individual observer or the population of neurons.

4.4.1 Mechanisms and Signatures of Perceptual Learning in the PTM

The predicted performance signatures of perceptual learning[9, 10] reflect different ways that learning might affect the parameters of the PTM. Stimulus enhancement (modeled equivalently as reduction in internal additive noise), external-noise filtering, and reduction in internal multiplicative noise and/or change in nonlinearity each make unique predictions about the effect of learning (figure 4.3). In fact, the same framework has also been applied to attention, adaptation, or other observer changes.

The mechanism of *stimulus enhancement*, or relative amplification of the stimulus compared to internal noise, is implemented as a multiplier on internal additive noise $A_{add}(t)$. This leads to improved performance (reduced thresholds) in zero or low external noise (figure 4.3a). There will be no improvement when high external noise is the primary limiting factor, because relative amplification benefits the external noise as well as the signal in the stimulus.

The mechanism of *external-noise exclusion*, or improved filtering of external noise, is implemented as a multiplier on external noise $A_{ext}(t)$. This predicts improved performance

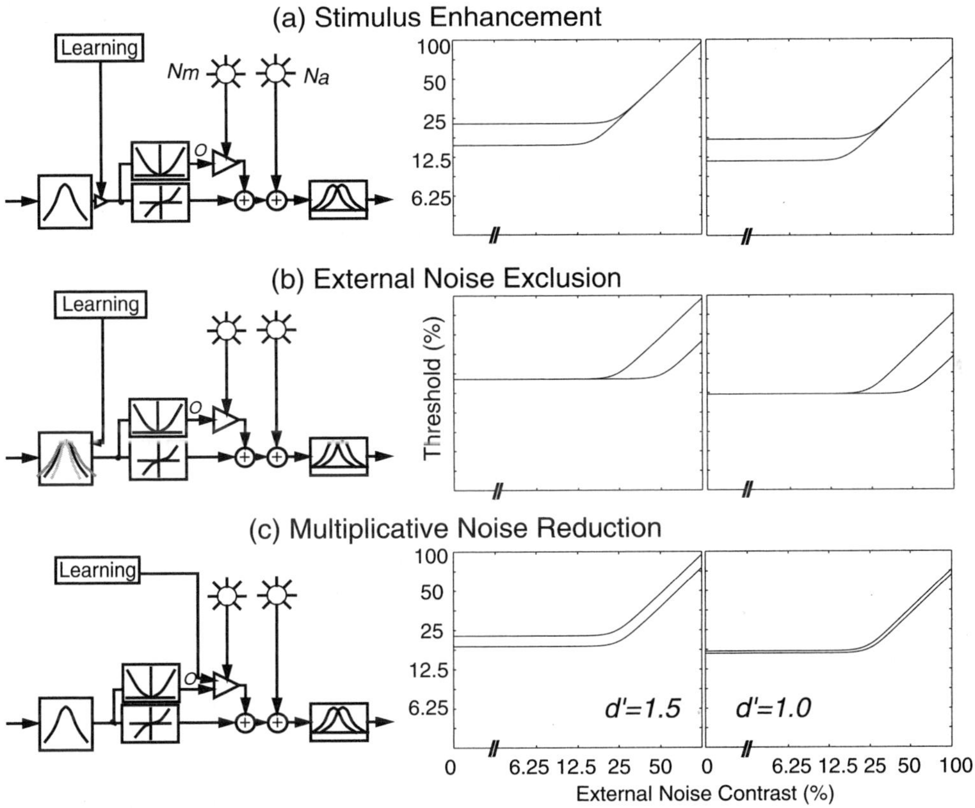

Figure 4.3
Signature mechanisms of perceptual learning in the perceptual template model (PTM): (a) stimulus enhancement (amplification) improves performance in low external noise; (b) external-noise exclusion (filtering) improves performance in high external noise; (c) gain control reduces internal multiplicative noise. Mixtures of (a) and (b) can be distinguished from (c) by considering two criterion performance levels. Modified from Dosher and Lu,[10] figure 3.

in high external noise (figure 4.3b). This has no impact in the absence of external noise in the stimulus. This mechanism might reflect improvements in focus on the appropriate time, spatial region, and/or contents of the signal stimulus.

The third mechanism is *reduction in internal multiplicative noise*. The magnitude of multiplicative noise is proportional to the contrast in the stimulus. This mechanism of change may correspond with a multiplier on multiplicative noise $A_{mult}(t)$. This mechanism improves performance in both high and low levels of external noise in a way that depends on the accuracy level or threshold requirement (figure 4.3c).

These distinct mechanisms in perceptual learning are captured as changes over practice or time t in the parameters of the PTM model: reduction in internal additive noise

$A_{add}(t)$, external-noise filtering $A_{ext}(t)$, or changes in internal multiplicative noise $A_{mult}(t)$ (and/or nonlinearity γ).

$$d' = \frac{(\beta c)^{\gamma}}{\sqrt{A_{ext}^{2\gamma}(t)\sigma_{ext}^{2\gamma} + A_{mult}^{2}(t)N_{mult}^{2}\left[A_{ext}^{2\gamma}(t)\sigma_{ext}^{2\gamma} + \frac{(\beta c)^{2\gamma}}{2}\right] + A_{add}^{2}(t)\sigma_{add}^{2}}}. \tag{4.6}$$

This can be rewritten to make predictions about TvC curves:

$$c_{\tau} = \frac{1}{\beta}\left[\frac{(1+(A_{mult}(t)N_{mult})^{2})(A_{ext}(t)\sigma_{ext})^{2\gamma} + (A_{add}(t)\sigma_{add})^{2}}{\left(\frac{1}{d'^{2}} - (A_{mult}(t)N_{mult})^{2}\right)}\right]^{\frac{1}{2\gamma}}. \tag{4.7}$$

These equations were first developed from a simple PTM model used for detection or discrimination that is limited by contrast rather than stimulus similarity (orthogonal or nearly orthogonal), though corresponding signature patterns have been developed for tasks in which very similar stimuli are being discriminated.[19, 20, 21]

The mechanisms at work can be deduced from changes in PTM parameters and from their corresponding signature changes in performance. If perceptual learning improves performance only in conditions of low external noise, this identifies the mechanism as stimulus enhancement; if it improves performance only in conditions of high external noise, this identifies the mechanism as external-noise exclusion or filtering. Often, however, learning improves performance in both low and high external noise. In this case, the pattern may reflect a mixture of both mechanisms or might reflect reduction in internal multiplicative noise or changes in nonlinearity. These two interpretations are discriminated by examining performance at two different threshold levels (e.g., d' of 1.5 and 1.0): on a log contrast axis, performance improvements will be the same at several threshold levels for stimulus enhancement, external noise exclusion, or their mixture—a property called the equivalent *shift relationship*. If either multiplicative internal noise or nonlinearity changes, however, the improvements (on the log scale) are larger at higher performance accuracies, and the equivalent shift relationship fails (see figure 4.3 for the shift relationship). Measurement at two or more criterion performance levels identifies the mechanisms.[9, 10] While in principle any mechanism could be involved, changes in multiplicative noise and nonlinearity have in fact never been observed. (As we will see in chapter 6, reweighting models can produce several of these mechanisms and combinations of them.)

4.4.2 A Typical External-Noise Study of Perceptual Learning

The PTM observer model and external-noise tests have been deployed in several task domains. In this section, we introduce a basic (hypothetical) study design that uses external noise. In this example, learning is studied for letter identification. A 10-alternative forced-choice identification of spatial-frequency filtered Sloan letters is tested in multiple levels of external noise. The performance measure is the stimulus (letter) contrast that produces threshold performance levels for multiple conditions of white external noise: the standard deviation σ_{ext} is set to one of eight levels (e.g., 0, 0.02, 0.04, 0.08, 0.12, 0.16, 0.25, and 0.33), where $\sigma_{ext}=0$ is no external noise and $\sigma_{ext}=0.33$ is the largest Gaussian standard deviation of noise contrasts that can be presented on the display device (i.e., about one-third of the achievable luminance range from its midvalue). In this hypothetical experiment, contrast thresholds might be measured with adaptive staircases, perhaps a one-down, one-up (1:1) staircase and a one-down, two-up (1:2) staircase tracking about 50% and 30% correct, respectively. In this 10-alternative forced-choice task, these accuracies correspond with d' of 1.5 and 0.9. Performance is measured for each session over, say, five days, with 16 thresholds measured in each session. If each 1:2 staircase is measured in 80 trials and each 1:1 staircase is measured in 60 trials, this requires 1,120 trials per session, or 5,600 trials per observer over the five days. Some features of this design and possible results are illustrated in figure 4.4.

The resulting threshold-versus-contrast (TvC) curves at two different criterion accuracies are traditionally graphed on log-log axes (often $\log_2$). This is done primarily because contrast

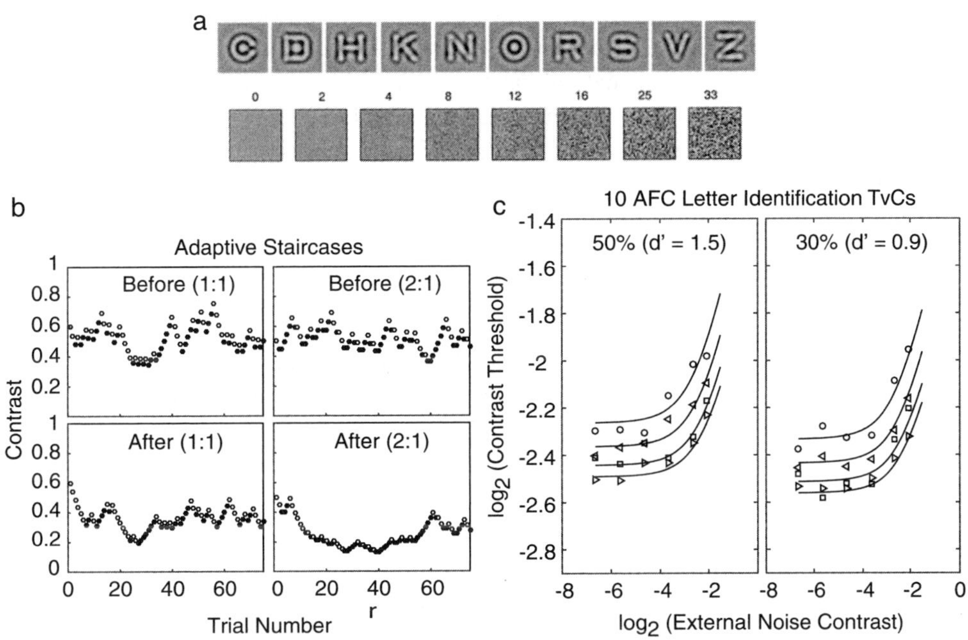

Figure 4.4
A hypothetical experiment measuring the mechanisms of perceptual learning using an external-noise method and the PTM model. (a) Filtered letters for 10 AFC identification (illustrated at high contrast). (b) Staircases (1:1 and 1:2) for measuring thresholds at 50% correct and 30% correct before and after learning. (c) Hypothetical TvC functions and corresponding PTM curves. (Note that these computations ignored stimulus similarities and are for illustration only.)

thresholds in high external noise can be 10 times or more higher than thresholds in zero external noise; the log scale also tends to equate the standard deviations of the threshold measures, which often are proportional to the threshold value. The x-axis also traditionally graphs the external-noise contrast on a log scale. This log-log graph reveals the classic TvC shape.

A PTM model is fit to the data to evaluate the mechanisms of perceptual learning over sessions of practice, using three sets of noise multipliers to describe the changes: $A_{add}(t)$ for stimulus enhancement (reduction in internal additive noise); $A_{ext}(t)$ for external-noise filtering; and $A_{mult}(t)$ for relative reduction of internal multiplicative noise. These parameters are set to 1 by definition for the first session. Estimated multipliers (<1) quantify improvements in performance by reducing the limiting noises. Although in principle nonlinearity γ could change, in practice it rarely or never has.

When external noise tests were used in the prior literature, before the introduction of the PTM (and sometimes even after), researchers generally measured only a single threshold function. As a result, they could only use a linear amplifier model (LAM) (without nonlinearity or multiplicative internal noise) to interpret their results. While the LAM can be informative in some ways, mechanisms of learning cannot be fully disambiguated without measuring thresholds at multiple performance levels. This is because, unfortunately, the LAM model estimates different parameter values for every measured performance level, and it fails to provide a cohesive (consistent across accuracy levels) description of the observer. It is our recommendation that researchers always measure performance for at least two threshold levels in studying perceptual learning or attention.

Further technical details related to the use of staircases (e.g., setting starting values and step sizes), how to fit and compare models, and how to carry out power analyses for some related experimental designs were treated in an earlier book,[31] including some example programs.

4.5 Mechanisms of Perceptual Learning in Visual Tasks

For years, perceptual learning research almost always focused solely on the presence or absence of learning or on transfer and specificity (phenomena treated in chapters 2 and 3, respectively). Only with the development of the external-noise paradigm along with observer models could the *mechanisms* of perceptual learning be further examined. In this section, we review the now considerable perceptual learning literature that demonstrates the existence (in data) of stimulus enhancement (relative amplification) and external-noise exclusion (filtering), as well as their mixture; changes in nonlinearity or internal multiplicative noise have yet to be observed. Performance accuracy will be our focus, although in principle the framework could be extended in a variety of ways.[5]

4.5.1 Using External Noise to Understand Perceptual Learning

One of the first applications of external noise to the study of perceptual learning actually predates the observer model. In 1995, Saarinen and Levi[32] used critical band masking to estimate the orientation sensitivity of Vernier judgments and how they changed with learning. Thresholds were shown to improve (be reduced) with training, with tuning of orientation sensitivity sometimes also occurring.

Later, the external noise paradigm was used in conjunction with the PTM observer model to identify the mechanisms of learning (figure 4.5).[9, 10] In these external noise experiments, two TvC functions were measured in a two-alternative forced-choice (2AFC) Gabor orientation identification task in the periphery; concurrent letter identification at the fovea was used to control fixation. External noise ranged from zero to high (0 to 0.33 in the eight steps); contrast thresholds at 79.4% and 70.7% correct were measured using adaptive staircases for each session (figure 4.5). In this study, practice shifted thresholds down in both low and high external noise, albeit with slightly different magnitudes, reflecting a mixture of improvements through stimulus enhancement and external-noise exclusion or filtering (as assessed by the PTM model fits). There was a remarkably strong shift relationship between the changes observed in the two threshold levels (the magnitudes of improvement resulting from perceptual learning are equivalent, in the log, at the two performance levels). Such strong shift properties rule out changes in multiplicative internal noise and/or nonlinearity.[10] Whether and to what degree these two perceptual learning mechanisms may be expressed separately has been an important question in several subsequent studies. In addition, it was in the context of these results that we were first prompted to postulate the reweighting hypothesis of perceptual learning.[9, 10]

Gold et al. extended our analysis to different stimuli and multialternative judgments (figure 4.6).[33, 34] Their observers identified which of 10 initially novel faces or which of

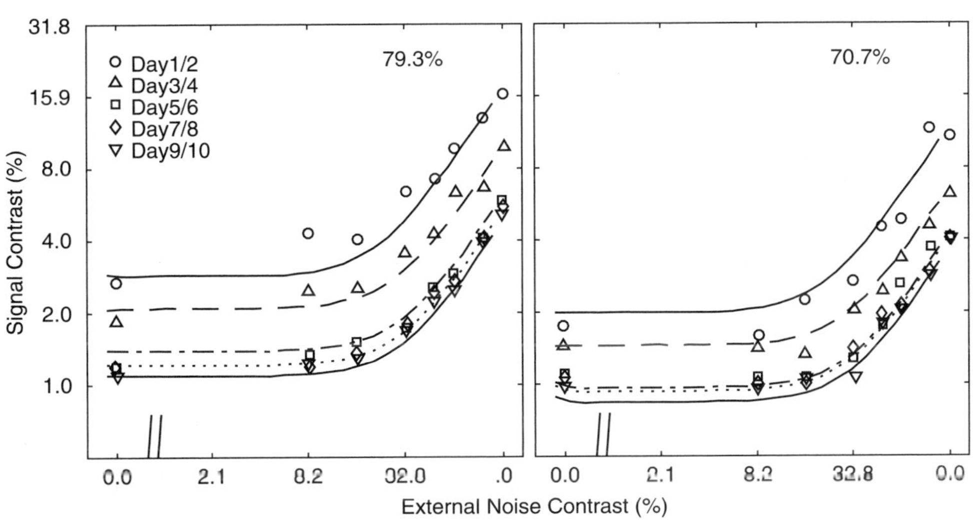

Figure 4.5
Measuring mechanisms of perceptual learning of orientation discrimination using the external-noise method and PTM model. Threshold improvements in low and high noise reflect a combination of stimulus enhancement and external-noise exclusion, consistent with the shift relationship between two threshold levels. After Dosher and Lu[9], part of figure 1.

10 novel exemplars of band-pass-filtered noise patterns was shown on each trial. The data closely paralleled our previous results,[9, 10] with performance improving with practice across conditions of both low and high external noise, here shown in figure 4.6 in the opposite way, as learning curves. (There is other work on learned face recognition.[35]) Because threshold was only measured at one criterion accuracy level (standard at the time), an LAM was used in the analysis. On this basis, a conclusion was drawn by Gold et al. that differed from our own. It was asserted that perceptual learning enhances processing efficiency for the signal stimuli (the filtered textures and the faces).[34] The LAM is almost surely too simple to draw such a strong conclusion, however, as it cannot consistently model performance at different accuracy levels, instead estimating a different set of parameters at each one.[16, 37] Several papers have compared the LAM and the PTM and shown that the PTM provides a more complete account of performance.[8, 16] (Gold et al. also applied a double-pass method to estimate different noise contributions using the LAM. The double-pass method along with the PTM analysis is described in subsection 4.6.3.) Although it would require measurement

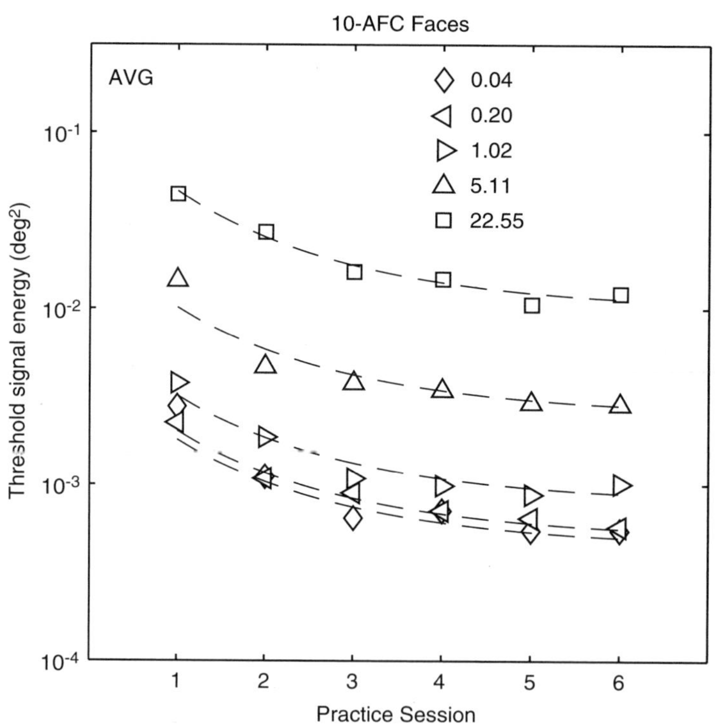

Figure 4.6
Perceptual learning in a 10-alternative forced-choice face-discrimination task in an external-noise experiment, in this case showing exponential learning curves at each external-noise level (corresponding with the TvC). Redrawn from the average of data in Gold et al.,[33] figure 1.

of at least two threshold levels to be sure, our interpretation is that the learning reported by Gold et al. occurred through a mixture of two mechanisms, stimulus enhancement plus external-noise exclusion, analogous to the perceptual learning of orientation judgments.[9, 10, 38] Another related study also found learned improvements at all levels of external noise in peripheral letter identification.[39] Perceptual learning improved stimulus enhancement and improved external-noise exclusion. (Here, too, the LAM failed to provide an internally consistent model of performance, while the PTM was successful, as in other studies.[8, 16, 37])

External-noise methods have also been used to study learning in a range of other task domains. In the case of motion-direction discrimination,[38, 40] improved performance was shown to occur through a combination of improved stimulus enhancement and external-noise exclusion. A separate assessment examined the eye specificity of the learning, finding almost complete transfer of monocular learning to the second eye in high-external-noise tests and about half transfer and half specificity in low-external-noise tests in the second eye; this demonstrated decoupled transfer in low and high external noise.[44]

A different kind of external-noise manipulation was used in a study about learning in a line-offset judgment task.[45] Observers chose which line of three (defined by horizontal Gabor elements) had an offset between the left and right portions in the presence of different levels of position noise (figure 4.7). The offset-threshold curves across different levels of position noise (figure 4.7b) were improved with practice. The weight given to each element location, estimated by regression methods (similar to reverse correlation, a method described in section 4.7), changed slightly with practice. Elements near the center (i.e., where the

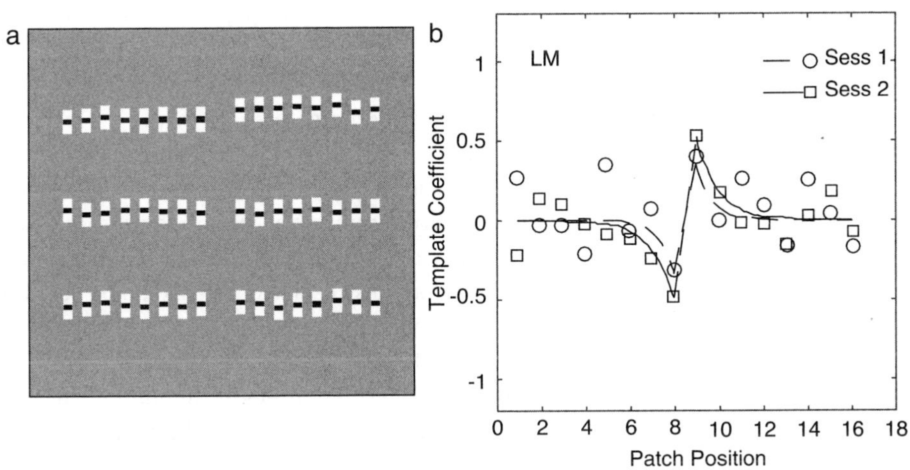

Figure 4.7
Perceptual learning of offset detection in noisy lines. (a) The observer chooses which line created by Gabor patches with different amounts of position noise has an offset (the top one). (b) Template weights for positions 1 to 16, as estimated from regression analysis, improve with perceptual learning. After Li, Levi, and Klein,[41] figure 5.

left and right halves converge) ultimately received the highest weighting. A regression analysis of this kind requires large amounts of data, so the estimated weights (template weights) were quite variable. Even though position noise is not directly conformable with the PTM model, the paradigm and potential models are nonetheless homologous.

In the auditory domain, the external-noise framework[9, 10] has been applied to learning in frequency discrimination.[42] Using psychometric functions, model fits, double-pass consistency, and classification boundary analysis analogous to that described here, these researchers concluded that perceptual learning predominantly reduced internal additive noise, showing primary learning in low-external-noise regions. The same conclusion was extended in a separate experiment to auditory frequency discrimination learning with unchanging stimuli.[43] (Further applications to audition and other modalities are described in chapter 10.) In a memorable phrasing, Hurlbert[36] likened the external-noise studies to "listening to an unfamiliar recording on an old LP, marred by fuzzy crackling sounds.... Over repeated hearings, you come to know... every warble and barely notice the static." She goes on to say that "learning to love a scratchy, old song is a complex, multilevel process." From this analogy it follows that perceptual learning may also involve "learning to extract the signal, learning to filter out external noise, and learning to reduce internal noise"[36] (p. R231).

4.5.2 Separate Expressions of Different Mechanisms of Perceptual Learning

As the preceding examples suggest, perceptual learning often reflects a mixture of improved stimulus enhancement and external-noise exclusion. But are there circumstances in which learning differentially trains one or the other mechanism? Can one mechanism occur without the other? There is some evidence that the answer to these questions is yes. By focusing on certain specific kinds of tasks, it has been possible to document nearly pure expressions of either mechanism. These studies focused on domains where it would be intuitively likely that a particular mechanism might dominate.

One such example of nearly (essentially) "pure" learning by external-noise exclusion was found in a relatively precise foveal (noncardinal) orientation identification task (45°±8° measured at two accuracy levels) (figure 4.8).[37] Learning improved performance significantly only in conditions of higher external noise, a pattern that was identified by PTM analysis as learning through the single mechanism of external-noise exclusion. (It should be noted that this observed pattern violates LAM-based accounts, because efficiency accounts require that improvements be of equal log magnitude at all levels of external noise.)[44]

On the other hand, an example of nearly "pure" learning through stimulus enhancement was documented in a foveal texture-defined orientation task in which observers were asked to discriminate between an alphabetic letter and its mirror-reversed form drawn as strokes of checkerboard, or second-order, texture.[45] In this case, learning improved performance only for tests in conditions of low external noise (figure 4.9). A similar experiment testing first-order luminance letters at the fovea, however, showed no learning. Because processing second-order patterns is thought to first require a rectification stage to extract the letter

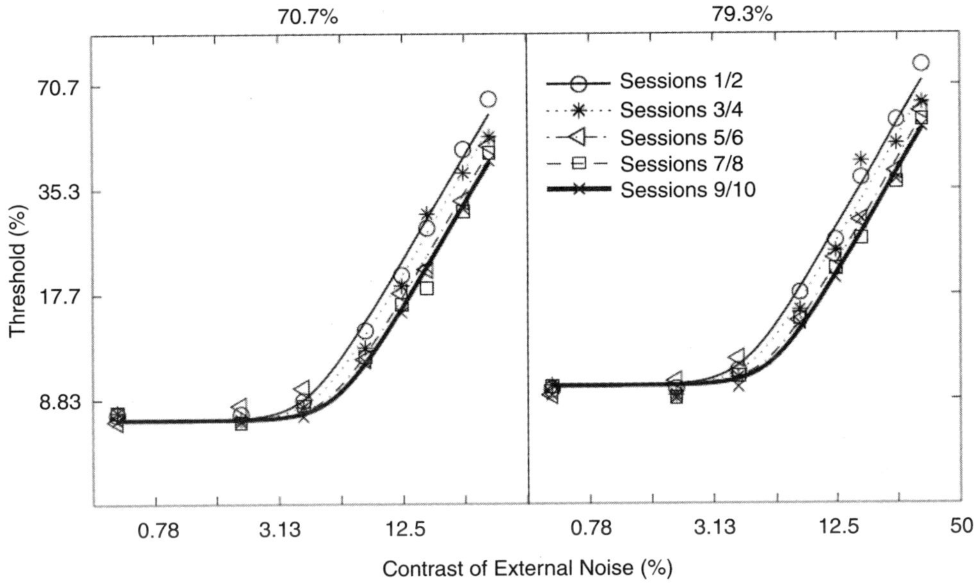

Figure 4.8
A pure case of perceptual learning by external-noise exclusion (filtering) occurs in orientation discrimination at the fovea (45°±8°). TvC functions at 70.7% and 79.3% accuracy. Performance improves only in higher external noise. After Lu and Dosher,[37] figure 4a.

pattern, learning appears to work by amplifying the stimulus relative to the limiting internal noise in the intrinsically noisy second-order stimulus representations.

Another series of studies showed that the two mechanisms—stimulus enhancement and external-noise exclusion—could be trained separately. One of these studies trained Gabor orientation identification separately in low or high external noise, finding an asymmetric pattern of transfer to the other external-noise level (figure 4.10).[46] Training in low noise alone improved performance in tests not just in low external noise but also in high external noise. On the other hand, training in high external noise alone failed to improve performance in low-noise tests. From these results, it follows that training in clear (no noise) displays seems to have unique advantages. Yet another study used differential pretraining to show that stimulus enhancement and external-noise exclusion could be trained separately in a motion-direction task at the fovea.[38, 40] In this experiment, one group of observers was pretrained in high external noise and another was pretrained in low external noise, while a third group received no pretraining. After this initial pretraining, all groups completed a main study that trained using intermixed external-noise stimuli, but showed different patterns of learning depending on the pretraining (figure 4.10). For the group without pretraining, the subsequent training improved performance at all external-noise levels (e.g., corresponding with the usual mixture of stimulus enhancement and external-noise exclusion). For the

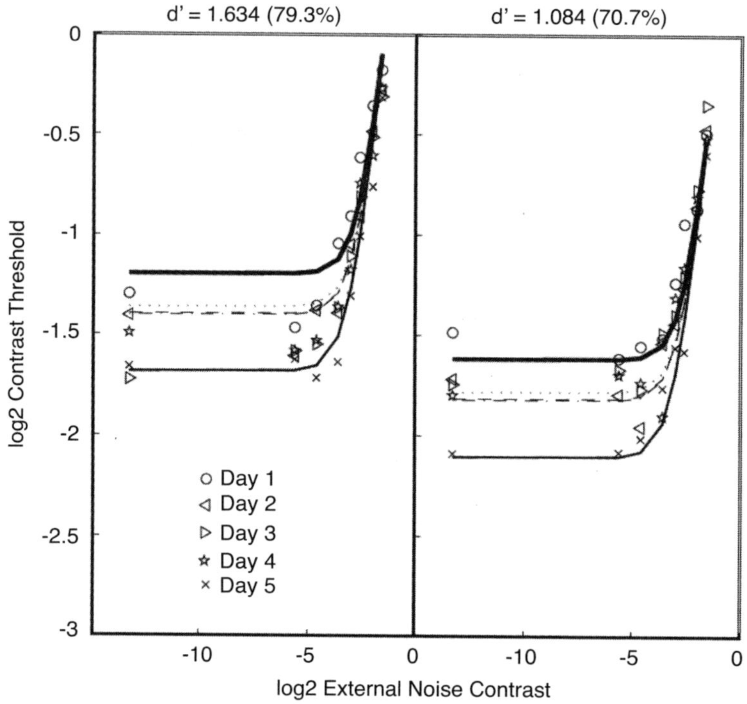

Figure 4.9
A pure case of perceptual learning through improved stimulus enhancement occurred in a letter-texture orientation task at the fovea, seen in TvC functions at two accuracy criteria. Performance improvements are restricted to zero or low external noise, corresponding to stimulus enhancement. After Dosher and Lu[45], figure 4.

group pretrained in high external noise, the subsequent learning improved low-noise conditions (corresponding with stimulus enhancement). For the group pretrained in low external noise, there was little subsequent learning visible at any external-noise level. These results were exactly consistent with the earlier results showing that training in low noise was sufficient to improve performance over all the external-noise conditions.[46] Related effects of training in different external-noise conditions have also been studied with regard to aging.[47]

4.5.3 Applications of the PTM and External Noise Methods

Initially developed to study visual attention,[8, 16, 18–20, 48–50] external-noise methods and the PTM model have also been assessed for use in practical applications.[51–55] Chapter 11 will discuss a broader range of practical applications, but several examples that use the external-noise approach deserve mention here. One recent study used the external-noise paradigm in conjunction with a PTM model to compare TvC curves for orientation discrimination before and after video game training.[55] As in many of the other studies we

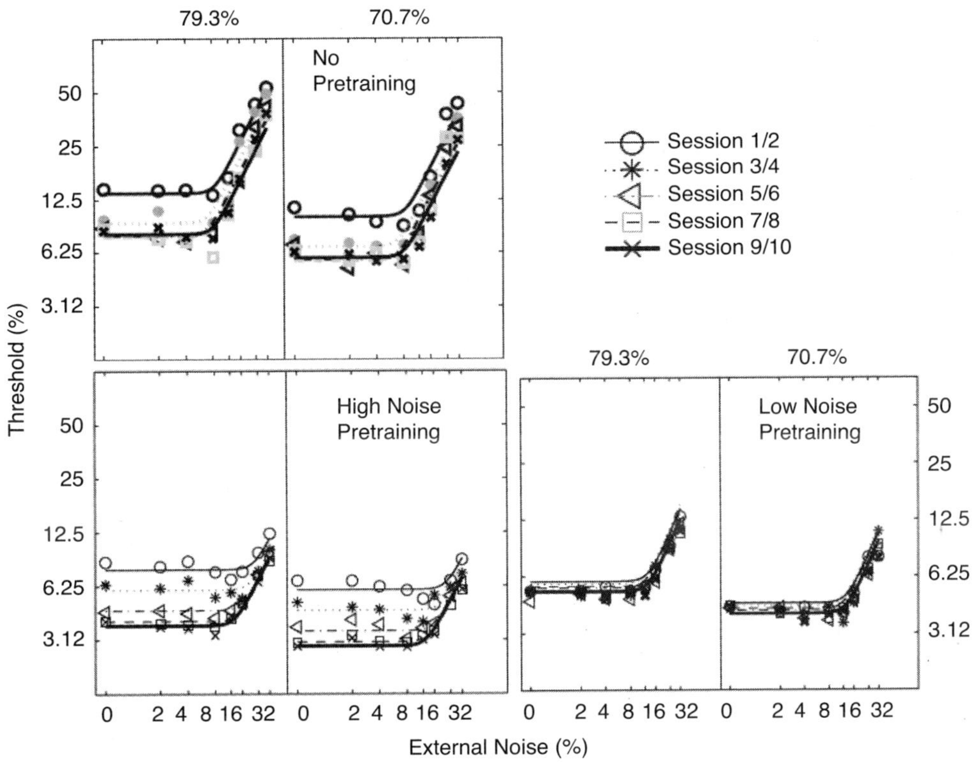

Figure 4.10
Perceptual learning through practice in all levels of external noise following different pretraining experiences in motion-direction discrimination, measured in TvC functions at two accuracy levels in three groups receiving (a) no pretraining, (b) pretraining in high external noise, and (c) pretraining in low external noise. After Lu, Chu, and Dosher,[40] figures 4 and 5.

cited, video game training improved performance through a mixture of stimulus enhancement and external-noise exclusion, while a no-training control group showed few or no improvements.

This approach has also been used to study and compare the mechanisms of perceptual learning in the context of aging. Using external-noise methods and the PTM model, one study examined the differences between younger and older populations.[56] Learning improved external-noise exclusion and stimulus enhancement in both age groups, with the performance of trained older adults approaching the performance of untrained younger adults. These findings were consistent with previous reports comparing learning in both groups,[53, 57] but this study went beyond the earlier literature by identifying the mechanisms of learning. Another study of aging compared different forms of training in an older population by measuring orientation-discrimination thresholds.[47] In this experiment, more precise orientation discriminations exhibited larger improvements. This study also examined the asymmetry of transfer between low and high external noise, which was found to be generally consistent with earlier reports in young adults. In these tests, there was less transfer from high to low external noise and more transfer from low to high external noise; training in high external noise led to more transfer to nontrained orientations. One generalization common to all these studies was that older adults were more limited by higher internal noise (possibly related to failures of inhibition, which are widely attributed to aging populations).[47] On the other end of the age distribution, developmental performance improvements have been measured (without explicit learning manipulations) in groups of five-, seven-, and nine-year-old children and then compared to those of adults.[58] Performance at all external noise levels improved with age, with the largest improvements from five to seven years old. (These shifts in the TvCs, when measured on a log scale, were larger at higher accuracy levels, a finding that requires a more complex explanation, potentially involving changes in gain as well.) The external-noise methods and PTM model have proven a fruitful framework within which to compare the signal-to-noise properties of different age populations and to examine the mechanisms of learning at different ages.

The PTM framework has also been used to study learning in special populations, including observers with relatively common visual deficits, such as amblyopia and myopia. (Some of these are described in more detail in chapter 11.) In one study, learning in adult amblyopes involved the two usual mechanisms, while normal observers, already quite good at the beginning of training, showed no significant improvements.[51] In another study, aspects of contrast-sensitivity detection were improved following training in adults with mild forms of myopia.[59] Other studies examined populations with more exceptional or extreme forms of visual deficits, such as cortical blindness (CB), which is associated with physical damage to the primary visual cortex. These studies found that the largest post-training improvements occurred in low external noise.[60] Yet another study examined learning of an orientation-identification task in patients with Wilson's disease,[61] a condition related to damage in basal ganglia, which is known to impact category learning. While exhibiting deficits in both rule-based and information-based types of category learning, as well as larger deficits in high external noise than in low external noise, the only correlation was between learning in high external noise and deficits in information-integration categorization. In the aggregate, these applied studies in special populations reinforce the notion that the two primary mechanisms of perceptual learning observed are stimulus enhancement and external-noise exclusion, and that these mechanisms of learning are at least partially separable.

4.5.4 Summary

External-noise paradigms and observer models, embodied in the PTM, provide a unique approach to understanding the mechanisms of perceptual learning. Across a number of task domains, learning has been shown to be associated with a combination of two mechanisms: stimulus enhancement and external-noise exclusion. These two mechanisms—working either in conjunction or separately—explained almost all the reported cases of perceptual learning studied so far. There is evidence for partially decoupled improvements as well as for learning primarily restricted to conditions of either low or high external noise, corresponding to stimulus enhancement or external-noise exclusion. (The demonstrations of strong decoupling involve training at the fovea; training in the periphery almost always results in improvements across levels of external-noise masking.) The exact pattern of learning was also found to depend on the initial state of the observer and on the details of the training protocol in each experiment. The method has furthermore been used to characterize learning in a small number of applied domains. These applications tend to characterize either the effects of a special kind of training, such as that found in video games, or the deficits in aging or in special populations. They also sometimes suggest ways in which training might mitigate these deficits.

4.6 Conclusions

What are the mechanisms by which perceptual learning results in improved performance? This chapter considered this question through the lens of a signal-to-noise analysis. If perceptual learning improves performance, then it must have either improved the processing of signal, reduced or excluded noise (either internal or external), or both. After describing the PTM model and external-noise methods, we examined the mechanisms of perceptual learning found in existing experimental studies. In this literature, training improved one or both of two mechanisms. Either it excluded (filtered) external noise or it improved (enhanced) the stimulus. Though in the majority of cases learning improved performance through both mechanisms, we also reported specific training protocols in which learning occurred *separately* in conditions of either low or high external noise. In these tasks, learning was focused almost entirely on one or the other mechanism.

All the observer models described here were *single-channel* implementations: they considered the visual system and decision processes of the observer as a whole. In reality, the visual system has many sensory channels (or representations), processed by different visual areas, arranged in a hierarchy of information and decision. In such *multichannel* models, each channel has its own signal and noise properties, with the cascade of information through an entire network of involved channels and modules generating the input-output behavior. In the first cases extending beyond the single-channel implementations, more complete network models implemented the same principles as the single-channel model in a multichannel architecture. In fact, the patterns of learning found in the external-noise experiments described (and analyzed with single-channel PTM models) are generally compatible with such multichannel implementations.

We will show how this PTM analysis is compatible with the reweighting hypothesis of perceptual learning. In a multichannel architecture, reducing the weights on irrelevant channels improves performance by reducing the effect of external noise and by reducing additive internal noise. In chapter 6, we present an augmented Hebbian reweighting model of perceptual learning that is an implemented form of multichannel architecture, along with other earlier computational models.[62, 63] Chapter 8 then presents an extended multilevel version, the integrated reweighting theory (IRT).[64] Both these models are compatible with the PTM observer approach. Taken together, all these models aim to transition toward the goal of a fully specified signal-and-noise model of the visual system and visual performance. While not identifying the "hardware" at the physiological level, they may provide guides of what to look for in the physiology (cellular recording, EEG, or fMRI responses; see chapter 5).

4.7 Appendix

This appendix points to some extensions and elaborations of the perceptual template model (PTM) and its applications. The topics treated include the methods of estimating PTM model parameters in several forms of experiments; methods to estimate the sensitivity of the perceptual template in several dimensions; important extensions of the model to the discrimination of similar (nonorthogonal) stimuli; elaborations that permit differential processes and parameter values in the signal and noise pathways; and equivalent gain-control formulations.

At the end, we provide simulated results of all the interrelated mechanism signatures of perceptual learning (i.e., perceptual state change) and their connected patterns in terms of psychometric functions, threshold-versus-contrast functions, contrast-ratio functions, and double-pass functions. The intention is simply to summarize these technical methods and to point the reader toward relevant theoretical or empirical developments and potential applications. Detailed derivations can be found in the source papers that we cite.

4.7.1 Specifying the PTM

What kinds of data are needed to specify the PTM observer model experimentally? Specifying the model requires checking the consistency of the model form with the pattern of the experimental data and estimating likely values of the model parameters. We briefly describe three types of experiments that have been used to specify the PTM models: the *multiple (triple) TvC* experiment, the *endpoint* method, and the *quick-TvC* method.

Performance, measured by observed discriminability d' values, in a single condition in a single experimental task for highly dissimilar stimuli (i.e., orthogonal or nearly orthogonal targets) or for detection, can be predicted with the simple PTM using four parameters: the gain of the template for a matching stimulus β, the exponent of the nonlinear transducer γ, the coefficient of multiplicative internal noise N_{mult}, and the magnitude of additive internal noise σ_{add}. The relationship between d' and percentage correct is predicted by standard SDT equations.

Measurement at three levels of performance accuracy, the so-called *triple-TvC* experiment, is sufficient to estimate all the parameters, including the nonlinearity parameters, in the PTM.[8] Measuring performance at three fairly widely separated percentage correct (or d') levels is a proxy for measuring the full psychometric functions. Sometimes this involves measuring contrast psychometric functions at multiple external-noise levels, from which three thresholds are interpolated from each psychometric function. Alternatively, thresholds may be estimated using, for example, adaptive (staircase) methods.[8, 10]

One hypothetical experiment would measure contrast thresholds in eight levels of external noise σ_{ext} (0, 0.02, 0.04, 0.08, 0.12, 0.16, 0.25, and 0.33) (see subsection 4.4.2). If psychometric functions were measured at eight levels of contrast using the method of constant stimuli (using contrasts selected for each external-noise level), the experiment would then have 64 primary conditions (8×8). A sample size of 60, for example, would require 3,840 test trials to fully quantify a single task condition. Such an experiment would lead to a family of psychometric functions at different external noises and a corresponding graph of three TvC curves. Each point on the three TvC curves (or equivalently each point on each psychometric function) is predicted by the PTM equations at a corresponding d'.

The spacing between the three TvC curves (graphed on a log contrast threshold versus log external noise contrast scale) reveals the nonlinearity in the system. This spacing can be summarized as the ratio between the threshold signal contrasts at two performance accuracy levels (d'_1 and d'_2):

$$\frac{c_{\tau_1}}{c_{\tau_2}} = \left[\frac{\frac{1}{d_2'^2} - N_{mult}^2}{\frac{1}{d_1'^2} - N_{mult}^2} \right]^{\frac{1}{2\gamma}}. \tag{4.8}$$

The PTM model makes strong predictions: that these ratios will be the same independent of external-noise level and will be a nonlinear function of the corresponding d'. In the traditional literature, external-noise experiments measured performance at only one performance level. In these cases, simpler models (i.e., the linear amplifier model, LAM, consisting of a single template, additive internal noise, and decision) were used to account for performance.[12] These simpler models cannot account for nonlinearities in the visual system; they are not internally consistent across different performance levels and require different parameter values to account for each performance level.

A triple-TvC experiment can fully specify the PTM and its parameters for a single condition. If several conditions are investigated together, measuring TvCs at two different threshold accuracies is often sufficient to specify the PTM because the data from the several conditions combine to constrain the estimates of shared parameters. Indeed, perceptual learning experiments have generally measured thresholds at only two performance accuracies, combining data from different stages of practice to test the model.[9, 10]

For practical reasons, it is sometimes inconvenient to carry out such a full assessment, because of the large number of test trials required, especially in experiments with many conditions. An *endpoint* method that measures performance only in zero and high external noise has sometimes been used instead. In this method, changes in performance in zero external noise are used to estimate stimulus enhancement, while changes in performance in high external noise are used to estimate external-noise exclusion or filtering. Performance must still be measured at several performance levels, or across psychometric functions, to constrain the PTM parameters.

Another approach uses quick adaptive testing methods to estimate a TvC. One of these is the quick-TvC (qTvC).[65] It is a Bayesian adaptive estimation method that uses information from the response on each trial to decide which stimulus and external-noise contrasts to test on the next trial to best constrain parameters of the model, distributing test trials to efficiently estimate the underlying TvC.

Programs for fitting and estimation of the PTM model can be found in chapter 7 of our book on laboratory methods.[66] Examples of experimental designs and corresponding power analyses, carried out by simulation, appear in chapters 9 and 12 of that book.

4.7.2 Specifying the Template

The PTM estimates the functional form and parameters of the observer for a given task. Additional collateral masking methods can be engaged to specify properties of the perceptual template and its sensitivity to different stimulus features in more detail. Two methods that have been used to estimate behavioral sensitivity to different features are critical band masking and the classification image. *Critical band masking* employs different masks to discover the observer's perceptual template sensitivity to different stimulus features, including spatial frequency, orientation, spatial location, and temporal location for a specific task. *Classification images* estimate spatial features of the template by categorizing noise samples for different observer responses. Although these methods have generally been used to provide heuristic and descriptive estimates of template sensitivity, in some cases they may also be integrated into the PTM model framework.

The principle of *critical band masking* is that external-noise energy (masking in certain frequencies, orientations, spatial positions, or temporal periods) impacts observer performance if and only if the perceptual template is sensitive to that energy. To measure the spatial-frequency sensitivity of the template, for example, the external noise added to the signal is manipulated by filtering white noise into different frequency bands and then measuring the contrast thresholds required to detect or discriminate a target. If the spatial frequencies in the external noise fall outside the template's sensitivity, this noise will not affect the thresholds, while external noise that includes energy in spatial frequencies to which the template is sensitive will elevate behavioral thresholds.

In the simple (single-channel) PTM model, the response gain of the perceptual template to a signal stimulus is captured by parameter β. In an elaborated model, the template response $T_S(v)$ may be further specified as a function of any variable v, such as spatial frequency f, orientation o, time t, or spatial location l (or some combination of these). Then the output of the template through the signal path for the stimulus, including signal and external noise, is[67]

$$S_1 = \alpha c \int T_s(\nu) S(\nu) d\nu, \tag{4.9}$$

$$\sigma_{N_1}^2 = \sigma_{ext}^2 \int T_s^2(\nu) F^2(\nu) d\nu. \tag{4.10}$$

The amplitude of the signal stimulus is $S(\nu)$, the amplitude of the external noise is $F(\nu)$, and the parameter α is the gain of the template to a signal stimulus relative to external noise. The template response in the gain control (multiplicative noise) path to ν is $T_N(\nu)$. The output of the gain-control template for signal and external noise is

$$S_2 = \alpha c \int T_N(\nu) S(\nu) d\nu, \tag{4.11}$$

$$\sigma_{N_2}^2 = \sigma_{ext}^2 \int T_N^2(\nu) F^2(\nu) d\nu. \tag{4.12}$$

As in the simple PTM, the ability to discriminate the target d' reflects the output of the signal path compared with the total noise:

$$d' = \frac{S_1^\gamma}{\sqrt{\sigma_{N_1}^{2\gamma} + N_{mult}^2 \left[\sigma_{N_2}^{2\gamma} + S_2^{2\gamma} / 2 \right] + \sigma_{add}^2}}. \tag{4.13}$$

For example, the template's sensitivity to different spatial frequencies can be measured through variations in threshold for signals embedded in a series of low-pass and high-pass external-noise images, defined by their cutoff frequencies—the so-called Tvf. Figure 4.11 shows samples of external noise passed through a series of high- and low-pass spatial-frequency filters, and the estimated sensitivities for three observers derived from their measured threshold profiles.[67] A similar approach has been used in other investigations to estimate the spatial-frequency tuning, orientation tuning, spatial footprint, and temporal window of the perceptual template. (For a more detailed review, see chapter 9 of our book on visual psychophysics.[66])

Investigating any stimulus dimension (e.g., spatial frequency, orientation) requires a relatively large psychophysical experiment. Future developments might use adaptive methods to measure several dimensions simultaneously. The advantage of characterizing templates by using the PTM observer framework is that it estimates the contributions of internal noises and nonlinearities as well as separating these perceptual factors from decision factors—rather than making qualitative conclusions from visual examination of the data. On a different point, the behavioral estimates of the perceptual template using band-pass masking could, in principle, be compared with neurophysiological measures of the tuning properties of visual neurons or visual areas.

Another related method designed to visualize visual features of a *spatial* template controlling perceptual judgments is the *classification-image method.* Developed from earlier work in audition,[68] classification images were first applied in vision by Ahumada in 1996.[69] The idea is that the observer's responses will be sensitive to random features of the

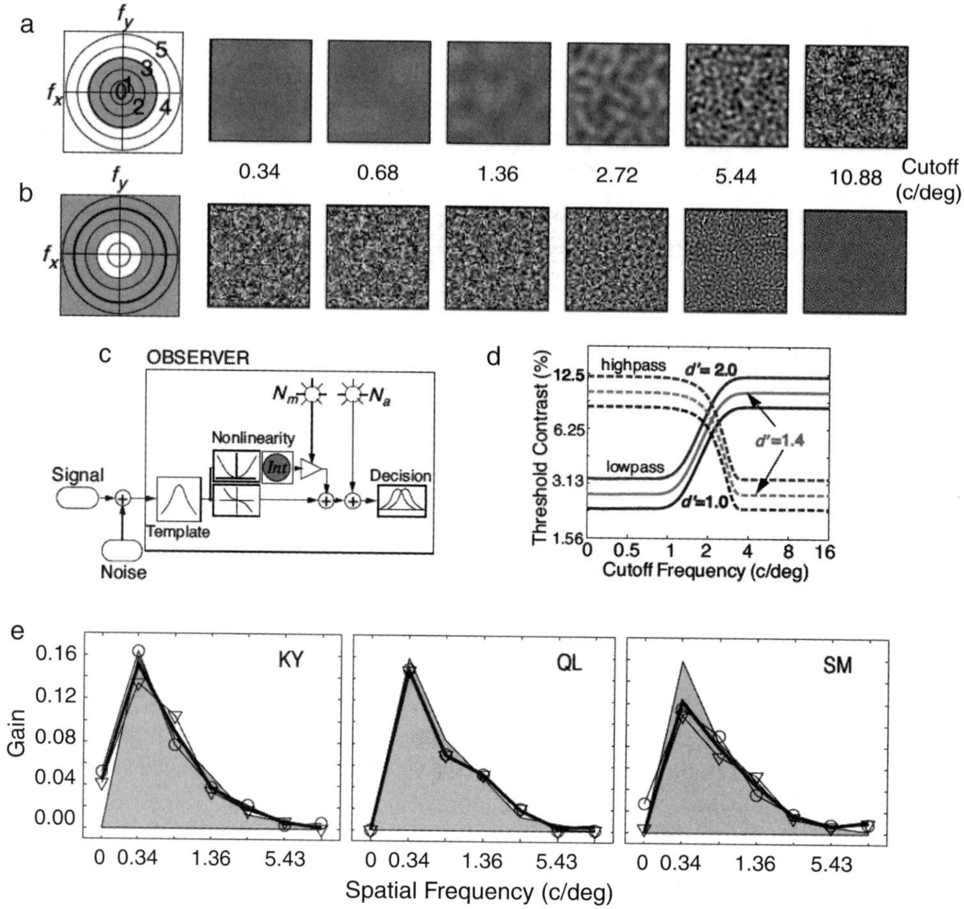

Figure 4.11
Specifying the perceptual template sensitivity of the spatial frequency by using critical band masking. (a) Low-pass spatial-frequency filters in Fourier space and external-noise examples. (b) High-pass spatial-frequency filters in Fourier space and external-noise examples. (c) Perceptual template model. (d) Threshold versus cutoff frequency curves for the low-pass and high-pass conditions. (e) Estimated template gains for three observers as a function of frequency (symbols) and matched for the stimulus (gray region). After Lu and Dosher,[37] parts of figures 1, 3, and 7.

external noise at different locations in the stimulus image. If the template is sensitive to a white patch in a given spatial location, then observers will be more likely to say "signal" when the image luminance is high (white) rather than low (black) in that location. In short, the correlation of the observer's *response* with many noisy stimulus features is used to infer the (positive or negative) weights on pixels, or groups of pixels, leading to a behavioral response.[70–72] This method is based on the same principles as *reverse-correlation* methods, often used to estimate the receptive fields of neurons in the visual cortex.[73–75] It can also be used to assess the spatial template's sensitivity at different points in time.[76] The original applications of the reverse-correlation method depended on the assumptions of the linear amplifier model (which generally are violated).[77] Later research has investigated more complete models in order to include multiplicative noise[77] and decision uncertainty.[78]

The details of the time course of the changes resulting from perceptual learning may be too fast to allow tracking by classification-image methods, because of the requirement for a very large number of trials (sometimes tens of thousands) to perform a reliable estimate of the classification-image template. Nonetheless, several authors have applied classification images or related methods to perceptual learning, often by examining tasks where it is possible to assume that the template may be rotationally symmetric.[79, 80]

4.7.3 Detailed Properties of Mechanisms of Perceptual Learning

The quality of performance in visual tasks reflects the fundamental limits of the signal-to-noise ratio in visual processing. When perceptual learning, or any other manipulation that alters the state of the observer, improves performance, these changes can improve the filtering of external noise, reduce internal additive noise (equivalent to amplifying the stimulus), or alter internal multiplicative noise or nonlinearity. As described in the main part of this chapter, these mechanisms make different signature predictions, as illustrated in TvC functions (see figure 4.3). In this section, we illustrate and discuss the predictions made by the PTM for a complete set of performance measures using simulated results.

Figure 4.12 shows the interrelated predictions of the PTM model for different mechanisms of learning (or other state change) on several measures of performance. From top to bottom, it shows the mechanisms of stimulus enhancement via amplification; stimulus enhancement via reduction in internal additive noise; external-noise exclusion; reduction in internal multiplicative noise; and a mixture of stimulus enhancement and external-noise exclusion. The predicted performance patterns include (left to right) effects on psychometric functions; signature shifts in TvC functions; predicted contrast ratio functions; and double-pass percentage correct versus percentage agreement P_c *v* P_A functions.

Double-pass experiments are a complementary method for constraining estimates of internal noise in relation to external noise. In a double-pass experiment, the same sequence of stimuli is presented to the observer twice (or sometimes more than twice, for n-pass).[13, 39, 44, 81–86] The observer's responses on the two identical trials can be scored as correct or incorrect, and as in agreement (the same response) or not. Each ratio between the total internal noise and the external noise yields a function relating percentage correct (PC) to percentage agreement (PA). For each (PC, PA) function, percentage correct goes from chance at 50% to very good near 100%. As percentage correct approaches 100%, percentage agreement also approaches 100%. As percentage correct becomes poor, the agreement in responses between the two copies of the same trial depends on whether the errors reflect independent samples of internal noise, and thus are not related, or the same sample of external noise, which controls the (error) response in the same way. Example (PC, PA) functions appear in the last column of figure 4.12. The external-noise level and signal contrasts are known variables, so the total internal noise can be estimated from the curve on which the data point lies. Relating the estimated total internal noise to different stimulus variables allows estimation of the constant additive noise and the contributions to multiplicative noise from the external noise and the signal. With this review of the double-pass method, each of the mechanisms of change is considered in turn (from top to bottom).

Stimulus enhancement, whether implemented as *relative amplification* (top row in figure 4.12) or as *reduction in internal additive noise* (second row), improves performance in conditions of low external noise. This is expressed as a leftward shift in the contrast psychometric functions in low external noise (first column) and the corresponding signature improvements in contrast thresholds restricted to low external noise (second column). Because of the enhancement, the contrast-threshold ratios are unaffected (third column). The (PC, PA) double-pass functions (last column) differ because the relative value of

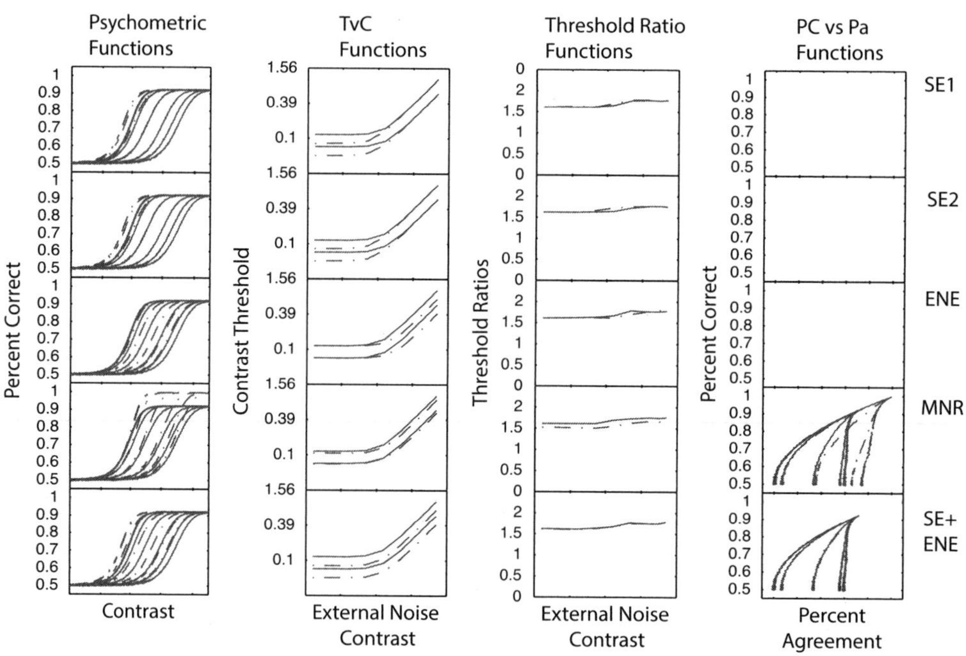

Figure 4.12
The PTM makes predictions for different perceptual learning mechanisms for multiple measures. Measures shown from left to right: psychometric functions, TvC functions, contrast threshold ratio tests (ratios of two criteria), and double-pass (percentage correct versus percentage agreement) functions. Mechanisms shown from top to bottom: stimulus enhancement (SE1, relative amplification), stimulus enhancement (SE2, reduction in internal additive noise), external-noise exclusion (ENE), internal-multiplicative-noise reduction (MNR), and a mixture of stimulus enhancement and external-noise exclusion (SE+ENE). Each panel compares pretraining (solid line) to posttraining (dashed line) performance. Simulated predictions.

internal noise changes for the same external noise and signal contrasts, but they converge at the top, where signal and external noise dominate performance.

External-noise exclusion (third row in figure 4.12) improves performance in conditions of high external noise. This is seen as a leftward shift in the contrast psychometric functions (first column) in high external noise and the corresponding signature improvements in thresholds in higher levels of external noise (second column). Here, too, the threshold ratios are unaffected for the same reasons (third column), and the (PC, PA) double-pass functions (last column) differ because the relative value of the internal noise changes for the same external noise and signal contrasts.

Internal-multiplicative-noise reduction (fourth row in figure 4.12) improves performance in the asymptotes of the psychometric functions (first column), which shifts the threshold functions across all levels of external noise (second column). This also serves to change the threshold ratios for the same two d' values because the psychometric functions are stretched (third column), and the (PC, PA) double-pass functions (last column) are shifted for the same reason.

Lastly, a combination of *stimulus enhancement plus external-noise exclusion* (fifth row in figure 4.12) shows the combined effects of these two mechanisms on the psychometric functions, thresholds, threshold ratios, and double-pass functions.

The results shown in figure 4.12 are the predictions of a simulated experiment with a particular (fairly typical) set of PTM parameters. A corresponding dataset would be the results of a particularly complete psychophysical experiment. In that case, a researcher could choose which form of the data to fit (sets of psychometric functions, the TvC curves derived from those psychometric functions, etc.). The PTM model could then be fit to the observed data, and its parameters could be estimated. The methods and procedures comparing models, including different mechanisms and the issues of model selection, were treated in our book on visual psychophysics.[66]

The double-pass predictions generated are based on a generic signal detection theory (SDT) model, constrained by PTM models. The PTM observer model provides an equation for the total internal noise as a function of external noise, and model parameters N_{mult}, N_{add}, β, and γ, which can then be used to derive predictions for each contrast and external-noise condition in the (PC, PA) space. This process can be inverted to recover PTM parameters from (PC, PA) data as long as there is sufficient stimulus variation, as detailed in equations and discussions in several papers.[16, 66] The double-pass method can be combined with the triple-TvC method to provide added constraints on estimation.[16] Empirically, most of the observed functions relating PC (percentage correct) to PA (percentage agreement) in the literature are quite similar across external-noise conditions—suggesting that the ratio of internal to external noise approaches a constant.[13, 16] This in turn implies the dominance of multiplicative noise over additive noise, consistent with Weber's law.[87] (Some prior research failed to recognize that multiplicative internal noise needs to be included in modeling the double-pass data, leading those researchers to draw erroneous conclusions from the approximate constancy of the internal to external noise ratios estimated by the (PC, PA) functions in perceptual learning.)[33] The predictions in figure 4.12 use PTM model predictions for contrast-limited discrimination or detection, but related predictions can be generated by elaborating the PTM for discrimination between very similar stimuli, in which similarity as well as contrast limits performance.[19, 30, 88]

4.7.4 Elaborations of the PTM

The PTM models and methods presented in this chapter were in their simplest form. A number of elaborations and extensions have been developed that allow the model to be applied in different situations. We consider several of these elaborations in turn.

One major extension elaborates the PTM to discrimination between *similar (nonorthogonal) stimuli.* The original PTM was developed for simple detection or discrimination in which the template(s) for stimuli are orthogonal or nearly so. In discriminating very dissimilar stimuli, performance accuracy is limited by contrast or visibility. In many situations, however, observers may be required to discriminate between very similar stimuli, in which the templates for the stimuli to be distinguished are not orthogonal but instead have high degrees of overlap. In these situations, performance accuracy is limited not just by contrast or visibility but also by similarity. Templates tuned to quite similar and quite dissimilar stimuli are illustrated in figure 4.13.

The response gain of the template that best matches the stimuli is $\beta_{Matched}$, and the response gain of the template for the other, unmatched template to that same stimulus is

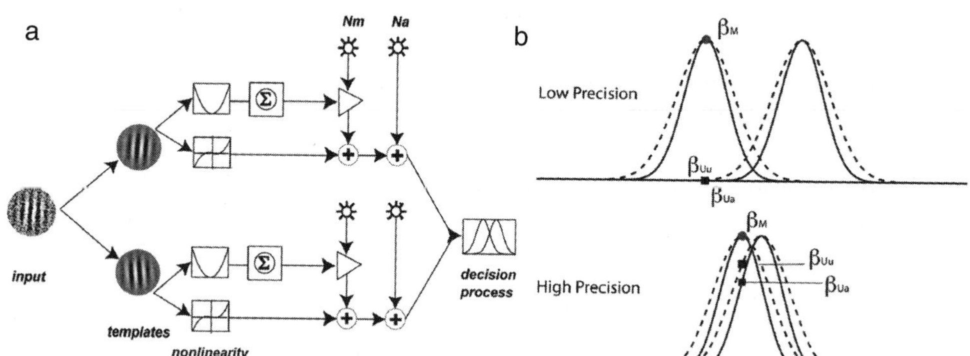

Figure 4.13
The elaborated perceptual template model (ePTM) models discrimination between two nonorthogonal stimuli and the resulting signal and noise distributions. (a) The ePTM computes the signal and noise properties when a stimulus input is processed by two possibly similar stimuli (e.g., orientations differing by a small amount). (b) Illustrations of the templates for (nearly) orthogonal stimuli (top) and very similar stimuli (bottom). The signal difference is large for dissimilar stimuli, and discrimination is limited by stimulus contrast and noise; the signal difference is small for similar stimuli, and discrimination is largely limited by similarity, in addition to contrast and noise. After Hetley, Dosher, and Lu,[19] figure 2.

$\beta_{Unmatched}$. Then, the relative response strength for a given stimulus is $(\beta_{Matched}\ c)^{\gamma}$—$(\beta_{Unmatched}\ c)^{\gamma}$ (numerator in the PTM d' equation). If the task is to discriminate stimuli that are very different—orthogonal or nearly orthogonal—then the response of the unmatched template to the signal stimulus is zero, so this simplifies to $(\beta_{Matched}\ c)^{\gamma} - 0$, which is the simple form for orthogonal stimuli. And since the response gain through the unmatched template is zero, this makes no contribution to the multiplicative noise. We presented this simplified model in subsection 4.2.3.

However, if the task is to discriminate stimuli that are very similar to each other, then the elaborated model estimates two template parameters, $\beta_{Matched}$ and $\beta_{Unmatched}$, rather than one. In addition, since both templates have nonzero responses, both contribute to multiplicative noise, resulting in a multiplicative-noise part of the overall noise equation that reflects the responses of both templates:[21]

$$N_{mult}^{2}\left[\sigma_{ext}^{2\gamma} + \frac{\beta_{Matched}^{2\gamma}c^{2\gamma} + \beta_{Unmatched}^{2\gamma}c^{2\gamma}}{2}\right]. \tag{4.14}$$

This nonorthogonal form of the PTM generates predictions of the joint effects of external noise, signal contrast, and stimulus similarity. For experiments in which observers discriminate between clearly different (orthogonal or nearly orthogonal) stimuli, performance will be limited by stimulus contrast and external noise. For experiments in which the observers discriminate between similar (nonorthogonal) stimuli, stimulus similarity, as well as signal contrast and external noise, limits performance. The nonorthogonal model is required to account for any experiment in which the similarity of the stimuli, rather than (or in addition to) contrast and external noise, is manipulated.

The nonorthogonal PTM has been used to evaluate the effects of stimulus similarity on discrimination.[21] This model has also been extended to account for mechanisms of change resulting from attention.[19, 30, 88] Model fitting in such cases requires the estimation of more parameters, which usually involves comparing performances for different levels of similarity of discrimination in a joint model fit.

Another model generalization includes potentially *different templates in the signal path and the multiplicative-noise or gain-control path.* The motivation for this elaboration follows the intuition that the signal templates, especially for well-practiced observers, can be narrowly tuned for the stimuli to be identified, while the template in the multiplicative-noise or gain-control pathway may be much broader. In the most general form, the signal and gain-control pathways may also differ in nonlinearity. The corresponding parameters are labeled β_1 and γ_1 in the signal path (numerator), and the generalized form then uses β_2 and γ_2 in the noise path (denominator) of the d' equation. When applied to similar (nonorthogonal) stimuli, the elaborated PTM parameters include $\beta_{1-Matched}$, $\beta_{1-Unmatched}$, and γ_1 in the signal path (numerator) and $\beta_{2-Matched}$, $\beta_{2-Unmatched}$, and γ_2 in the noise path (denominator).

The original PTM developed equations for orthogonal stimuli but allowed different nonlinearities for signal and noise processing.[8] A form that permitted different nonlinearities and different gains in the gain-control pathway was also examined in more detail in one study.[16] The simplest form of the PTM, which equates the parameters in the signal and noise pathways and assumes orthogonal discrimination, requires four parameters for one condition in a given task. Allowing the parameters in the signal and noise paths in the orthogonal model requires six parameters, and the model for nonorthogonal discriminations with different parameters in the signal and noise paths is specified in eight parameters. Although it is possible that a future dataset will require the complication, essentially all the data currently available have been very well fit using the same parameters in the signal and noise pathways. Still, some experiments that focused on teasing apart forms of internal noise may require these model elaborations, especially for designs that are more complicated.[16]

The original PTM formulation is mathematically equivalent to a development in which system nonlinearities are recast as *contrast gain control*.[50] In contrast gain control, the magnitude of the internal representation is scaled or normalized by the total contrast energy in the input stimulus. At the level of the overall observer, gain-control variants of the PTM can be equivalently rewritten to the original formulation. (That is, we can rewrite one set of equations into the other.) In the gain-control PTM, the signal-to-noise equation (for the orthogonal form) is

$$d' = \frac{(\beta c)^{\gamma} / \sqrt{b+E}}{\sqrt{(\sigma_{ext}^{2\gamma} + N_1^2)/(b+E) + N_2^2}}, \tag{4.15}$$

where $E = \beta_2^{2\gamma} c^{2\gamma} + \sigma_{ext}^{2\gamma} + N_1^2$.

This is exactly equivalent to the original (orthogonal) PTM, with the following rewritten equivalences:

$$N_{mult} = N_2, \tag{4.16}$$

$$\sigma_{add} = \sqrt{N_1^2 + N_2^2(b + N_1^2)}. \tag{4.17}$$

A recent paper has fully detailed and tested this gain-control formulation.[50] The PTM observer model is also generally consistent with a significant parallel set of observer models, several of which are in gain-control form, developed in the context of pattern-masking experiments in which a pattern stimulus rather than a noise stimulus is combined with the signal stimulus.[89–94] Although the gain-control formulations have not yet been extended to nonorthogonal templates or distinct parameters in the signal and the multiplicative-noise pathways, this should be straightforward.

The original PTM and its many extensions generate equations, such as those shown here and in the rest of the chapter, are *analytic approximations* of a *stochastic form* of the model. In the analytic formulations, random variables resulting from noises are replaced with their expected values, and a number of cross product terms are eliminated.[8, 10, 18, 49] The stochastic form of the model has been simulated and compared to the analytic forms.[10] The analytic formulas provide a good approximation to the key properties of the fully stochastic model. For example, the analytic formulas and stochastic form are consistent in their predictions for the ratio tests for nonlinearity and in the signature predictions of the different mechanisms of state change, such as perceptual learning.[10] The approximation is exact if the nonlinearity parameter γ is 1; the approximation is very good if the nonlinearity parameter γ is in the neighborhood of 2; and it is strained if the nonlinearity parameter γ is 3 or higher. In many applications, the empirical estimates of γ have often been in the range 1.7–2.5.

Another elaboration examined the model forms' sensitivity to *where* internal noises are introduced. In the original PTM, internal multiplicative noise is introduced, followed by internal additive noise. However, in principle, internal additive noise could occur before the template, after the template but before multiplicative noise, after multiplicative noise, or in all three places en route to the decision. These multiple potential sources of additive noise can be referred to a single late additive-noise source by rescaling and cumulating all sources.[10] However, the empirical observations of pure external-noise exclusion mechanisms found in both attention and perceptual learning suggest that the dominant internal additive noise occurs late in the process; if there is any internal additive noise before the template, it must be very small.

Another class of elaborations of the PTM expands the number of channels involved in processing the stimulus and the decision. The *multichannel model* illustrated in Dosher and Lu[11] provides one such example. In further elaborations, different neurons may operate as a "channel," each with its own noise and nonlinearity, and groups of neurons may contribute to a behavioral decision based on a population code.[95, 96] These *neural population models* can mimic the TvC functions and other properties of the PTM under perceptual learning.[97] See chapter 9 of our previous book[66] for further discussion.

References

1. Green D, Swets J. *Signal detection theory and psychophysics.* Wiley; 1966.

2. Macmillan NA. Signal detection theory as data analysis method and psychological decision model. In Karen G and Lewis C, eds. *A handbook for data analysis in the behavioral sciences: Methodological issues.* Lawrence Erlbaum Associates;1993:21–57.

3. Geisler WS, Albrecht DG. Visual cortex neurons in monkeys and cats: Detection, discrimination, and identification. *Visual Neuroscience* 1997;14(5):897–919.

4. Geisler WS, Albrecht DG, Salvi RJ, Saunders SS. Discrimination performance of single neurons: Rate and temporal-pattern information. *Journal of Neurophysiology* 1991;66(1):334–362.

5. Petrov AA, Van Horn NM, Ratcliff R. Dissociable perceptual-learning mechanisms revealed by diffusion-model analysis. *Psychonomic Bulletin & Review* 2011;18(3):490–497.

6. Ball K, Sekuler R. A specific and enduring improvement in visual motion discrimination. *Science* 1982;218(4573):697–698.

7. Watanabe T, Náñez JE, Koyama S, Mukai I, Liederman J, Sasaki Y. Greater plasticity in lower-level than higher-level visual motion processing in a passive perceptual learning task. *Nature Neuroscience* 2002;5(10):1003–1009.

8. Lu Z-L, Dosher BA. Characterizing human perceptual inefficiencies with equivalent internal noise. *Journal of the Optical Society of America A* 1999;16(3):764–778.

9. Dosher BA, Lu Z-L. Perceptual learning reflects external noise filtering and internal noise reduction through channel reweighting. *Proceedings of the National Academy of Sciences* 1998;95(23):13988–13993.

10. Dosher BA, Lu Z-L. Mechanisms of perceptual learning. *Vision Research* 1999;39(19):3197–3221.

11. Sperling G. *Three stages and two systems of visual processing*. New York University Department of Psychology; 1989.

12. Pelli DG. *Effects of visual noise*. PhD dissertation, University of Cambridge; 1981.

13. Burgess AE, Colborne B. Visual signal detection: IV. Observer inconsistency. *Journal of the Optical Society of America A—Optics Image Science and Vision* 1988;5(4):617–627.

14. Pelli DG. Uncertainty explains many aspects of visual contrast detection and discrimination. *Journal of the Optical Society of America A* 1985;2:1508–1532.

15. Eckstein MP, Ahumada AJ Jr, Watson AB. Visual signal detection in structured backgrounds: II. Effects of contrast gain control, background variations, and white noise. *Journal of the Optical Society of America* 1997;14(9):2406–2419.

16. Lu ZL, Dosher BA. Characterizing observers using external noise and observer models: Assessing internal representations with external noise. *Psychological Review* 2008;115(1):44–82.

17. Dosher BA, Han S, Lu Z-L. Perceptual learning and attention: Reduction of object attention limitations with practice. *Vision Research* 2010;50(4):402–415.

18. Dosher BA, Lu Z-L. Mechanisms of perceptual attention in precuing of location. *Vision Research* 2000;40(10):1269–1292.

19. Hetley R, Dosher BA, Lu Z-L. Generating a taxonomy of spatially cued attention for visual discrimination: Effects of judgment precision and set size on attention. *Attention, Perception, & Psychophysics* 2014;76(8):2286–2304.

20. Han S, Dosher BA, Lu Z-L. Object attention revisited: Identifying mechanisms and boundary conditions. *Psychological Science* 2003;14(6):598–604.

21. Jeon ST, Lu ZL, Dosher BA. Characterizing perceptual performance at multiple discrimination precisions in external noise. *Journal of the Optical Society of America A* 2009;26(11):B43–B58.

22. Foley JM, Legge GE. Contrast detection and near-threshold discrimination in human vision. *Vision Research* 1981;21(7):1041–1053.

23. Foley JM. Human luminance pattern-vision mechanisms: Masking experiments require a new model. *Journal of the Optical Society of America A* 1994;11(6):1710–1719.

24. Barlow HB. Retinal noise and absolute threshold. *Journal of the Optical Society of America* 1956;46(8):634–639.

25. Hillyard SA, Mangun GR, Woldorff MG, Luck SJ. Neural systems mediating selective attention. In Gazzaniga MS (ed.), *The cognitive neurosciences*. MIT Press;1995:665–681.

26. McAdams CJ, Maunsell JH. Effects of attention on orientation-tuning functions of single neurons in macaque cortical area V4. *Journal of Neuroscience* 1999;19(1):431–441.

27. Moran J, Desimone R. Selective attention gates visual processing in the extrastriate cortex. *Science* 1985;229(4715):782–784.

28. Treue S, Trujillo JCM. Feature-based attention influences motion processing gain in macaque visual cortex. *Nature* 1999;399(6736):575–579.

29. Heeger DJ. Normalization of cell responses in cat striate cortex. *Visual Neuroscience* 1992;9(2):181–197.

30. Dosher B, Lu Z-L. Object attention: Judgment frames, perceptual learning and mechanisms. In Raaijmakers JGW, Criss AH, Goldstone RL, Nosofsky RM, Steyvers M. *Cognitive modeling in perception and memory: A festschrift for Richard M. Shiffrin*. Psychology Press;2015:35–62.

31. Lu Z-L, Hua T, Huang C-B, Zhou Y, Dosher BA. Visual perceptual learning. *Neurobiology of Learning and Memory* 2011;95(2):145–151.

32. Saarinen J, Levi DM. Perceptual learning in Vernier acuity: What is learned? *Vision Research* 1995;35(4):519–527.

33. Gold J, Bennett P, Sekuler A. Signal but not noise changes with perceptual learning. *Nature* 1999;402(6758):176–178.

34. Gold JM, Sekuler AB, Bennett PJ. Characterizing perceptual learning with external noise. *Cognitive Science* 2004;28(2):167–207.

35. Peterson MF, Eckstein MP. Learning optimal eye movements to unusual faces. *Vision Research* 2014;99:57–68.

36. Hurlbert A. Visual perception: Learning to see through noise. *Current Biology* 2000;10(6):R231–R233.

37. Lu Z-L, Dosher BA. Perceptual learning retunes the perceptual template in foveal orientation identification. *Journal of Vision* 2004;4(1):5,44–56.

38. Lu Z-L, Chu W, Dosher BA, Lee S. Independent perceptual learning in monocular and binocular motion systems. *Proceedings of the National Academy of Sciences* 2005;102(15):5624–5629.

39. Chung STL, Levi DM, Tjan B. Learning letter identification in peripheral vision. *Vision Research* 2005;45(11):1399–1412.

40. Lu Z-L, Chu W, Dosher BA. Perceptual learning of motion direction discrimination in fovea: Separable mechanisms. *Vision Research* 2006;46(15):2315–2327.

41. Li RW, Levi DM. Characterizing the mechanisms of improvement for position discrimination in adult amblyopia. *Journal of Vision* 2004;4(6):7,476–487.

42. Jones PR, Moore DR, Amitay S, Shub DE. Reduction of internal noise in auditory perceptual learning. *Journal of the Acoustical Society of America* 2013;133(2):970–981.

43. Micheyl C, McDermott JH, Oxenham AJ. Sensory noise explains auditory frequency discrimination learning induced by training with identical stimuli. *Perception & Psychophysics* 2009;71(1):5–7.

44. Gold J, Bennett PJ, Sekuler AB. Signal but not noise changes with perceptual learning. *Nature* 1999;402(6758):176–178.

45. Dosher BA, Lu Z-L. Level and mechanisms of perceptual learning: Learning first-order luminance and second-order texture objects. *Vision Research* 2006;46(12):1996–2007.

46. Dosher BA, Lu Z-L. Perceptual learning in clear displays optimizes perceptual expertise: Learning the limiting process. *Proceedings of the National Academy of Sciences* 2005;102(14):5286–5290.

47. DeLoss DJ, Watanabe T, Andersen GJ. Optimization of perceptual learning: Effects of task difficulty and external noise in older adults. *Vision Research* 2014;99:37–45.

48. Dosher BA, Lu Z-L. Noise exclusion in spatial attention. *Psychological Science* 2000;11(2):139–146.

49. Lu Z-L, Dosher BA. External noise distinguishes attention mechanisms. *Vision Research* 1998;38(9):1183–1198.

50. Dao DY, Lu Z-L, Dosher BA. Adaptation to sine-wave gratings selectively reduces the contrast gain of the adapted stimuli. *Journal of Vision* 2006;6(7):739–759.

51. Huang C-B, Lu Z-L, Zhou Y. Mechanisms underlying perceptual learning of contrast detection in adults with anisometropic amblyopia. *Journal of Vision* 2009;9(11):24,1–14.

52. Andersen GJ, Atchley P. Age-related differences in the detection of three-dimensional surfaces from optic flow. *Psychology and Aging* 1995;10(4):650–658.

53. Andersen GJ, Ni R, Bower JD, Watanabe T. Perceptual learning, aging, and improved visual performance in early stages of visual processing. *Journal of Vision* 2010;10(13):4,1–13.

54. Jurs BS. *Learning to "see through the noise": A training study on the development of fingerprint expertise.* PhD dissertation, Indiana University;2009.

55. Bejjanki VR, Zhang R, Li R, Pouget A, Green CS, Lu, ZL, Bavelier D. Action video game play facilitates the development of better perceptual templates. *Proceedings of the National Academy of Sciences* 2014;111(47):16961–16966.

56. Bower JD, Andersen GJ. Aging, perceptual learning, and changes in efficiency of motion processing. *Vision Research* 2012;61:144–156.

57. Ball K, Sekuler R. Direction-specific improvement in motion discrimination. *Vision Research* 1987;27(6):953–965.

58. Jeon ST, Maurer D, Lewis TL. Developmental mechanisms underlying improved contrast thresholds for discriminations of orientation signals embedded in noise. *Frontiers in Psychology.* 2014;5(977):1–10.

59. Yan F-F, Zhou J, Zhao W, Li M, Xi J, Lu ZL, Huang CG. Perceptual learning improves neural processing in myopic vision. *Journal of Vision* 2015;15(10):12,1–14.

60. Cavanaugh MR, Zhang R, Melnick MD, Das A, Roberts M, Tadin D, Carrasco M, Huxlin KR. Visual recovery in cortical blindness is limited by high internal noise. *Journal of Vision* 2015;15(10):9,1–18.

61. Xu P, Lu Z-L, Wang X, Dosher B, Zhou J, Zhang D, Zhou Y. Category and perceptual learning in subjects with treated Wilson's disease. *PLoS One* 2010;5(3):e9635.

62. Petrov AA, Dosher BA, Lu Z-L. The dynamics of perceptual learning: An incremental reweighting model. *Psychological Review* 2005;112(4):715–743.

63. Petrov AA, Dosher BA, Lu Z-L. Perceptual learning without feedback in non-stationary contexts: Data and model. *Vision Research* 2006;46(19):3177–3197.

64. Dosher BA, Jeter P, Liu J, Lu Z-L. An integrated reweighting theory of perceptual learning. *Proceedings of the National Academy of Sciences* 2013;110(33):13678–13683.

65. Lesmes LA, Jeon S-T, Lu Z-L, Dosher BA. Bayesian adaptive estimation of threshold versus contrast external noise functions: The quick TvC method. *Vision Research* 2006;46(19):3160–3176.

66. Lu Z-L, Dosher B. *Visual psychophysics: From laboratory to theory.* MIT Press; 2013.

67. Lu Z-L, Dosher BA. Characterizing the spatial-frequency sensitivity of perceptual templates. *Journal of the Optical Society of America A—Optics, Image Science, and Vision* 2001;18(9):2041–2053.

68. Ahumada AJ, Lovell J. Stimulus features in signal detection. *Journal of the Acoustical Society of America* 1971; 49(6, Pt. 2):1751–1756.

69. Ahumada A Jr. Perceptual classification images from Vernier acuity masked by noise. *Perception* 1996;25(1_suppl):2.

70. Eckstein MP, Ahumada AJ Jr. Classification images: A tool to analyze visual strategies. *Journal of Vision* 2002;2(1):i.

71. Abbey CK, Eckstein MP. Estimates of human-observer templates for a simple detection task in correlated noise. In *Medical Imaging 2000: Image Perception and Performance.* 2000;3981:70–77. International Society for Optics and Photonics.

72. Gold JM, Murray RF, Bennett PJ, Sekuler AB. Deriving behavioural receptive fields for visually completed contours. *Current Biology* 2000;10(11):663–666.

73. Jones JP, Palmer LA. The two-dimensional spatial structure of simple receptive fields in cat striate cortex. *Journal of Neurophysiology* 1987;58(6):1187–1211.

74. Ohzawa I, DeAngelis GC, Freeman RD. Encoding of binocular disparity by simple cells in the cat's visual cortex. *Journal of Neurophysiology* 1996;75(5):1779–1805.

75. Ringach DL, Hawken MJ, Shapley R. Dynamics of orientation tuning in macaque primary visual cortex. *Nature* 1997;387(6630):281–284.

76. Ahumada AJ Jr. Classification image weights and internal noise level estimation. *Journal of Vision* 2002;2(8):121–131.

77. Murray RF, Bennett PJ, Sekuler AB. Optimal methods for calculating classification images: Weighted sums. *Journal of Vision* 2002;2(1):6,79–104.

78. Tjan BS, Nandy AS. Classification images with uncertainty. *Journal of Vision* 2006;6(4):8,387–413.

79. Dobres J, Seitz AR. Perceptual learning of oriented gratings as revealed by classification images. *Journal of Vision* 2010;10(13):8,8–11.

80. Dupuis-Roy N, Gosselin F. Perceptual learning without signal. *Vision Research* 2007;47(3):349–356.

81. Levi D, Klein S. Noise provides some new signals about the spatial vision of amblyopes. *Journal of Neuroscience* 2003;23(7):2522–2526.

82. Green DM. Consistency of auditory detection judgments. *Psychological Review* 1964;71:392–407.

83. Ahumada AJ. *Detection of tones masked by noise: A comparison of human observers with digital-computer-simulated energy detectors of varying bandwidths.* Technical Report No. 29, University of California, Los Angeles;1967.

84. Gilkey RH, Frank AS, Robinson DE. Estimates of internal noise. *Journal of the Acoustical Society of America* 1978;64:S36.

85. Gilkey RH, Hanna TE, Robinson DE. Estimates of the ratio of external to internal noise obtained using repeatable samples of noise. *Journal of the Acoustical Society of America* 1981;69(S1):S23.

86. Spiegel MF, Green DM. Two procedures for estimating internal noise. *Journal of the Acoustical Society of America* 1981;70:69–73.

87. Weber EH. *De Pulsu, resorptione, auditu et tactu: Annotationes anatomicae et physiologicae*. CF Koehler; 1834.

88. Liu S-H, Dosher BA, Lu Z-L. The role of judgment frames and task precision in object attention: Reduced template sharpness limits dual-object performance. *Vision Research* 2009;49(10):1336–1351.

89. Nachmias J, Sansbury RV. Grating contrast: Discrimination may be better than detection. *Vision Research* 1974;14(10):1039–1042.

90. Fredericksen RE, Hess RF. Temporal detection in human vision: Dependence on stimulus energy. *Journal of the Optical Society of America* 1997;14(10):2557–2569.

91. Gorea A, Sagi D. Disentangling signal from noise in visual contrast discrimination. *Nature Neuroscience* 2001;4(11):1146–1150.

92. Klein SA, Levi DM. Hyperacuity thresholds of 1 sec: Theoretical predictions and empirical validation. *Journal of the Optical Society of America A* 1985;2(7):1170–1190.

93. Kontsevich LL, Chen CC, Tyler CW. Separating the effects of response nonlinearity and internal noise psychophysically. *Vision Research* 2002;42(14):1771–1784.

94. Legge GE, Foley JM. Contrast masking in human vision. *Journal of the Optical Society of America* 1980;70(12):1458–1471.

95. Watson AB, Solomon JA. Model of visual contrast gain control and pattern masking. *Journal of the Optical Society of America* 1997;14(9):2379–2391.

96. Goris RL, Putzeys T, Wagemans J, Wichmann FA. A neural population model for visual pattern detection. *Psychological Review* 2013;120(3):472–496.

97. Bejjanki VR, Beck JM, Lu Z-L, Pouget A. Perceptual learning as improved probabilistic inference in early sensory areas. *Nature Neuroscience* 2011;14(5):642–648.

5

Physiological Basis

Changes in physiological responses as a result of visual perceptual learning may identify the loci and modalities of brain plasticity, leading to improvements in behavioral responses. Cellular recording in early visual cortices sometimes finds small but potentially important changes in V1 or MT (middle temporal area) after training; evidence for changes higher in the visual cortex, in V4 or LIP (lateral interparietal cortex), is stronger, especially when measured during active task performance. Early cortical response changes account for a small amount, but by no means all, of the full behavioral improvements, while plasticity in higher cortical levels is more closely related to behavior. Along with measures using fMRI and EEG in humans, these findings in animals highlight the involvement of complex brain networks, multiple loci of visual plasticity, and a likely role for top-down attention and decision in perceptual learning.

5.1 Biological Substrates of Perceptual Learning

Visual perception engages many brain regions. Even the simplest perceptual task depends on an integrated set of processes, starting with the initial sensory registration of the stimulus and progressing through higher cortical regions, ultimately leading to a decision and sometimes a corresponding action. Systems for attention, expectation, and reward may also be engaged. All these interacting brain regions and processing modules work together to achieve any given behavioral goal.

Just as the experience of seeing a given stimulus involves a network of brain processes, *getting better* at seeing that stimulus will also involve a distributed set of processes. Improvements in perceptual ability could, in principle, reflect plasticity anywhere along the neural pathway. Indeed, plasticity observed in one brain region may not in itself signal the primary locus of plasticity. A change in an early sensory area could furthermore have cascading consequences throughout the system. A fuller picture would likely admit that learned plasticity more probably occurs at multiple levels in emergent ways. What is clear, however, is that whether the goal is to better understand visual perception or visual perceptual learning, an analysis of the biological substrates undergirding either of these functions will necessarily involve consideration of the entire brain system and its modules.[1]

In an ideal world, scientists would be able to monitor physiological activity simultaneously in multiple brain regions at a range of spatial and temporal resolutions. If this were possible (which it currently is not), then we could presumably identify the primary site or sites of plasticity with precision and document exactly how any observed changes affect system response. Given the inherent limitations of any measurement apparatus, and the limitations of our current technological methods, claims about brain plasticity have, by necessity, been more modest. Any physiological measurements have typically been made either from specific brain regions or at selected spatial or temporal resolutions, with trade-offs involved in each.

There currently exists a menu of possible methods from which researchers can choose. On the local, cellular level, single-unit or multiunit recordings measure the responses of individual neurons or groups of neurons in a small number of locations at a fine temporal resolution. EEG measures the activity of larger brain regions or networks at good temporal resolution, while fMRI measures responses of select brain regions or the whole brain at higher spatial resolution, but at lower temporal resolution, than EEG. Researchers may also use active methods such as transcranial magnetic stimulation (TMS) to either enhance or disrupt activity in superficial brain regions in order to assess its causal relationship to the target perceptual behavior (or assess the disruption or enhancement of learning). Although at present we cannot measure neural activities for the whole brain simultaneously at high temporal and spatial resolution, each experiment provides a potentially important glimpse of the physiological substrates of perception or of learning. These pieces of information must be knitted together to form an integrated view of system plasticity.[2]

One especially important goal for researchers is to trace the causal connection between the physiological responses and the behavioral responses—traversing the multiple levels of the brain. This would encompass the physiological responses to the stimulus, the involvement of higher regions in attention and expectation, the regions computing decision, and the recruitment of behavioral responses—as well as how each of these levels (and their interaction) might be altered by practice or training.

Such an aspirational goal is perhaps in sight but is still far off. Linking patterns of brain activity to behavioral responses can be complicated. To navigate this complexity, a computational model or theory is almost always critical. Without a predictive quantitative model, the relationship between physiological response and behavior cannot really be assessed. The best current work uses theories, rules, or algorithms to meaningfully connect changes in a local physiological response to changes in decision and behavior. Many single-cell recording studies, for example, apply signal detection, pattern classifiers, or Bayesian models to connect neural responses to the behavioral choices.[3] In addition, raw physiological data are also sometimes interpreted in terms of changes in derived measures such as tuning functions, response magnitude, or topology of the responses.

At the same time, the current investigations leave open certain avenues of investigation. All these physiological measures, but especially those focused on localized regions of the early visual cortex, implicitly see the neural responses as representations of different characteristics of the stimulus or at least as information encoded from the stimulus that is passed along to other brain regions.[4] At this point, drawing an analogy between physiological measures and learning in neural networks is useful. Learning in neural networks is embodied in the connection weights. If these local neural responses in particular brain areas are akin to the activation of representation units in a neural network model, this leaves open the question of how and where this information is connected up to make a decision, and where these connection weights are stored during learning. Chapters 6–9 detail the success of neural network models in which learning *reweights* the connections between relevant representation units and decision. These models, and general principles on which they are based, lead to the following questions about the physiology: If properties of the stimulus are represented in activities in visual cortical areas, where are the weights connecting these activities to decision and action, and where is the learning embodied in the brain? Is it, as hypothesized by some, embodied in the low-level representation neurons? Or is learning embodied in changes of connectivity in the larger brain network? And how can these phenomena be measured in physiology? We return to these questions at the end of the chapter.

In what follows, we begin by providing a brief snapshot of some of the most relevant brain regions examined in the literature. Although this information will be well known to vision scientists, this short treatment is meant to highlight the cortical regions that have been the focus of current studies—in short, to provide background for other readers. The chapter then goes on to analyze the evidence from physiological studies of perceptual learning and what these studies say about visual plasticity. Our treatment is organized by the technology used, starting with cellular recording in animals (usually monkeys), followed by different modalities of brain imaging studies in humans. The field's investigation of the physiological substrates of perceptual learning is just beginning. Nevertheless, as we will see, a number of fascinating conclusions, with profound implications, can already be sketched.

5.2 Physiological Substrates

The study of the neuroscience underlying brain function is one of the most prolific scientific enterprises of our time. Thousands of books and millions of articles have been written about it. This section provides just a brief view of the most relevant brain regions for visual perceptual learning. We start by reviewing the eye and the visual cortex, with a focus on those cortical regions that have been the primary targets of physiological research; we also touch on the cognitive, reward, decision, and motor areas possibly involved in learning. This brief treatment is meant to serve as immediate, if partial, background for the investigations treated later in the chapter. It is meant for nonexperts. Interested readers should of course consult relevant textbooks and the referenced papers for details.

5.2.1 Functional Areas of the Brain

The brain is a complex connected network with many modules;[1] different regions have been identified with different sensory, motor, and mental functions.[5–8] Figure 5.1 shows a

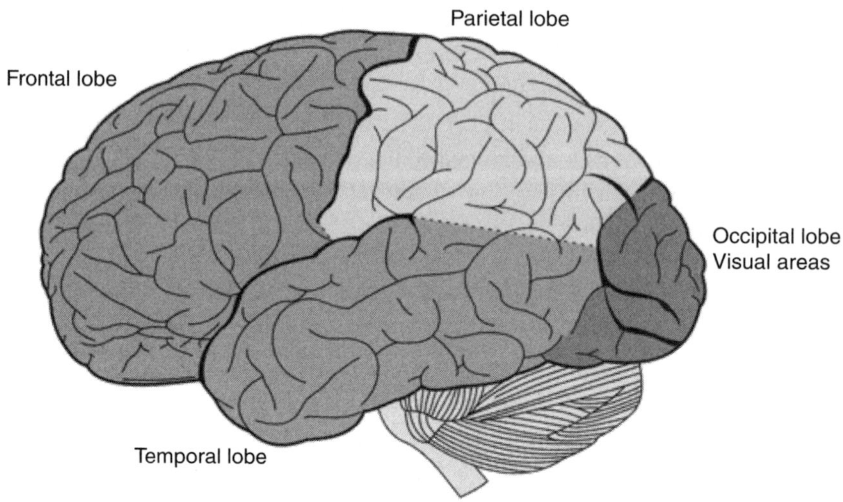

Figure 5.1
Functional regions of the human brain.

classic illustration of the large functional regions of the human brain. These regions include smaller subdivisions for the visual cortex, auditory cortex, olfactory areas, and the sensory and somatosensory areas that support vision, hearing, smell, and tactile senses. Other regions represent sensorimotor associations, motor activity, and all the many processes that support humans' ability to make decisions, remember past experiences, and carry out higher mental functions, such as language and speech, thinking, and planning. Any individual perception, thought, or motor action reflects integrated activities of many brain regions working together in concert.

The visual system and associated visual cortical areas are among the largest functional areas in the human brain, rivaling in size the brain territory specialized for processing of language. In humans, this system has many submodules, which carry out a complex set of computations on the sensory inputs (see figure 1.4). For this reason, visual perceptual learning embodies many of the complexities of learning in any sensory or motor domain. Since the focus of this book is on visual learning, we begin with a brief overview of visual areas, starting with the eye.

5.2.2 The Visual System

Images of the outside world arrive through the eye by way of photons reflected off some objects or emitted by others. Light images project onto the the retina through the lens at the front of the eye, where they are detected by light-sensitive cells. This system (cornea, lens, and muscles controlling the lens) focuses the images on the retina, while the iris, pupil, and eyelids control the amount of light, like the aperture and shutter of a camera. The quality of the eye's optics and the resolution of coding at the back of the retina are believed to have evolved to be approximately consistent with one another.[9]

Readers will of course be familiar with this precis of vision. Of special importance to the early physiological study of visual perceptual learning, however, is the pathway from the eyes, through the lateral geniculate nuclei (LGN), and then to the visual cortices. Also of special interest for a number of perceptual learning studies is the retinotopic mapping from the visual fields of the two eyes to the visual cortex.

Light arriving at the retina, where the image is upside down and reversed from left to right, is converted to neural firing by light-sensitive photoreceptors in the rods and cones.[10] Activity of the cones, sensitive to the long, medium, and short wavelengths (color), and the rods, sensitive to low light, in spatially localized receptive fields, drive the activity of the retinal ganglion cells.

The LGN, an important way station between the retina and the visual cortex, is a layered structure with specialized cell types (M, P, and K cells), each conveying different kinds of information.[11–14] The majority of neurons in the optic nerve follow a pathway to the LGN through the optic radiation to the primary visual cortex (geniculostriate path). About 10% of the neurons in the optic nerve go a different route (tectopulvinar path), which may be used to integrate audition and vision with motor systems. The LGN has a *retinotopic* arrangement; spatial regions of the LGN represent spatial regions of the visual field. The representations of the left hemifields (right visual field) of both eyes project to the left LGN, and vice versa for the right LGN (see figure 5.2).[15] (This fact has been exploited by some studies that train stimuli presented in one hemifield and measure physiological responses in the corresponding cortical representation in the other hemifield as an untrained control.) Some researchers have actually proposed that learning may reach as far down as the LGN; indeed one computational model proposed that learning reweights information in the LGN.[16, 17]

The sensory representations in the visual cortex, however, have been the primary focus of physiological studies on perceptual learning. These physiological studies have asked whether learning changes the responses in the early visual cortex—effectively asking whether learning changes these early representations. The motivating enterprise has been "how low can you go" in documenting learned plasticity at the earliest possible levels in these visual representations. For this reason, studies have focused on V1, V2, and V4 for pattern tasks and on MT and MST for motion tasks.

V1, or the primary (striate) visual cortex in the occipital lobe of each hemisphere, is the primary end point of the path from the LGN.[5] Like the LGN, V1 has a *retinotopic* organization, with adjacent regions of V1 representing adjacent regions of the visual world[18, 19] and different layers receiving inputs from different kinds of LGN cells.[20, 21] Within this, alternating regions or ocular dominance columns are differentially sensitive to inputs from the two eyes.[19, 22] Neurons in subregions of V1 called *blobs* are color sensitive, monocular, and have small receptive fields (inputs from P and K cells).[19, 23] Neurons in the *interblob* regions may be sensitive to orientation, motion, and form (inputs from M and P cells); many are binocular, receiving inputs from both eyes. V1 is certainly a central input hub for visual information. Some estimates place the number of neurons in each hemisphere of V1 at 140

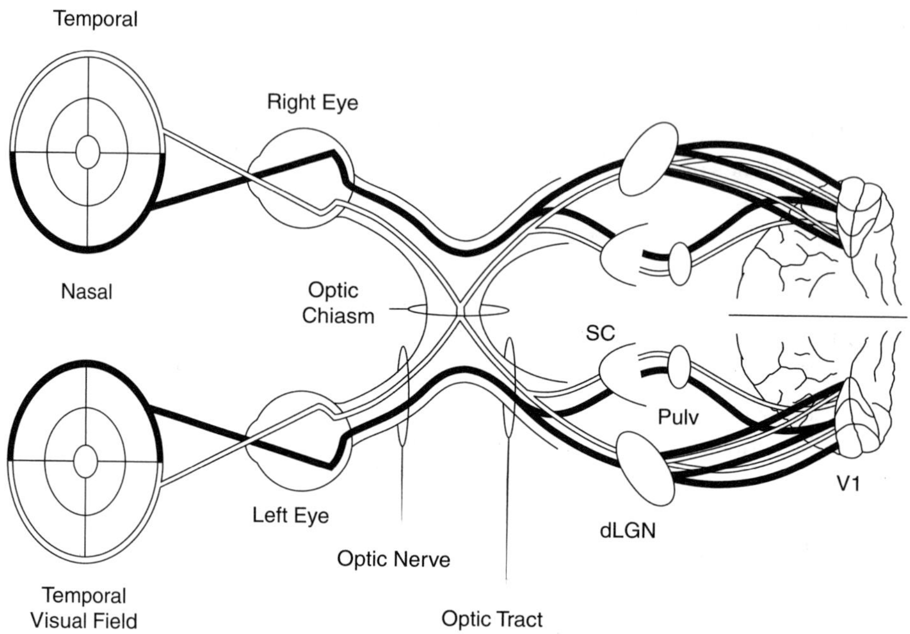

Figure 5.2
The visual pathways show the connections from the eye to the primary visual cortex via the lateral geniculate nuclei (LGN). The left LGN takes inputs from the right visual field from both eyes and vice versa (SC=superior colliculus; Pulv=pulvinar). Figure adapted from Burnat,[17] figure 1. Creative Commons, copyright 2015 Kalina Burnat.

million, or about 40 V1 neurons per LGN neuron.[24, 25] This provides a vast computational resource for image processing.

V1 passes on information to a cascade of other higher visual areas in the extrastriate visual cortex.[5, 26–31] Neurons sensitive to color, shape, depth, or motion send information upstream, either directly or indirectly, to many other visual areas: V2, V3, V4, V5 (MT), and elsewhere (figure 5.3).[32] Each of these areas may serve as a way station to higher-level processing but also code specific features of the visual input. In turn, these regions feed information back to V1, which also receives modulatory inputs from nonvisual areas. (Feedback and modulatory connections are omitted in the figures.) Major lesions of V1 cause loss of vision in the visual hemifield opposite to the side of the lesion. While damage to LGN or V1 qualitatively disrupts vision, damage to the extrastriate visual cortex has more complicated and subtler impacts on perception.[33–37] V4 neural responses (and above) can also be modulated by cognitive factors such as attention or object salience. Indeed, the nature of task influence on responses in early visual areas is still actively debated. It seems likely that the higher visual areas may code many visual properties relevant to object recognition and may incorporate top-down influences.[37, 38]

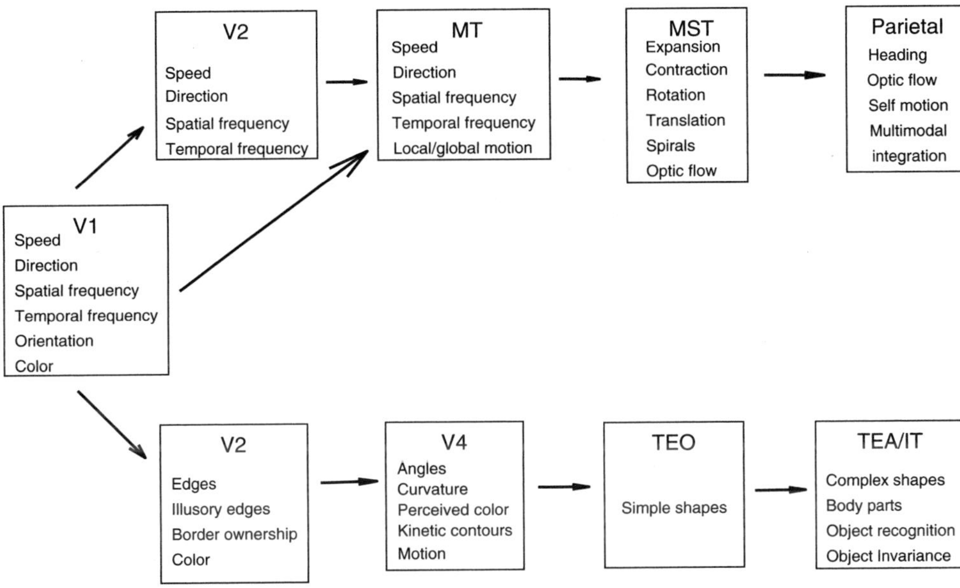

Figure 5.3
Feed-forward pathways of the visual system. The parietal (dorsal) pathway processes motion, depth, and spatial information. The inferior temporal (ventral) pathway to the inferior temporal cortex processes form and color. Both pathways take input from V1 projected via the LGN. After Perry and Fallah,[32] figure 1. Creative Commons, copyright 2014 Perry and Fallah.

Much research has focused on describing the exact nature of the neural receptive fields in different visual areas. For example, retinal and LGN cells are sometimes characterized in terms of center-surround receptive fields (either on- or off-cell) (figure 5.4)[39] whose sizes increase with distance from the fovea. V1 neurons are sometimes modeled with receptive fields that are elongated and oriented (figure 5.4), coding properties such as oriented edges. One analysis of V4 sees it as representing the contours and locations of object parts, while the inferior temporal cortex (ITC) may represent object categories (figure 5.5).[40] In other analyses, V4 has regions sensitive to color, orientation, shape, depth, and motion.[38] MT (V5), which receives inputs from directionally sensitive neurons in V1, represents primary motion information, with individual neurons being selectively sensitive to different directions of motion (figure 5.6).[41] It passes information to code for integrated motion patterns thought to be represented in MST, where some cells are sensitive to properties of optical flow motion (expansion, contraction, rotation, etc.).[42–45]As we will see, some research in visual perceptual learning has focused on some of these higher-level visual cortical regions.

The many visual areas of the human brain are organized in two broad pathways or streams of visual processing (see the schematic illustration in figure 5.3): the ventral and the dorsal. They have a common origin in V1 and extend through the extrastriate visual cortex to regions of either the temporal or posterior parietal cortex.[27, 46] The *ventral*

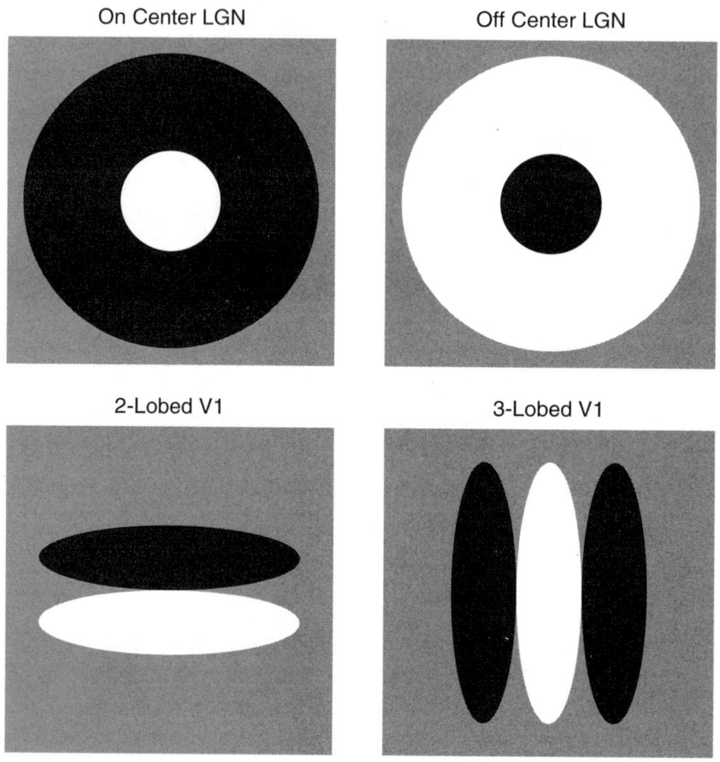

Figure 5.4
The center-surround receptive fields of retinal and LGN neurons and the oriented receptive fields typical of V1 simple-cell neurons: on- or off-center cells in the LGN or retina excited by light surrounded by dark stimuli or darkness surrounded by light stimuli. Many cells in V1 have oriented receptive fields that either respond to edges (two-lobed) or bars (three-lobed), with horizontal and vertical orientations being more common.

(bottom) stream takes inputs from V1 to V2 and V4 and then on to the inferior temporal (IT) cortex. The receptive field sizes increase from V1 (< 1°), to V4, to IT (20°), depending on retinal eccentricity and the complexity of the stimuli that are used to measure them. The ventral stream seems to be involved in the representation of shape, size, orientation, form, and objects. It has been called the "what" pathway—these features help you decide what you are seeing. The *dorsal* (upper) stream takes inputs from V1, passes through V2 and on to area MT (V5), and then to the posterior parietal cortex, eventually providing inputs to the motor cortex. Unlike the ventral stream, the dorsal stream functions primarily in spatial awareness, motion, and actions such as reaching, and has been called the "where" pathway, or thought of alternatively as an "action" system.[47] One current view is that the ventral and dorsal processing streams are parts of a complex network that coordinates both kinds of processing in order to orchestrate behavior. Overall, a network of up to 40 different areas that participate in visual processing has been identified (see

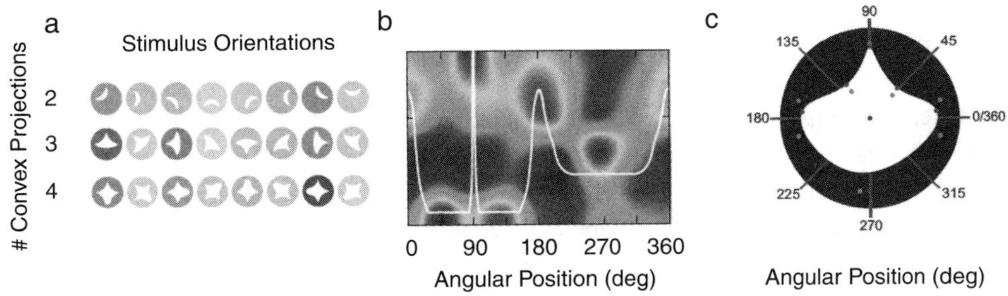

Figure 5.5
Receptive fields of V4 neurons may code for spatial contours. (a) Examples of convex contours with two, three, or four vertices and gray level indicating cell response. (b) Composite shapes coded by activities over several V4 neurons identify curvature and angular position; hot spots reflect different V4 neurons that together code an object shape. (c) A corresponding object shape. From Kourtzi and Connor,[40] figure 1a, c, and d, with permission. (See plate 3.)

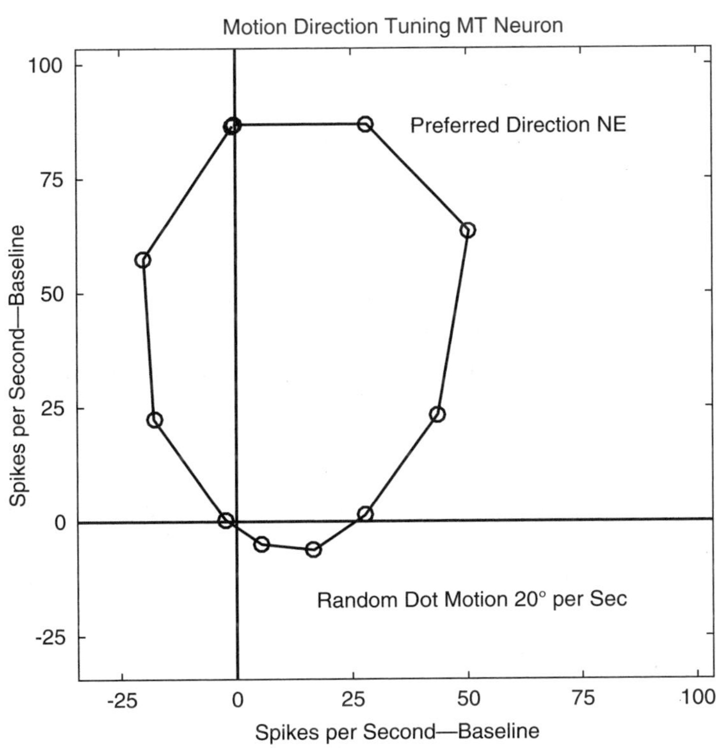

Figure 5.6
Motion-direction selectivity of an MT neuron to random dot motion, in this case with a preference for motion to the upper right. The connected shape in polar coordinates represents the summed spike rates to 16 different directions of motion. Redrawn from data in Albright, Desimone, and Gross,[41] figure 1.

figure 1.4), though this understanding continues to develop thanks to analysis of human brain activity using fMRI.

Yet another—and quite different—approach to characterizing the coding of different visual areas cites analogies between the responses of visual areas and the responses of units in different layers in a deep-learning network trained to discriminate object categories.[48–51] (Deep-learning networks are those with many layers, which are typically trained with images of objects using supervised labeling; see chapter 8.) Several studies, for example, have related the responses in layers of the computational network to fMRI activation in the IT cortex.[48, 51, 52] The various computations carried out in V1 and other visual areas, and the details of the outward projections and feedback connections, are still an active focus of research in macaque monkeys[53] and in functional magnetic resonance imaging (fMRI) in humans.

5.2.3 Circuits of Perceptual Decision-Making, Reward, and Attention

Perceptual decisions do not simply respond to sensory evidence; they must also be informed by task context, expectations, rewards, and other cognitive factors, including attention, which almost surely plays an important role in the selection of inputs and responses, as we will detail. In order to accomplish this complex set of activities, the human system involves a number of pathways connecting the prefrontal cortex to sensory- and motor-control areas. Figure 5.7 illustrates some of the connections and circuitry that could be involved in making perceptual decisions.

Visual perceptual decisions begin with the visual signals in V1, processed through the dorsal and ventral streams and then integrated in the dorsolateral prefrontal cortex (dlPFC). Computations of reward expectation occur in a network of brain areas, including the substantia nigra par reticulata (SNr) and the ventral tegmental area (VTA), which also project to the prefrontal cortex (PFC), where reward information is integrated with visual information, prior experience, and cognitive evaluation in order to assign values to potential outcomes.[54] From the PFC, information is sent to *decision and response selection* mechanisms in a loop that includes the cerebral cortex (including the premotor cortex), the thalamus, and the basal ganglia. The basal ganglia are integral to the control of voluntary motions. The thalamus is involved in regulation of excitation and inhibition. The selected response is communicated to the relevant motor systems in order to execute a response behavior. If the response is an eye movement, for example, it is communicated to the frontal eye fields (FEFs) and to the lateral interparietal cortex (LIP), which are associated with eye movements and working memory related to them.[54] (See figure 5.7 for a schematic of these areas based on an analysis in monkeys.)

Reward is a common instrument of learning. The physiological *reward systems* were originally identified primarily with dopamine neurons in regions such as the nucleus accumbens (NAcc) and the ventral tegmental area (VTA)[55–59] (see Haber and Knutson[60] for a review). The list of areas thought to be involved in reward processing has expanded, notably including the ventral striatum (VS) and the dopamine neurons of the substantia nigra

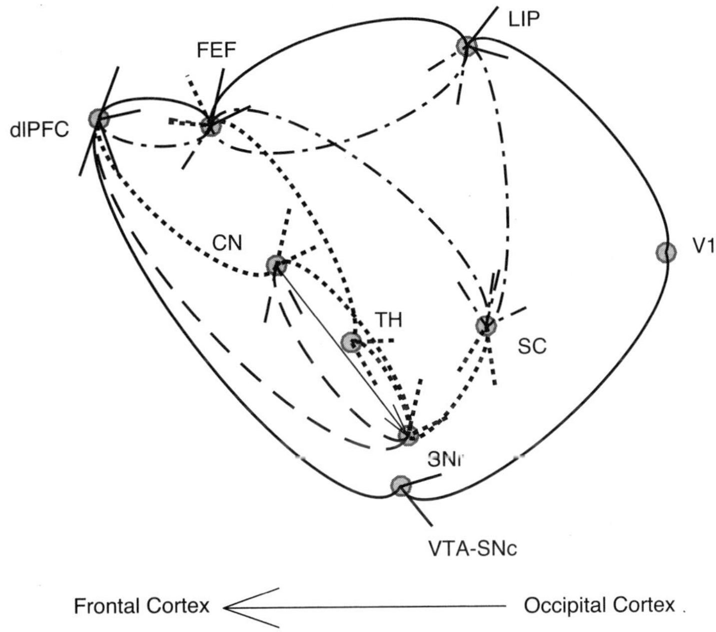

Figure 5.7
Neurocircuits that connect vision with decision-making and action. Visual signals from the dorsal and ventral streams are integrated in the dorsolateral prefrontal cortex (dlPFC) (solid arrows). Reward and reward expectation (dashed arrows), processed in the ventral tegmental area (VTA) and substantia nigra pars compacts (SNc), are integrated in the PFC. Response selection (dotted arrows) engages a loop that includes the basal ganglia, thalamus, and cortex: caudate nucleus (CN), substantia nigra par reticulata (SNr), thalamus (TH), and superior colliculus (SC). Eye responses are represented in the frontal eye field (FEF), the lateral interparietal cortex (LIP), and superior colliculus (SC) (dot-dash arrows), which also sends messages to the brain stem. Based on analysis of macaque monkeys by Opris and Bruce,[54] figure 3.

(SN), with a reward circuit embedded within a cortico–basal ganglia network. Basal ganglia, originally associated with their roles in motor and sensory function, are now thought to contribute more widely to the coding of reward value, motivation, and early decision. In short, the reward system involves a network of complex circuitry that interacts with frontal cortical regions in understanding and selecting behaviors.

By selectively orienting the observer toward the relevant visual stimuli, attention is also an integral part of the processing of sensory inputs. In current theories of spatially selective attention, the putative subcortical attention circuit includes some of the same elements: the superior colliculus (SC) in the midbrain and the pulvinar nucleus of the thalamus (TH). The ventral pulvinar maintains a topographic representation of the sensory space and receives information from and feeds information back to the visual cortex; it has been associated with the function of a saliency map. Covert attention and eye-movement systems also share circuits in the frontal eye fields (FEFs), lateral interparietal cortex (LIP), and prefrontal cortex (PFC) that help to select and orient attention to salient or cued visual stimuli.[61–63]

As we will see, these higher regions that are likely involved in visual task performance have not yet played as central a role as they perhaps could have in the physiological analysis of visual perceptual learning. Chapter 9 considers the possible relations to learning of task structure, attention, and reward.

5.2.4 Discussion

Visual perception and perceptual learning occur in the context of a set of interwoven brain modules. Any long-lasting change—even a local change in one module—could theoretically also change the processing elsewhere in the network. If visual perceptual learning were associated with plasticity in the earliest visual cortex in V1, for example, this might then feed different information directly or indirectly to many other visual areas, whose subsequent responses would thus also be altered. Changing the responses of V1 would almost certainly also require changes in the interpretation of the sensory evidence used for making a decision. More generally, it should be stressed that observing a change in one brain area does not necessarily imply that this change is the causal factor behind changed behavior—it may instead be a *consequence* of a change in another brain area that mediates the observed behavioral effects. Conversely, a change in response in an early visual area such as V1 could in principle be a consequence of feedback from higher visual areas where plasticity originated. Changes in early responses caused by the top-down processing are one way to instantiate learned plasticity that can be multiplexed depending on the task context. Given all these considerations and the possible system hazards of too much early plasticity, researchers may wish to more fully contextualize the interpretation of local measurements.

5.3 Using Biology to Understand Learning

The first investigations into the physiology underlying visual perceptual learning focused on neuroplasticity in the early sensory cortices. This choice was largely motivated by many physiological studies from the 1990s that reported substantial plasticity in the earliest sensory cortices in the somatosensory and auditory domains.[64–66] It was also motivated by the specificity of behavioral improvements to the trained retinal location and features, a feature of learning that led some researchers to infer a central role for early visual cortical plasticity.[67, 68] Though initial physiological investigations focused on changes in the responses in V1 and other early visual areas, more recent work has shifted focus to higher-level regions of the visual cortex and decision areas.

Regardless of the brain area studied, a number of techniques have been used to investigate the substrates of learning. The most common method is cellular recording (usually in monkeys), while other work has used technologies such as EEG, fMRI, and TMS (usually in humans). In cellular recordings, researchers have tended to look for correlations between perceptual learning and changes of tuning, topology, and/or magnitude of neuronal responses. In EEG and fMRI brain imaging, the effects researchers look for involve changes in the response properties of specific cortical regions.

Correlation in either cellular recording or EEG and fMRI does not, of course, prove causation. Even when physiological changes correlate quite directly with behavioral improvements, the causal relationship between physiology and learning is difficult to pin down,[69] as the data may reflect not only changes in cortical responses to specific stimuli but also an ongoing cascade of processes in other brain regions that jointly modulate physiological responses in the recorded sensory areas (figure 5.8). In a pure bottom-up mode, changes in the visual cortical responses would lead directly through subsequent processes to changes in the perceptual behavior. In a pure top-down mode, attention, alertness, or a task-modulated goal would induce changes in both the visual cortical responses and the perceptual behavior, inducing a correlation (but not causation) between the sensory response and the behavior. More likely, however, changes in the physiological responses reflect some mixture of bottom-up and top-down influences. Given this likely reality, it again bears noting that observed changes in any single cortical region do not mean that this site is the only or even the primary substrate of learning.[4, 69]

Whatever the technology, the measurements of brain responses reflect the functioning of the system under the demands of a behavioral task—and thus may also be influenced by task expectation, stimulus processing, attention, decision-making, motor response, and feedback or reward, among other factors. Disentangling these respective influences is quite

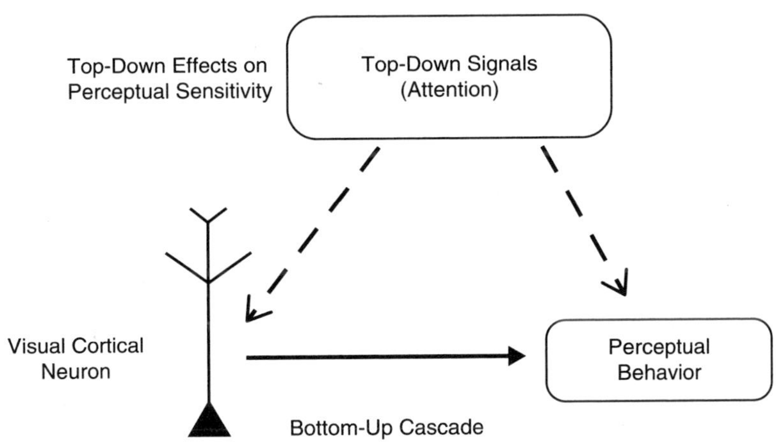

Figure 5.8
Two neural mechanisms accounting for the correlation of responses of a visual cortical neuron and perceptual behavior via either bottom-up processing of signals or shared top-down influences (attention, alertness, goal direction) on both cortical responses and behavior. Modified from elements of Smith et al.,[69] figure 1 (open access).

challenging. Heuristically, researchers have assumed that brain responses occurring closer in time to the stimulus reflect bottom-up effects, while brain responses occurring later reflect top-down influences, an assumption that has driven the interpretation of data in single-unit recording and EEG (fine temporal data are largely unavailable in the sluggish fMRI signals). Though intuitively compelling, even this heuristic is a simplification, because expectation, attention, and decision bias are known to induce anticipatory effects well before any stimulus is presented.

Distinguishing causation from correlation is difficult in any field, but in the case of physiology and perceptual learning this is doubly so, thanks to the dual timescale at work in any given experiment. In the basic study of attention or decision, for example, neuronal responses are measured while the animal is performing the task, and the relevant data are precisely those perhaps momentary changes in visual cortical responses that occur during task performance. In perceptual learning, by contrast, plasticity may involve *persistent effects of training* on visual responses that alter the system even when the trained task is not being performed. Alternatively, perceptual learning might alter the visual responses solely during the performance of the trained task, or very similar tasks, reflecting *task-induced effects of training* on visual responses. While physiological studies have so far tended to focus on either persistent changes or task-induced changes in early visual cortical responses, a rigorous account of the effects of perceptual learning should evaluate both in concert with one another.

Other interpretive issues have arisen that are more specific to fMRI and EEG, where both attention and task difficulty are hypothesized to also affect visual-cortical responses. A number of experimental paradigms exist. In fMRI, for example, some investigators have measured brain responses during the task performance itself, while others have measured responses to some set of stimuli before and after perceptual learning. Still others have tested a standard task during pre- and posttraining imaging sessions to assess the effects of training in a different but related task (keeping the stimuli and performance the same in pre- and posttraining imaging sessions, so that the results will likely depend on the similarity of the assessment task to the training task). A final group has studied connectivity during a resting state in order to avoid performance-level confounds. Each of these methods differs so widely from the others that we must be careful in interpreting any given dataset.

The injunction to take analytic care, however, should not in itself be discouraging. Complexity is often as much an invitation as a warning, and based on the modest number of existing physiological studies, a remarkably clear pattern of results is already beginning to emerge, especially when each individual and often localized experimental snapshot is knitted into a larger framework.

This kind of meta-analysis of the existing literature led us to the following provisional conclusions. Although modest changes in the tuning of the earliest level of the visual cortex (V1) after learning have been reported in some cases, the changes in higher visual areas (V4, IT cortex) are more substantial. Changes in V1 seem to account at most for a very small proportion of the variation in behavior, while neural responses in higher areas seem to account for more. Additionally, while some small, persistent changes in physiological responses have been reported, altered neural responses have been more tightly coupled to behavior only when measured during active task performance. In these cases, the improved predictions of behavior almost always also reflect changes in readout used by the researcher from before to after learning. Finally, and perhaps more controversially, we conclude that the changes in these low- or mid-level representations—the weights that connect them to decision or that embody the learning about the tasks—quite likely reside elsewhere in the brain system.

Our provisional conclusions should have implications for the involvement of whole-brain systems in perceptual learning and for the plasticity/stability dilemma (or the relative balance of retuning and reweighting). A further consideration of these issues appears at the end of the chapter. First, in the following sections, we aim to systematically evaluate the evidence obtained from the physiological studies, organized according to measurement technology. As in previous chapters, we provide initial summaries, followed by a discussion of specific exemplar studies.

5.4 Evidence from Single-Cell Recording

A growing body of research has studied the relationship between perceptual learning and neural responses using single-unit recording methods. We group these studies based on the nature of the training task: focused on features, patterns, or objects or scenes. This partition roughly corresponds with learning in low-, mid-, and high-level vision.

5.4.1 Perceptual Learning of Features

The first substantial single-cell recording studies of mechanisms of visual perceptual learning in nonhuman primates appeared in the first decade of this century. They were largely modeled on single-cell recording studies of learning in the somatosensory cortex and auditory cortex. So far, there have been studies using judgments of *low-level* visual features and physiological measures in V1, V2, and V4 of monkeys and areas 17 and 18 of cats. The largest training effects occur on the tuning or contrast responses of neurons in V4 or higher, with some relatively small changes in V1. In some studies, quantitative models were used to link observed changes in neuronal responses to the task behavior. In these low-level visual feature tasks, changes in neural responses in early visual areas were generally far too small to account for the large behavioral improvements from training—in some cases too small by an order of magnitude.

This literature only begins to investigate the territory. Most of the studies examined learning in fine orientation-discrimination tasks, with only one using coarse discrimination. In many cases, changes in neural response were measured during fixation or control tasks, with a few measuring neural responses during active performance of the trained task. The former seeks to measure persistent cortical changes from training, while the latter includes effects of training that are mediated by top-down programming during task performance. It is unknown whether plasticity is the same in fine and coarse discrimination tasks or how these results would generalize to other visual features. Furthermore, other neural properties, such as the correlation between responses of populations of neurons, measured with multielectrode array technology, which may be important, have yet to be widely assessed.[16] The differences between persistent and top-down transient changes in responses also require further systematic investigation. In what follows, we consider key examples of these single-cell recording studies in the literature.

In a seminal study, Schoups et al. reported that training produced a slight retuning of neurons in V1 of macaque monkeys, although the distribution of neural responses remained essentially unchanged.[70] The task was fine orientation discrimination between grating patches rotated away from 45° either clockwise or counterclockwise, and was trained for thousands of trials. Specificity of such learning to retinal location in humans suggested V1 as the relevant representation, because of its small receptive fields.[71, 72] Extensive training reduced the monkey's behavioral thresholds by 80%–90%, from more than 10° to about 1° in difference thresholds, and these improvements were largely specific to the trained location and orientation. To find evidence of a *persistent* change in the early visual cortex resulting from perceptual learning, V1 responses after training were measured while the monkey was performing a central fixation task or passively viewing oriented patterns (i.e., while *not* performing the trained task). The distributions of preferred orientations in neurons were essentially the same in the trained and untrained V1 locations. There was no overrepresentation of tuned neurons for the trained orientation, nor were the responses stronger in the trained field. There were subtle changes in the *slope* of the tuning functions in a subset of neurons with selectivity near the trained orientation and location (figure 5.9), with the authors stating that "the slope of the orientation tuning curve that was measured at the trained orientation increased only for the subgroup of trained neurons most likely to code the orientation identified by the monkey"[70] (p. 550). Changing the slope of the tuning functions of neurons with preferred orientations just slightly off from the trained orientation was seen as one way to improve performance by changing the slope of the tuning functions of neurons sensitive to the small differences in the task stimuli.

Schoups et al. used a Bayesian model (an "ideal observer analysis of the population response") to relate the neuronal responses to the behavioral responses.[70] The neuronal discrimination was estimated from the population response from a randomly selected set of 20 trained or untrained cells sensitive to the trained orientation. The curve for the trained cells was shifted toward smaller orientation differences, with a difference in slope between trained and untrained neurons of about 12%, corresponding to a 7% improvement in angular resolution,[73] while leaving the overall variability of the neuronal responses unchanged. According to the model, changes in the neural population responses accounted for a small part (about one-tenth) of the behavioral improvement, leading the authors to conclude that learning must occur at many levels in addition to V1, or in the correlations between V1 neurons.[70, 73]

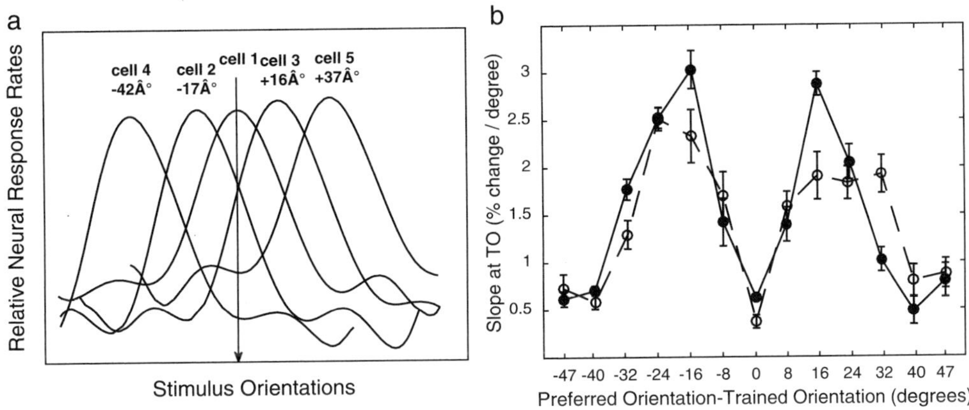

Figure 5.9
The perceptual learning in rhesus monkeys induced small changes in the slopes of receptive fields of V1 neurons. (a) Examples of orientation tuning of neurons with preferences for the trained orientation (cell 1) and for adjacent orientations. (b) Slopes for cells tuned near the trained orientation for trained (dark circles) and untrained cells (light circles). These small changes are not sufficient by themselves to account for the substantial behavioral effects of learning. Adapted from Schoups et al., [70] parts of figures 2b and c, with permission.

A subsequent study examined the consequences of perceptual learning in fine orientation discrimination in V1 and V2 and found essentially no changes in the neural responses.[74] Monkeys trained for thousands of trials in a delayed match-to-sample task requiring discrimination of small orientation differences near 45°, with different spatial frequencies in the sample and the test. Training improved the orientation thresholds from about 30° to less than 5°, and these improvements were specific to orientation but not to location. Neural responses were assessed for *persistent* changes during passive peripheral viewing of irrelevant stimuli while the monkeys performed an easy match-to-sample orientation task (i.e., horizontal versus vertical) at the fovea. Response amplitude, orientation tuning, and response variability were essentially identical in trained and untrained neurons, although there was a slight underrepresentation of neurons tuned near the trained orientations in the trained locations. A population model based on the signal and noise properties of the neural responses accounted for perhaps one-tenth of the behavioral improvements. These researchers concluded that persistent changes in V1 or V2 were not the basis of perceptual learning, which might have occurred in other brain areas.[74] Overall, then, both these studies report that any changes in the persistent properties of V1 or V2, if they did occur, were too small to account for the substantial behavioral effects of perceptual learning.

The next step for experimenters was to examine perceptual learning a bit further along in the visual pathway in V4, where slightly larger effects were found.[75] In one study, monkeys discriminated orientations near 45° in the delayed match-to-sample task.[74] Orientation thresholds improved from about 30° to 2°–5°, and learning was substantially location-specific. V4 neural responses during an easy orientation task at the fovea were used to assess the persistent impacts of training: neurons tuned near the trained orientation and location had 14% stronger responses and 13% narrower orientation tuning than untrained control neurons, leading the authors to conclude that perceptual learning induced persistent plasticity in intermediate levels of the visual cortex. However, the authors also noted that the changes were modest and could have reflected modifications either within the V4 circuitry or in connections between earlier visual areas up to V4, similar to reweighting proposals.[75] A 24% increase in *d′* (33% reduction in threshold) was estimated from the population signal detection model, so these modest persistent changes in V4 were still unable to account for more than a fraction of the very large changes in behavioral thresholds.[74]

Monkeys in another study were trained in the Schoups task but in 10% pixel noise.[70, 73] Behavioral thresholds improved to about 2° from 10° in one monkey and from 30° in another. V4 neurons with orientation tuning offset (25°–65°) from the trained orientation in the trained location showed slightly reduced variance and modest narrowing of orientation tuning. The improvement in orientation discrimination from the neural responses was estimated at 28%, compared to 7% in the V1 data.[70, 73] These V4 changes were still an order of magnitude too small for the corresponding behavioral improvements.

All these studies share several characteristics that may prove important. First, almost all of them trained fine orientation-discrimination thresholds (Type I tasks), though, as previously noted, the choice of training protocol could have significant consequences for plasticity.[76, 77] Second, recording in V1, V2, and V4 occurred during passive viewing of orientation stimuli, not during the task itself, so the conclusions concerned *persistent* changes in neural response after learning, though there may also have been top-down effects during task performance.

Several other experiments measured neural responses differently, using coarse discrimination training and/or task-engaged neuronal activity. One innovative study trained coarse orientation discrimination and recorded neural responses during active task performance together with passive controls (figure 5.10).[78]

Monkeys discriminated different oblique Gabors in masking noises. The effects of training were largest primarily in intermediate noise masking levels, while the information in neural firing rates was measured as the area under the receiver operating characteristic (AUROC). Training increased the range of spike rates for the best compared to the worst stimuli and reduced the ratio of the variance in the spike rate to its mean (the Fano factor) at intermediate noise-masking levels—increases in the signal and decreases in the noise. The estimated "neurometric" thresholds late in training were shown to be about half the behavioral thresholds (24% and 12% SNR, respectively).

In this same study, Bayes classifiers computed on populations of neurons linked the responses of V4 neurons during active task performance to the behavioral choice. Different classifiers were trained for early and late in training. The thresholds computed from the classifier and behavioral thresholds both showed about 50% improvements. It should be noted that these improvements in classifier performance reflected *both* changes in V4

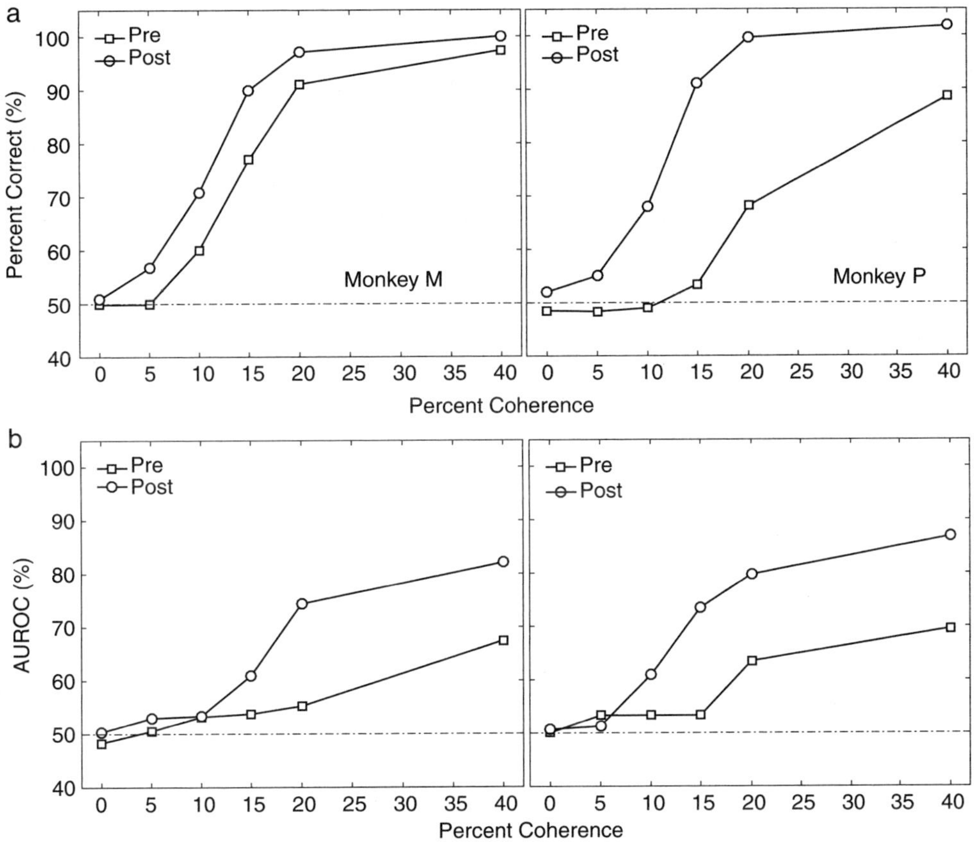

Figure 5.10
Perceptual learning in coarse orientation discrimination with masking noise in two rhesus monkeys increases (a) behavioral percentage correct psychometric functions of percentage coherence and (b) corresponding changes in the area under the receiver operating characteristic (AUROC) measures of discriminability based on neural responses in V4. These behavioral and neural functions both increase with coherence, but the neural AUROC accounts for only part of behavioral performance. Psychometric functions estimated from data in Adab and Vogels, [78] figure 1.

neural responses and changes in the readout (*reweighting*) of the responses. With this in mind, the parallel between the performance of the Bayes classifier and behavior may have included significant contributions from improved reweighting. Significantly, the tuning functions showed no changes under *passive viewing*, although perhaps there were small changes in the AUROC and the Fano factor.[78]

This important study of perceptual learning was the first to measure both *persistent* changes in neural response that could impact other tasks alongside *transient task-induced* changes in the early visual cortical responses that occur while actively carrying out the trained task. These transient changes in the absence of persistent changes can be thought of as a form of "multiplexing" (reweighting of connections within the visual cortex, depending on the task) or perhaps an improved ability to use top-down attention signals to alter responses.[79–86] The dataset thus supports our general hypothesis regarding the prevalence of reweighting.

It should also be noted in this survey of the physiological literature that the substrates of learned plasticity may depend on the animal used in the experiment. For example, when cats were rewarded for nose-press responses for coarse discriminations, changes in V1 under anesthesia gave a better account of the behavior.[87] In that study, monocular behavioral contrast sensitivity curves (CSFs) were measured by assessing contrast thresholds for the discrimination of sine waves oriented at either 45° or 135° over a range of spatial frequencies, while monocular perceptual training was carried out at a high spatial frequency and then the CSF was measured again. Contrast-response functions for a preferred orientation were measured for individual neurons in area 17 (early visual cortex) under anesthesia and then used to construct population CSFs for the trained or untrained eye. Training was shown to improve behavioral contrast sensitivity, with some specificity to spatial frequency and the eye of training. Training also improved the contrast sensitivity of V1 neurons tuned to the trained spatial frequency by increasing the neuronal contrast gain. Having said this, the magnitude of the change in the behavioral CSF before and after training was essentially the same as the magnitude of the change in the neuronal CSF measured under anesthesia.

To summarize, for perceptual learning of low-level visual features, the *persistent* visual plasticity in early visual areas V1 and V2 appears far too small to account for the large behavioral improvements. This pattern becomes especially clear when all the relevant studies are considered together (see table 5.1 for a summary).

Moderate persistent visual plasticity following perceptual learning in visual area V4 has been reported; however, the computed effect of these neural changes is generally far smaller than the changes in behavioral thresholds. Changes in V4 neural responses measured during active task performance in coarse discrimination tasks close the gap, but we do not know whether this would generalize to fine discrimination. Even there, however, the population response functions estimated for V4 neurons showed changes that only accounted for a small proportion of the changes in the perceptual judgment behavior. However, the neural changes came closer to predicting the behavior when *different* optimal population classifiers were used to predict the neural responses before and after learning. That is, the optimal classifiers for making predictions about responses reflect both any changes in neural response and optimized changes in readout of those responses (reweighting). The changed weighting is likely one critical factor in accounting for perceptual learning. One interpretation of these results is that V1 and V2 are not especially plastic, while later stages of visual representation are (except perhaps in cats). Another interpretation is that all levels of the visual cortex are plastic, but since essentially everything in the visual world

Table 5.1
Perceptual learning effects on single-cell physiology in feature tasks

Source	Training task	Neural task	Neural responses	Model analysis	Accounts for behavior?
Schoups et al.[70]	Fine orientation discrimination (parafoveal)	V1, passive, fixation task	No overall changes, select small slope changes	Bayesian population model	Accounts for less than one-tenth of behavioral improvement
Ghose et al.[74]	Fine orientation discrimination, delayed match to sample (parafoveal)	V1 and V2, passive, easy fixation task	No significant changes in trained vs. untrained	Ideal observer model of neural responses	Accounts for small fraction of behavioral improvement
Yang and Maunsel[75]	Fine orientation discrimination, delayed match to sample	V4, passive, easy fixation task	Modest 13% to 14% tuning change in relevant neurons	Ideal observer model of neural responses	Accounts for less than one-third of behavioral improvement
Raiguel et al.[73]	Fine orientation discrimination in pixel noise	V4, passive, fixation task	Small changes in tuning for some relevant neurons	Bayesian population models with different pre- and post-inputs	Accounts for less than one-third of behavioral improvement
Adab and Vogels[78]	Coarse orientation discrimination	V4, active task, easy passive controls	Moderate changes in intermediate S:N ratios under active viewing, inconsistent effect in passive viewing	Neural area under the ROC	Accounts for approximately half of behavioral improvement under active task
Hua et al.[87] (cats)	Coarse orientation discrimination, CSF	Area 17, under anesthesia		Population neuronal contrast-sensitivity functions	Substantially accounts for behavioral improvements

activates V1, experiences outside the experimental training context mask any changes that might have occurred during an experimental session.[75]

These conclusions are clear from the existing data, yet there are several factors that might have influenced the pattern of results. First, fine discrimination tasks and coarse discrimination tasks could, in principle, rely on somewhat different learning mechanisms, or at least different learning regimes. Coarse discrimination tasks are limited by low contrast and internal noise (but not by similarity), while fine discriminations are limited by the bandwidth and correlation between templates (as discussed in chapter 4, they would also be influenced by contrast and noise, except that they are typically carried out at high contrast and zero external noise). Observers should weight evidence differently in the coarse and fine tasks. In coarse discrimination tasks, optimal use of evidence and therefore optimal weight structures can be very broad and inclusive. In fine discriminations, the differences in stimulus evidence are by definition smaller, such that evidence in just a few units may distinguish between close alternatives (e.g., orientation differences). Additionally, it is plausible that the correlation between firing rates in neurons is likely to be more important in fine discrimination tasks than in coarse ones. But since so few studies of perceptual learning of low-level visual tasks involved coarse discrimination, and most used orientation tasks, further research is needed in order to substantiate existing theories.

A second factor that may have influenced the data involves a distinction between passive viewing and active task performance. The differences between the observed changes in the neural responses during the two conditions were profound. Changes in the early visual cortex measured under passive viewing persist outside the context of task performance and thus would affect performance in other tasks. By contrast, changes in neural responses in the early visual cortex measured under active viewing are likely associated with changes anywhere in the complex brain networks engaged in the task, including decision, reward, and attention. These changes are presumably transient and task induced—permitting multiplexing of tasks over the same early cortical representations—and so may be more compatible with system stability.

Third, it seems that there may be important species differences between the majority of the measurements in monkeys and the measurements in cats, with the monkey's visual system being more similar to that of humans. Indeed, the changes in area 18 of the cat may be more analogous to changes in V4 in monkeys and humans.

Finally, essentially all these single-cell recording studies of perceptual learning for features have relied on measurements of the (orientation) tuning functions of neurons. Other aspects of the responses of populations of neurons, such as synchronization and correlation—or indeed other aspects of neural coding—may prove to be just as important in explaining plasticity and final behavioral performance. Future research is needed to explicate their contributions.

5.4.2 Perceptual Learning of Patterns

In this section, we turn to single-unit recording and perceptual learning of mid-level visual tasks such as visual motion, depth, or long-range interactions over space, coded in mid-level visual areas. Like the single-unit investigations of low-level visual features, studies of perceptual learning and plasticity in the mid-level brain have just begun. The same issues are relevant (e.g., the difference between persistent and task-induced plasticity, training of coarse or fine discriminations). All existing studies suggest that perceptual learning of patterns in mid-level visual tasks, as with single features in early vision, may largely reflect reweighting or changed readout with greater reliance on higher visual areas. A study in pattern detection, however, suggested that the lateral connections within V1 were reweighted during active task performance only after training, while leaving the classical receptive fields of neurons unchanged.

One very influential study found that perceptual learning altered responses to random dot-motion stimuli further along in the motion processing pathway in the LIP (lateral intraparietal area), associated with integration of motion that leads to classification and response selection, but not in the earlier MT (middle temporal area), which performs early coding of local motion.[88] Neurons were measured while monkeys actively performed coarse visual random dot motion discrimination (left versus right); meanwhile, behavioral performance was measured for random-dot-motion coherence levels from 0 to 99.9% and for display durations up to 1 s. Coherence thresholds improved from about 80% to about 20% for one monkey and from near 99% (basically immeasurable) to about 30% for another. Error rates at 99% coherence also improved. The sensory responses in MT remained largely unchanged, while the responses of LIP neurons were sensitive to training (figure 5.11). The neural responses and the behavior were modeled using regression and signal detection models.[88] Behavior correlated with LIP activity (*r* near 0.6) but not with MT activity (*r* near 0), though it should be noted that an optimal classifier model would have been an alternative model to fit the data.[78] The results of this study have been widely cited as demonstrating perceptual learning through changed readout to decision, [89, 90] though it is also possible that attention had some influence.[91]

Another interesting study found a complex effect of training fine depth discrimination on performance during transient lesions of the MT.[77] Temporary MT deactivation (through muscimol injection) devastated performance of a coarse depth task (at +45 min, +1 day after injection) in an untrained monkey, while both coarse and fine depth discriminations were unaffected in a monkey previously trained in fine depth discrimination; deactivation of the MT damaged both coarse and fine motion direction judgments. Disparity tuning of MT neurons during a fixation task before and after fine-disparity training was essentially

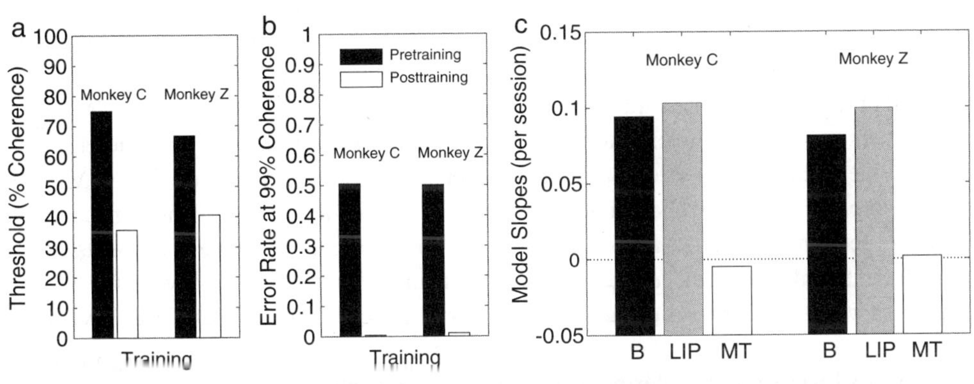

Figure 5.11
Training coarse motion discrimination in monkeys is related to neural response changes in the LIP but not the MT. Behavioral motion coherence thresholds (a) and percentage error at 99% coherence (b) improve as a function of session. (c) Slopes reflecting changes over a session from a model are positive for LIP and behavior but not MT activity. Redrawn from selected data in Law and Gold,[88] figures 2 and 4.

unchanged, so changed MT neural responses did not mediate the training effect. Apparently, experience with fine judgments did not change the signals in MT but rather seemed to down-weight the importance of MT signals in favor of other signals, likely from ventral areas.[92, 93] These results, then, are consistent with reweighting explanations of perceptual learning in motion,[88] including the original reweighting claims.[89, 94] In this case, the reweighting is among different brain areas.

Several similar studies investigated other kinds of potential effects, such as the change in internal noise correlation between neurons. For example, learning of motion flow heading judgments in at least one case affected the *correlations* between neurons in the dorsal medial superior temporal area (MSTd), a brain area involved in the perception of heading from optic flow and vestibular signals.[95–97] Reduced noise correlations between neurons coding the same stimuli can improve classification performance by making them more statistically independent. Monkeys trained to report the motion heading (left or right) reduced their behavioral thresholds from more than 10° to about 1°–3° after extensive training. Training reduced correlations during a fixation task as compared to correlations in untrained animals—although tuning curves, response variability, and individual neuron discrimination functions were unchanged. Using a population-coding model and only the most relevant neurons, observed changes in correlation at best accounted for an 8% improvement in threshold compared to behavioral improvements of 80% or more. Given this small contribution to the behavioral improvement, changed readout to decision, or reweighting, seemed to be the dominant explanation. A related computational study,[98] reanalyzing previous data,[88] found no changes in correlation between MT responses after learning, concluding that learning changed readout weights from MT to LIP. It should be noted that altered correlations between neurons[95] may be very small relative to other factors limiting fine discrimination, such as the similarity of the templates for the discrimination, the amount of internal noise, and the correlation of the noise (see the discussion of elaborated PTM in chapter 4).

In addition to motion and depth tasks, perceptual learning has also been examined in *contour-detection* tasks.[99, 100] In these tasks, observers detect collinear lines or closed contours in a field of randomly oriented and positioned lines or Gabor elements. Sensitivity of V1 neurons to contours, initially interpreted as an extended association field beyond the classical receptive field, is now understood as a contour-sensitive response occurring first in V4 and then in a broader brain network that is subsequently fed back into V1.[101–103] Training generally improved performance, especially for contours with an intermediate number of elements,[103–106] while learning influenced the *late* responses of V1 neurons (e.g., those sensitive to the central element in a contour display) during active task performance. These late responses have been attributed to top-down effects, however.[104, 105] They may also reflect attention, since negligible changes in V1 were observed during fixation or attention control tasks.[105]

Gilbert et al. refer to the task-specific changes in V1 as multiplexing; we have called this task-selective reweighting. A few examples help to give a sense of this phenomenon. In one study, in V1, neural responses to contour patterns were changed while monkeys performed the contour-detection task.[105] Before training, V1 responses were unaffected by the presence of a contour compared to a random pattern in all tasks, including fixation and attention control tasks. Training in contour detection led to improvements of about 15% accuracy for intermediate or longer contours. Following training, late V1 responses also changed during active contour detection for longer contours. Neural accuracy functions based on late spike rates of individual neurons roughly paralleled, but were significantly lower than, behavioral accuracy. Overall, these data suggested that changes in late V1 responses occurred only while performing the contour task. These neural changes have also been interpreted by some as task-specific attention effects.[107]

Another study used chronically implanted multiple electrode arrays in V1 and found that perceptual training increased V1 responses during a contour-detection task for fixed location and orientation of the contours.[108] Late responses of V1 neurons with receptive fields along the contour increased as training progressed and behavioral accuracy improved, while those sensitive to pattern elements away from the contour decreased. (Here, "late" means responses occurring later in the response interval to each stimulus.) Artificial intelligence classifiers trained to discriminate contours from the late V1 responses of individual neurons and their interactions at different points during training showed that improvements in behavioral accuracy lagged behind improvements in the neural classifier, especially earlier in training. Thus, while the late V1 response during active task performance increased with training, performance improvements may have been primarily impacted by improved readout of this evidence for the behavioral response.

The studies of perceptual learning in all these mid-level tasks provided one of the largest sets of physiological evidence from single-cell recording of the visual cortex and perceptual learning. These studies generally were consistent with some form of selective reweighting as a dominant mechanism in perceptual learning. The idea here is that responses in higher cortical areas may change, while the earliest levels of the visual cortex remain largely unaltered by training, even during active task performance.

The conclusion of learning through reweighting is also consistent with the predominant effects of learning in the LIP rather than the MT in coarse motion discrimination and with the temporally delayed effects in V1 in contour detection that reflect top-down influence from V4 or higher regions. Another reason to support the reweighting hypothesis is that many or all the changes in neural response have been seen to occur during active task performance, while the persistent properties of neurons measured under fixation or attention control tasks are weak to nonexistent. Because the bulk of the studies use either detection or coarse discrimination, with only a single study involving fine discrimination, the generality of these conclusions needs to be assessed, though reweighting still presents itself as the most compelling explanation of plasticity in these mid-level tasks. See table 5.2 for a summary.

The concept of reweighting can cover a multitude of potential influences. Although we tend to think of reweighting as combining evidence from lower-level cortical representations to influence or create higher-level ones, reweighting may also change weights that

Table 5.2
Perceptual learning effects on single-cell physiology in pattern tasks

Source	Training task	Neural task	Neural responses	Model analysis	Accounts for behavioral improvements?
Law and Gold[88]	Coarse motion discrimination	MT, LIP, active task	No effects in MT, significant changes in LIP	Regression analysis	LIP accounts for 60% of variance
Chowdhury and DeAngelis[77]	Fine depth discrimination	Behavior with and without transient MT lesion	Training changes sensitivity to lesion	None	Not estimated
Gu et al.[95]	Coarse motion-heading discrimination	MSTd, active task	Tuning unchanged, reduced interneuronal correlations	Population coding model	Accounts for about one-tenth of behavioral improvement
Li et al.[104]	Long-range contour detection	V1, active task, some fixation controls	Late response changes from V4 or higher	None	Not estimated
Li et al.[105]	Long-range contour detection in the periphery	V1, active task, some fixation controls	Late response changes originating from V4 or higher	Average neural classification accuracy	Neural performance worse but effects similar to behavioral accuracy; not estimated
Yan et al.[108]	Long-range contour detection in the periphery	V1, active task performance, chronic multiarray	Late response changes from V4 or higher	Classifier analysis, with changing classifiers	Improved neural classifications reflect readout changes as well as late V1 response changes; not estimated

control the interactions of neurons within a single cortical area. Likewise, changes in earlier cortical areas can also be influenced at a delay later in the response cycle by reweighted feedback connections from higher cortical areas, or reweighting may change the feed-forward connections between lower areas. Reweighting can occur both within and between brain areas. It can thus reflect feed-forward, feedback, or intra-area connectivity. Our hypothesis is that unique task-dependent weighting of neural activities in multiple brain areas, possibly through the top-down programming from attention, decision, and reward, may be one way of balancing plasticity and stability in the system. Such top-down task-dependent effects can leave early visual cortical responses relatively stable—allowing the system to be plastic to the demands of the individual task while still preserving the stability of performance in other tasks and contexts.

5.4.3 Perceptual Learning of Objects and Scenes

Perceptual learning occurs not only in low- or mid-level tasks but also in higher-level visual tasks such as object or face identification, and this, too, has been selectively studied using single-cell recording. Perceiving objects and faces—especially from different viewpoints—is an important visual function in normal daily life.[109] Motivated by observations that damage to the IT cortex can disrupt object or face processing, researchers have primarily focused on measuring responses in IT and adjacent regions, though a range of studies exist and complex objects are of course processed for identification throughout the ventral pathway, from V1, V2, and V4 all the way to the prefrontal cortex.[109]

Most of these higher-level studies—primarily in the IT cortex but also in V4 and the PFC—have focused on finding some neurons that are responsive to specific trained objects. Our synoptic interpretation is that perceptual learning served primarily to recruit neurons to represent objects, which through reweighting become sensitive to the most diagnostic stimulus features coded in early visual areas. With extensive training, the two-dimensional features coded in early brain areas are connected to IT neurons that represent three-dimensional objects, which in turn are connected to neurons in the PFC that are used in memory and decision. In the language of chapter 2, this is less a process of winnowing from among preexisting representations than training neurons to represent the unique combination of features (among the millions of possible combinations) that best represents the object.

One such study examined the responses of V4 neurons as monkeys were trained to identify objects embedded in different amounts of external noise (figure 5.12).[110] In this delayed match-to-sample task, a test stimulus of 100% coherence appeared 1 s after a sample stimulus, which was between 0% coherence (phase-randomized noise) and 100% coherence (relatively clear). Monkeys trained for 20 sessions on four repeated objects (the familiar set) and four new objects that were different in every session (the novel set). Following practice, the accuracy for the familiar set exceeded that for the novel set, especially for intermediate levels of coherence. Measures of mutual information were computed between the neural response patterns to each of the four familiar stimuli and to each of the four novel stimuli (measured while the monkey was performing the task). Responses to 100% coherence samples showed the same response rates, variability, and mutual information at all levels of training for both familiar and novel stimuli. When the sample stimuli were at intermediate levels of stimulus noise, mutual information was better for familiar stimuli than for novel stimuli. The conclusion was that "basic response properties of V4 neurons … appear not to be altered by learning, similar to findings in V1 that … receptive field size or orientation tuning remain unchanged even after extensive training …"[110] (p. 280). Instead, V4 neurons may be "specifically recruited for difficult discriminations … [with] indeterminate visual inputs"[110] (p. 281) (e.g., for those stimuli with external noise).

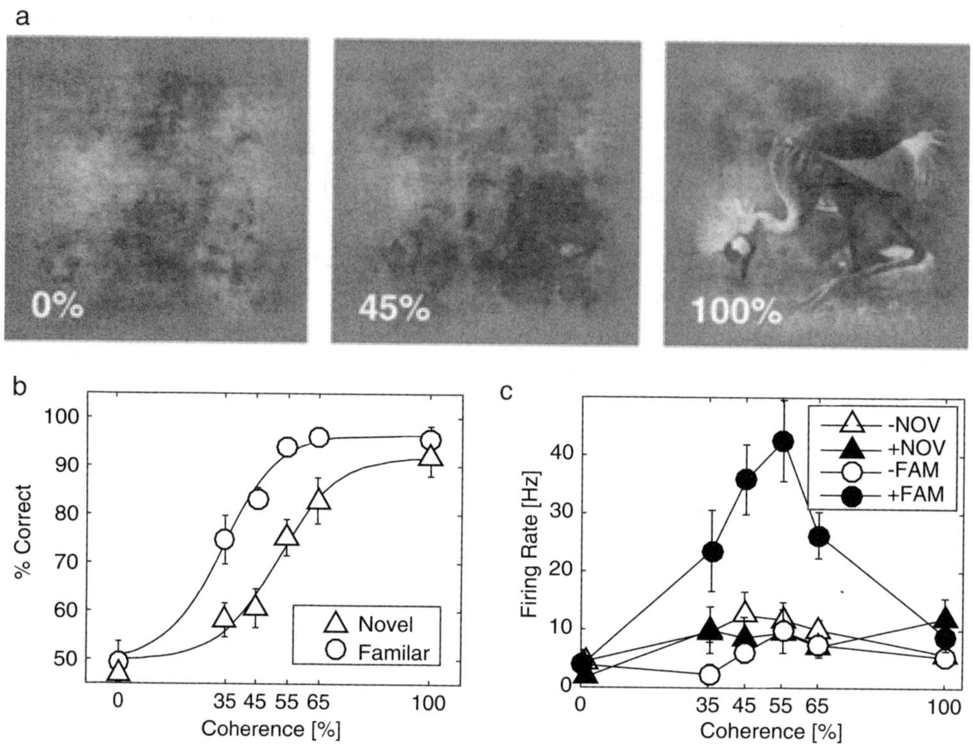

Figure 5.12
Perceptual training in a delayed match-to-sample task of objects in noise of various coherence (a), behavioral accuracy (b), and corresponding changes of V4 responses to noisy stimuli (c). Firing rates in V4 neurons increase for familiar trained stimuli in intermediate noise levels. After Rainer, Li, and Logothetis,[110] parts of figures 1, 2, and 4. Creative Commons, copyright 2004, Rainer, Li, and Logothetis. (See plate 4.)

Responses in lateral PFC neurons after training yielded an opposite but related pattern of results.[111] The lateral PFC receives inputs from the IT cortex, which in turn receives inputs from other visual areas. The PFC is involved in discriminating, remembering, or making decisions about visual stimuli.[112, 113] Neurons in the PFC responded more to novel stimuli than to familiar ones, especially when degraded by the presence of external noise, and showed a direct relationship between the neural changes during active task performance and behavioral improvements. This was interpreted as follows: "As a stimulus becomes more familiar, neurons coding features not essential for recognizing it reduce their responses, leaving . . . a smaller number of more selective neurons that optimally represent the familiar stimulus"[111] (p. 181).

One early study of the IT cortex showed differential responses of neurons to familiar objects.[114] Monkeys were trained to identify a small number of wire-frame or irregular spheroidal objects by showing them a rotating three-dimensional view for several seconds. The test then required the discrimination of rotated target views from random views of distractor objects. In a fixation task following object training, a small number of IT neurons were found that each responded selectively to different views of the objects. For most, rotating the image away from the trained orientation reduced responding; however, a few seemed to respond to two views, and a very few seemed to have developed view-invariant responses to the trained object. For the five extensively trained objects that achieved view-independent behavioral performance, some of these also showed invariance to the size and location. In a related study of trained monkeys, more IT neurons (region TE) responded under anesthesia to the trained two-dimensional shape stimuli relative to untrained controls.[115] In yet another study, familiarity affected selectivity in the neural responses (the difference in response between the most and least preferred objects for trained objects in upright orientation), measured in a fixation task.[116]

There was also some evidence that anterior IT neurons become sensitive to the visual features most relevant to the trained classifications.[117] Monkeys were taught to classify schematic faces or fishes into the corresponding categories; two of four stimulus dimensions varied systematically with category, while the two others varied randomly. After training, IT neurons showed differential sensitivity to the features important for categorization. This occurred later in the response interval, likely reflecting top-down effects of recognition. In a separate but related study, sensitivity to diagnostic compounds of perceptual features increased over the course of training during active task performance.[118]

Perceptual learning of objects and faces, then, involves a network of visual processes with inputs from the early visual cortices (V1 through V4) feeding into object representations in IT cortex and then in the PFC. The cellular recording studies generally suggest that selected individual neurons in the IT cortex or the PFC develop responses to trained objects, or at least some view of those objects. One way in which this literature differs from that on tasks involving low- or intermediate-level visual tasks is that these trained neurons were sometimes found under passive task conditions. None of these studies attempts to estimate whether the responses of these few neurons could account for the changes in behavioral accuracy in performance. See table 5.3 for a summary of results.

We believe that for these high-level tasks, the primary function of perceptual learning is to recruit neurons to represent specific objects, which through reweighting become sensitive to the most diagnostic stimulus features of the object coded in early visual areas. Each complex object is defined by a unique combination of possible features. The idea that neural representations will exist for that unique combination, from among millions of possible combinations, is quite implausible. Therefore, we have proposed that such complex representations must be learned, or created, by recruiting and training neurons or sets of neurons to represent each new object. Extensive training may connect two-dimensional image features in early brain areas to neurons in IT, which are then connected to neurons in PFC used in memory and decision. Similarly, many two-dimensional representations may converge to a single higher-level three-dimensional object representation. Learning high-level tasks logically requires the creation of new representations for new unique feature combinations.

Table 5.3
Perceptual learning effects on single-cell physiology in objects and natural scene tasks

Source	Training task	Neural task	Neural responses	Model analysis	Accounts for behavioral improvements?
Rainer, Lee, and Logothetis[110]	Object recognition in noise, delayed match to sample	V4, active task	Slightly stronger responses in intermediate noise	Mutual information analysis	Not estimated
Rainer and Miller[111]	Object recognition in noise, delayed match to sample	Lateral PFC, active task	Reduced responses for familiar stimuli in noise	Mutual information analysis	Not estimated
Logothetis, Pauls, and Poggio[114]	Object identification under rotation	IT cortex, passive, fixation control	Responses to familiar objects in select neurons	Neurometric discrimination	Not estimated
Kobatake, Wang, and Tanaka[115]	Object recognition	IT cortex, under anesthesia	More neurons tuned to trained objects	None	Not estimated
Freedman et al. [116]	Category judgments in successive same vs. different	IT cortex, passive, fixation control	Increased selectivity for upright trained objects	Selectivity analysis (max to min response)	Not estimated
Sigala and Logothetis[117]	Two category judgments	IT cortex, active task	Late neural responses to relevant features	Feature analysis	Not estimated
Baker, Behrmann, and Olson.[118]	Objects defined by feature combinations	IT cortex, passive, fixation control	Responses to trained feature combinations in select neurons	Feature analysis	Not estimated

5.4.4 Summary of Perceptual Learning in Single-Cell Experiments

Cellular recording studies of visual perceptual learning have looked for changes in neural responses in many areas of the visual cortex. Researchers have investigated learning in tasks with single features, intermediate patterns, and higher-level stimuli such as objects, faces, or scenes. Neural recordings carried out with passive viewing and/or fixation tasks assessed the persistent or more permanent changes caused by learning, while recordings taken as the animal performs the training task reveal possibly transient task-induced or top-down changes in neural response induced by perceptual learning, as well as possible persistent changes.

Overall, the cellular recording studies indicate a remarkable level of stability in neural response in the earliest levels of the visual cortex during learning. Generally speaking, the changes in neural responses in areas below V4 occur only under active task conditions, and often with delayed latency, suggesting top-down influences. Only the experience-dependent changes in cortical response at higher levels of the visual cortex during active task performance come close to accounting for the substantial improvements in behavioral performance with training. By contrast, the minor changes in the performance of discriminant modeling that derive from changes in V1 or V2 are an order of magnitude too small compared to behavioral improvements. There is, however, some evidence to suggest that with extensive practice persistent changes do occur in representations along the visual hierarchy. These are in the development of representations of newly experienced objects and corresponding neurons that develop sensitivity to them in the IT cortex or the PFC. These studies tend to isolate a few neurons that express selectivity to a particular object, without considering whether the neural responses can account for behavioral improvements.

Virtually all these cellular recording investigations of perceptual learning measure neural responses in a single cortical area, yet the real story of plasticity likely occurs within the context of a *network* of relevant brain regions acting in a coordinated way. Simultaneous measurement in multiple brain regions is challenging given the current technology, but such measurements could lead to important discoveries regarding both bottom-up and top-down processing during active and passive viewing. One practical consequence of this state of affairs is that both top-down transient and persistent changes in neural responses of the visual cortex should be measured if possible. Although persistent changes measured under passive viewing or fixation tasks are presumably also present during active task performance, top-down changes may in fact overwhelm these small changes. The precision of the required judgment (i.e., fine versus coarse judgments) is another underused experimental manipulation that may also help to better determine the precise expression of visual learning.

Transient changes in neural responses in early visual cortices that occur only during active performance of the trained task, or very similar tasks, embody one means by which experience-driven plasticity can occur while leaving early cortical responses relatively stable and calibrated for use in many tasks. Yet visual learning that is retained, in some cases for years, must reflect persistent plasticity. The fact that (at least in the cases investigated so far) persistent changes in the early visual cortex have tended to be small—too small to account for the bulk of the learning—suggests that important sites of plasticity occur elsewhere in the brain network. The fact that learning is better expressed in the early visual cortex during top-down directed activity also reinforces this conclusion.

One possibility is that the task frame originates in the prefrontal cortex, which in turn coordinates a task-specific decision unit (or units) in decision areas, which in turn are activated by the relevant sensory representations through a learned weight structure. In reality, however, storage of these persistent traces of the learned visual task would likely involve strengthening the connections between different regions. Alternatively, they could be stored in the synapses or could be epigenetically modulated, though measuring both these processes would require different kinds of technologies.

5.5 Evidence from Brain Imaging

Brain imaging technologies, including PET, fMRI, and EEG, provide alternative measures of brain activity that may reveal the broader substrates of plasticity in visual perceptual learning in humans. Although there are relatively few such studies, we continue to cluster tasks involving features, patterns, and objects historically associated with low-, mid-, and high-level vision. At this point, practically every study seems to point to something different from the others. In principle, however, evidence from brain imaging could provide critical information about the simultaneous changes over multiple brain regions and networks complementary to the localized information provided by cellular recording studies.

Some initial comments are useful here. The existing imaging studies often compare task performance before and after training. Some studies use fixed stimuli, so the performance level improves over the course of training. In others, the researchers perform before-and-after assessments with very easy stimuli that lead to very high and therefore unchanging performance or use passive viewing of trained stimuli during the imaging sessions. These researchers have been motivated to assess brain responses with different, often easier, stimuli (compared to the training task) as a control for task difficulty, which can itself affect brain activity. Using passive viewing or control tasks parallels the use of fixation tasks to assess persistent changes in single-cell recording. These choices are made knowing that many aspects of the task, including changes in expectation, variations in the stimulus, changes in attention, task difficulty, and task set may all influence the activation measured in different brain regions.

The main dependent measure in these studies has been the amplitude of the activation in particular brain regions. Another technique is the multivariate pattern analysis (MVPA). MVPA uses machine-learning methods such as the support vector machine to decode patterns of activity in multiple fMRI voxels to predict task classifications, such as presence or absence of a target in a detection experiment, or the identity of the target in an identification experiment. Typically, different pattern analyzers are used to classify the patterns of brain activation before and after behavioral training in the perceptual task (which intrinsically assumes reweighting of evidence to optimize readout). The prediction accuracy of such a pattern classifier is used to estimate the relevant information available in a particular region in the brain. The use of MVPA analysis is a logical parallel to population response analyses in single-unit recording. However, the quality of the signal in fMRI can limit decoder performance, often leading to marginal increases relative to the near-chance levels of MVP classification performance.

5.5.1 Perceptual Learning of Features

Learning in low-level visual feature tasks has been studied using PET, fMRI, and EEG, with a variety of results. In some of these studies, the amplitudes of brain responses decrease with training; in others, the amplitudes increase; and still others show no changes, especially in early visual cortical areas.

One fMRI decoding study found changes in decoding accuracy only in high visual areas. Another study, by contrast, found improvements in decoder performance in lower visual areas. Such variability in results likely reflects the diversity of experimental choices: fine versus coarse discrimination tasks, passive viewing of easier stimuli versus active task performance for the trained stimuli, and so on. The extent of training also has varied considerably from study to study. This variation can make experimental results challenging to interpret and understand. Having said this, however, each study has its own logic, and the localization of learning-induced modulations may nonetheless provide helpful clues to the brain network involved in perceptual learning.

One early imaging study worth noting examined brain responses during orientation discrimination using PET imaging ($_{15}$O-water labeled positron emission tomography) and found reductions in activations in early visual cortical areas.[119] Pre- and posttraining sessions measured brain activity while observers performed one of three tasks: orientation discrimination around the trained orientation (+45°±10°), around the untrained orientation (−45°±10°), and in a control task. The discrimination task (±10°) was at or near ceiling accuracy (>95% correct) for both trained and untrained orientations before and after training—selected to equate performance across conditions. Training in a threshold orientation-discrimination task (which occurred between the imaging sessions) led to a 66% reduction in the behavioral threshold (JND) for the trained orientation and a 39% reduction for the untrained orientation. Following training, brain activity (regional cerebral blood flow, rCBF) was reduced in V1, V2, and V3, more so for the trained orientation. The inference was that training reduced the number of neurons involved in the task, reflecting bottom-up selection of neurons in the representation as well as reduced requirements for top-down attention to the stimulus location.

A related fMRI study of perceptual learning in orientation discrimination found increased activity measures in V1 after training for noncardinal stimuli.[120] Orientation discrimination was examined for a cardinal direction (0°±2.4°) and an oblique direction (45°±6.7°), with these two tasks (for different precision judgments) yielding approximately equivalent near ceiling accuracy performance both before and after training. Training these judgments for the oblique stimuli outside the scanner reduced the contrast threshold by 39% and increased the relative fMRI response to the oblique stimuli in V1 but not in V2 or V3. So, in contrast to the previous PET study, this study found that training *increased* the response to the trained orientation in V1, and not at all in two other visual areas. These authors attribute the differences between the two tasks to the use of fine orientation training (JND angular difference) in the PET study but coarse discrimination training in this fMRI study.

An EEG study also found increases in signature responses associated with early visual cortical activity.[121] Observers improved their contrast thresholds for detecting a peripheral sine patch after training. EEG responses to ("easy") high-contrast oriented stimuli were measured before and after training. The C1 component of the visual evoked potential (VEP) (70 to 100 ms after stimulus onset) is typically associated with responses from V1 or from a mixture of V1, V2, and V3 sources. After training, the C1 response amplitude *increased* for the trained orientations in the trained location. Early visual cortical responses may have been modulated directly through training, although plasticity at higher cortical areas may also have contributed through top-down influence. In contrast, a related study[122] found a *reduction* in C1 amplitude as a result of training in a texture-discrimination task (see discussion of pattern tasks in subsection 5.5.2).

Another study innovatively set out to induce activation modulation in V1 by using fMRI-based neurofeedback to train an oriented pattern by providing positive feedback when activation in V1 regions that were known to code the relevant pattern increased and also showing that there were corresponding improvements in behavioral accuracy.[123]

A related cluster of studies focused on the effects of training on the successful decoding of patterns of activities in different regions of interest. One notable fMRI study found that training increased the successful decoding of activity patterns at higher cortical levels but found no changes in early visual cortical responses.[124] This study used a multivoxel pattern analysis (MVPA) on BOLD signals in early visual cortical areas and higher areas such as the lateral parietal cortex and the angular cingulate cortex (ACC), estimated on the first and fourth days of training in an orientation-discrimination task. Stimulus orientation could be decoded in some degree from activity in both early visual areas and higher areas; however, pattern-decoding performance in early visual areas was unaffected by training, while the decoded activity in the ACC was correlated with behavioral improvements. The authors argued that learning occurred only in higher-order areas and not in the early sensory areas. Another fMRI decoding study found no change in the overall level of the BOLD responses in early visual areas V1–V4,[125] unlike another study,[120] although the researchers did find an attention-dependent improvement that suggested the influence of top-down processes. Modulations of higher-level decision processes were also reported in a recent EEG study of learning to discriminate faces and cars embedded in external noise,[126] paralleling a number of single-cell studies.[77, 88, 104] All these findings were generally consistent with the reweighting theory of perceptual learning.[89, 94]

As mentioned previously, several factors may play a role in the apparent inconsistency of results concerning changes in early visual cortices. Variation in experimental factors as well as active versus passive performance context could very well have influenced outcomes. Furthermore, as will become apparent in several higher-level tasks, the effects of training on early visual cortical areas may change depending on the degree of training, with increases in activity likely associated with increased use of attention, especially early in the dynamic process of learning.

The sparseness of the brain imaging studies and the variation in the results challenge the creation of a theoretical structure to organize them. The fMRI results in particular may be consistent with top-down induction of changes in early levels of the cortex that derive from higher regions of the brain network because of the lower temporal resolution of fMRI responses.

5.5.2 Perceptual Learning of Patterns

Imaging studies of learning in pattern tasks (mid-level vision) include those for motion discrimination, texture discrimination, visual search, and discrimination of glass patterns. Here, too, there has been a diversity of outcomes. Some studies found increases in the responses in early visual areas, others found an increase followed by a decrease, and still others showed decreases. Several studies found evidence for more improvements in decoding activations in higher-level brain areas, leading to the association of training with either increases or decreases of activity in early visual areas as a consequence of perceptual learning. Interestingly, in a few cases, the results are at variance with the results of related single-cell recording studies. Furthermore, this collection of studies (unlike those involving low-level features) has typically involved the measurement of brain activations at quite different behavioral accuracy levels before and after perceptual training, which suggests that at least some of the results may reflect consequences of a shift in the task difficulty or in attention.

One of the earliest fMRI studies of perceptual learning examined the consequences of training in a motion task and interpreted the results both as an increase in the activity of motion representations and the complex effects of attention.[127] In this experiment, performance increased from near chance to nearly 100% correct after four blocks in a motion-direction discrimination task using a 20% coherent two-frame random dot-motion stimulus. The fMRI response in MT increased, while activation in the cerebellum and in other areas related to attention was reduced with practice, possibly reflecting either recoding of the sensory motion information or changes in attention or decision associated with improved accuracy of performance.

Another fMRI study also examined perceptual learning in a motion task, and the researchers found relatively significant changes in the decoding from a mid-level visual area.[128] A single motion direction was trained in a task detecting 15% coherent motion from random motion, preceded and followed by assessment of motion performance in nine motion directions, while detection improved from about 70% to about 88% correct, with the largest improvements focused near the trained direction. Distinct multivoxel pattern classifiers were created for each of the tested motion directions before and after behavioral training, using run-normalized fMRI BOLD responses in various regions of interest for 50% coherent motion in the nine directions versus random stimuli. The size of the behavioral improvements and size of the range-normalized decoder improvements for V3A were similar—although in this case the overall accuracy of the decoders was not reported, so this comparison is difficult to assess. One possible interpretation was that higher areas leading to behavioral choice read out information from V3A, a motion-integration area.

Several studies have examined perceptual learning in texture-discrimination tasks (TDTs), with a few implicating changes in V1. In one single-session training study, performance was tested during fMRI scanning under active task conditions following monocular training.[129] Although the accuracy of behavioral performance in the scanner was strangely low (57.75% compared to >80% during training outside the scanner), a single activation cluster that differed between the eyes was identified, and this was interpreted as a change in activation in the monocular cells of V1. A related study of training in the TDT found first an increase and then a decrease in activation in V1 under active task conditions as training went on.[130] Activation of a subregion of V1 corresponding to the trained visual quadrant increased along with improvements in behavioral task performance. As training continued, however, V1 activation levels returned to their original levels. The researchers suggested a form of plasticity "in which different patterns of synaptic activity occur at different stages."[130] (Other examples of stage-dependent returns to baseline during learning have been found in other modalities; see chapter 10.) One interpretation may be that changes in V1 activation early in training reflected heightened top-down effects of attention.

The temporal ambiguity of the comparatively sluggish fMRI response has been addressed in EEG studies of training and the TDT task, which have found reductions in responses associated with early visual cortices.[122, 131] In one case, EEG responses to easy TDT stimuli with stimulus onset asynchrony of about five times threshold were measured before and after training in one visual quadrant. Here, training reduced the amplitude of the C1 responses, which was interpreted as learned increases in inhibitory responses to the mask in the early visual cortex.[122] Another recent study, which measured performance using high-density EEG before and after training in a TDT task, also found a decrease in the C1 responses, but also in N1 amplitide and latency, as well as increases in P3. The researchers concluded that perceptual plasticity occurs at several levels, from early visual responses to higher-order responses of attention or cognition.[131]

The effects of training in visual search tasks have been examined using fMRI.[132] BOLD activation was measured for blocks of trained and untrained targets (rotated Ts of different orientations) among brief 12-element displays. Target present-versus-absent judgment accuracy was 90% for trained targets and 20% for untrained ones, while BOLD activation increased in early visual areas but decreased in attention networks while performing the task. Comparing BOLD responses across radically different levels of behavioral performance, including the paradoxically below-chance performance on the untrained targets, challenges this interpretation. Still, the researchers conclude that training alters responses throughout the brain network for trained tasks compared to untrained tasks.

Finally, the effects of training on the perception of Glass patterns have also been measured.[133] Glass patterns are created when a dot pattern is duplicated and shifted slightly, leading to the perception of paired dots with offsets. Observers discriminated between Glass patterns shifted concentrically around fixation or radially away from it, at six levels of radial sheer with either 45% or 80% of signal dots. The fMRI BOLD activations in early visual areas and higher visual areas V3a, V3b/KO, V7, and LOC were processed separately by decoders that learned to classify the six levels of radial sheer, with separate decoders for pre- and posttraining in each area (figure 5.13). For most visual areas, decoder classification was about 20%, with chance at 16.7%. Training improved classification only in higher visual

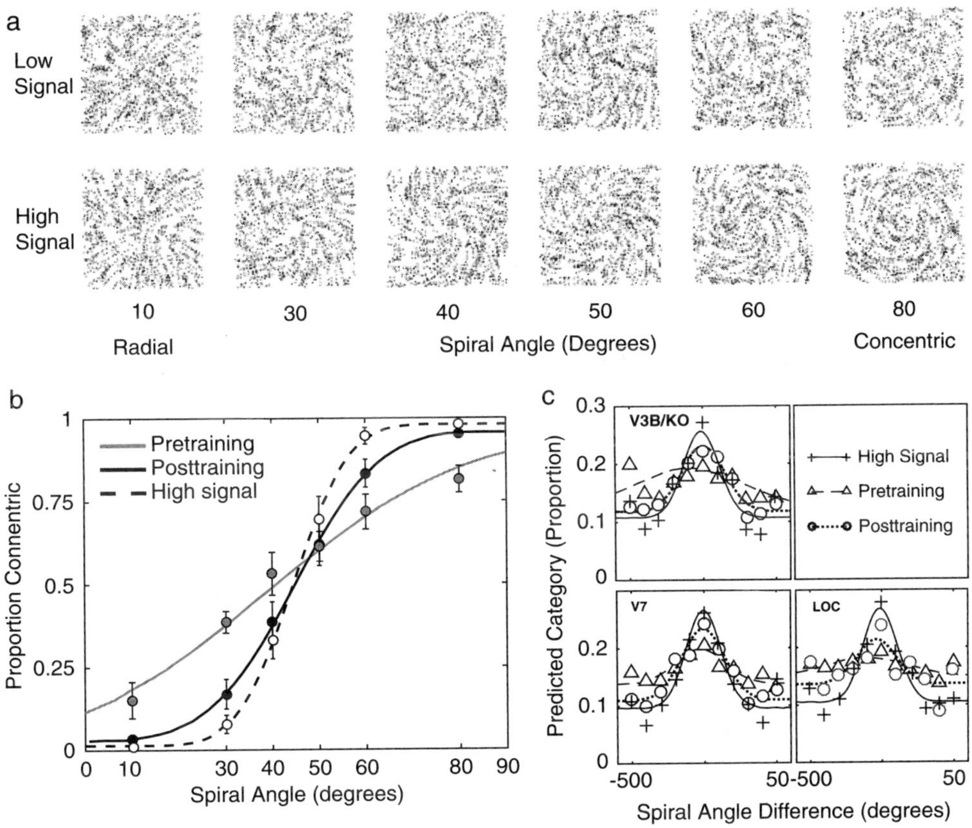

Figure 5.13
Learning to discriminate radial and concentric Glass patterns, and corresponding changes in multivariate pattern classifiers in LOC, V7, and KO after training and compared to responses to very high signal stimuli (chance classification at 16.7%). There were few changes after training in V1, V2, V3a, V3v, V3d, or V4v. After Zhang et al.,[133] parts of figures 1 and 2. Creative Commons, copyright 2010 Zhang et al.

areas, which led these researchers to the conclusion that learning occurred in these higher visual areas. Improved classifier performance may reflect changes in the representations or reductions in noisiness of the responses. The classifiers provide evidence that information is available to readout at higher cortical levels, however the relation between classifier performance and corresponding behavioral choices on individual trials was not reported.

To summarize, each imaging study of training in a mid-level pattern task reports interesting changes in the activation somewhere in the brain. Some studies found changes in brain activations in relevant sensory areas, while others did not. These studies generally held the stimuli constant, which required comparisons at very different behavioral performance levels. For this reason, inferring a causal relationship is complicated. Do the observed changes in brain activation before and after perceptual training cause the changes in the accuracy of behavioral performance? Or, do the changes in behavioral performance mediate the observed brain activations? Other studies in this series focused instead on the performance of voxel pattern decoders applied to different regions of interest before and after training in the visual task. These studies tended to find improvements in higher-level visual areas whose representations correspond more closely with those of the training task; this is similar to the conclusion from the cellular recording studies. At the same time, it should be noted that the fact that a researcher finds evidence of improved classification using an MVPA pattern decoder indicates that there may be better information to read out from this location, not that the readout takes place there. We return to this point in the general discussion at the end of this chapter.

5.5.3 Perceptual Learning of Objects

There have been only a handful of brain imaging studies examining the effects of training on categorizing or naming objects. In one, observers identified which object on either side of fixation contained a symmetric contour among oriented elements, finding that the pattern of changes in different brain regions depended on the contour salience.[134] In this study, learning was compared in a low-salience condition (in which contours were embedded in a field of Gabors with random orientations) and a high-salience condition (in which contours were embedded in fields of Gabors with the same orientation). Training increased detection of the contour from about 60% to 85% for low-salience objects and from about 73% to 95% for high-salience objects. Training in the low-salience task led to *increases* in fMRI BOLD responses throughout early visual areas and higher areas (V1, V2, Vp, V4, LOC, pFs) where pFs is the posterior fusiform sulcus. Training in the high-salience task led to *decreases* in higher areas (LOC, pFs), while leaving responses in early visual areas unchanged. The size of the change in fMRI responses was correlated over subjects with increases in behavioral accuracy. The different patterns in the low- and high-salience regimes suggest a flexible locus of plasticity that depends on the nature of the task.

A second study trained observers to judge slanted-line objects made from oriented Gabors with axes either collinear with or orthogonal to the direction of the slant.[135] Exposure without feedback was sufficient to improve slant judgments for easier collinear displays but not for the more difficult orthogonal contours. With feedback, learning occurred for both kinds of stimuli, with orthogonal contours increasing from near chance (50%) to about 80% accuracy. fMRI scans measured response amplitudes during active task performance from the visual cortex through the motor cortex (V3A, V3B, LOC, and other higher areas) for contours and random displays. Training primarily changed brain responses in higher areas, while visual areas through LOC remained largely unchanged. This suggested that plasticity occurred in higher brain areas, especially in conditions requiring supervised training.

5.5.4 Summary of Brain Imaging Studies of Perceptual Learning

Brain imaging studies have the potential to provide a more holistic view of the physiological substrates of perceptual learning. These methods could reveal a network of areas active in perceptual tasks and how responses in these areas change as a consequence of training. Most cellular recording studies, by contrast, have tended to measure responses in a single brain area. In section 5.5, we reviewed the results of imaging studies organized by the three levels of perceptual training tasks: features, patterns, and objects (low-, mid-, and high-level vision). The cumulative results from the existing studies are interesting but inconsistent. Some studies found more activation in a given region following training, while others found less, and still others found small or no changes. In several cases, the data from either fMRI or EEG differed from data in corresponding single-cell recording studies.

As indicated previously, training coarse versus fine discriminations would likely have significant consequences for brain activity, as would measurements from active task performance compared to passive control tasks. In addition, many of the early brain imaging studies based their conclusions on the change in responses to the same stimulus before and after training, which in some cases corresponded to different behavioral performance levels. In this case, either the altered brain responses led to improved behavioral accuracy, differences in performance accuracy may have influenced the observed changes in neural responses, or both. To avoid comparing task performances at different accuracy levels, several researchers opted to compare responses to easier stimuli (near ceiling in behavioral accuracy) or during passive viewing before and after training on a different task. Determining which approach is best will only become clear as further imaging studies of learning are carried out, allowing us to compare the outcomes of studies with these different design features.

One possible interpretation of mixed observations of activation in the early visual cortex after training builds on the singular observation in an fMRI study of dynamic changes between such patterns early and late in training.[130] One possibility is that attention to the stimulus is engaged during active task performance early in learning. Attention can alter (often enhance) the responses as early as V4, potentially affecting V1 via top-down feedback.[82] Then, as learning improves connections from relevant stimulus representations to decision, the trained performance ceases to rely on attention.[136] (See chapter 9 for several examples.) This explanation may also be consistent with claims that video game players, who can experience broad improvements in performance associated with attention, show faster learning than those who are not video game players only in the early stages of visual learning.[137] In this view, at least some of the evidence for increased activation in early visual areas may in fact be a function of attention that is engaged early in training and then fades away as the task becomes more practiced. Validating this kind of hypothesis could motivate the design of future studies.

A different approach evaluates not the activation in visual areas per se but rather the ability to classify the stimuli based on MVPA analyzers (multivoxel pattern analyzers) of fMRI activations in different regions of interest. These analyzers or decoders have been used to quantify the signal-relevant information that may lead to behavioral classifications. Studies using this approach reported increases in decoder accuracy after training. Two issues complicate the interpretation, however. First, conclusions may be limited because the decoder performance can be very close to chance, far below behavioral classification accuracy (e.g., in one case, decoder classification increased from 50%, or chance, to 55% after training, while behavioral accuracy was > 80%). Voxels include many neurons and may haphazardly differ in their collective differential sensitivity to the relevant stimulus characteristics, which limits the information. Second, different decoders are used before and after perceptual training—so improvements reflect a mixture of changes in the response pattern and/or their noisiness, together with the newly optimized readout of the classifier itself. The methods used to train the classifiers can be enormously sensitive to even the smallest changes in the quality of input information. Finally, the ability to predict behavioral responses with optimized machine learning readout does not imply that the stimulus patterns are classified in the corresponding brain regions.

Overall, while some studies have suggested changes in the activity in early visual areas after training, others are consistent with the reweighting of evidence into higher visual and decision areas, and still others are consistent with the influence of top-down processes of expectation and attention.[69] There are relatively few imaging studies of perceptual learning, so research in this area is in its early stages. Developments in imaging technologies, the invention of new and improved experimental designs, and an increase in the number of studies will all likely influence interpretations and improve our understanding. In principle, whole-brain imaging technologies should provide important insights into the brain networks involved in perception and decision, and their plasticity.

There are a number of other promising new technologies for investigating the substrates of learning. Transcranial magnetic stimulation (TMS) and transcranial direct current stimulation (tDCS) could potentially lead to new insights about the localization or mechanisms of learning. This research has generally focused on other memory phenomena, such as working memory (e.g., findings that tDCS stimulation either improves or damages the effectiveness of working-memory training depending on the task and the location of stimulation).[138, 139]

Another new technique measures the connection between brain regions, relating the integrity of white matter tracts in specific brain areas (measured by diffusion tensor imaging, DTI) to success in various forms of learning, from speech to general memory tasks.[140–142] For example, visual perceptual learning has been related to a thickening of white matter tracts under the early visual cortex following training in older but not in younger adults.[143]

There are also new methods of measurement using multicellular recording in animals or ECoG (electrocorticography) or iEEG (intracranial electroencephalography) in humans.[144, 145] These new technologies may ultimately prove useful in evaluating multiple points along the cascade of visual processing, measuring correlations between firing patterns in sets of neurons, in the role of synchrony in neural firing, and in how these properties may be altered by training.

Other new forms of brain imaging may be used to better understand different kinds of changes in brain responses. For example, GABA imaging is used for special sensitivity to inhibitory processes. A brain region could either increase or decrease in the expression of GABA, a molecule involved in inhibitory modulation of neurons. A recent study of perceptual learning found effects in opposite directions in two different visual tasks.[146] Individuals for whom GABA decreased in the visual cortex after training did better in a target-detection task, while individuals for whom GABA increased after training did better in a feature-discrimination task—roughly in coarse and fine tasks. Future innovations may improve techniques of GABA imaging, which refers changes in GABA to those of other metabolites. Using this kind of approach may further elucidate the roles of inhibitory and excitatory processes in visual perceptual learning. This list of new methods and directions only begins to detail how future research may advance our understanding of the nature of plasticity in learning.

5.6 Discussion

This chapter examined what physiological studies have to tell us about the substrates of visual perceptual learning. From a synoptic analysis, especially in light of the literature from cellular recording studies, several insights emerged.

The enterprise so far largely went looking for changes in early visual cortical responses to trained stimuli. The primary goal, whether stated or unstated, seems to have been to establish the existence of experience-dependent changes as early in the visual cortex as possible. Over a number of studies, changes in several early brain regions were measured either under active task performance or in passive or control tasks; the visual tasks themselves also differed, first in relation to the given domain (e.g., orientation, stereopsis, and motion) but also in their choice to use either fine or coarse judgment tasks.

Even as we acknowledge that the methodological diversity on offer might lead to condition-dependent interpretations, certain broad if tentative hypotheses can be sketched. The bulk of the evidence supports general stability of visual representations, especially for those in the earliest visual cortical areas. Against this backdrop, there have been several reports of subtle coding changes in neurons tuned slightly away from the relevant trained stimulus feature (e.g., slightly away from a trained orientation in an orientation task) as early as V1. But even in these cases, experience-dependent response changes tended to be significantly stronger in higher visual areas during active task performance (e.g., V4 versus V1 or V2; or the IT cortex or MST versus MT or V1 during task performance).

The evidence for altered responses during passive viewing or in control tasks accounted for only very small amounts of behavioral improvements, often less than one-tenth (estimated by various classification models). Learning therefore almost surely integrates effects of task-specific contexts or goals that in turn would specify top-down factors during active task performance. In addition, the important functional changes might have occurred upstream in any event.

The exception to the emphasis on active task performance came in high-level object-recognition tasks, where a number of cases showed learned tuning effects in the absence of active task performance (e.g., with fixation tasks or under anesthesia). In these, learning was associated with the emergence of a few neurons in the IT cortex or the PFC that came to represent individual familiar objects (sometimes dependent on view selection). These cases, however, simply represent the responses of a few selected neurons; no computations were carried out to estimate how much of the overall behavior was accounted for by the neural responses.

5.6.1 Where Is the Reweighting?

Whether there are significant modulations of early visual cortical responses to input stimuli—essentially modulations in the patterns of activity that represent the stimulus—or not, the classifications of these patterns and the subsequent decision and translation into action almost surely occur further along the processing pathways. That these modulations translate into a connection to behavior primarily under active task performance also suggests that what is learned resides in weighted connections between the representations and a decision that is at least jointly specified by information about the task stored in the prefrontal cortex and invoked in top-down activities. A strong claim based on these observations must place a significant amount of the *learning* involved in visual perceptual learning elsewhere, upstream, where things about the task are remembered and can be deployed, in some cases years later.

In other words, a full investigation of the physiological substrates of visual learning should focus not only or even primarily on representations but rather on tracing the weighted connections that route information from those representations to the point of decision. (One potentially relevant study traced how initially inactive cells are recruited to participate in an experience-dependent network of active units in rodent V1 using optical and epigenetic approaches.)[147] It should also seek to understand how the task context or goal structure is invoked by top-down influences, perhaps by involving task-context contingent reinstatement of decision structures, reward structures, and/or the involvement of attention. In the language of the reweighting versus representation dichotomy introduced in chapter 1, if the first stages of visual cortical response are the representations, where is the reweighting or the readout?

The focus on how well the possibly complex model "decoders" read the information in the stimulus "encoding" in the early cortex may be misleading. As we alluded to, the fact that the researcher can measure the cortical responses and then use mathematical tools to extract information from these responses that to some degree can predict the behavioral response does not indicate that such a process is taking place in those early cortical areas that were measured (see figure 5.14). Indeed, a more appropriate interpretation would be that there is adequate information in one or more representational areas to support the observed behavior—the question is how much or how well that information is used in the behavioral decision. These machine learning analyses of the physiological observations should perhaps

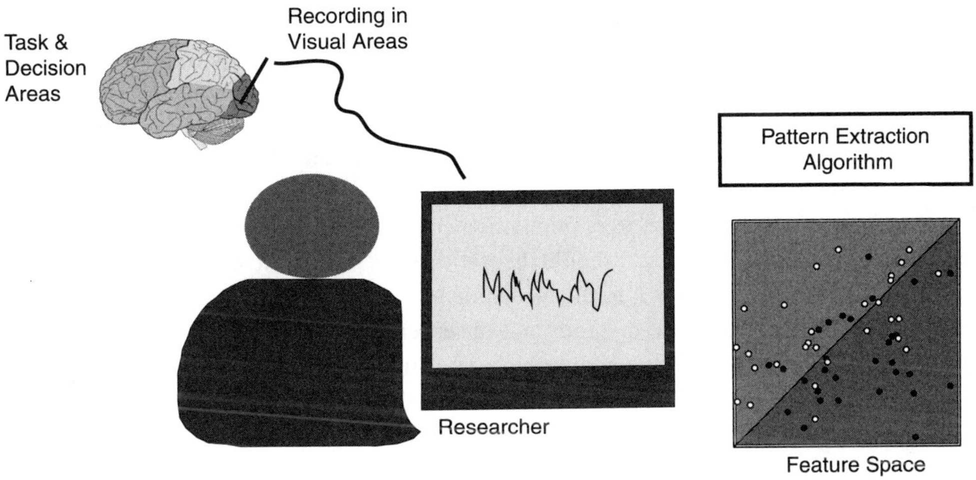

Figure 5.14
Where in the brain are the learned weights that convert activity in visual areas into a decision? A researcher, with the help of a computer and sophisticated pattern-recognition algorithms, statistically categorizes two-alternative stimuli into categories with some success. This indicates that evidence in that location could support some level of categorization—likely carried out in higher decision areas. Much perceptual learning may lie in the connections between the evidence and the decision that do the same work as the pattern-extraction algorithms of the researcher.

be understood in a different spirit: as observer calculations that take noiseless copies of stimulus images and compute how well an ideal observer could perform the task.

If the changes in representation are relatively small and the learning resides primarily in the changes in weights connecting these representation activities to decision, then the new challenge for scientists will be to devise new methods that can measure these weights as well as those in the other brain areas active in task performance, such as decision, learning, and goal setting. At this point, given where the field currently stands, a number of questions emerge: Is it possible to reveal the areas and weight changes involved in the brain network during active performance? If so, what is the correct index of the network? Will the network's location and weights be revealed in the connections during a resting-state task or will they be indexed by integrity changes in white matter tracts, as with fMRI and EEG? Will the connection diagrams estimated from functional imaging help to identify relevant brain areas? Can new imaging modalities be devised that would better reveal weights and weight changes? What about optical imaging? Given the existing technology, are we even at a point where we can adequately measure these aspects of learning?

5.6.2 Relation to Internal Noise and the Observer Model

Improvements in behavioral performance observed in a visual learning task must reflect changes in the intrinsic signal and noise properties of the observer. Each physiological study can thus be seen as measuring some aspect of inherently noisy physiological brain responses that could, in principle, be related to observed behavioral changes. This chapter considered some of the relevant evidence revealed in single-cell responses in monkeys and fMRI and EEG activation across brain areas in humans.

Perceptual learning can increase or decrease neural activity. It can also change tuning curves or result in improvements in quantitative estimates of the information in the neural responses derived from estimates of population codes in cellular responses or decoding or classification in brain imaging. It should theoretically be possible to connect these changes in physiological responses back to the mechanisms of learning discussed in chapter 4 and explored using external-noise methods and observer models such as the perceptual template model (PTM). Given this analogy, changes in physiological response amplitudes might be related to stimulus enhancement, while changes in the neural tuning would plausibly be related to changes in the perceptual template. In the context of physiology, the goal would be to understand the response properties of the population of neurons in different brain regions and to link these population responses to the changes in signal-to-noise ratio and the behavioral outcomes. Furthermore, as we will see in the multichannel computational models discussed in subsequent chapters, retuning at a single neural level or reweighting evidence from lower levels to higher levels can alter the internal noises that limit the signal-to-noise ratio in performance by implementing template change.

Further analogies that relate the physiology of learning to the three mechanisms specified in the PTM model are worth pursuing. Observer models have the advantage of fully characterizing the system, including the template for the relevant signal information, the nonlinearities in response, and the intrinsic noises that limit accuracy. The corresponding analysis in physiology would look to characterize the neural response to signal, the noise in those responses, and the correlation between neural responses—all those properties that determine the population code for a given task.

It comes as a surprise, then, that although noise in the neural responses—whether measured by single-cell recording or fMRI activation—is one of the fundamental properties of neural responses, these noise properties have yet to be significantly examined in relation to visual perceptual learning. One measure of neural response noisiness in single neurons is the so-called Fano factor, defined as the ratio of the variance to the mean of the neural spikes during a measured time interval. This is a kind of noise-to-signal ratio, a measure of dispersion relative to the mean. A reduction in the Fano factor indicates that the noisiness in the response is reduced relative to the mean firing rate. The Fano factor and neural correlations have been studied experimentally in several recent single-cell recording studies of visual attention.[148–150] Nevertheless, such explicit noise analyses have yet to be widely integrated into studies of visual perceptual learning.

We suggest that along with the focus on changes in the amplitude of neural responses following learning, the changes in the response variability in single-cell recording, EEG, and fMRI signals should be systematically studied. One important approach would be to characterize the signal and noise, and their ratio, in different brain regions, as well as the covariance structure between the responses of different neurons or different regions. Advancements in multiarray neural recording,[108] and technical improvements in the spatial and temporal resolution of brain imaging modalities, have the potential to transform the quality of these important analyses of the signal, noise, and correlation properties of neural population responses and their relation to behavioral choice.

5.6.3 Elaborated Computational Studies

Future advances may also come not only through experimentation but also through theoretical or computational studies of neural population responses. One recent theoretical study in this vein used a computational model to argue that perceptual learning could reflect improved probabilistic inference (e.g., readout) of the neural population responses to the stimulus.[16] The primary simulations in this study sought to understand the role of changing correlations between neural responses in predicting behavior as focused on predicting the external-noise manipulations or threshold versus contrast (TvC) functions (see chapter 4). A simulated neuronal model of orientation discrimination included layers of the visual system from retinal ganglion cells, to LGN, to V1. The V1 layer provided the stimulus representation that was read out to determine the behavioral decision. Model predictions were compared to the behavioral effects of perceptual learning on TvC (threshold versus external-noise contrast) functions, where feed-forward reweighting of evidence from LGN to V1 could simulate the observed effects of perceptual learning.[89, 94] In contrast, changes in recurrent connections within V1 corresponding to retuning of V1 neurons could not. Changes in V1 recurrent connections between V1 neurons typically increased the correlation between the neurons, which decreased the information available that translated to the behavioral choice in the simulation. Feed-forward reweighting did not increase, and sometimes decreased, the correlation between the responses of neurons in the simulation. The amplification and slight sharpening of tuning functions observed in a few single-cell recording studies were shown to be neither necessary nor sufficient to account for the changes in the TvC functions in perceptual learning. Furthermore, the simulations showed that estimated neural TvC functions were relatively robust and did not depend very much on the number of neurons being recorded or simulated.

Although the study[16] modeled a multilevel neural system from the retina to V1, the same points apply to multilevel neural systems at higher levels of the visual system. The study highlights the potential value of measuring external-noise functions for neural responses. Modeling the relationship between the properties of physiological responses and the behavioral outcomes could also connect the physiological level of analysis and the systems analysis of signal and noise, a connection that could illuminate both analyses. Other similar computational studies, perhaps those that incorporate entire observer systems, may provide further insights that could directly affect future evaluations of the functional physiology.

5.7 Conclusion

This chapter reviewed the evidence of plasticity in different brain areas for perceptual tasks at different levels of analysis, including those focused on low-level individual features, patterns represented in mid-level vision, and higher-level objects and scenes, using the available studies in cellular recording and in brain imaging. Though there are some reports of modest changes in the responses of early visual areas, by far the bulk of the evidence points to the importance of readout or reweighting from higher-level visual areas. Measuring connection weights and weight changes remains a major challenge. The development of new technologies involved in physiological assessment of brain activity promises to enable new measurements of neural populations in multiple brain regions simultaneously throughout the course of perceptual learning. The evidence collected by these future experiments may very well resolve the open questions of the field more conclusively, as they promise to yield new and refined insights regarding the plastic changes in the entire processing system that supports perceptual learning.

References

1. Striedter GF. *Neurobiology: A functional approach.* Oxford University Press; 2015.
2. Horwitz B, Poeppel D. How can EEG/MEG and fMRI/PET data be combined? *Human Brain Mapping* 2002;17(1):1–3.
3. de-Wit L, Alexander D, Ekroll V, Wagemans J. Is neuroimaging measuring information in the brain? *Psychonomic Bulletin & Review* 2016;23(5):1415–1428.
4. Wandell BA, Smirnakis SM. Plasticity and stability of visual field maps in adult primary visual cortex. *Nature Reviews Neuroscience* 2009;10(12):873–884.
5. Brodmann K. *Vergleichende lokalisationslehre der grosshirnrinde in ihren prinzipien dargestellt auf grund des zellenbaues.* Barth; 1909.
6. von Economo CF, Koskinas GN. *Die cytoarchitektonik der hirnrinde des erwachsenen menschen.* Springer; 1925.
7. Parent A, Carpenter MB. *Human neuroanatomy.* Williams and Wilkins; 1995.
8. Thompson RF. *The brain: A neuroscience primer.* Macmillan; 2000.
9. Kaufman P, Alm A. *Adler's physiology of the eye: Clinical application.* Mosby; 2003.
10. Cornsweet T. *Visual perception.* Academic Press; 2012.
11. Kaplan E, Shapley R. X and Y cells in the lateral geniculate nucleus of macaque monkeys. *Journal of Physiology* 1982;330(1):125–143.
12. Nealey T, Maunsell J. Magnocellular and parvocellular contributions to the responses of neurons in macaque striate cortex. *Journal of Neuroscience* 1994;14(4):2069–2079.
13. Merigan W, Katz LM, Maunsell J. The effects of parvocellular lateral geniculate lesions on the acuity and contrast sensitivity of macaque monkeys. *Journal of Neuroscience* 1991;11(4):994–1001.
14. Hendry SH, Reid RC. The koniocellular pathway in primate vision. *Annual Review of Neuroscience* 2000;23(1):127–153.
15. Malpeli JG, Baker FH. The representation of the visual field in the lateral geniculate nucleus of *Macaca mulatta. Journal of Comparative Neurology* 1975;161(4):569–594.
16. Bejjanki VR, Beck JM, Lu Z-L, Pouget A. Perceptual learning as improved probabilistic inference in early sensory areas. *Nature Neuroscience* 2011;14(5):642–648.

17. Burnat K. Are visual peripheries forever young? *Neural Plasticity* 2015;15:1–13.

18. Tootell RB, Switkes E, Silverman MS, Hamilton SL. Functional anatomy of macaque striate cortex. II. Retinotopic organization. *Journal of Neuroscience* 1988;8(5):1531–1568.

19. Hubel DH, Wiesel TN. Ferrier lecture: Functional architecture of macaque monkey visual cortex. *Proceedings of the Royal Society of London Series B: Biological Sciences* 1977;198(1130):1–59.

20. Gilbert CD. Laminar differences in receptive field properties of cells in cat primary visual cortex. *Journal of Physiology* 1977;268(2):391–421.

21. Martinez LM, Alonso JM, Reid RC, Hirsch JA. Laminar processing of stimulus orientation in cat visual cortex. *Journal of Physiology* 2002;540(1):321–333.

22. Tootell R, Hamilton S, Silverman M, Switkes E. Functional anatomy of macaque striate cortex. I. Ocular dominance, binocular interactions, and baseline conditions. *Journal of Neuroscience* 1988;8(5):1500–1530.

23. Horton JC, Hubel DH. Regular patchy distribution of cytochrome oxidase staining in primary visual cortex of macaque monkey. *Nature* 1981;292(5825):762–764.

24. Leuba G, Kraftsik R. Changes in volume, surface estimate, three-dimensional shape and total number of neurons of the human primary visual cortex from midgestation until old age. *Anatomy and Embryology* 1994;190(4):351–366.

25. Wandell BA. *Foundations of vision*. Sinauer Associates; 1995.

26. Essen Dv, Zeki S. The topographic organization of rhesus monkey prestriate cortex. *Journal of Physiology* 1978;277(1):193–226.

27. Mishkin M, Ungerleider LG, Macko KA. Object vision and spatial vision: Two cortical pathways. *Trends in Neurosciences* 1983;6:414–417.

28. Van Essen DC, Maunsell JH. Hierarchical organization and functional streams in the visual cortex. *Trends in Neurosciences* 1983;6:370–375.

29. Lennie P. Single units and visual cortical organization. *Perception—London* 1998;27:889–936.

30. Maunsell JH, Newsome WT. Visual processing in monkey extrastriate cortex. *Annual Review of Neuroscience* 1987;10(1):363–401.

31. Felleman DJ, Van Essen DC. Distributed hierarchical processing in the primate cerebral cortex. *Cerebral Cortex* 1991;1(1):1–47.

32. Perry CJ, Fallah M. Feature integration and object representations along the dorsal stream visual hierarchy. *Frontiers in computational neuroscience* 2014;8,84:1–17.

33. Damasio AR, Damasio H, Van Hoesen GW. Prosopagnosia: Anatomic basis and behavioral mechanisms. *Neurology* 1982;32(4):331–341.

34. Zihl J, Von Cramon D, Mai N, Schmid C. Disturbance of movement vision after bilateral posterior brain damage: Further evidence and follow up observations. *Brain* 1991;114(5):2235–2252.

35. Vaina LM. Functional segregation of color and motion processing in the human visual cortex: Clinical evidence. *Cerebral Cortex* 1994;4(5):555–572.

36. Zeki S. A century of cerebral achromatopsia. *Brain* 1990;113(6):1721–1777.

37. Agrawal P, Stansbury D, Malik J, Gallant JL. Pixels to voxels: Modeling visual representation in the human brain. *arXiv preprint 14075104.* 2014.

38. Roe AW, Chelazzi L, Connor CE, Conway BR, Fujita I, Gallant JL, Lu H, Vanduffel W. Toward a unified theory of visual area V4. *Neuron* 2012;74(1):12–29.

39. Kuffler SW. Discharge patterns and functional organization of mammalian retina. *Journal of Neurophysiology* 1953;16(1):37–68.

40. Kourtzi Z, Connor CE. Neural representations for object perception: Structure, category, and adaptive coding. *Annual Review of Neuroscience* 2011;34:45–67.

41. Albright TD, Desimone R, Gross CG. Columnar organization of directionally selective cells in visual area MT of the macaque. *Journal of Neurophysiology* 1984;51(1):16–31.

42. Duffy CJ. MST neurons respond to optic flow and translational movement. *Journal of Neurophysiology* 1998;80(4):1816–1827.

43. Orban GA, Lagae L, Raiguel S, Xiao D, Maes H. The speed tuning of medial superior temporal (MST) cell responses to optic-flow components. *Perception* 1995;24(3):269–285.

44. Wurtz RH. Optic flow: A brain region devoted to optic flow analysis? *Current Biology* 1998;8(16):R554–R556.

45. Graziano MS, Andersen RA, Snowden RJ. Tuning of MST neurons to spiral motions. *Journal of Neuroscience* 1994;14(1):54–67.

46. Van Essen DC, Gallant JL. Neural mechanisms of form and motion processing in the primate visual system. *Neuron* 1994;13(1):1–10.

47. Milner AD, Goodale MA. *The visual brain in action*. Oxford University Press; 1995.

48. Cadieu CF, Hong H, Yamins DL, Pinto N, Ardila D, Solomon EA, Majaj NJ, DiCarlo JJ. Deep neural networks rival the representation of primate IT cortex for core visual object recognition. *PLoS Computational Biology* 2014;10(12):e1003963.

49. Yamins DL, DiCarlo JJ. Eight open questions in the computational modeling of higher sensory cortex. *Current Opinion in Neurobiology* 2016;37:114–120.

50. Yamins DL, DiCarlo JJ. Using goal-driven deep learning models to understand sensory cortex. *Nature Neuroscience* 2016;19(3):356–365.

51. Yamins DL, Hong H, Cadieu CF, Solomon EA, Seibert D, DiCarlo JJ. Performance-optimized hierarchical models predict neural responses in higher visual cortex. *Proceedings of the National Academy of Sciences* 2014;111(23):8619–8624.

52. Cadieu CF, Hong H, Yamins D, Pinto N, Majaj NJ, DiCarlo JJ. The neural representation benchmark and its evaluation on brain and machine. *arXiv preprint 13013530*. 2013.

53. Van Essen DC. Organization of visual areas in macaque and human cerebral cortex. *Visual Neurosciences* 2004;1:507–521.

54. Opris I, Bruce CJ. Neural circuitry of judgment and decision mechanisms. *Brain Research Reviews* 2005;48(3):509–526.

55. Hikosaka O, Sesack SR, Lecourtier L, Shepard PD. Habenula: Crossroad between the basal ganglia and the limbic system. *Journal of Neuroscience* 2008;28(46):11825–11829.

56. Rolls ET. The orbitofrontal cortex and reward. *Cerebral Cortex* 2000;10(3):284–294.

57. Schultz W. Multiple reward signals in the brain. *Nature Reviews Neuroscience* 2000;1(3):199–207.

58. Schultz W, Tremblay L, Hollerman JR. Reward processing in primate orbitofrontal cortex and basal ganglia. *Cerebral Cortex* 2000;10(3):272–283.

59. Wise RA. Brain reward circuitry: Insights from unsensed incentives. *Neuron* 2002;36(2):229–240.

60. Haber SN, Knutson B. The reward circuit: Linking primate anatomy and human imaging. *Neuropsychopharmacology* 2010;35(1):4–26.

61. Shipp S. The brain circuitry of attention. *Trends in Cognitive Sciences* 2004;8(5):223–230.

62. Kastner S, Pinsk MA. Visual attention as a multilevel selection process. *Cognitive, Affective, & Behavioral Neuroscience* 2004;4(4):483–500.

63. Petersen SE, Posner MI. The attention system of the human brain: 20 years after. *Annual Review of Neuroscience* 2012;35:73–89.

64. Recanzone GH, Merzenich MM, Jenkins WM, Grajski KA, Dinse HR. Topographic reorganization of the hand representation in cortical area 3b owl monkeys trained in a frequency-discrimination task. *Journal of Neurophysiology* 1992;67(5):1031–1056.

65. Recanzone GH, Schreiner C, Merzenich MM. Plasticity in the frequency representation of primary auditory cortex following discrimination training in adult owl monkeys. *Journal of Neuroscience* 1993;13(1):87–103.

66. Weinberger NM, Ashe JH, Metherate R, McKenna TM, Diamond DM, Bakin JS, Lennartz RC, Cassady JM. Neural adaptive information processing: A preliminary model of receptive-field plasticity in auditory cortex during Pavlovian conditioning. In Gabriel M, Moore J, eds., *Learning and computational neuroscience: Foundations of adaptive networks*. MIT Press;1990:91–138.

67. Karni A, Sagi D. Where practice makes perfect in texture discrimination: Evidence for primary visual cortex plasticity. *Proceedings of the National Academy of Sciences* 1991;88(11):4966–4970.

68. Ahissar M, Hochstein S. The reverse hierarchy theory of visual perceptual learning. *Trends in Cognitive Sciences* 2004;8(10):457–464.

69. Smith JE, Chang'an AZ, Cook EP, Masse NY. *Linking neural activity to visual perception: Separating sensory and attentional contributions.* INTECH; 2012.

70. Schoups A, Vogels R, Qian N, Orban G. Practising orientation identification improves orientation coding in V1 neurons. *Nature* 2001;412(6846):549–553.

71. Schoups AA, Vogels R, Orban GA. Human perceptual learning in identifying the oblique orientation: Retinotopy, orientation specificity and monocularity. *Journal of Physiology* 1995;483(3):797–810.

72. Vogels R, Orban GA. The effect of practice on the oblique effect in line orientation judgments. *Vision Research* 1985;25(11):1679–1687.

73. Raiguel S, Vogels R, Mysore SG, Orban GA. Learning to see the difference specifically alters the most informative V4 neurons. *Journal of Neuroscience* 2006;26(24):6589–6602.

74. Ghose GM, Yang T, Maunsell JH. Physiological correlates of perceptual learning in monkey V1 and V2. *Journal of Neurophysiology* 2002;87(4):1867–1888.

75. Yang T, Maunsell JH. The effect of perceptual learning on neuronal responses in monkey visual area V4. *Journal of Neuroscience* 2004;24(7):1617–1626.

76. Law C-T, Gold JI. Reinforcement learning can account for associative and perceptual learning on a visual-decision task. *Nature Neuroscience* 2009;12(5):655–663.

77. Chowdhury SA, DeAngelis GC. Fine discrimination training alters the causal contribution of macaque area MT to depth perception. *Neuron* 2008;60(2):367–377.

78. Adab HZ, Vogels R. Practicing coarse orientation discrimination improves orientation signals in macaque cortical area v4. *Current Biology* 2011;21(19):1661–1666.

79. Motter BC. Neural correlates of feature selective memory and pop-out in extrastriate area V4. *Journal of Neuroscience* 1994;14(4):2190–2199.

80. Luck SJ, Chelazzi L, Hillyard SA, Desimone R. Neural mechanisms of spatial selective attention in areas V1, V2, and V4 of macaque visual cortex. *Journal of Neurophysiology* 1997;77(1):24–42.

81. Desimone R, Duncan J. Neural mechanisms of selective visual attention. *Annual Review of Neuroscience* 1995;18(1):193–222.

82. McAdams CJ, Maunsell JH. Effects of attention on orientation-tuning functions of single neurons in macaque cortical area V4. *Journal of Neuroscience* 1999;19(1):431–441.

83. McAdams CJ, Maunsell JH. Attention to both space and feature modulates neuronal responses in macaque area V4. *Journal of Neurophysiology* 2000;83(3):1751–1755.

84. Cohen MR, Maunsell JH. Using neuronal populations to study the mechanisms underlying spatial and feature attention. *Neuron* 2011;70(6):1192–1204.

85. Maunsell JH, Treue S. Feature-based attention in visual cortex. *Trends in Neurosciences* 2006;29(6):317–322.

86. Williford T, Maunsell JH. Effects of spatial attention on contrast response functions in macaque area V4. *Journal of Neurophysiology* 2006;96(1):40–54.

87. Hua T, Bao P, Huang C-B, Wang Z, Xu J, Zhou Y, Lu Z-L. Perceptual learning improves contrast sensitivity of V1 neurons in cats. *Current Biology* 2010;20(10):887–894.

88. Law C-T, Gold JI. Neural correlates of perceptual learning in a sensory-motor, but not a sensory, cortical area. *Nature Neuroscience* 2008;11(4):505–513.

89. Dosher BA, Lu Z-L. Perceptual learning reflects external noise filtering and internal noise reduction through channel reweighting. *Proceedings of the National Academy of Sciences* 1998;95(23):13988–13993.

90. Gold JI, Shadlen MN. The neural basis of decision making. *Annual Review of Neuroscience* 2007;30:535–574.

91. Gold JI, Law C-T, Connolly P, Bennur S. The relative influences of priors and sensory evidence on an oculomotor decision variable during perceptual learning. *Journal of Neurophysiology* 2008;100(5):2653–2668.

92. Uka T, DeAngelis GC. Linking neural representation to function in stereoscopic depth perception: Roles of the middle temporal area in coarse versus fine disparity discrimination. *Journal of Neuroscience* 2006;26(25):6791–6802.

93. Colby CL, Duhamel J-R, Goldberg ME. Ventral intraparietal area of the macaque: Anatomic location and visual response properties. *Journal of Neurophysiology* 1993;69(3):902–914.

94. Dosher BA, Lu Z-L. Mechanisms of perceptual learning. *Vision Research* 1999;39(19):3197–3221.

95. Gu Y, Liu S, Fetsch CR, Yang Y, Fok S, Sunkara A, DeAngelis GC, Angelaki DE. Perceptual learning reduces interneuronal correlations in macaque visual cortex. *Neuron* 2011;71(4):750–761.

96. Angelaki DE, Gu Y, DeAngelis GC. Multisensory integration: Psychophysics, neurophysiology, and computation. *Current Opinion in Neurobiology* 2009;19(4):452–458.

97. Britten KH. Mechanisms of self-motion perception. *Annual Review of Neuroscience* 2008;31:389–410.

98. Kumano H, Uka T. Neuronal mechanisms of visual perceptual learning. *Behavioural Brain Research* 2013;249:75–80.

99. Kovacs I, Kozma P, Feher A, Benedek G. Late maturation of visual spatial integration in humans. *Proceedings of the National Academy of Sciences* 1999;96(21):12204–12209.

100. Li W, Gilbert CD. Global contour saliency and local colinear interactions. *Journal of Neurophysiology* 2002;88(5):2846–2856.

101. Gilbert CD, Li W. Adult visual cortical plasticity. *Neuron* 2012;75(2):250–264.

102. Kapadia MK, Ito M, Gilbert CD, Westheimer G. Improvement in visual sensitivity by changes in local context: Parallel studies in human observers and in V1 of alert monkeys. *Neuron* 1995;15(4):843–856.

103. Chen M, Yan Y, Gong X, Gilbert CD, Liang H, Li W. Incremental integration of global contours through interplay between visual cortical areas. *Neuron* 2014;82(3):682–694.

104. Li W, Piëch V, Gilbert CD. Perceptual learning and top-down influences in primary visual cortex. *Nature Neuroscience* 2004;7(6):651–657.

105. Li W, Piëch V, Gilbert CD. Learning to link visual contours. *Neuron* 2008;57(3):442–451.

106. McManus JN, Li W, Gilbert CD. Adaptive shape processing in primary visual cortex. *Proceedings of the National Academy of Sciences* 2011;108(24):9739–9746.

107. Thiele A. Perceptual learning: Is V1 up to the task? *Current Biology* 2004;14(16):R671–R673.

108. Yan Y, Rasch MJ, Chen M, Xiang X, Huang M, Wu S, Li W. Perceptual training continuously refines neuronal population codes in primary visual cortex. *Nature Neuroscience* 2014;17(10):1380–1387.

109. Hoffman K, Logothetis N. Cortical mechanisms of sensory learning and object recognition. *Philosophical Transactions of the Royal Society of London B: Biological Sciences* 2009;364(1515):321–329.

110. Rainer G, Lee H, Logothetis NK. The effect of learning on the function of monkey extrastriate visual cortex. *PLoS Biology* 2004;2(2):e44.

111. Rainer G, Miller EK. Effects of visual experience on the representation of objects in the prefrontal cortex. *Neuron* 2000;27(1):179–189.

112. Fuster JM. The PFC: Anatomy, physiology and neuropsychology of the frontal lobe. Raven Press; 1997.

113. Fuster JM. Network memory. *Trends in Neurosciences* 1997;20(10):451–459.

114. Logothetis NK, Pauls J, Poggio T. Shape representation in the inferior temporal cortex of monkeys. *Current Biology* 1995;5(5):552–563.

115. Kobatake E, Wang G, Tanaka K. Effects of shape-discrimination training on the selectivity of inferotemporal cells in adult monkeys. *Journal of Neurophysiology* 1998;80(1):324–330.

116. Freedman DJ, Riesenhuber M, Poggio T, Miller EK. Experience-dependent sharpening of visual shape selectivity in inferior temporal cortex. *Cerebral Cortex* 2006;16(11):1631–1644.

117. Sigala N, Logothetis NK. Visual categorization shapes feature selectivity in the primate temporal cortex. *Nature* 2002;415(6869):318–320.

118. Baker CI, Behrmann M, Olson CR. Impact of learning on representation of parts and wholes in monkey inferotemporal cortex. *Nature Neuroscience* 2002;5(11):1210–1216.

119. Schiltz C, Bodart J-M, Michel C, Crommelinck M. A pet study of human skill learning: Changes in brain activity related to learning an orientation discrimination task. *Cortex* 2001;37(2):243–265.

120. Furmanski CS, Schluppeck D, Engel SA. Learning strengthens the response of primary visual cortex to simple patterns. *Current Biology* 2004;14(7):573–578.

121. Bao M, Yang L, Rios C, He B, Engel SA. Perceptual learning increases the strength of the earliest signals in visual cortex. *Journal of Neuroscience* 2010;30(45):15080–15084.

122. Pourtois G, Rauss KS, Vuilleumier P, Schwartz S. Effects of perceptual learning on primary visual cortex activity in humans. *Vision Research* 2008;48(1):55–62.

123. Shibata K, Watanabe T, Sasaki Y, Kawato M. Perceptual learning incepted by decoded fMRI neurofeedback without stimulus presentation. *Science* 2011;334(6061):1413–1415.

124. Kahnt T, Grueschow M, Speck O, Haynes J-D. Perceptual learning and decision-making in human medial frontal cortex. *Neuron* 2011;70(3):549–559.

125. Jehee JF, Ling S, Swisher JD, van Bergen RS, Tong F. Perceptual learning selectively refines orientation representations in early visual cortex. *Journal of Neuroscience* 2012;32(47):16747–16753.

126. Diaz JA, Queirazza F, Philiastides MG. Perceptual learning alters post-sensory processing in human decision-making. *Nature Human Behaviour* 2017;1:35,1–9.

127. Vaina LM, Sundareswaran V, Harris JG. Learning to ignore: Psychophysics and computational modeling of fast learning of direction in noisy motion stimuli. *Cognitive Brain Research* 1995;2(3):155–163.

128. Shibata K, Chang L-H, Kim D, Náñez JE Sr, Kamitani Y, Watanabe T, Sasaki Y. Decoding reveals plasticity in V3A as a result of motion perceptual learning. *PLoS One* 2012;7(8):e44003.

129. Schwartz S, Maquet P, Frith C. Neural correlates of perceptual learning: A functional MRI study of visual texture discrimination. *Proceedings of the National Academy of Sciences* 2002;99(26):17137–17142.

130. Yotsumoto Y, Watanabe T, Sasaki Y. Different dynamics of performance and brain activation in the time course of perceptual learning. *Neuron* 2008;57(6):827–833.

131. Ahmadi M, McDevitt EA, Silver MA, Mednick SC. Perceptual learning induces changes in early and late visual evoked potentials. *Vision Research* 2017;152:101–109.

132. Sigman M, Pan H, Yang Y, Stern E, Silbersweig D, Gilbert CD. Top-down reorganization of activity in the visual pathway after learning a shape identification task. *Neuron* 2005;46(5):823–835.

133. Zhang J, Meeson A, Welchman AE, Kourtzi Z. Learning alters the tuning of functional magnetic resonance imaging patterns for visual forms. *Journal of Neuroscience* 2010;30(42):14127–14133.

134. Kourtzi Z, Betts LR, Sarkheil P, Welchman AE. Distributed neural plasticity for shape learning in the human visual cortex. *PLoS Biology* 2005;3(7):1317–1327.

135. Zhang J, Kourtzi Z. Learning-dependent plasticity with and without training in the human brain. *Proceedings of the National Academy of Sciences* 2010;107(30):13503–13508.

136. Dosher BA, Han S, Lu Z-L. Perceptual learning and attention: Reduction of object attention limitations with practice. *Vision Research* 2010;50(4):402–415.

137. Kim Y-H, Kang D-W, Kim D, Kim H-J, Sasaki Y, Watanabe T. Real-time strategy video game experience and visual perceptual learning. *Journal of Neuroscience* 2015;35(29):10485–10492.

138. Ferrucci R, Marceglia S, Vergari M, Marceglia S, Cogiamaniam F, Barvieri S, Scarpini E, Priori A. Cerebellar transcranial direct current stimulation impairs the practice-dependent proficiency increase in working memory. *Journal of Cognitive Neuroscience* 2008;20(9):1687–1697.

139. Au J, Katz B, Buschkuehl M, Bunarjo K. Enhancing working memory training with transcranial direct current stimulation. *Journal of Cognitive Neuroscience* 2016;28(9):1419–1432.

140. Schlegel AA, Rudelson JJ, Tse PU. White matter structure changes as adults learn a second language. *Journal of Cognitive Neuroscience* 2012;24(8):1664–1670.

141. Flöel A, de Vries MH, Scholz J, Breitenstein C, Johansen-Berg H. White matter integrity in the vicinity of Broca's area predicts grammar learning success. *Neuroimage* 2009;47(4):1974–1981.

142. Bennett IJ, Madden DJ, Vaidya CJ, Howard JH Jr, Howard DV. White matter integrity correlates of implicit sequence learning in healthy aging. *Neurobiology of Aging* 2011;32(12): 2317,1–12.

143. Yotsumoto Y, Chang L-H, Ni R, Pierce R, Andersen GJ, Watanabe T, Sasaki Y. White matter in the older brain is more plastic than in the younger brain. *Nature Communications* 2014;5:5504,1–8.

144. Geller AS, Burke JF, Sperling MR, Sharan AD, Litt B, Baltuch GH, Lucas TH II, Kahana MJ. Eye closure causes widespread low-frequency power increase and focal gamma attenuation in the human electrocorticogram. *Clinical Neurophysiology* 2014;125(9):1764–1773.

145. Parvizi J, Kastner S. Promises and limitations of human intracranial electroencephalography. *Nature Neuroscience* 2018;21:474–483.

146. Frangou P, Correia M, Kourtzi Z. GABA, not BOLD, reveals dissociable learning-dependent plasticity mechanisms in the human brain. *eLife* 2018;7:e35854.

147. Marshel JH, Kim YS, Machado TA, Quirin S, Benson B, Kadmon J, Raja C, Chibukhchyan A, Ramakrishnan C, Inoue M, Shane JC, McKnight DJ, Yoshizawa S, Kato HE, Ganguli S, Deisseroth K. Cortical layer–specific critical dynamics triggering perception. *Science* 2019;365:558,1–12.

148. Mitchell JF, Sundberg KA, Reynolds JH. Differential attention-dependent response modulation across cell classes in macaque visual area V4. *Neuron* 2007;55(1):131–141.

149. Mitchell JF, Sundberg KA, Reynolds JH. Spatial attention decorrelates intrinsic activity fluctuations in macaque area V4. *Neuron* 2009;63(6):879–888.

150. Cohen MR, Maunsell JH. Attention improves performance primarily by reducing interneuronal correlations. *Nature Neuroscience* 2009;12(12):1594–1600.

IV

Models of Perceptual Learning

6

Models

By efficiently capturing empirical findings and making testable predictions, quantitative models play a critical role in understanding the phenomena of perceptual learning. In this chapter, we review several early classical models, all of them essentially reweighting models, and consider the application of neural networks to perceptual learning. The core of the chapter shows how neurally inspired stimulus representations together with computational networks can account for a wide range of learning phenomena. The augmented Hebbian reweighting model (AHRM) and others like it are also extensible to other tasks and domains by introducing the appropriate new representation, decision, or learning subsystems. Future models may be needed to implement the creation of new concepts using reweighting.

6.1 The Goals of Modeling

One important ambition of many fields within cognitive neuroscience is to develop and test computational models. The case of perceptual learning should be no different, yet the field as a whole has often pursued an observational approach to theory. This gap between observed phenomena and modeling has left a fertile if somewhat underexplored middle ground: while fully comprehensive models are still a long way off, the development of even partial models can lead to fresh insights, advancing our theoretical understanding while also potentially playing a central role in optimizing training paradigms for practical applications.

Previous chapters documented a sweeping set of phenomena in visual perceptual learning that could, in principle, be modeled. These include the extent, range, and nature of learning in tasks at different levels of visual encoding; the degree of specificity or transfer of training; and the consequences of using different training protocols. Subsequent chapters will consider the influence of feedback, reward, and attention. As we will see, current models exist for specific domains and certain tasks, and can implement many (but not all) of manipulations and can account for many (but not all) of these phenomena, while at the same time further work is required to develop the next generation of increasingly robust and comprehensive models.

In order to account for perceptual learning, a successful predictive model will necessarily incorporate several key functions. It will need to encode the stimuli, specify how the task-relevant decision is made, and implement the training and test paradigms and learning. Each of these functions may be instantiated in a distinct *module*. A representation module, for example, would specify the sensory encoding and the resulting representations. The decision module would specify the way decisions are made. The learning module would specify learning rules. The model as a whole might also specify the top-down influences of attention and the effects of feedback and reward. Of course, any model of behavior must also incorporate internal noises.

A quantitative or computational model of this kind should generate precise and testable predictions about observed phenomena in specific experiments. Testing the accuracy of the model's predictions in turn helps us determine whether the proposed principles of representation, decision, and learning in fact operate in the expected ways (see figure 6.1). This three-way dialogue between modeling, theory, and experiment will be one important avenue for advancement in our understanding of perceptual learning.

Modeling is rewarding but not easy. In visual perceptual learning, the development of accurate models faces a number of key theoretical challenges.

Challenge 1: Even the simplest visual task involves circuits from a number of levels, from the early visual system to higher-level decision and top-down task-relevant processing. The challenge here is to specify the levels most relevant for any given task, including perceptual, decision, and learning processes, as well as the connections between and within them. A fully validated specification that relates relevant computational modules to brain function is currently unavailable. The functions of visual cortical areas such as V1, V4, or MT and IT are still under investigation, as are the relevant learning and decision mechanisms. Nevertheless, future models should attempt to specify the functions and connectivity consistent with current knowledge of the relevant brain systems.

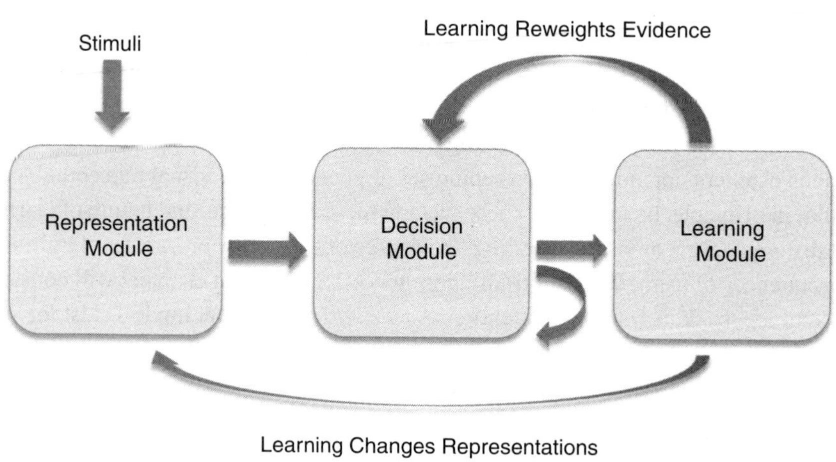

Figure 6.1
Key modules of perceptual learning models, and two mechanisms of learning: reweighting and representation change.

Challenge 2: By necessity, a computational learning model will specify the relevant sensory representation(s) and connectivity. Learning could, in principle, modify the tuning of existing representations; it could occur through reweighting the connections in any of the feed-forward, feedback, or recurrent connections between and within multiple levels of representation; or it could do a combination of both. Physiological investigations and specificity observed in psychophysical studies can play a role in constraining the level and the nature of plasticity in the model.

Challenge 3: A computational model must specify the rule or algorithm used in learning as well as the observer's prior knowledge. The initial state of the system prior to learning will reflect how much is known about the task ahead of time. This a priori knowledge will then determine how much remains to be learned and the conditions required for learning. With minimal prior knowledge, the learning process may require an explicit teacher; with enough prior knowledge, the teacher may in fact become unnecessary.

Challenge 4: A meta-level challenge (and a recurring theme of this book) concerns how to strike the right balance between plasticity and stability. Does learning (and plasticity) go on indefinitely within any given task, or does the learning system become stable at some point? A successful model must specify if and when the system stops learning (perhaps because the internal noise is limiting further improvement) and how learning will be retained over longer time periods. These choices will relate to predictions about observed specificity and/or transfer between learning tasks, as well as the preservation of previously learned tasks.

Challenge 5: Another constraint on models of learning is the biological feasibility of their mechanisms. On the one hand, purely computational models often proceed from abstract properties; on the other hand, perceptual learning models have historically been connected to claims about brain systems and plasticity. For neural network modeling in particular, biological implausibility has been one consideration in the evaluation of certain learning rules (such as fully supervised back propagation), although more biologically feasible versions have also been proposed.[1] Ideally, models should aim for biological plausibility and work within the known properties of brain systems.

Challenge 6: A final important challenge facing researchers hoping to model learning is to specify the nature of the experiments that would be required to evaluate the models they develop. While not properly a component of the model itself, the need to define how a model may be seriously tested is an important challenge. The PTM model, for example, was specified by manipulations of the observer (attention, learning, etc.) and of external noise, as well as measurement at several levels of accuracy or across the psychometric function, all of which are useful in measuring the template, internal noises, and nonlinearities of the model (indeed, many of our experiments include some combination of external noise and multiple contrasts or multiple criteria) (see chapter 4). Similarly, specific manipulations and tests must be developed to test any perceptual learning model.

The challenges listed in table 6.1 delineate the theoretical terrain within which most classical models have been situated and out of which we developed our model, the augmented

Table 6.1
Theoretical challenges in modeling perceptual learning

1. To specify the relevant brain modules and noises and their connections for perception
2. To identify the level(s) of learning and the appropriate learning rules
3. To specify prior knowledge in the starting state and the task environment
4. To consider the balance between plasticity and stability
5. To consider the biological plausibility of all model components
6. To specify the constraining experiments used in testing

Hebbian reweighting model (AHRM).[2, 3] The AHRM is the primary focus of this chapter. In what follows, we show that the AHRM accounts for a considerable range of observed empirical data while striking a balance between plasticity and stability. Likewise, as we will see, by using sufficiently good (but still somewhat simplified) implementations of representations, decision, and learning, the AHRM may have a number of advantages over models of greater or lesser abstraction. (Neither the classical models nor the AHRM, however, have implemented the recruiting or creation of new nodes and new weighting structures. We return to this point in the discussion in section 6.7.) Before describing the AHRM and its applications, however, it is useful to set the stage by considering some of the classical models of visual perceptual learning. Our survey aims to give a sense of the fundamental choices—possible network architectures, different learning rules, and top-down influences—that must be made when constructing a successful model of perceptual learning.

6.2 Classical Models of Perceptual Learning

Early classical models of perceptual learning were developed to account for learning in specific perceptual tasks. These included hyperacuity,[4–7] motion-direction discrimination,[8] contrast discrimination,[9] and orientation discrimination.[10] Over many variations (in tasks, forms of representation, and learning rules), the classic models generally left the sensory representations unchanged, so learning occurred through changes in information integration. The majority of computational models, as a result, have largely focused on reweighting as the theoretical frame for understanding visual learning.

One of the earliest classical models, the *hyper basis function* (HBF) model, was designed to account for learning in a visual hyperacuity task in which observers judge whether the top line was offset slightly left or right of the bottom one.[5] It was among the first to use a three-layer feed-forward network architecture with an input layer, an intermediate layer of representations (radial basis functions), and an output layer consisting of a single decision unit (see figure 6.2). Because of its canonical form and seminal status in the field, a technical discussion of the model will be useful.

In the first network layer of the HBF, the input image was converted into activities in different receptive fields by matching (convolving) the input image with Gaussian filters

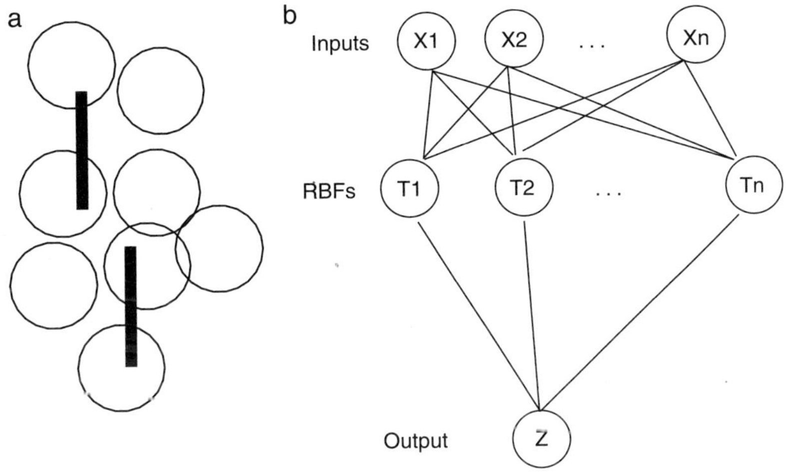

Figure 6.2
A network model of a visual hyperacuity task by Poggio, Fahle, and Edelman.[5] (a) A Vernier offset stimulus overlaid on circles representing radial basis functions in different locations. (b) The three-layer feed-forward network consisting of an input layer, nonlinear radial basis functions, and an output or decision unit. Redrawn from Poggio, Fahle, and Edelman,[5] figure 2, with permission.

centered on a set of retinal locations $x_i = G_\sigma(r - r_i) * I(r)$, where $I(r)$ is the input at location r, r_i is the center of a given receptive field, and G_σ is a two-dimensional Gaussian. In the second layer, the model computed the similarity (distance) between the input vector and a set of templates $t_a : Y_a = B(\| X - t_a \|_w)$, where $\| X - t_a \|_w$ is the weighted distance between the input vector and the circular template, and the vector w contains the weights. (These spatially circular functions were called radial basis functions, or RBFs.) Together, the input and RBF layers made up the *representation module*. In the final layer, the single-unit *decision module* computed a linear combination of the activations in the RBFs: $Z = \Sigma_a c_a y_a$. Negative and positive values of Z corresponded to a top left or a top right response, implicitly assuming a bias of zero. Later implementations added decision noise.[6] Finally, there was a *learning module* that updated the weights as the model experienced the stimulus on each trial. The observer "learned" the spatial locations of the RBF units and the weights to decision using inverse methods, a form of supervised learning. The authors speculated that, in the absence of feedback, internal training signals might be available if the experimenter included large offset stimuli.

This pioneering model used a linear-nonlinear-linear "sandwich," in which nonlinear computations in the middle internal representation layer were combined with linear input and decision layers. (The first version of the model also added new basis function units in an unsupervised learning stage to account for fast perceptual learning, with subsequent supervised learning accounting for slower perceptual learning over the long course of training.) Although the HBF was designed primarily to account for task learning, the improvements were also found to be specific to aspects of the trained stimuli, such as the orientations of the lines, their lengths, and the gap between them.

As one of the first computationally implemented models of perceptual learning, the HBF was a watershed accomplishment. Nevertheless, it was the subject of several early criticisms: that it lacked interactions between basis functions; used biologically implausible mathematical inverse methods; predicted near-chance initial performance, which rarely occurs in human performance; failed to incorporate noise in either the representations or decision as necessary to predict stochastic performance; and, finally, it was unclear how the model could account for other empirical phenomena, such as different forms of feedback.[11]

A number of subsequent modifications of the HBF model aimed to address various limitations.[6] Newer variants held the number of relevant input representations constant (at eight) while adding random input units to simulate low-level noise in internal representations; radial basis functions were replaced with oriented basis functions in the intermediate layer; and decision noise was added at the output layer. The resulting simulations could then account for the effects of some stimulus manipulations, such as the size of the offset between the lines, the length of the lines, and the gap between them, on learning and transfer. Another simulation study investigated the mode of learning by comparing two supervised and two unsupervised/self-supervised rules.[6] For the supervised rule, the correct response was known; a self-supervised rule assumed the existence of internal feedback on trials with large offsets; and an exposure-dependent rule simply constrained the initial connection weights to be either positive or negative as appropriate. The authors favored self-supervised or unsupervised learning rules, on the grounds that learning had sometimes been shown to occur without external feedback. Overall, this series of papers delineated many of the issues any model of visual perceptual learning must address, thus foreshadowing a number of design choices in the field.

Following in the footsteps of the HBF and its offshoots, an analogous model was developed for learning global-motion judgments in random dot motion (figure 6.3).[8] Taking motion vectors of the dots as the input, the activities of MT-like units were tuned to different motion directions integrated over large receptive fields consistent with global motions based on a weighted average of activities and a noisy threshold function (i.e., left or right). (In different versions, integration occurred either by multiplying or summing over local responses.) Finally, learning was implemented either through an exposure-based rule or a self-supervised rule, both basically Hebbian in nature. This model predicted learning but also the effects on performance of the proportion of signal dots and the total number of dots. It also predicted full specificity to the trained direction. The researchers characterized these improvements as learning to ignore the noise.

Several other variants of these models explored the use of recurrent (feedback) as well as feed-forward weights. In one model, "preprocessing" of input stimuli by recurrent network connections was used to "clean up" the input prior to the linear-nonlinear-linear feed-forward classification, which led to the conclusion that learning serves not "to better encode the stimuli… [but rather to] modify the neural responses in a task specific manner that is unlikely to

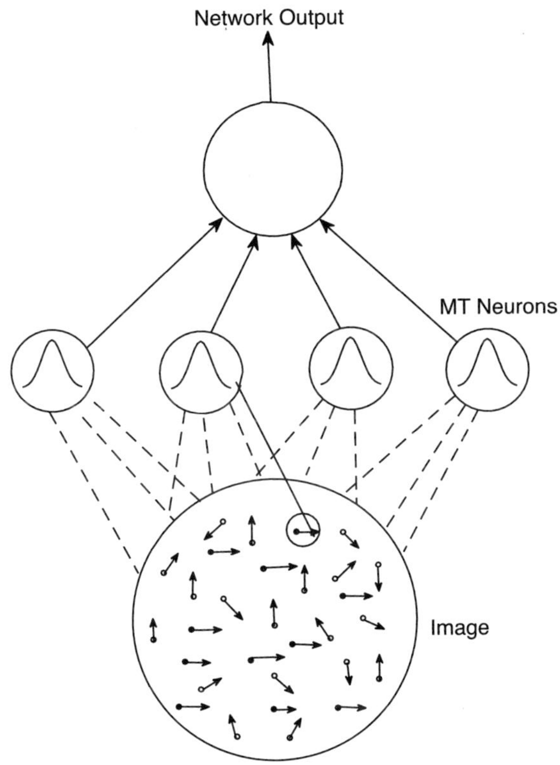

Figure 6.3
A model for judgments of global-motion direction. Arrows show the motion direction of signal dots (dark circles) and the random directions of noise dots (light circles). MT neurons code for motion in different directions, and the output decision reflects the weighted average of all MT neurons. After Vaina, Sundareswaran, and Harris,[8] figure 4, with permission.

improve the coding or representation of the stimuli for other tasks"[7] (p. 244). In these variants, the decision was based on evidence from feed-forward connections from representation to decision, while learning changed recurrent weights. Another model also proposed that learning occurred through recurrent connections that were guided by attention and feedback (figure 6.4),[4] with the goal of explaining differential learning with trial-by-trial feedback, block feedback, uncorrelated feedback, biased feedback, and no feedback.[12] This also involved a three-layer feed-forward architecture (an input layer, a hidden layer, and an output layer of two competing units that determined the response), combined with supervised or self-supervised teaching signals that drive task-specific top-down inhibition that recurrently modifies the weights from the input to the hidden layer. In essence, this sketched model learns through task-dependent reweighting of top-down inhibitory connections.[13, 14]

These classic models explored the theoretical foundations of learning through reweighting within a three-layer architecture with several learning rules but also left ample room

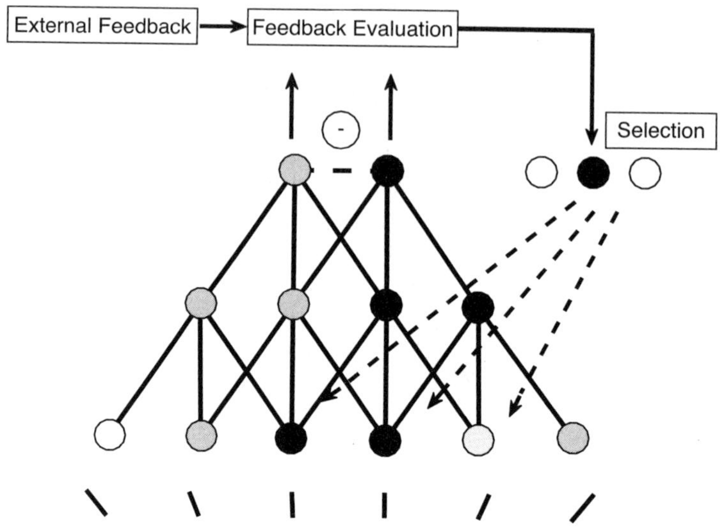

Figure 6.4
An extended model for perceptual learning of hyperacuity with top-down modification of weights based on feedback and a winner-take-all decision competition in a three-layer feed-forward model. After Herzog and Fahle,[4] figure 5, with permission.

for future models to better address the challenges described at the beginning of the chapter. Regarding challenge 1, the sensory representations of these models were often abstract and simplified, and internal noise, which is necessary to capture stochastic aspects of behavior, was only included in a few cases. Regarding challenge 2, the previous studies were designed to predict learning, with little concern for more complex outcomes. (This focus on simple behavioral phenomena is understandable, as at the time perceptual learning experiments were very simple as well.) Regarding challenge 3, learning rules were generally tested simply for learning, while other manipulations were rarely considered.[4, 6] Many of the models used supervised learning rules, while some hybrid form combining unsupervised and supervised learning is almost surely required to account for the literature on feedback (see chapter 7 for a further discussion of this problem). Furthermore, the potential role of prior knowledge was generally ignored. Regarding challenge 4, some aspect of stability was an implicit feature of these models, because almost all used some form of feed-forward reweighting that left low-level representations largely unchanged.[4, 6] Regarding challenge 5, biological plausibility was sometimes mentioned but rarely systematically examined (and many of the learning rules would be considered biologically implausible). Finally, most models failed to define the requirements for verification and testing as expressed in challenge 6. The experimental paradigms used to test the model would almost surely require many of the methods used to specify internal noises, as were subsequently used in tests of the PTM (see chapter 4).

6.3 The Reweighting Hypothesis and the AHRM Model

Many hundreds of experimental studies of visual perceptual learning have been carried out since the early classical models were first developed. With these growing datasets, we now have a broader range of phenomena against which models can be tested. In this section, we turn to our own model, the AHRM, and how it has accounted for a range of empirical phenomena while simultaneously navigating the basic design challenges.

6.3.1 Perceptual Learning through Channel Reweighting

Despite developments in classic computational modeling, or at least independent from them, the primary theoretical position in the behavioral and physiological literature has focused on explaining visual perceptual learning as a result of plastic retuning of early cortical sensory representations. Since roughly the mid-1990s, the behavioral and physiological literature has been motivated by groundbreaking observations of unusual specificity in tasks associated with properties coded in V1. These observations led to the dominant proposal that the primary substrate of visual perceptual learning is plasticity of the tuning of early cortical sensory representations.

Our own approach focused on an alternate hypothesis. Our goal was to consider how much of visual perceptual learning could in fact be accounted for by reweighting alone, essentially by improving the "readout" of precoded information to a decision unit. Our proposal initially grew out of an early analysis of visual learning through an external and internal noise analysis (as developed in chapter 4), but it also derived from a recognition that plasticity should be kept in balance with stability. We thought that if early sensory representations were constantly changing in response to experience with new stimuli, the result would be an unstable sensory apparatus. Our theory of reweighting expressed the belief that maintaining stable calibration of sensory systems must be one overall system goal.

The experiments that led us to a reweighting counterproposal to explain perceptual learning were some of the first external-noise studies in this area. They used the perceptual template model (PTM) to analyze how learning altered the signal and noise properties of the observer (see subsection 4.4.1). Perceptual learning substantially reduced the contrast thresholds by a factor of about two in both high and low external-noise exclusion conditions (measured using TvC functions at two accuracy levels).[15] This in turn led to our initial proposal of learning through reweighting within a multichannel observer (see figure 6.5).[13, 14] We concluded that "perceptual learning primarily serves to select or strengthen the appropriate channel and prune or reduce inputs from irrelevant channels. The connections between the most closely tuned visual channel(s) and a learned categorization structure are maintained or strengthened, while input from other channels is reduced or eliminated"[13] (p. 13992).

Learning could have been accounted for without presuming any retuning of individual sensory channels, but this logically does not preclude the possibility of individual channel retuning. This multichannel reweighting also provided an explanation of specificity to

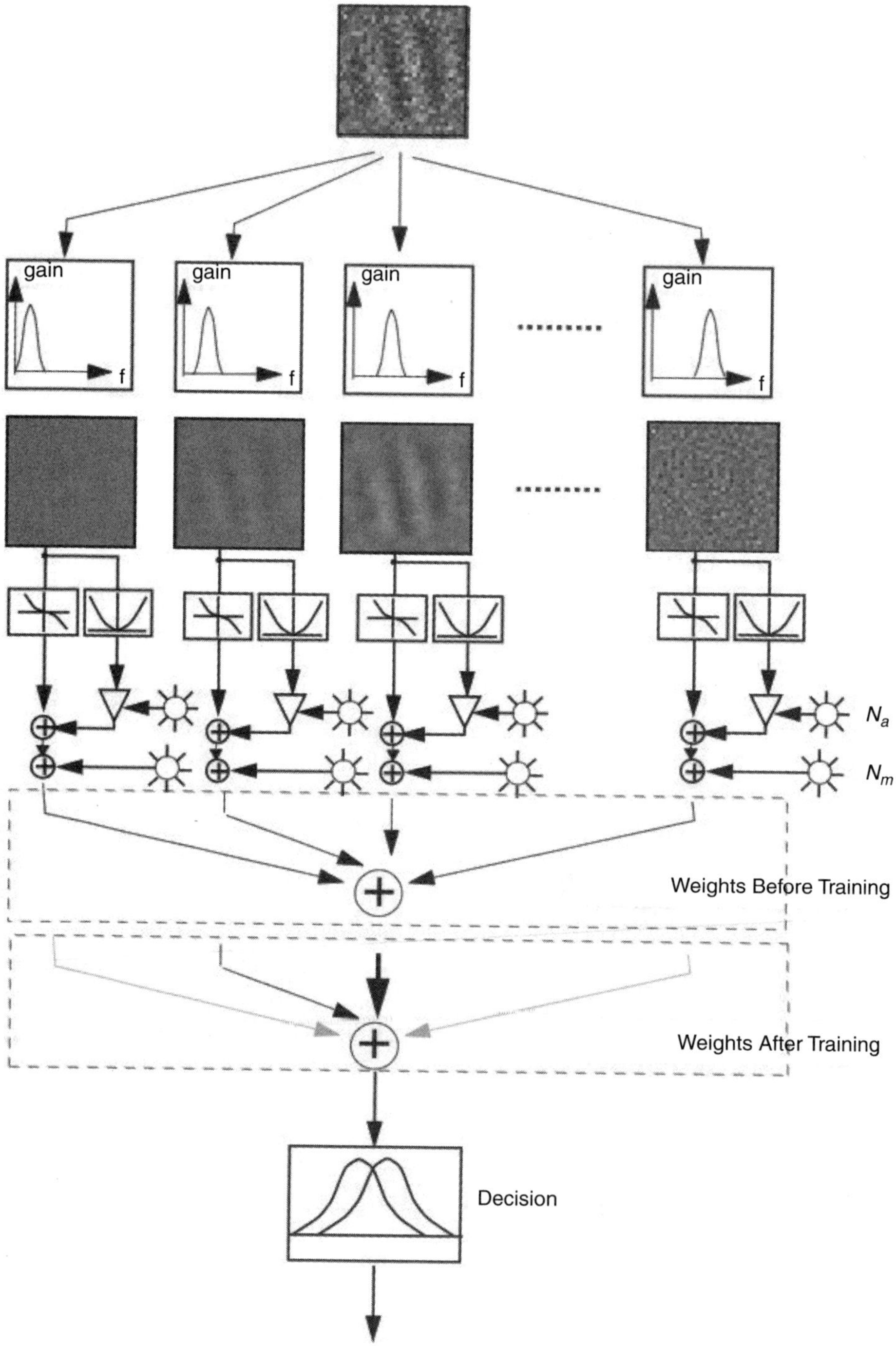

Figure 6.5
Perceptual learning through reweighting in a multichannel observer model. The input image is processed through multiple sensory channels (here shown as being sensitive to different spatial frequencies and orientations) with nonlinearity and internal noises. Adapted from Dosher and Lu,[13] figure 3.

retinal location, spatial frequency, and orientation, because these features have retinotopic representations in the early visual cortex. Because of this, reweighting (selective readout) explains specificity just as well as retuning does. (The position that specificity requires only readout from low-level channels, and not their retuning, was independently proposed by Mollon and Danilova.[16])

In our original reweighting proposal, we sketched the multichannel model of perceptual learning (figure 6.5) in which the nonlinear and noisy responses of multiple channels were weighted to decision. The next step was to implement the multichannel model and the reweighting hypothesis—and to carry out perceptual learning experiments that would generate a rich dataset for testing internal noise, nonlinearity, and especially the learning rule. The AHRM was the response to this challenge.

6.3.2 The Development of the AHRM

The AHRM grew directly out of the proposal that learning reflected multichannel reweighting. Like many classical models, the AHRM included a sensory representation module, a decision module, and a learning module suitably defined for each task. The sensory representation module was designed to mimic characteristic neuronal properties of the early visual cortex (V1); it computes noisy activations for the representation units. These activations were then weighted to make a decision. Finally, the model learns by updating weights using a hybrid or semisupervised Hebbian learning rule. Throughout, internal noise operates as one limiting factor in both performance and learning. As we will see, this hybrid approach has proved largely successful. While retuning was previously the dominant theoretical explanation, most recent reviews of perceptual learning now include reweighting mechanisms, often along with retuning.

The original AHRM, developed with Alex Petrov, implemented the multichannel model of perceptual learning for applications involving spatial pattern judgments such as orientation discrimination (figure 6.6).[2, 3] The *representation module* codes activations in orientation and spatial-frequency units from input patterns. The *decision module* makes orientation judgments, for example, based on the activations of the representations. A nonlinear decision unit combines weighted evidence from the representation activations with input from a bias unit that aims to balance the responses in a two-alternative choice. The *learning module* reweights the connections from representation units to a decision unit using Hebbian learning augmented with bias control and feedback. After the simulated response, explicit feedback, if available, shifts activation in the decision unit toward the correct response before Hebbian learning updates the weights. In the absence of feedback, the learning is unsupervised, extracting correlations between representation activations and the decision. In fitting data, the model reprises the experiment exactly: it takes in a stimulus image seen by the observer, produces a predicted decision on that trial, learns over trials, and uses the same data analyses.

The processing embodied in the representation module is more realistic than in most classical models, with nonlinearities, limiting internal noises, and normalization that are

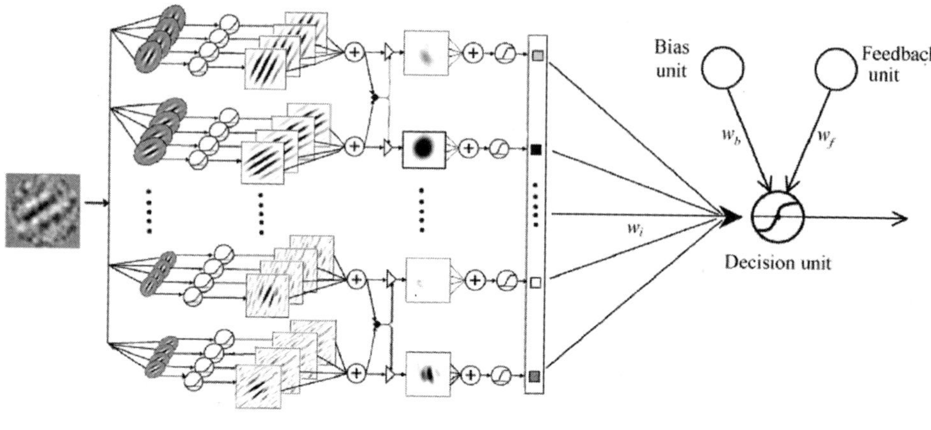

Figure 6.6
The augmented Hebbian reweighting model (AHRM). The model takes a stimulus image (far left) and processes it in a *representation module* that mimics early visual system coding (left) to generate representation activations (vertical rectangle) that are weighted in the *decision module* (right). At the end of the trial, the *learning module* updates the weights with Hebbian learning augmented by feedback and bias control. Simulations re-create experimental sequences of trials to make predictions. Adapted from Petrov, Dosher, and Lu,[2] figures 6 and 8.

characteristic of early visual responses. This front-end module was meant to be consistent with the noise properties of the perceptual template observer model (PTM), which previously had been used to characterize change in observer state (e.g., learning, attention, adaptation; see chapter 4).[13, 14, 17–24] The internal noise introduces stochastic properties to the predictions. Initial weights are set to reflect general prior knowledge. In these ways, the AHRM aimed to address many of the modeling challenges specified earlier. As we will see, the design of the AHRM allows it to make predictions consistent with the mechanisms measured in external-noise studies, testable predictions about feedback (see chapter 7), and predictions about the specificity of learning to the stimulus. Further details of the implementation, including equations, are provided in the chapter's appendix (section 6.8).

Subsequent variants have substituted the representation module with alternative representation modules for the particular stimulus domains (e.g., for motion direction or point Vernier tasks). Finally, the AHRM was elaborated into an integrated reweighting theory (IRT) that involves additional hidden layers of (location-)invariant representations in order to account for aspects of location transfer (see chapter 8).

Although inspired by features of the human brain, the AHRM architecture is still simplified and abstract. It reduced the key modules to their essentials for the purposes of investigating the explanatory power of reweighting, or changed "readout," as a learning principle. The framework also may provide a structure within which to incorporate more elaborate, neutrally inspired representation, decision, or learning modules developed in the future.

6.4 Tests and Applications of the AHRM

The AHRM model has been applied to a range of empirical phenomena of visual learning. In most of these applications, the model has been fit to experimental data (in many model applications, a similarity between predicted patterns and observed patterns of behavioral data is simply noted).

Accurately fitting a model to data requires estimating the values of its parameters. This has generally been done using hierarchical grid-search methods, first evaluating a matrix of spaced parameter values and then narrowing in on regions of the parameter space that are more promising. Each model fit can be modestly computationally intensive, because of the processing of many different samples of external noise added to the stimuli, internal noise in the simulated representations, and decision. It also depends on the number of estimated parameters and the intrinsic complexity of the model. Simulations have typically been run many times (hundreds to thousands) to generate average predictions and confidence bands. Each run of the model leads to a different sequence of responses and somewhat different weight changes for each simulated observer, because of stochastic trial-by-trial variations resulting from internal and external noises and different random trial sequences. Variations in performance from one simulated run to the next may suggest something about differences in outcomes between individual observers.

Although the AHRM has a number of parameters, we set many values a priori based on the physiology (e.g., bandwidths of the orientation and spatial-frequency representations of the representation module), while the others were estimated in initial applications and were then held constant.[2, 3] Of approximately 15 core parameters, the majority (usually 9 or 10) were fixed, while the values of the remaining four or five were varied slightly to match observed human data in a particular experiment. These variable parameter values include the internal noises, model learning rate, weights on feedback and bias control, and sometimes decision nonlinearity. In addition, initial weights have typically been set to include some knowledge of the stimulus domain and the task instructions (e.g., initial weights on activities of units tuned to clockwise or counterclockwise orientations have been set as negative and positive, respectively). This is required to account for initial above-chance performance in experimental data, which cannot be matched with random initial weights. Exploratory simulations have indicated that modest changes in representation bandwidths or starting weights were far less important than the other estimated parameters, such as internal noises and the learning rate parameter, in accounting for the data.

As the following sections describe, the original AHRM model has done quite a good job of accounting for human data so far.

6.4.1 Perceptual Learning in Nonstationary Environments

Our initial development and test of the AHRM was based on a relatively complex experiment by the field's standards, one designed to challenge the stability of learning by repeatedly switching the task context.[2, 3] Observers judged the orientation of Gabor stimuli (tilted top right or left) embedded in external noise that was itself dominantly oriented either left or right. (The relationship of the two tasks is of class D; see chapter 3.) The Gabors were of low, medium, or high contrast (figure 6.7). Variations in contrast—which manipulated the accuracy of responses—also constrained estimates of system nonlinearities and the signal and noise properties of the system (see chapter 4).

A repeated alternation design was selected in order to test the architecture and rules used in learning. After a single block in the first external-noise orientation, the external-noise

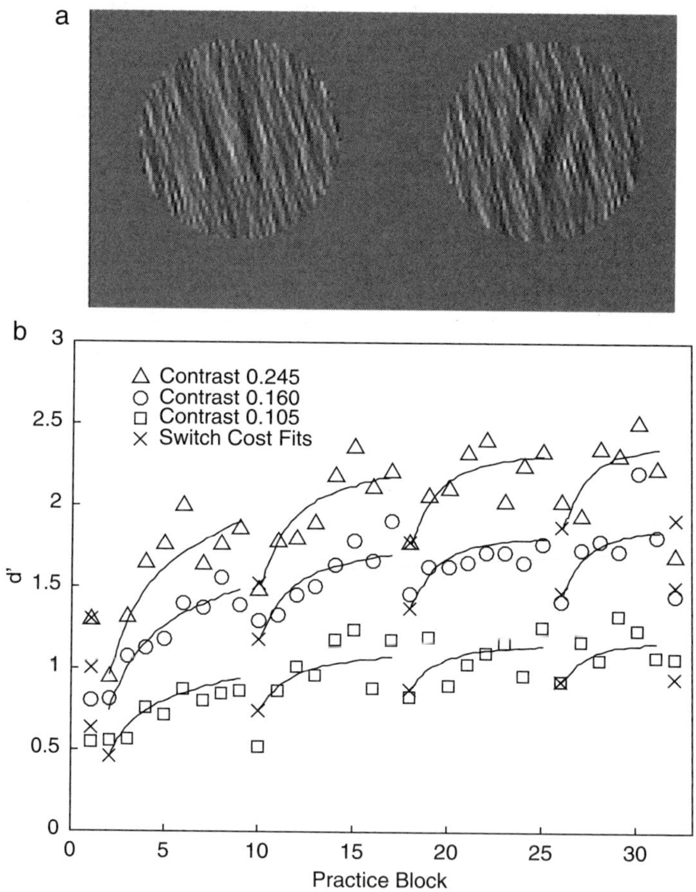

Figure 6.7
Sample stimuli tested in alternating external-noise contexts, (a) shown here in left-oriented external noise congruent (tilted top left) or incongruent (tilted top right) with the orientation of the external noise (a), and (b) discriminability for Gabors of three contrasts as a function of practice block. After Petrov, Dosher, and Lu,[2] figures 1 and 3.

orientation was alternated after every eight blocks of trials for a total of five switches (300 trials per block). In addition, a separate groups of observers performed this task with and without feedback (in separate experiments) in order to more fully understand the nature of the learning rule.[2, 3] A fully supervised back-propagation learning rule used in a network with hidden layers in principle should be able to overcome switching costs by learning distinct weight structures for the two external-noise contexts, as in an "exclusive or" (XOR) problem.[25] In contrast, a Hebbian learning regime in a less complicated network, even one augmented by feedback supervision, should continue to show persistent switching costs.

The human data revealed many expected results but also some surprising ones. The observed discriminability (d') for feedback-trained observers showed learning over training blocks but also persistent switching costs whenever the orientation of the external noise was changed (see figure 6.7). Higher-contrast stimuli of course led to higher response accuracies. What these discriminability curves (put together in the d' computation) failed to show, however, was an apparently counterintuitive interaction between contrast and congruency (figure 6.8). Remarkably, increasing the contrast in a congruent condition (e.g., a right-tilted Gabor in right-tilted external noise) had, if anything, a slightly negative effect on response accuracy. The higher contrasts improved performance only for incongruent stimuli (e.g., a left-tilted Gabor in right-tilting external noise). The AHRM predicted this unusual pattern of data (see the curves in figure 6.8). A second experiment trained observers without feedback and found the same patterns of learning, with persistent switching costs and a similar effect of congruency, and so learning differed primarily in the overall level of bias.[3] The net bias in the direction of the external-noise orientation was 57% in the group trained with feedback, compared with 64% for the group trained without it.

The AHRM provided natural explanations for this pattern of data (see the predictive curves in figure 6.8). Several features of the model contributed to its success. The systematic effect of contrast was accounted for by stimulus transformations in the representation module, which included nonlinear gain control and internal noises. Like the data, the model showed large effects of contrast for incongruent stimuli only and a very small or slightly reversed impact on response accuracy for congruent stimuli. Although at first glance this finding is counterintuitive, it is logical; it makes sense to downweight representation units whose activation is primarily driven by irrelevant orientation energy in the external noise. This, in turn, also downweights evidence from the congruent stimuli. As a result, the model learned across blocks of training but also suffered switching costs at the swap of external-noise orientations. The push and pull between orientations competing for differential weighting alternated in the separate external-noise contexts, while the weights favoring the relevant orientations and spatial frequencies increased through learning.

The operation of the model—in the changing values of the weights connecting representations to decision—provides a window into how learning unfolds, with the weights from the best-fitting model shown in figure 6.9. In order to account for above-chance performance at the beginning of training, initial weights in the model included prior knowledge about

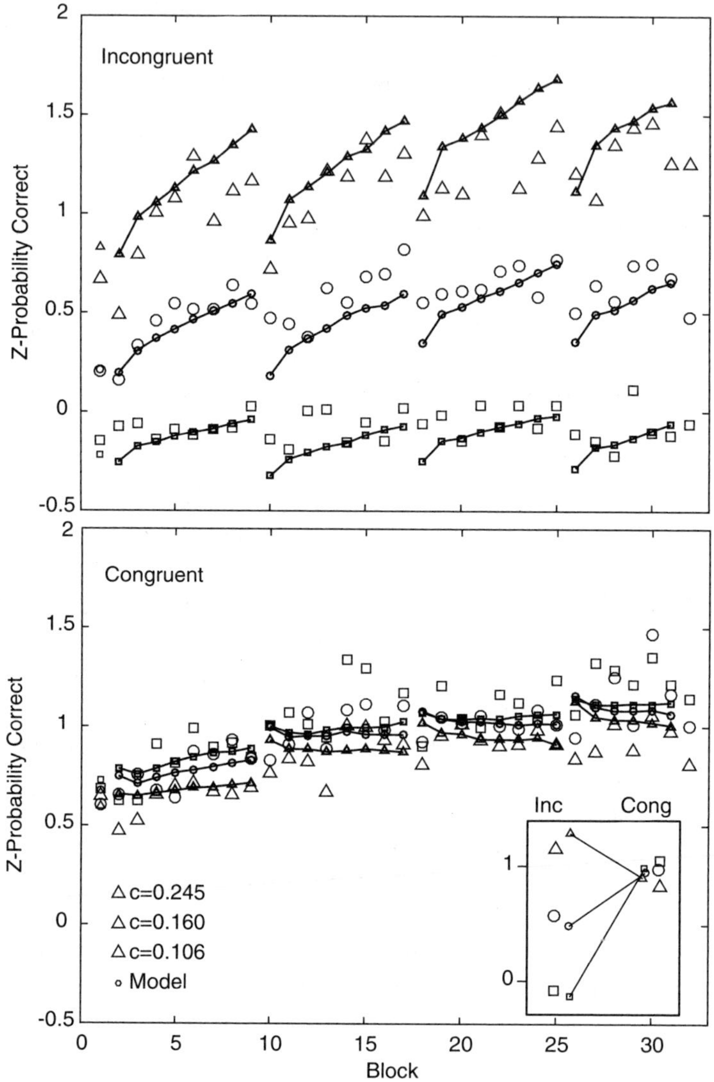

Figure 6.8
Performance accuracy, shown as the Z-score of the probability of response over training blocks for three levels of Gabor contrast for incongruent (a) and congruent (b) stimuli. Data are light symbols, and AHRM fits are dark symbols with lines. After Petrov, Dosher, and Lu,[3] figure 7.

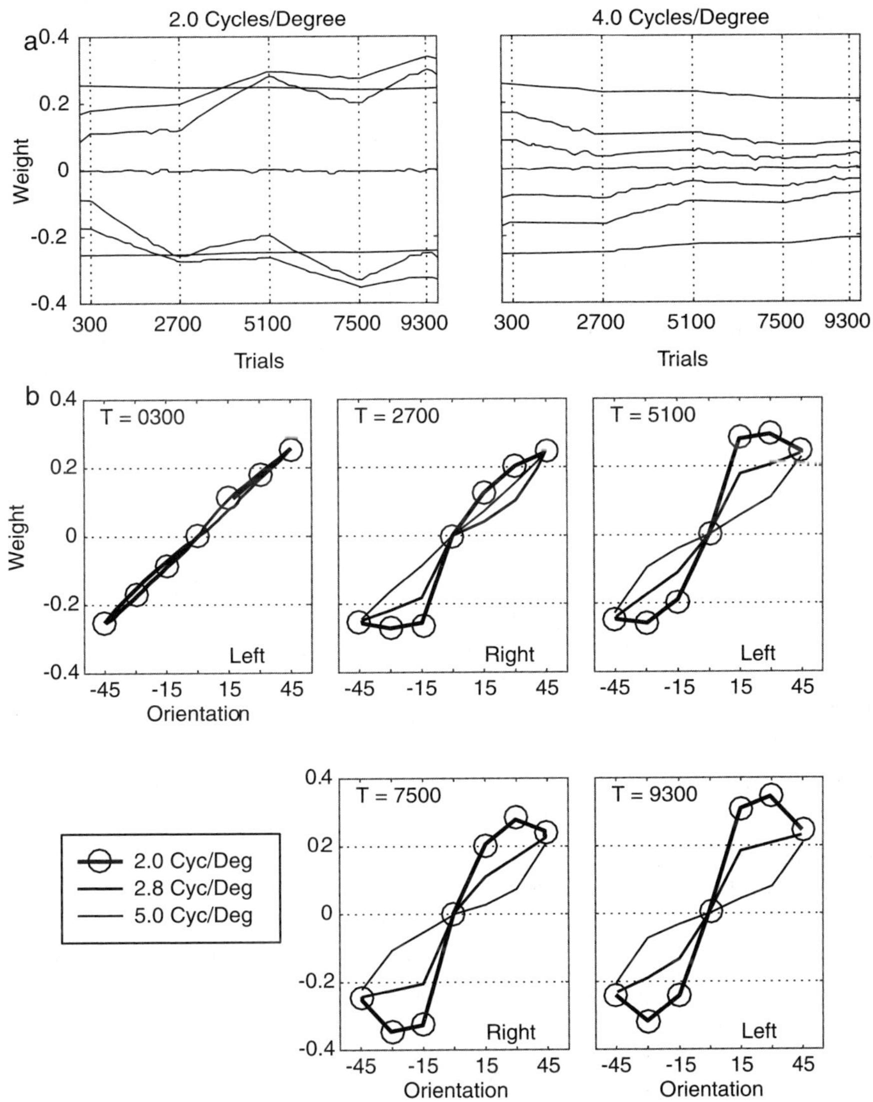

Figure 6.9
Weights change during learning with feedback in alternating external-noise contexts from the best-fitting AHRM simulations. (a) Weight traces for units centered on the target frequency (2 cycles per degree) and another frequency (4 cycles per degree), with lines for each orientation. (b) Weights at the end of each training context epoch (T). The weights near the most diagnostic orientations shift with each context change, showing shift costs. After Petrov, Dosher, and Lu,[2] figure 11.

orientation, with counterclockwise orientations set negative and clockwise ones set positive. Over the course of training, weights on representations most sensitive to the spatial frequency and orientations of the Gabor targets increased, while others decreased. In the model, the switching cost occurred because weights adjusted for the *other* external-noise context were no longer optimal, and the incongruent weights changed more than the congruent weights.[2, 3]

To further understand the optimal weight structure for the task, we performed a separate analysis and found that the optimal decision boundaries in the two-dimensional spatial-frequency/orientation space of the stimuli were approximately linear, though different in the two external-noise contexts. Furthermore, the model predicted that learning with and without feedback would be similar because the high-contrast stimuli provided such good information even without feedback. On the other hand, bias correction was found to be essential to learning without feedback whenever the external noise was changed. In this case, learning and performance often became unstable shortly after the first switch of external noise, then leading to a loss of accuracy and an inability to recover performance in the absence of either bias control or feedback. When feedback is available, it usually dominates the influence of bias control.

The AHRM used a simple architecture without hidden layers, and an augmented Hebbian learning rule. This was no accident but rather was an adaptation to fit the observed data. Because we viewed the structure of the experiment as an XOR problem, we first fit a network model with hidden layers and back-propagation learning rules. Learning with this model, however, was too powerful to fit the behavioral data; it adjusted too quickly and ultimately reduced switching costs. This led us to a network model with fewer layers and rules that combined unsupervised and supervised learning. Fortunately, this form of the AHRM has subsequently been shown to account for many other experiments. The hybrid learning rules, in particular, have proven essential to predicting the behavioral effects of feedback (see chapter 7).

While designed to learn by reweighting from stable early representations, the AHRM also provided a framework for evaluating early sensory retuning. In separate simulations, we tested various retuning schemes (i.e., narrowing the orientation bandwidths for units tuned near the target, for all units, and other schemes). The resulting simulations showed only relatively small improvements in performance, with the best schemes yielding on the order of 10% improvements in d' compared to the observed behavioral improvements, which were an order of magnitude larger.[2] Interestingly, though, this 10% estimate is in line with the estimated contributions of cellular retuning in V1 to behavioral responses in monkeys (see subsection 5.4.1). From a purely theoretical point of view, the ability to translate information carried in newly retuned representations into improved decision accuracy almost surely would also require changes in readout, such that newly learned weights could capitalize on the altered representations. Changed encoding requires changed decoding. (This simulation analysis did not incorporate correlated internal noises or their potential impacts on performance, as discussed elsewhere.)[26]

These experiments that deliberately altered context were designed to pose a challenge to the reweighting framework by exercising the modules for representation, decision, bias control, and feedback as well as the learning rule and network architecture. Nevertheless, reweighting has been able to account for this complex pattern of perceptual learning remarkably well.

6.4.2 Basic Mechanisms of Perceptual Learning

In order to further test the AHRM model, we examined its ability to account for earlier external-noise studies on mechanisms of learning. Three possible mechanisms were identified in the perceptual template model (PTM) and external-noise analyses: *stimulus enhancement*, *external-noise exclusion*, or *multiplicative noise reduction/gain control change*.[13, 14] The previous datasets have not found evidence for the third mechanism. (See subsection 4.4.2 for a description of these mechanisms.)

One such simulation study found that reweighting in the AHRM could indeed account for the typical pattern of combined stimulus enhancement (improvements in zero or low external noise) and external-noise exclusion or filtering (improvements in high external noise). The model was fit to data from the experiment in which orientation discrimination (±12° Gabors) was trained in eight levels of external noise, with staircases at two accuracy criteria, resulting in reductions in contrast threshold of about 65% across all external-noise levels.[13, 14] The model provided an excellent fit to the data (figure 6.10), accounting for 95.3% of the variance. Likewise, reweighting accounted for improvements in both low and high levels of external noise, corresponding to the mixture of stimulus enhancement and external-noise exclusion. It also predicted the threshold shift between the two accuracy criterion levels (a shift invariance on the log scale that rules out changes in gain control or nonlinearities in the PTM analysis).[14] After some internal-noise parameters were selected to fit the initial thresholds, the learning rate was selected to fit the data; all the remaining features of the data simply fell out of the model structure.

An examination of changes in model weights over training again showed increases for representations tuned closer to the spatial frequency and orientations of the Gabor (2 cycles per degree, ±12°) and modest decreases for units tuned to irrelevant spatial frequencies and orientations (see details in Lu, Liu, and Dosher[27]). The final model weights were close to optimal, as estimated by the weights after extended training of the AHRM at high contrast and zero external noise.[27]

6.4.3 Asymmetric Transfer of Learning in High and Low Noise

In section 6.4.2, the AHRM was shown to be consistent with improvements in performance when an observer had been trained by a mixture of all external-noise levels. Other training protocols, however, have yielded curious asymmetries in learning and transfer between tests with zero and high external noise.[28] It turns out that the AHRM can predict these experimental results as well. Following pretests in zero and high external noise, one group

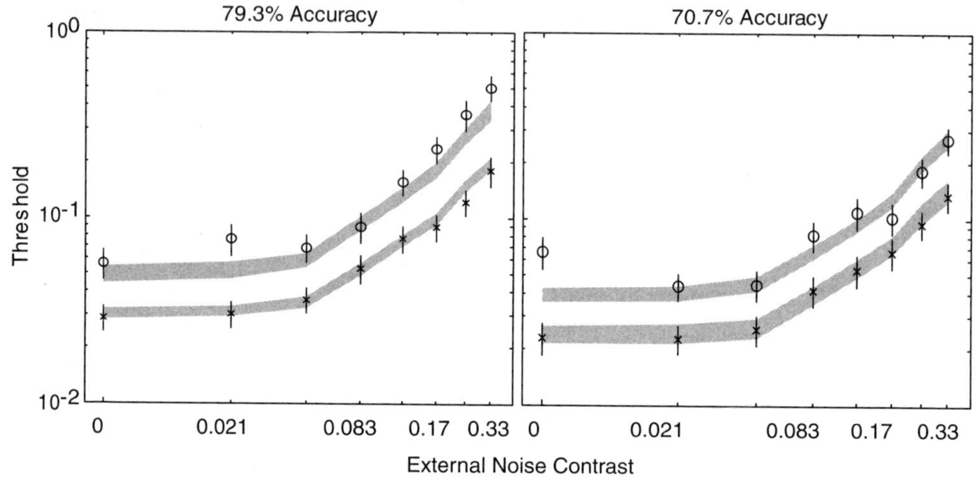

Figure 6.10
The AHRM model fits to perceptual learning in orientation discrimination seen in threshold versus external-noise contrast (TvC) curves at two accuracy levels, comparing early (higher thresholds) and late (lower thresholds) levels in training, showing the improvements in low and high external noise and at different threshold accuracies found in the behavioral data (see subsection 4.4.2) (data in symbols, model predictions in gray bands). Redrawn from Lu, Liu, and Dosher,[27] figure 4; data from Dosher and Lu.[13]

trained in zero external noise, while the other group trained in high external noise. Each group then continued training in the switched condition of a peripheral orientation discrimination task (±8° of vertical in the periphery measuring the contrast threshold at 75% correct, with an RSVP letter task at the fovea, as described in subsection 3.4.3; see figure 6.11). Training in either zero or high external noise produced learning (negative slopes of $\log_{10}$ contrast threshold versus $\log_{10}$ practice blocks), yet training in zero noise transferred almost completely to testing in high external noise, while training in high external noise showed little benefit for performance in zero external noise. This asymmetric transfer might intuitively seem to challenge a reweighting account, yet the AHRM did an excellent job of handling it.

For both training groups, weights on task- and stimulus-relevant representations increased with practice, while other weights decreased (see the weights in the source paper[27]). Weights changed more quickly for training in zero external noise; following this, the weights were damaged slightly by subsequent training in high external noise. Extensive training in zero external noise should come to approximate an optimal weight structure, while training in external noise can never approach more than a very rough approximation of the optimal weight structure, because external noise continues to move the weights in random directions away from the optimal weight state. These destabilizing effects of external noise produce the asymmetries in training and transfer.

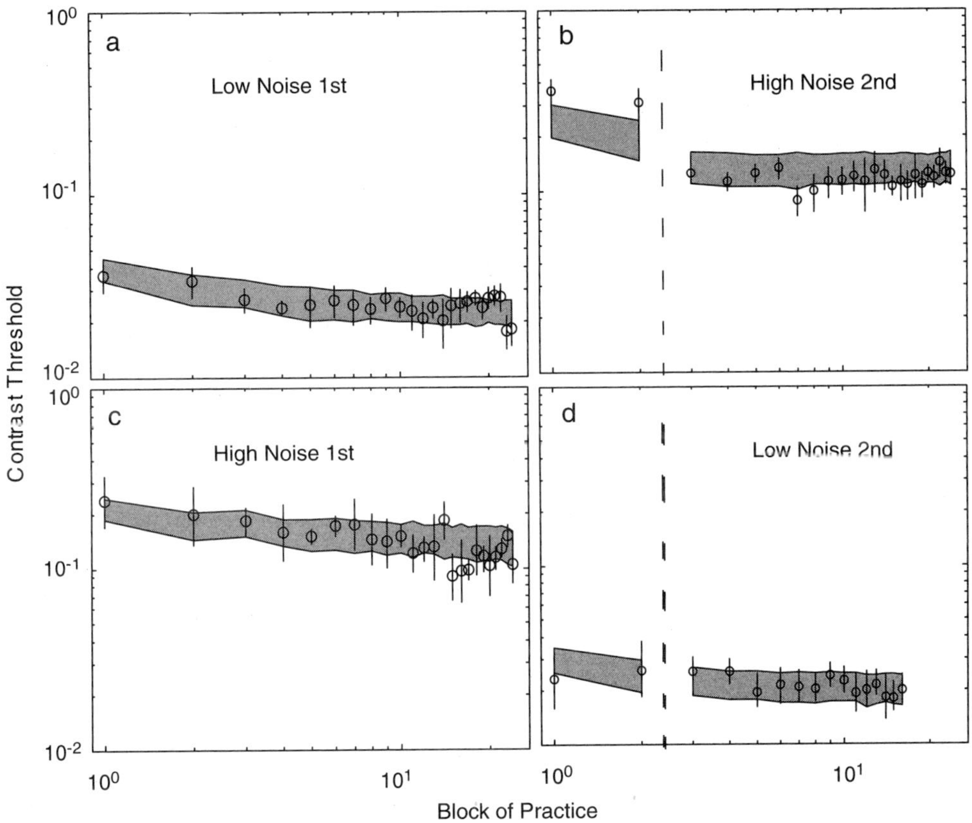

Figure 6.11
The AHRM model (gray bands) accounts for asymmetries in transfer in orientation judgments trained in low and high external noise, measured as contrast thresholds (symbols). Learning first in low noise (a) transfers to high external noise (b) in one group, while learning in high noise (c) does not improve performance in low noise (d) in another. Redrawn from Lu, Liu, and Dosher,[27] figure 8, for data taken from Dosher and Lu.[28]

6.4.4 Effects of Pretraining Mechanisms

The AHRM also provided an excellent account of learning following pretraining in contexts of zero or high external noise, as demonstrated in left-right motion-direction discrimination.

In this study, three groups received different training histories: no pretraining, pretraining in high external noise, and pretraining in zero external noise.[29] Pretraining in high or low external noise reduced the corresponding contrast thresholds for left-right motion discrimination by about 37% and 44%, respectively. Then, all three groups were trained in multiple levels of external noise in the main experiment (10,000 trials total). This subsequent main-experiment training yielded about 41% threshold reduction across all external-noise levels for the no-pretraining group; about 55% threshold reduction in low external noise and 25% in (pretrained) high external noise; and only about 5% threshold reduction

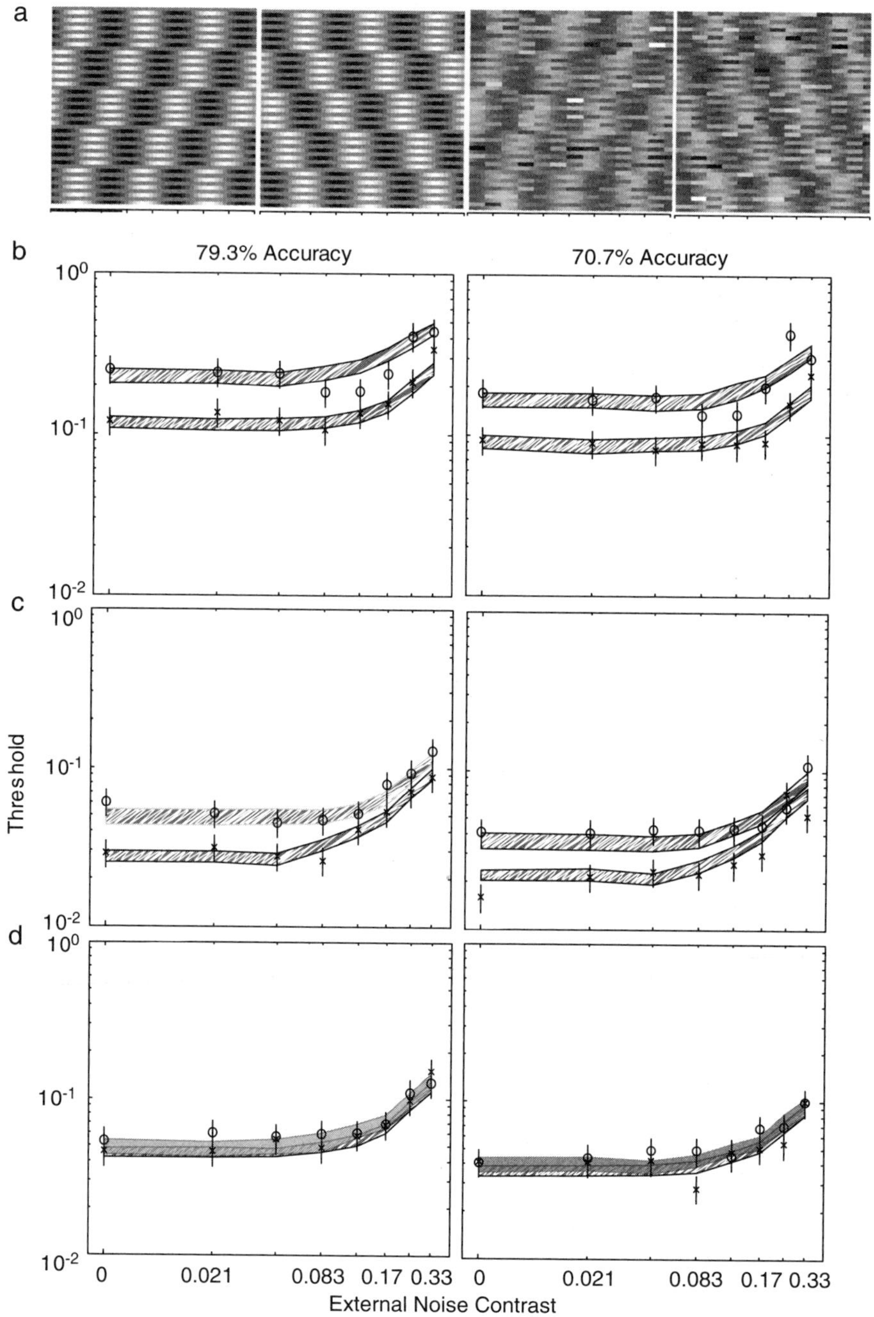

Figure 6.12

Learning to discriminate left-right sine motion direction in different external noises after different pretraining. Sample motion stimuli (a) and contrast-threshold data (symbols) before and after learning at two accuracy levels in a main task without pretraining (b), after pretraining in high external noise (c), and after pretraining in zero external noise (d); fits of the AHRM are shown as gray bands. Redrawn from Lu, Liu, and Dosher,[27] figures 7 and 8; data from Lu, Chu, and Dosher.[29]

in any external-noise level after zero-external noise pretraining. Pretraining in zero noise led to nearly complete learning. (It should be noted that the initial levels also differ in threshold between the three groups of three observers each.) The AHRM provided excellent fits to data of the main experiment, as seen in figure 6.12.

Pretraining in zero external noise was more efficient than training in high external noise at increasing weights on relevant units, while pretraining in high external noise was more effective in reducing the weights on irrelevant units. This application of the AHRM used the analogy between the motion in space and time and orientation in x and y to code sine-wave motion stimuli.[27] (Spatial-frequency and orientation-tuned representations processed five frames of sinusoidal luminance motion, with 90° phase shifts between frames, as if they were x-y instead of x-t images). A model of the sensory inputs to motion, analogous to inputs at the levels of V1 and MT/MST, was subsequently developed to handle dot motion[30] and has been applied to other learning phenomena in the motion domain.

6.4.5 Colearning Analysis of Multiple Tasks

Another important empirical phenomenon found in perceptual learning is the specificity of learning to the task. If perceptual learning occurs through retuning, then learning in different tasks that share the same sensory representations should interact because the initial training in one would alter the sensory representations used in the other. By contrast, separate decision structures for two different tasks are generally assumed by reweighting explanations, thus leading to predictions of independent learning.

In order to adjudicate these questions, the AHRM model was tested in an experiment that alternated training in a bisection and Vernier offset task every 10 blocks (see chapter 3).[31] The task used the same stimuli (in bisection, a middle dot was judged closer to the top or the bottom of two outer reference dots, while in Vernier a middle dot was judged left or right of the reference dots, with initial threshold offsets set to yield a criterion of 70.7% correct). The human data showed essentially independent learning in the two tasks, similar to a previous training study with only one phase of training in each task (see figure 6.13).[32] Not surprisingly, the AHRM was consistent with independent perceptual learning, since separate decision structures were required for the two tasks. This application used a new representation module that coded location in radial basis functions with divisive gain control between the location units and added internal noise; the model was otherwise equivalent to the AHRM.[2, 3] This implementation was related to early radial basis function models of Vernier tasks but different in that it also incorporates gain control, nonlinearity, internal noises, and a Hebbian learning rule.[5, 6]

6.5 Other Reweighting Models of Learning

A variety of other models with the same architecture but possibly different representation, decision, and learning modules—similar to the AHRM—have also been proposed recently. These represent only a small sample of the models that the framework could generate.

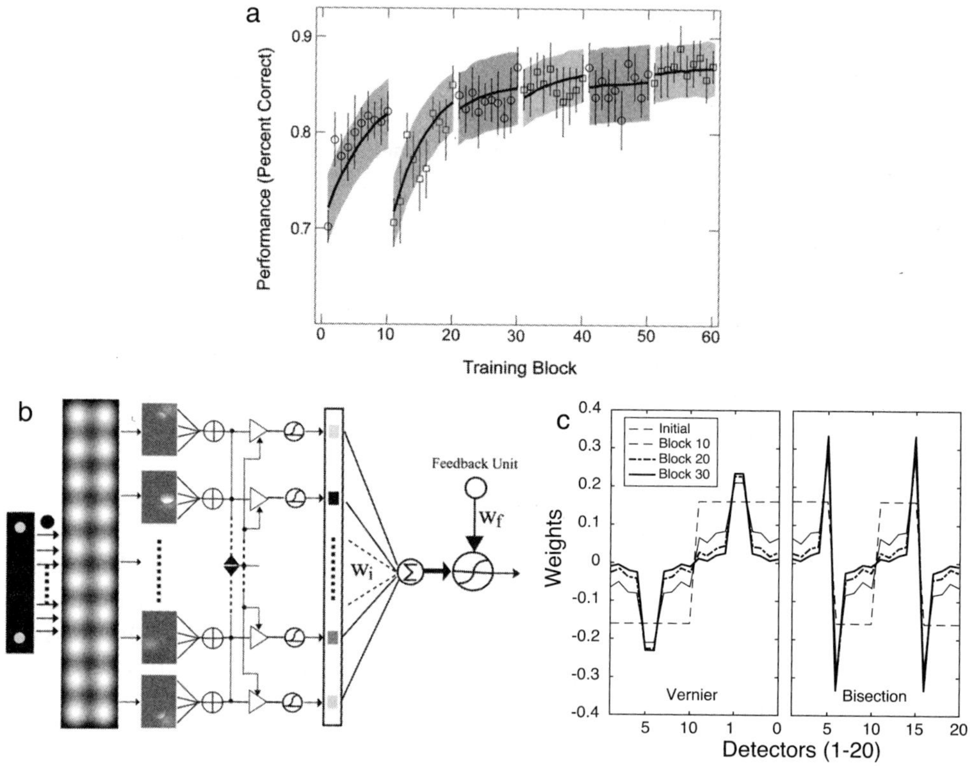

Figure 6.13
Colearning of bisection and Vernier tasks with dot stimuli using alternate task training measured with percentage correct. (a) An AHRM model, with a front-end coding spatial locations (radial basis functions with divisive gain control and internal noise, not shown) (b) and weight diagrams for the best-fitting model of the two tasks (c). Adapted from Huang, Lu, and Dosher,[31] figures 2 and 6.

One of these models basically modified the classic HBF model for hyperacuity.[6] It replaced units in the middle layer with oriented Gabor filters, added a simple nonlinearity in responses, and used a supervised rule (the Widrow-Hoff least-squares loss).[33] This model (figure 6.14) improved on the original HBF model by predicting the effects that varying line lengths and separations, mixed training, and transfer to other forms of hyperacuity have on performance.[34–38] The authors also fit their model to the alternating oriented external noise data (but not the congruency effect),[2, 3] stating that it was "a simplification of [the AHRM], which includes multiple stages of integration with respect to spatial phase and scale and features complex operations such as response normalization."[33] (p. 597).

Another model used the orientation and spatial-frequency representation module of the AHRM but substituted different decision and learning modules based on Bayesian adaptive precision pooling to predict tilt judgments.[39] Adaptive precision pooling is essentially

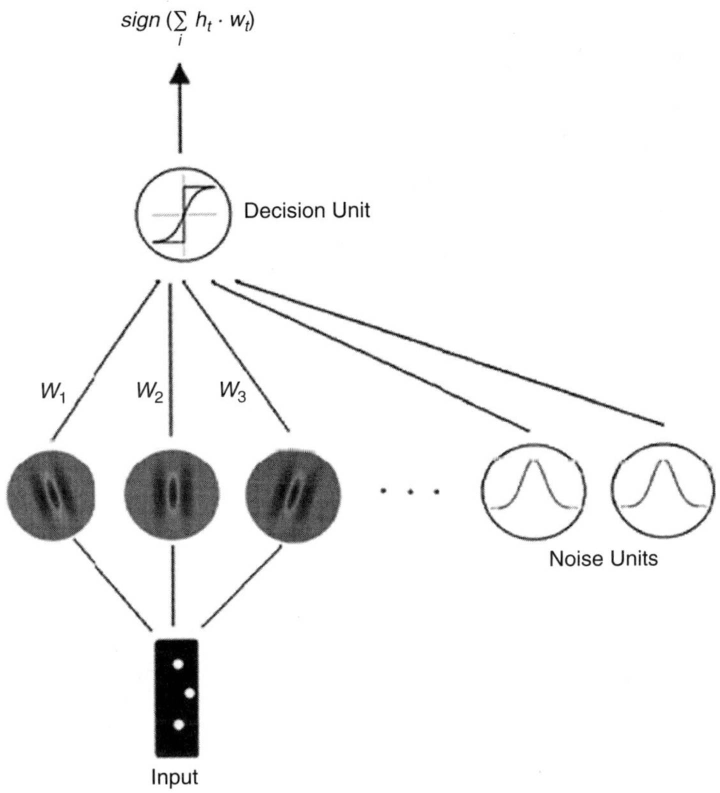

Figure 6.14
A modified reweighting model of hyperacuity tasks. A three-dot Vernier stimulus feeds into oriented Gabor basis functions whose output, with the outputs of noise units, are reweighted to a decision unit. Adapted from Sotiropoulos, Seitz, and Seriès,[38] figure 1, with permission.

a form of reweighting that identifies a small number of sensory inputs that drive decision while ignoring all others, leaving a sparse set of weights to decision. Precision pooling predicted accuracies that far exceeded actual human accuracy. The authors see the model as predicting an upper bound on behavior and emphasize the relative inefficiency of humans compared to the Bayesian norm.[39]

Yet another model based on reweighting was developed to account for learning in motion-direction judgments in monkeys.[40] The model architecture was analogous to the AHRM but substituted the representation module with one that approximated MT population motion responses, whose reweighted pooling is used to make a decision that mimics the pooling of neural evidence from MT in the LIP (figure 6.15). In this model, weights were updated based on reward expectation error (the difference between an actual and expected reward), labeled reinforcement learning. The model was fit to two types of data for

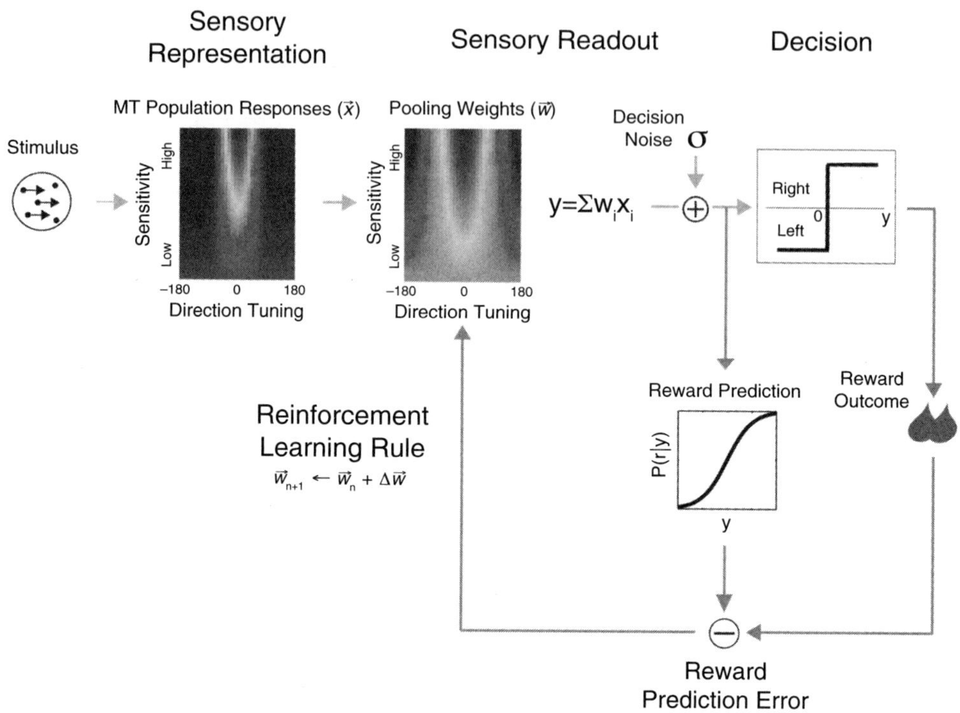

Figure 6.15
Reweighting model of perceptual learning for motion-direction discrimination in monkeys, from Law and Gold.[40] A motion stimulus activates an MT-like sensory representation and is passed through a weight structure to a decision unit. Reinforcement rules driven by a deviation between the expectation of reward and the actual reward were used for learning. After Law and Gold,[40] figure 1, with permission.

left-right motion-direction discrimination: error rates in trials with 99.9% motion coherence (labeled associative learning) and coherence thresholds in trials with weaker motion coherence (labeled perceptual learning). The model also predicted perceptual learning for finer direction discrininations (±10°), based on the most informative neurons, which were assumed to be tuned about 40° away from the true motion directions.

Despite their formal variations, all these new models were also based on the principle of perceptual learning by reweighting, and each used an AHRM-like network architecture, replacing one or another of the model modules. In each, the reweighting framework provided a successful account of the data.

6.6 Summary

This chapter examined many of the classic computational models of visual perceptual learning, with a focus on the structural predictions of the augmented Hebbian reweighting model (AHRM) and how it accounts for a range of empirical phenomena, and ended with a brief account of other similar recent models and their applications. Each of these models used a representation module specialized for the stimulus domain, a decision module specialized to the task judgment, and a learning module based on either supervised, unsupervised, and/or semisupervised algorithms from neural network theory. All the models accounted for learning with some form of reweighting, almost always from stable initial sensory representations. Where it has been tested, then, the reweighting principle has been successful in accounting for the human behavioral data. If learning actually combines reweighting with retuning (representation enhancement), it may be possible to incorporate this within the broader reweighting framework.

The AHRM moved beyond its predecessors in a number of ways, most notably in the design of its representation module and in the treatment of noise. Whereas the classic models often used symbolic or simplified characterizations of the inputs, the representation module of the AHRM was designed to mimic visual system responses to actual stimulus images. The representation modules of the AHRM and other related models, which incorporate nonlinearities and internal noises, thus have the power to make predictions for a wide variety of stimuli. Performance predictions for stimuli with different contrasts or external noises fall out directly from the "just complicated enough" representation computations. The AHRM (and some of its modified forms) was also shown to powerfully explain the quantitative phenomena of perceptual learning in multiple task domains. It predicted the nature of switching costs, mechanisms of perceptual learning revealed in different external-noise conditions, asymmetric transfer of training between zero and high external noise, and colearning of independent tasks. As we will see in the following chapters, the AHRM has also organized the existing literature and made new predictions about feedback in learning, the generalization across stimuli and judgments within a location, the impact of mixing training of multiple tasks, and other aspects of learning.

Several recent models with similar architectures implemented alternative representation or learning modules while leaving the fundamental architecture of the model intact. The alternative representation modules were required to make predictions in a new task domain (e.g., replacing the spatial pattern module with modules representing motion or dot location), while the variations in the learning modules seemed to have reflected preferences for other popular theoretical positions, such as reinforcement learning or Bayesian inference. Although these alternative forms were successful in accounting for perceptual learning, the corresponding experiments to which they were applied were not designed as strong tests of the learning rules.

Our original reweighting implementation and many other models that we reviewed have used feed-forward reweighting of evidence from multiple sensory channels, sometimes called channel reweighting or altered readout.[13, 14] The idea of reweighting, however, is a broad one—it is not restricted only to feed-forward connections. In principle, reweighting in a neural system could occur through many kinds of connectivity: bottom-up feed-forward connections from one module or level to the next; top-down feedback connections from higher-level modules; the interplay of feed-forward and feedback connectivity between networks of modules; and/or recurrent connections within modules or areas. The broad success of purely feed-forward reweighting models, however, suggests that it is generally adequate to account for the behavioral measures of learning, yet there is also evidence from the physiology that top-down signals influence processing. Taken together, all these possible configurations (bottom-up, top-down, and within-level) should permit implementations of the reweighting principle that are even more flexible.

At the same time, neither the classical nor the new models have systematically addressed those situations in which learning requires the recruitment or creation of units to represent complex or naturalistic objects composed of combinations of many different features. It was this form of recruitment or creation that we argued (in chapter 2) was required to solve the combinatoric explosion of all possible feature combinations. This is not unlike the approach to accounting for multiple tasks, in which the models simply assume that new decision units and weight connections exist for different tasks. Future models may be able to address these issues by implementing a process by which units are recruited or modified to represent feature combinations and subpart relationships inherent in high-level visual tasks while retaining the selective reweighting mechanisms that have been so powerful in explaining learning in early and mid-level visual tasks.

Lastly, the AHRM suggests new approaches to studying the physiology of learning. Though early qualitative theories of perceptual learning proposed modified cortical representations, as early as V1 (or even earlier in the LGN), the reweighting models keep the earliest layers of visual representation relatively stable. For those models with multilayer architectures, reweighting of evidence from unchanging lower layers to units in higher layers de facto alters the stimulus representations in those higher layers. Implicitly, too, new decision units must be recruited for new tasks, so new weight structures will connect to them from stimulus representations. The reweighting models also place a strong emphasis on the weights or connections between different processing modules. Future physiological studies may find a way to focus on measuring the changing weights. Although feed-forward reweighting in relatively shallow architectures has so far provided an excellent account of a range of perceptual learning phenomena, reweighting in a more complex, powerful, and flexible multilayer network is still to be fully investigated. Several efforts that follow this line of thought will be considered in chapters 8, 9 and 12.[41–43]

6.7 Future Directions

Intuition alone is inadequate to evaluate how proposed principles of learning might work together in a complex system. A formal computational model that generates quantitative predictions is critical. Only within a formal model can we determine whether the proposed principles of representation, decision, and learning, taken together, in fact operate in the expected ways and make the intuitively expected predictions.

In practice, theoretical statements about perceptual learning have often started from a small set of experimental investigations. Every empirical investigation involves the selection of a stimulus, task, and training paradigm. It must also specify the number of training trials and kind of feedback, among other things. A quantitative or computational model can save precious time and resources by helping to guide the lengthy and expensive empirical observations with human observers. It does this by summarizing a wide variety of observed data efficiently and by predicting how the system is likely to perform under novel testing conditions. The observed behavioral data may corroborate the model, but it may instead challenge the model, ultimately prompting improvements. In either case, researchers have a robust tool to help optimize discovery given limited time and research capacity.

Models can also make new predictions that drive future model validation, which occurs when a model accounts for performance in different situations with intuitively plausible changes in model parameters. Generative models that make predictions for many experimental protocols may also serve as the theoretical engine for computationally optimizing training protocols (as discussed in chapter 12).

There are a number of directions that future research might take to advance this initiative:

1. To challenge existing models and drive new developments, future work could test stimuli and tasks that are more varied. So far, learning models have been applied to single tasks, or to two quite different tasks that by definition require different decisions and decision weights. Furthermore, the vast majority of perceptual learning experiments have used two-alternative tasks, which divide the stimulus space in relatively simple ways. These are well described, at least approximately, as linearly separable discrimination problems. Novel experiments and models should begin to approach the complexity of the real world.
2. Future work is needed to more fully examine the nature of stochastic noise in the perceptual system and whether and how learning alters noise properties. Some classical models added a single decision noise or a few sources of representation noise. The AHRM incorporated internal noise in all representations and decisions, which can take the form of either internal additive or internal multiplicative noise in the perceptual template model (PTM).[13, 23] For model tractability, different sources of internal noises have generally been assumed to be independent, but if internal noises were correlated, this would have implications for the information available in the signals. If learning reduced the correlations between the noises in different sensory units, this could be one way to improve stimulus coding. Some simulation studies have looked at such mechanisms as changing the correlations between activities in representation units.[26] (There are also formulations of correlated noise in the PTM.)[44]
3. Future computational research could incorporate changes in top-down, feedback, or recurrent connections during learning. Architectures that are more complex have already been outlined in some schematic models of perceptual learning, but these remain to be

implemented. These proposals included networks with hidden layers and gating structures,[45] feedback and recurrence,[4, 26] and attention and other top-down influences.[46–48]

4. Future models could explore how new units are recruited to represent either new task judgments or, in high-level perceptual tasks, new multifeatured objects. This would require enriching the representation modules to include multiple kinds of inputs and perhaps addressing how these are incorporated to define complex new entities (see chapters 2 and 8).
5. Ideally, the models developed at different levels of description will dovetail to create an integrated understanding of the perceptual learning systems. The models considered in this chapter were necessarily abstracted away from biological implementations, yet they clarify the major functional components and systematize the findings in the field. The correspondence between these models and biologically plausible computational models of neurons or groups of neurons may lead to new insights that can guide research into the functional significance of neural computations and structures.

6.8 Appendix: Implementation Details of the AHRM

This appendix provides some implementation details of the augmented Hebbian reweighting model (AHRM). These equations are closely consistent with those initially developed in collaboration with Alex Petrov[2, 3] and extended in subsequent publications in collaboration with Jiajuan Liu.[27, 41, 49–54]

6.8.1 Representation Module

The representation module encodes the stimulus image into activations distributed over a set of characteristic representation units—here a bank of orientation and spatial-frequency tuned filters. It also includes subsequent processing stages that carry out normalization and gain control (see figure 6.6), as described in more detail in the original papers.[2, 3]

Briefly, representation units tuned to different orientations θ, spatial frequencies f, and spatial phases ϕ compute *phase-sensitive maps* $S(x, y, \theta, f, \phi)$ of the input image $I(x, y)$ at retinotopic location (x, y). In most applications, there were 35 *channels* corresponding with channels centered at seven orientations ($\theta \in [0, \pm 15, \pm 30, \pm 45]$ degrees) and at five spatial frequencies ($f \in [1, 1.4, 2, 2.8, 4]$ cycles per degree). For some experiments, additional orientations span the full 180°. Each channel is computed at four phases ($\phi \in [0, 90, 180, 270]$ degrees) (or phase quadrature). In general, experimental stimuli do not correspond exactly with any of the channels; instead, the representation is distributed in activity over a number of partially matching channels. The phase-sensitive maps $S(x, y, \theta, f, \phi)$ are computed with templates corresponding with two-dimensional Gabor receptive fields:

$$S(x, y, \theta, f, \phi) = \left| RF_{\theta, f, \phi}(x, y) \otimes I(x, y) \right|^2. \tag{6.1}$$

The ⊗ symbol denotes the convolution operator; $|*|^2$ represents rectification, similar to computations in other forms of normalization.[55] Receptive field bandwidth parameters were chosen to be similar to the tuning of parafoveal simple cells in the macaque striate cortex (full-bandwidth at half-amplitude of $h_\theta = 30°$ for orientation and $h_f = 1$ octave for spatial frequency).[56] While these values were typical of physiology, a sensitivity analysis of the model showed that modest variations in bandwidth had little effect on the fits to behavioral data.[2] The phase-sensitive maps are then combined into phase-invariant maps $E(x, y, \theta, f)$ by summing, as in this equation:

$$E(x, y, \theta, f) = \Sigma_\phi S(x, y, \theta, f, \phi). \tag{6.2}$$

Phase invariance often occurs in V1 complex cells,[56–58] and it has been used in other models of texture and motion perception.[59–61] After phase combination, the responses are converted into normalized maps $C(x, y, \theta, f)$ by using nonlinear divisive normalization, as in this equation:

$$C(x, y, \theta, f) = E(x, y, \theta, f)/(s^2 + N(f)). \tag{6.3}$$

The nonlinear divisive normalization term $N(f)$ is a normalization pool (sum over unit activations) that is independent of orientation and modestly tuned for spatial frequency. This is meant to approximate the shunting inhibition observed in the visual cortex, combining activations across all orientations and only modestly tuned for spatial frequency, corresponding with physiological and psychophysical evidence.[23, 62–65] The small semi-saturation constant s^2 prevents a division by zero, relevant in stimulus conditions of zero external noise.

The activation information is further aggregated over spatial location into an activation $A(\theta, f)$ for a single representation unit per channel by pooling the energy maps over space in the image roughly corresponding to the relevant stimulus. A radial symmetric Gaussian kernel W_r with full-width at half-height h_r approximated this weighted summation. In many applications, this corresponds roughly to 2°, but this should be altered depending on the stimulus. This spatial pooling is described in equations (6.4) and (6.5), which also introduce internal noise $\varepsilon_{\theta, f}$.

$$A'(\theta, f) = \Sigma_{x,y} W_r(x, y) C(x, y, \theta, f) + \varepsilon_{\theta, f}, \tag{6.4}$$

$$A(\theta, f) = \begin{cases} \dfrac{1 - e^{-\gamma_r A'}}{1 + e^{\gamma_r A'}} A_{max}, & A' \geq 0 \\ 0, & otherwise \end{cases} . \tag{6.5}$$

This produces activations in representation units that are positive and saturate for high-contrast inputs. These internal noises added to the representation units, together with decision noise, limit the accuracy of predicted behavioral performance even in the absence of external noise.[23, 66–68]

This representation module, then, takes the stimulus image and generates a corresponding distributed activation pattern, where the activation $A(\theta, f)$ of each unit encodes the (noisy) normalized spectral energy with sensitivity centered at the corresponding orientation and spatial frequency. Although used by us and others to model perceptual learning in a range of tasks, including orientation judgments,[2, 3, 27, 49] Vernier offset judgments,[49, 50] sinusoidal motion judgments,[27] and tilt judgments,[39] other kinds of perceptual tasks, such as other kinds of hyperacuity[31] or dot motion tasks, have required alternative front-end modules.[8, 30, 40] Although simplified, this computation of sensory representations has been sufficient for a good account of a range of data.

6.8.2 Task-Specific Decision Module

The decision module takes the pattern of activations over the representation units as the input and generates a response based on weighted summation. For example, if the task requires deciding whether the stimulus was tilted to the top left or right (counterclockwise or clockwise) of a reference angle, then activities for units tuned to the left or right received negative or positive weights at the decision unit. (Two decision units in competition with one another can also be used to replace the one-unit decision module, which also allows the decision dynamics to be driven by a winner-take-all competition.)[2]

The decision unit sums the representation activations using current weights w_i. The sum also includes an input from a top-down bias unit b, plus Gaussian decision noise ε (mean 0 and standard deviation σ_d), as in this equation:

$$u = \sum_{i=1}^{N\ channels} w_i a_i - w_b b + \varepsilon. \tag{6.6}$$

The output of the decision unit o', corresponding with the binary behavioral response, is a sigmoidal function of u, as in equations (6.7) and (6.8) (e.g., the model responds left if $o' < 0$ and right otherwise). The value $\pm A_{max}$ is the positive or negative value at which the decision unit saturates.

$$G(u) = \frac{1 - e^{-\gamma u}}{1 + e^{-\gamma u}} A_{max}, \tag{6.7}$$

$$o' = G(u) \text{ (early)}. \tag{6.8}$$

This decision computation in two-alternative tasks acts similarly to a single linear classification boundary whose orientation in representation space is set by the current weight vector.[2] (Note that we subsequently have developed a variant of the reweighting framework to carry out n-alternative forced choice.[69, 70])

It is common practice in perceptual learning experiments to instruct observers about the task, including showing them examples of the stimuli in some cases. Often, initial performance even before training is above chance. The knowledge from prior experience and instruction is implemented in the model as initial weights. These initial values, acting like priors, are an intrinsic aspect of model performance early in training, although different initial weight settings that embody some knowledge have often led to similar predictions.[2] For example, initial weights have sometimes been set proportional to the preferred orientation of the representation unit relative to the instructed standard, or $w_i = \left(\frac{\theta_i}{s}\right) w_{init}$ relative to a vertical standard. In the applications described in this chapter, the initial weights reflected knowledge of the instructed task-relevant dimension but not other dimensions (i.e., approximate specification of orientation if the task is orientation discrimination, but flat over spatial frequency). These initial weights are then modified on a trial-by-trial basis by learning.

6.8.3 The Learning Module

The learning module gradually upweights inputs from the most diagnostic representation units (channels) for the task and downweights others. On every trial, weights are updated using Hebbian rules; feedback, if available, is incorporated as a shift in the decision variable toward the correct response that yields a new late-phase decision variable o, as in this equation:

$$o = G(u + w_f F) \text{ (late).} \tag{6.9}$$

Feedback $F = \pm 1$ (for binary decisions) is added into the decision variable with weight w_f. This drives the late activation o toward the activation limit $\pm A_{max}$, which is often set at ± 0.5. If the weight on feedback is high, the decision variable will be shifted to the positive or negative maximum (whichever is correct), whereas if the weight on feedback is low, it may only slightly shift the decision variable, which is often in an intermediate range in the absence of feedback, where $o = o'$. Pure Hebbian learning occurs without feedback. Incorporating feedback into the late activation at the decision unit before Hebbian learning associates input activations with decision variables that are more accurate, which operates as a de facto form of semisupervision.

Weight changes are computed in equations (6.10)–(6.12):

$$\delta_i = \eta a_i (o - \bar{o}), \tag{6.10}$$

$$\Delta w_i = (w_i - w_{min})[\delta_i]_- + (w_{max} - w_i)[\delta_i]_+, \tag{6.11}$$

$$\bar{o}(t+1) = \rho o(t) + (1 - \rho)\bar{o}(t). \tag{6.12}$$

The weight change δ_i depends on the presynaptic activation a_i, the difference of the postsynaptic activation o from its weighted long-term average $\bar{o}$ and the learning rate of the model η. Equation (6.11) bounds the weights by scaling the weight changes in proportion to their distance from the upper or lower limit (e.g., O'Reilly and Munakata[71]). The addition of a comparison between the postsynaptic activation and its weighted long-term average is used in some Hebbian models and has some basis in physiology.[72] The long-term average weights recent trials more heavily, as described in equation (6.12). All these details together are a form of normalization that constrains the weights. Together, this learning module reweights the evidence from sensory representations to decision to improve the accuracy of the response classification.

6.8.4 Adaptive Bias or Criterion Control

Adaptive bias or criterion control shifts the decision variable to compensate for biases in the immediate response history by adding a corrective input to the decision unit. This serves to guide (supervise) the learning process by counteracting response biases. When feedback is available, it will dominate, and the bias control becomes unimportant. However, in nonstationary learning conditions in the absence of feedback, this bias or criterion control has been shown to be critical for system stability and learning.[3]

Adaptive criterion control assumes that observers monitor their own response behavior and seek to equalize response frequencies, essentially trying to match stimulus probabilities that are balanced in many experiments (e.g., 50%:50% in two-alternative tasks). A running average exponentially discounts the past response history, as in equations (6.13) and (6.14):

$$r(t+1) = \rho R(t) + (1-\rho) r(t), \qquad 6.13)$$

$$b(t+1) = r(t). \qquad (6.14)$$

This control on the bias input to decision is a weak form of supervision. Different weighting on the bias input was used to model different levels of bias in behavioral data with and without feedback.[2, 3] It has also been used to model the effects of block feedback.[49, 50]

References

1. Lillicrap TP, Cownden D, Tweed DB, Akerman CJ. Random synaptic feedback weights support error backpropagation for deep learning. *Nature Communications* 2016;7(1):1–10.

2. Petrov AA, Dosher BA, Lu Z-L. The dynamics of perceptual learning: An incremental reweighting model. *Psychological Review* 2005;112(4):715–743.

3. Petrov AA, Dosher BA, Lu Z-L. Perceptual learning without feedback in non-stationary contexts: Data and model. *Vision Research* 2006;46(19):3177–3197.

4. Herzog MH, Fahle M. Modeling perceptual learning: Difficulties and how they can be overcome. *Biological Cybernetics* 1998;78(2):107–117.

5. Poggio T, Fahle M, Edelman S. Fast perceptual learning in visual hyperacuity. *Science* 1992;256(5059):1018–1021.

6. Weiss Y, Edelman S, Fahle M. Models of perceptual learning in Vernier hyperacuity. *Neural Computation* 1993;5(5):695–718.

7. Zhaoping L, Herzog MH, Dayan P. Nonlinear ideal observation and recurrent preprocessing in perceptual learning. *Network: Computation in Neural Systems* 2003;14(2):233–247.

8. Vaina LM, Sundareswaran V, Harris JG. Learning to ignore: Psychophysics and computational modeling of fast learning of direction in noisy motion stimuli. *Cognitive Brain Research* 1995;2(3):155–163.

9. Sagi D, Adini Y, Tsodyks M, Technion AW. Context dependent learning in contrast discrimination: Effects of contrast uncertainty. *Journal of Vision* 2003;3(9):173.

10. Teich AF, Qian N. Learning and adaptation in a recurrent model of V1 orientation selectivity. *Journal of Neurophysiology* 2003;89(4):2086–2100.

11. Tsodyks M, Gilbert C. Neural networks and perceptual learning. *Nature* 2004;431(7010):775–781.

12. Herzog MH, Fahle M. The role of feedback in learning a Vernier discrimination task. *Vision Research* 1997;37(15):2133–2141.

13. Dosher BA, Lu Z-L. Perceptual learning reflects external noise filtering and internal noise reduction through channel reweighting. *Proceedings of the National Academy of Sciences* 1998;95(23):13988–13993.

14. Dosher BA, Lu Z-L. Mechanisms of perceptual learning. *Vision Research* 1999;39(19):3197–3221.

15. Graham NVS. *Visual pattern analyzers*. Oxford University Press; 1989.

16. Mollon JD, Danilova MV. Three remarks on perceptual learning. *Spatial Vision* 1996;10(1):51–58.

17. Dosher BA, Lu Z-L. Noise exclusion in spatial attention. *Psychological Science* 2000;11(2):139–146.

18. Dosher BA, Lu Z-L. Mechanisms of perceptual attention in precuing of location. *Vision Research* 2000;40(10):1269–1292.

19. Han S, Dosher BA, Lu Z-L. Object attention revisited: Identifying mechanisms and boundary conditions. *Psychological Science* 2003;14(6):598–604.

20. Lu Z-L, Chu W, Dosher BA, Lee S. Independent perceptual learning in monocular and binocular motion systems. *Proceedings of the National Academy of Sciences* 2005;102(15):5624–5629.

21. Lu Z-L, Dosher BA. Spatial attention: Different mechanisms for central and peripheral temporal precues? *Journal of Experimental Psychology: Human Perception and Performance* 2000;26(5):1534–1548.

22. Lu Z-L, Dosher BA. External noise distinguishes attention mechanisms. *Vision Research* 1998;38(9):1183–1198.

23. Lu Z-L, Dosher BA. Characterizing human perceptual inefficiencies with equivalent internal noise. *Journal of the Optical Society of America A* 1999;16(3):764–778.

24. Lu Z-L, Liu CQ, Dosher BA. Attention mechanisms for multi-location first- and second-order motion perception. *Vision Research* 2000;40(2):173–186.

25. Rumelhart D, Hinton G, Williams R. Learning representations by back-propagating errors. *Nature* 1986;323(6088):533–536.

26. Bejjanki VR, Beck JM, Lu Z-L, Pouget A. Perceptual learning as improved probabilistic inference in early sensory areas. *Nature Neuroscience* 2011;14(5):642–648.

27. Lu Z-L, Liu J, Dosher BA. Modeling mechanisms of perceptual learning with augmented Hebbian reweighting. *Vision Research* 2010;50(4):375–390.

28. Dosher BA, Lu Z-L. Perceptual learning in clear displays optimizes perceptual expertise: Learning the limiting process. *Proceedings of the National Academy of Sciences* 2005;102(14):5286–5290.

29. Lu Z-L, Chu W, Dosher BA. Perceptual learning of motion direction discrimination in fovea: Separable mechanisms. *Vision Research* 2006;46(15):2315–2327.

30. Tlapale É, Dosher BA, Lu Z-L. Construction and evaluation of an integrated dynamical model of visual motion perception. *Neural Networks* 2015;67:110–120.

31. Huang C-B, Lu Z-L, Dosher BA. Co-learning analysis of two perceptual learning tasks with identical input stimuli supports the reweighting hypothesis. *Vision Research* 2012;61:25–32.

32. Fahle M, Morgan M. No transfer of perceptual learning between similar stimuli in the same retinal position. *Current Biology* 1996;6(3):292–297.

33. Sotiropoulos G, Seitz AR, Seriès P. Perceptual learning in visual hyperacuity: A reweighting model. *Vision Research* 2011;51(6):585–599.

34. Westheimer G, McKee SP. Spatial configurations for visual hyperacuity. *Vision Research* 1977;17(8):941–947.

35. Seitz AR, Yamagishi N, Werner B, Goda N, Kawato M, Watanabe T. Task-specific disruption of perceptual learning. *Proceedings of the National Academy of Sciences* 2005;102(41):14895–14900.

36. Webb BS, Roach NW, McGraw PV. Perceptual learning in the absence of task or stimulus specificity. *PLoS One* 2007;2(12):e1323.

37. Seitz A, Watanabe T. A unified model for perceptual learning. *Trends in Cognitive Sciences* 2005;9(7):329–334.

38. Sotiropoulos G, Seitz AR, Seriès P. Changing expectations about speed alters perceived motion direction. *Current Biology* 2011;21(21):R883–R884.

39. Jacobs RA. Adaptive precision pooling of model neuron activities predicts the efficiency of human visual learning. *Journal of Vision* 2009;9(4):22,1–15.

40. Law C-T, Gold JI. Reinforcement learning can account for associative and perceptual learning on a visual-decision task. *Nature Neuroscience* 2009;12(5):655–663.

41. Dosher BA, Jeter P, Liu J, Lu Z-L. An integrated reweighting theory of perceptual learning. *Proceedings of the National Academy of Sciences* 2013;110(33):13678–13683.

42. Sotiropoulos G, Seitz AR, Seriès P. Performance-monitoring integrated reweighting model of perceptual learning. *Vision Research* 2018;152:17–39.

43. Wenliang LK, Seitz AR. Deep neural networks for modeling visual perceptual learning. *Journal of Neuroscience* 2018;38(27):6028–6044.

44. Lu Z-L, Dosher BA. Characterizing observers using external noise and observer models: Assessing internal representations with external noise. *Psychological Review* 2008;115(1):44–82.

45. Mato G, Sompolinsky H. Neural network models of perceptual learning of angle discrimination. *Neural Computation* 1996;8(2):270–299.

46. Roelfsema PR, van Ooyen A. Attention-gated reinforcement learning of internal representations for classification. *Neural Computation* 2005;17(10):2176–2214.

47. Roelfsema PR, van Ooyen A, Watanabe T. Perceptual learning rules based on reinforcers and attention. *Trends in Cognitive Sciences* 2010;14(2):64–71.

48. Schäfer R, Vasilaki E, Senn W. Perceptual learning via modification of cortical top-down signals. *PLoS Computational Biology* 2007;3(8):e165.

49. Liu J, Dosher B, Lu Z-L. Modeling trial by trial and block feedback in perceptual learning. *Vision Research* 2014;99:46–56.

50. Liu J, Dosher BA, Lu Z-L. Augmented Hebbian reweighting accounts for accuracy and induced bias in perceptual learning with reverse feedback. *Journal of Vision* 2015;15(10):10,1–21.

51. Liu J, Lu Z, Dosher B. Augmented Hebbian learning accounts for the complex pattern of effects of feedback in perceptual learning. *Journal of Vision* 2010;10(7):1115.

52. Liu J, Lu Z-L, Dosher BA. Mixed training at high and low accuracy levels leads to perceptual learning without feedback. *Vision Research* 2012;61:15–24.

53. Liu J, Lu Z-L, Dosher BA. Augmented Hebbian reweighting: Interactions between feedback and training accuracy in perceptual learning. *Journal of Vision* 2010;10(10):29,1–14.

54. Liu L, Kuyk T, Fuhr P. Visual search training in subjects with severe to profound low vision. *Vision Research* 2007;47(20):2627–2636.

55. Heeger DJ. Normalization of cell responses in cat striate cortex. *Visual Neuroscience* 1992;9(2):181–197.

56. De Valois RL, Yund EW, Hepler N. The orientation and direction selectivity of cells in macaque visual cortex. *Vision Research* 1982;22(5):531–544.

57. Movshon JA, Thompson ID, Tolhurst DJ. Spatial summation in the receptive fields of simple cells in the cat's striate cortex. *Journal of Physiology* 1978;283:53–77.

58. De Valois RL, Albrecht DG, Thorell LG. Spatial frequency selectivity of cells in macaque visual cortex. *Vision Research* 1982;22(5):545–559.

59. Adelson EH, Bergen JR. Spatiotemporal energy models for the perception of motion. *Journal of the Optical Society of America A* 1985;2(2):284–299.

60. Knutsson H, Granlund GH. Texture analysis using two-dimensional quadrature filters. In *IEEE Computer Society Worshop Computational Architecture Pattern Analysis Image Database Management* 1983;206–213.

61. Pollen DA, Ronner SF. Phase relationships between adjacent simple cells in the visual cortex. *Science* 1981;212(4501):1409–1411.

62. Cannon MW, Fullenkamp SC. Spatial interactions in apparent contrast: Inhibitory effects among grating patterns of different spatial frequencies, spatial positions and orientations. *Vision Research* 1991;31(11):1985–1998.

63. Carandini M, Heeger DJ, Movshon JA. Linearity and normalization in simple cells of the macaque primary visual cortex. *Journal of Neuroscience* 1997;17(21):8621–8644.

64. Chubb C, Sperling G, Solomon JA. Texture interactions determine perceived contrast. *Proceedings of the National Academy of Sciences* 1989;86(23):9631–9635.

65. Graham N, Sutter A. Normalization: Contrast-gain control in simple (Fourier) and complex (non-Fourier) pathways of pattern vision. *Vision Research* 2000;40(20):2737–2761.

66. Ahumada A, Watson A. Equivalent-noise model for contrast detection and discrimination. *Journal of the Optical Society of America A* 1985;2(7):1133–1139.

67. Burgess A, Shaw R, Lubin J. Noise in imaging systems and human vision. *Journal of the Optical Society of America A* 1999;16(3):618–618.

68. Burgess A, Wagner R, Jennings R, Barlow HB. Efficiency of human visual signal discrimination. *Science* 1981;214(4516):93–94.

69. Liu J, Dosher B, Lu Z-L. Perceptual learning in n-alternative forced choice with response and accuracy feedback, and a reweighting model. *Journal of Vision* 2017;17(10):1078.

70. Dosher B, Liu J, Chu W, Lu Z-L. Dissimilarity and spatial separation enable perceptual learning in multitask visual training. *Journal of Vision* 2020; accepted.

71. O'Reilly RC, Munakata Y. *Computational explorations in cognitive neuroscience: Understanding the mind by simulating the brain.* MIT Press; 2000.

72. El B, Cooper L, Munro P. Theory for the development of neuron selectivity: Orientation specificity and binocular interaction in visual cortex. *Journal of Neuroscience* 1982;2(1):32–48.

7

Feedback

Although feedback is often incorporated in perceptual learning protocols, learning can still occur in its absence. What is the role of feedback in learning? In this chapter, we classify feedback into different categories and consider the influence of each. These include trial-by-trial feedback, intermittent or partial feedback, block feedback, false feedback, and exaggerated feedback, each relying more or less on unsupervised or supervised learning algorithms. The experimental literature consists of a complex set of feedback phenomena, which turn out to be well explained by the learning rules of the augmented Hebbian reweighting model (AHRM) and other semisupervised or hybrid models of learning.

7.1 Feedback in Perceptual Learning

Successful learning almost always requires practice. But is simple repetition enough? How important is it to know how well you are doing as you learn? Even if you can learn without knowing, how might it help? What kind of supervisory evaluation, or *feedback*, leads to the most effective learning?

The relationship between feedback, practice, and learning can be complicated. Manipulations of feedback have produced mixed findings in the literature. In some circumstances, explicit feedback has been shown to improve learning. It can increase learning in many cases and indeed may actually be necessary. In other circumstances, however, learning will occur in the absence of feedback. The question is, what distinguishes the two situations?

In the typical perceptual learning experiment, feedback is generally provided, even if the feedback is not an explicitly manipulated factor. To evaluate its role precisely, however, requires the comparison of several different feedback protocols in otherwise equivalent learning conditions. Such studies can be very informative. Discovering those forms of feedback that are most effective, and the circumstances that might determine this efficacy, has implications far beyond the experiment in question. Such studies could, in principle, help to reveal more general principles of learning, while aiding in the design of real-world training protocols.

Whether inside or outside the laboratory, the extent to which learning requires feedback seems to depend critically on the nature of the task. A person might easily learn to classify perceptual stimuli with full information (e.g., when an external teacher provides the desired response for each stimulus).[1, 2] However, a teaching signal of this kind may only be available some of the time, or only partial information may be provided. If a teacher is not available, people will have to learn in whatever way they can based on their general knowledge and the statistics of the stimuli themselves.[3]

Humans learn in many of these circumstances, but whether they learn turns out to depend also on the task's difficulty. An easy task may be learned without feedback, while feedback may be critical in a more demanding one.

Analogous distinctions in the related field of neural network learning theory provide a useful framework. This field makes a distinction between *modes* of learning that are purely supervised, those that are purely unsupervised, and those that are a mixture of the two.[4] Purely supervised learning denotes algorithms that use a full teaching signal, purely unsupervised learning denotes those for which no teaching signal is used, and mixed learning denotes those cases where the teaching signal provides either partial information or full information for selective trials.[4, 5]

Neural network theory makes further relevant distinctions. Supervised learning can involve an external teaching signal that fully specifies the correct response; it can also involve reinforcement learning, in which the signal conveys response accuracy (but not the correct response or the direction of the error). Another burgeoning field, machine-learning theory, further distinguishes between fully supervised learning, in which every instance is labeled by the correct response, and semisupervised learning, in which only a subset of instances are labeled.[6]

Neural networks learn by changing the connection weights between nodes based on specified rules or algorithms. Within each mode (supervised, unsupervised, hybrid, or semisupervised), several learning regimes have been proposed and investigated. These include back-propagation,[7] reinforcement,[8] unsupervised Hebbian,[9] modified Hebbian,[10, 11] and Kohonen rules,[12, 13] as well as various clustering algorithms,[14] alongside other options. Some learning rules, such as back-propagation of error signals, are fully supervised, while others, such as Hebbian rules, are unsupervised. Still other algorithms use some mixture.

At first glance, it might seem easy to identify which of these modes of learning a human observer is using, but in practice, identifying the specific rule or algorithm in play in a given perceptual learning experiment is difficult, because several algorithms might be consistent with the empirical observations.[15–18] Quite often, other experimental factors must be varied in order to rule out incompatible algorithms and thus narrow the set of possible learning modes. Feedback manipulations have been a powerful tool in the literature.

Beyond the questions mentioned earlier, there are additional questions about how the information that feedback provides interacts with reward or attention (topics considered in chapter 9) and how the feedback process operates in the human brain. To take one example, if the experimenter delivers a beep after a correct response and a buzz after an error, how might this engage distinct brain circuits to modify learning? Clearly, a complex neural system will be involved in any feedback-based learning. At the end of the chapter, we briefly speculate about the possible biological substrates that feedback may bring into play.

7.2 The Empirical Literature

Discovering the role that feedback plays in perceptual learning has recently become an active avenue of investigation. Historically, the vast majority of empirical studies were relatively simple. They used trial-by-trial feedback, and the perceptual task being learned almost always involved a binary classification (e.g., left/right, above/below, same/different), even though, outside the laboratory, many real-world situations require more complex or graded decisions. In binary tasks, any accuracy feedback not only tells the observer whether their response was accurate but, by definition, also indicates the correct response (among the two choices). It follows from this that feedback about response accuracy can only disentangle full supervision from reinforcement supervision in more complex tasks (a topic taken up later in the chapter).

Alongside trial-by-trial feedback, a few experiments have studied so-called block feedback, in which aggregate information about performance over a block of trials is provided.[19, 20] Both trial-by-trial and block feedback can produce successful learning in some circumstances, although, as we will see, trial-by-trial feedback is far more powerful. In one unusual demonstration, trial-by-trial feedback produced some evidence for learning even in the absence of a stimulus.[21, 22]

At the same time, we know that in some cases learning has occurred without external feedback[19, 20, 23–27] (e.g., the no-feedback variant of the alternating external-noise experiment described in chapter 6),[10, 11] though sometimes learning has failed in the absence of feedback, often with difficult stimuli or tasks.[19, 20] In some demonstrations, trial-by-trial feedback improved the learning *rate*, although learning as such could occur without it.[23, 25] Another study reported that learning without feedback was able to successfully achieve asymptotic performance, while the addition of feedback was shown to have little impact.[27]

The toll of dysfunctional or misleading feedback (e.g., random feedback uncorrelated with the observer's response or intentionally false feedback) has also been investigated. In one case, false feedback was reported to eliminate learning, though learning rebounded rapidly once accurate feedback was provided.[19] False feedback on a subset of near-threshold stimuli favoring one response over the other induced significant response biases that extended to suprathreshold stimuli.[28–30] Surprisingly, it has been suggested that exaggerated block feedback can actually improve the rate of learning.[31]

To review, learning can sometimes—and perhaps more often than we might think—occur without external feedback, even to the extent that it achieves asymptotic levels. Yet feedback can sometimes improve the rate of perceptual learning or enable learning that is otherwise impossible in very difficult tasks in which initial task performance is very low.[5, 32, 33] In visual learning, trial-by-trial feedback is usually better than block feedback, although block feedback sometimes supports learning. Finally, false or random feedback can disrupt learning.

Perceptual learning models can be very useful in explaining this complex set of empirical findings and then in generating new predictions. Broadly speaking, this modeling literature suggests that perceptual learning is neither purely supervised nor purely unsupervised.[34–36] As we have emphasized previously, theoretical progress is at best qualitative in the absence of models. In the following sections, we describe some prominent network learning theory rules, along with their requirements for supervision (section 7.3), the rules that guide our interpretation of the experimental results (section 7.4), and how these results relate more broadly to the role of feedback in perceptual learning.

7.3 Learning Rules and Feedback

As discussed earlier, models of perceptual learning often borrow the concepts, language, and algorithms of neural network models.[37, 38] Once the learning rules and system architectures have been specified, the most useful models can generate very useful quantitative predictions.

As seen in the examples in chapter 6, network models include at least an input layer, which represents the input from the stimulus, and a decision or output layer, which classifies the stimulus and therefore determines the response (figure 7.1). Hidden layers between the input and output layers, if present, allow more complex representations of stimulus features or feature combinations and enable more complex classifications. Weighted connections, analogous to neural connectivity in a biological system, send information from the input layer, through the hidden layers, to the output layer. The learning rule or algorithm learns on trial k by updating the weights between nodes (units) i and j: $W_{i,j}\ (k+1) = W_{i,j}(k) + \Delta W_{i,j}$, resulting in a change in weight $\Delta W_{i,j}$. Each different learning rule may interact with feedback during learning in different ways, so empirical data may help to constrain the choice.

One of the prominent learning rules used for learning in multilayer networks with hidden layers is *back-propagation*.[1] This fully supervised rule learns the relationships between the stimulus inputs and output targets provided by the teacher. In a perceptual task, this equates to providing feedback that specifies the correct response on each trial. Weight changes are driven by the error signals, the direction and size of the disparity between the current output layer and the target output, and error signals are propagated back through multiple layers of the network. The back-propagation rule defaults to a *delta* rule when the network has only an input layer and an output layer (a perceptron): $\Delta W_i = \eta(t-o)x_i$. Here, o is the output given the current weight state, t is the target output provided by the teacher, x_i is the activity in input unit i, and η is the learning rate. That is, the weight change is jointly proportional to the learning rate, the size of the error, and the activity of the driving input unit. With multiple output units, this corresponds to $\Delta W_{i,j} = \eta(t_j - o_j)x_i$. For example, $o_j = \phi(\Sigma_1^n W_{i,j}x_i)$, where $\phi(z) = 1/(1+e^{-z})$, in which the output values are weighted sums of the input activations, passed through a nonlinear

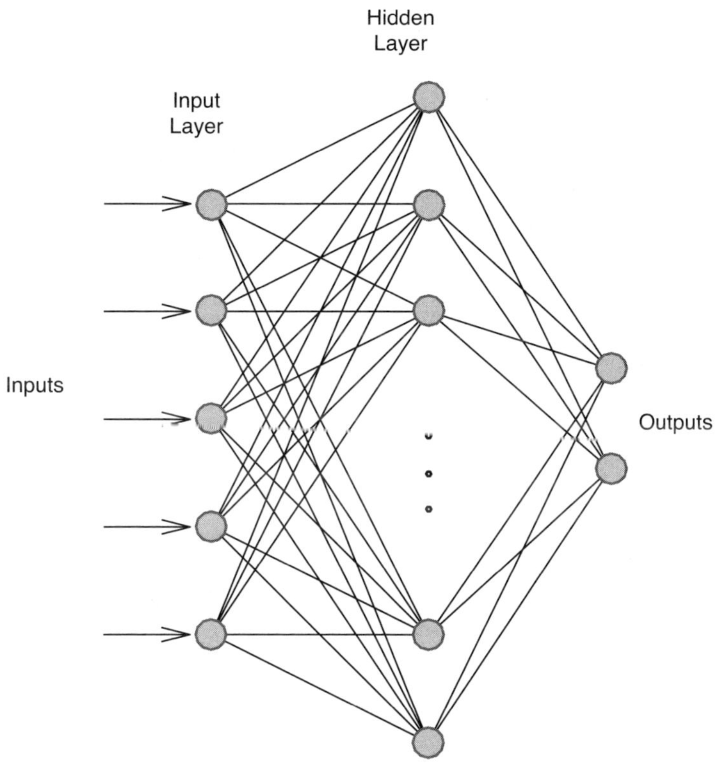

Figure 7.1
Neural network with an input layer, an output layer, and a hidden layer. Learning occurs by changing the weights between units with a learning rule or learning algorithm after each trial (same as figure 1.4).

activation function ϕ such as the logistic. This delta rule is generalized to networks with hidden layers by assigning an error to each (by differentiating the error function with respect to each weight).

The fully supervised back-propagation rule is very powerful, allowing systems to learn complex mappings in multilayer networks. It uses differentiation to compute the error assignment at each successive layer, which is why back-propagation is widely seen as biologically implausible, although researchers are developing variants that are more biologically relevant.[39,40] Some models of perceptual learning have steered away from back-propagation, partly because of biological implausibility but also because its need for explicit teaching signals seems to be incompatible with observations that people sometimes learn in the absence of any feedback.

The *Hebbian learning rule* is one of the standard purely unsupervised learning rules. It strengthens the connection weights between coactive units, extracting their correlations. Hebb explained the rule this way: “When an axon of cell A is near enough to excite a cell

B and repeatedly or persistently takes part in firing it, some growth process or metabolic change takes place in one or both cells such that A's efficiency, as one of the cells firing B, is increased."[41] Put differently, "what fires together wires together." The basic Hebbian rule is $\Delta W_{i,j} = \eta o_j x_i$, where x_i is the activity of the "presynaptic" unit, o_j is the activity or output of the "postsynaptic" unit, and η is the learning rate. Weight change depends on the learning rate and the correlation between x_i and o_j. Hebbian learning does not use a teaching signal, and it is seen as biologically plausible because weight changes could be locally computed. If the activities of the input and representation units and the activity of the response unit(s) are sufficiently correlated, then systematic learning occurs. This reflects initial above-chance performance (i.e., usually 70%–75% correct in two-alternative tasks).

Hebbian rules can have technical issues: weights can increase without bounds and can be driven disproportionately by a single dominant signal. Consequently, some form of normalization is often used in actual implementations, such as limiting the sum of all the weights, introducing nonlinear transformations on the presynaptic or postsynaptic activation, or placing bounds on either the weights or the activation levels.[42, 43] In the AHRM, the magnitude of the weight change depends on the difference between the current weight and a minimum or maximum value.[10, 11]

Although classic Hebbian learning is purely unsupervised, it can be augmented by information from feedback supervision. In the AHRM, activity in the decision unit ("postsynaptic") is driven toward the correct response when feedback is available before a cycle of Hebbian learning, increasing the accuracy of what is learned because the correlated response is the correct one. If performance accuracy is low at the beginning of learning, correlations are low, and supervisory inputs will be necessary to learning.

In *reinforcement learning*, sometimes also called weakly supervised learning, reinforcement signals (either positive or negative) provide supervision. Reinforcement is a process of exploration in which learning is driven by different reinforcement histories for different choices. In the computational literature, reinforcement learning is usually applied when several competing behaviors are initially produced with some probability. Rewards then increase the probability of the desired actions and decrease the probability of less desirable ones. In the presence of multiple possible actions, the delivery of reward or punishment (or the informational equivalent, feedback) provides information about whether the response is right or wrong, though it does not specify the direction or the magnitude of an error. This aligns naturally with some real-world situations. Like back-propagation, however, reinforcement learning operates only in the presence of a teaching signal and does not by itself provide a mechanism for unsupervised learning in its absence.

The feedback literature has a lot to tell us about the learning rules actually used by humans in the course of visual perceptual learning. Reports of learning in the absence of feedback, with trial-by-trial feedback, with block feedback, and with false or manipulated feedback all place important constraints on the choice of a learning rule to explain

Plate 1
The human perceptual system uses all perceptual senses as the interface to a complex world. From www.freeimages.com (#1240544).

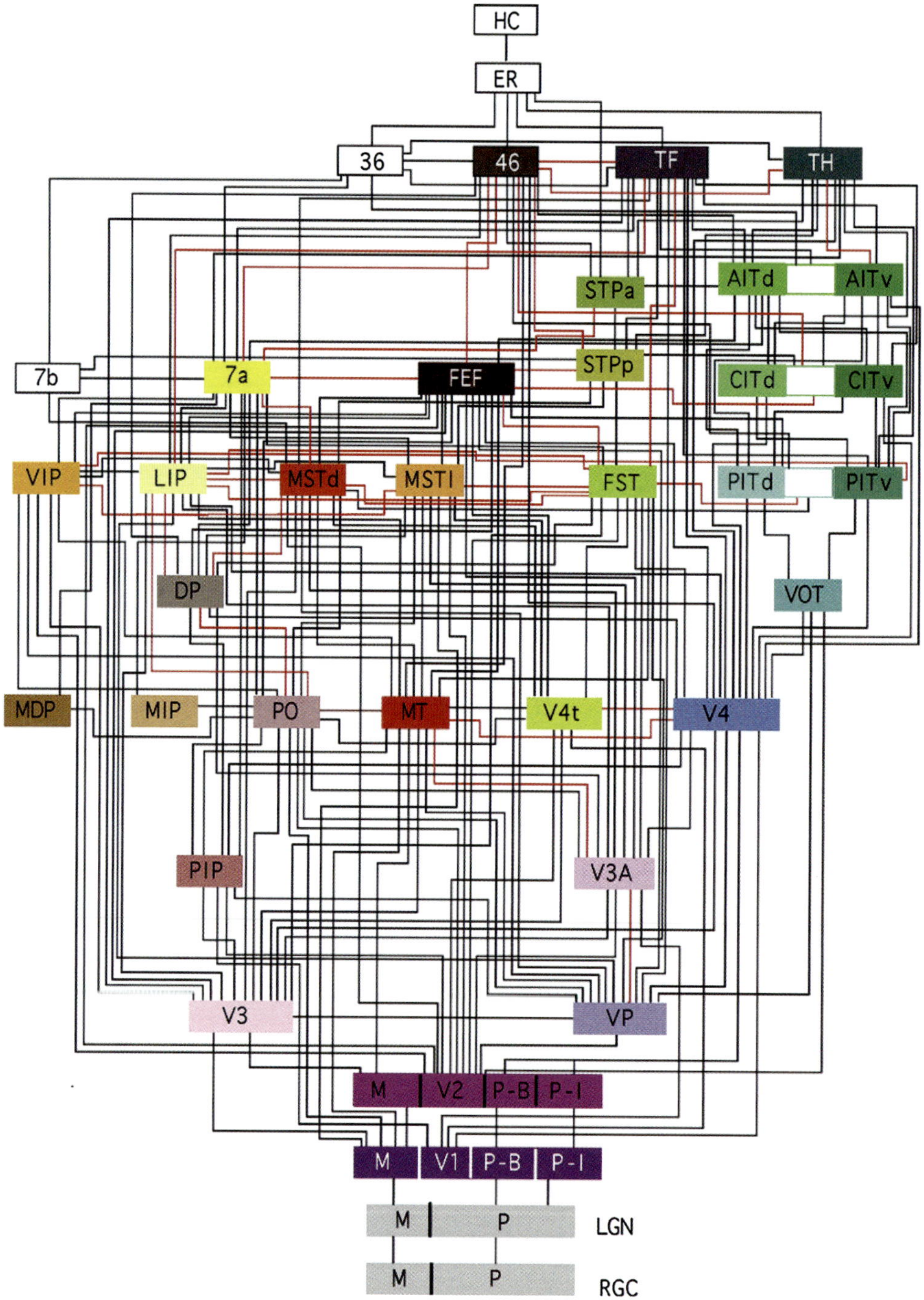

Plate 2

Diagram of the connected network of visual brain areas, based on monkey physiology. After Van Essen, Anderson, and Felleman, figure 2, with permission.

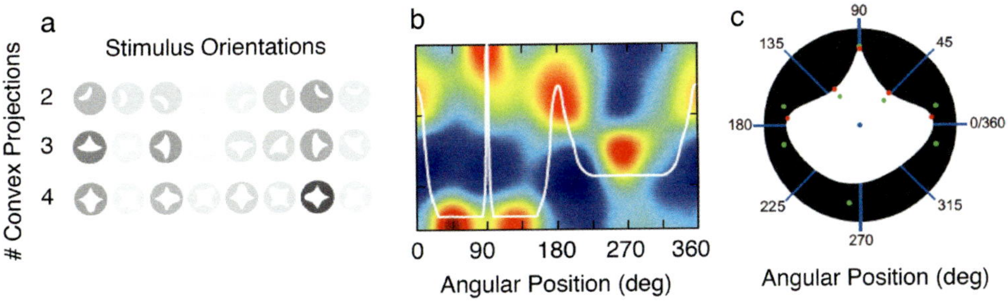

Plate 3
Receptive fields of V4 neurons may code for spatial contours. (a) Examples of convex contours with two, three, or four vertices and gray level indicating cell response. (b) Composite shapes coded by activities over several V4 neurons identify curvature and angular position; hot spots reflect different V4 neurons that together code an object shape. (c) A corresponding object shape. From Kourtzi and Connor, figure 1a, c, and d, with permission.

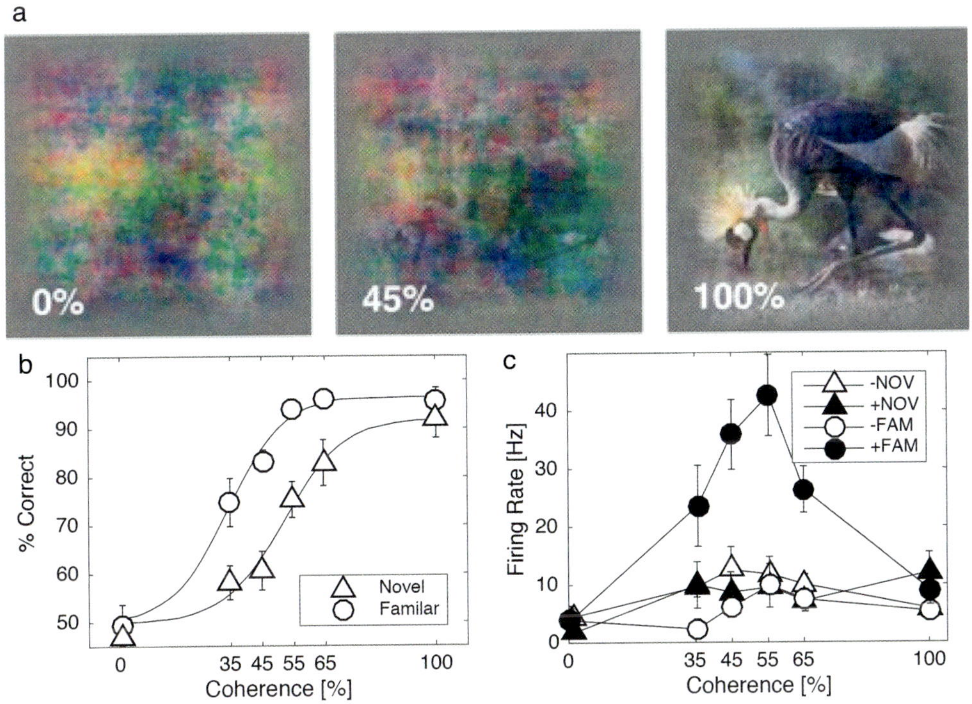

Plate 4
Perceptual training in a delayed match-to-sample task of objects in noise of various coherence (a), behavioral accuracy (b), and corresponding changes of V4 responses to noisy stimuli (c). Firing rates in V4 neurons increase for familiar trained stimuli in intermediate noise levels. After Rainer, Li, and Logothetis, parts of figures 1, 2, and 4. Creative Commons, copyright 2004, Rainer, Li, and Logothetis.

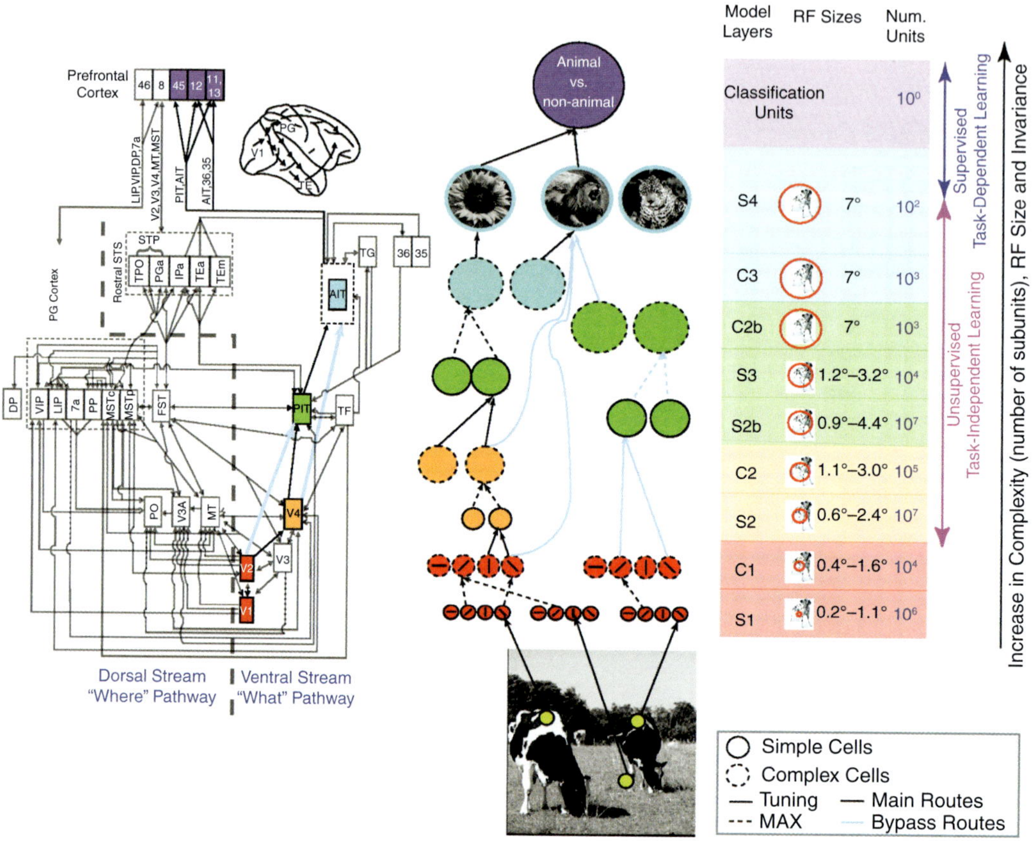

Plate 5

Illustration of a hierarchy of representations of a visual object, ranging from low-level orientation and spatial-frequency representations of the early visual cortex up to higher-level object representations. After Serre, Oliva, and Poggio, figure 1. Copyright (2007) National Academy of Sciences.

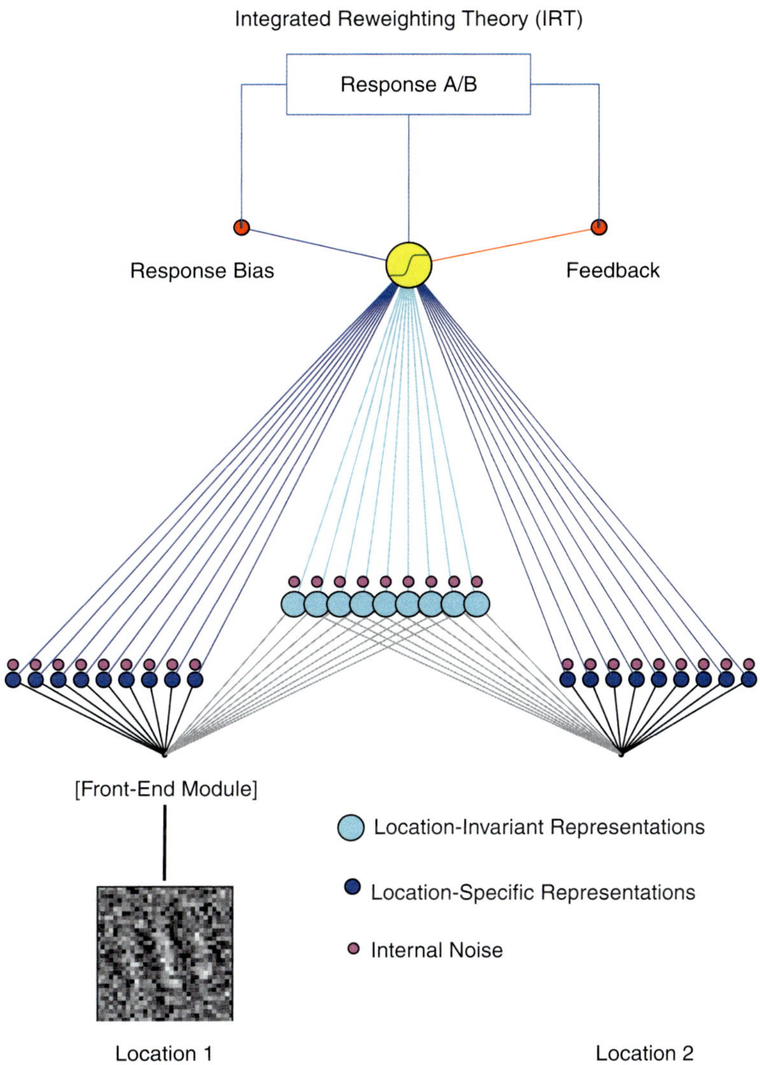

Plate 6
An integrated reweighting theory (IRT) designed to account for transfer over locations and to different stimuli. The architecture illustrated here includes two sets of location-specific representation units and one set of location-invariant representation units, each tuned for orientation and spatial frequency and computed by the front-end module. The weight structure connects each unit to the decision unit. A Hebbian learning rule, augmented with bias and feedback inputs, learns by reweighting the connections. After Dosher et al., figure 1.

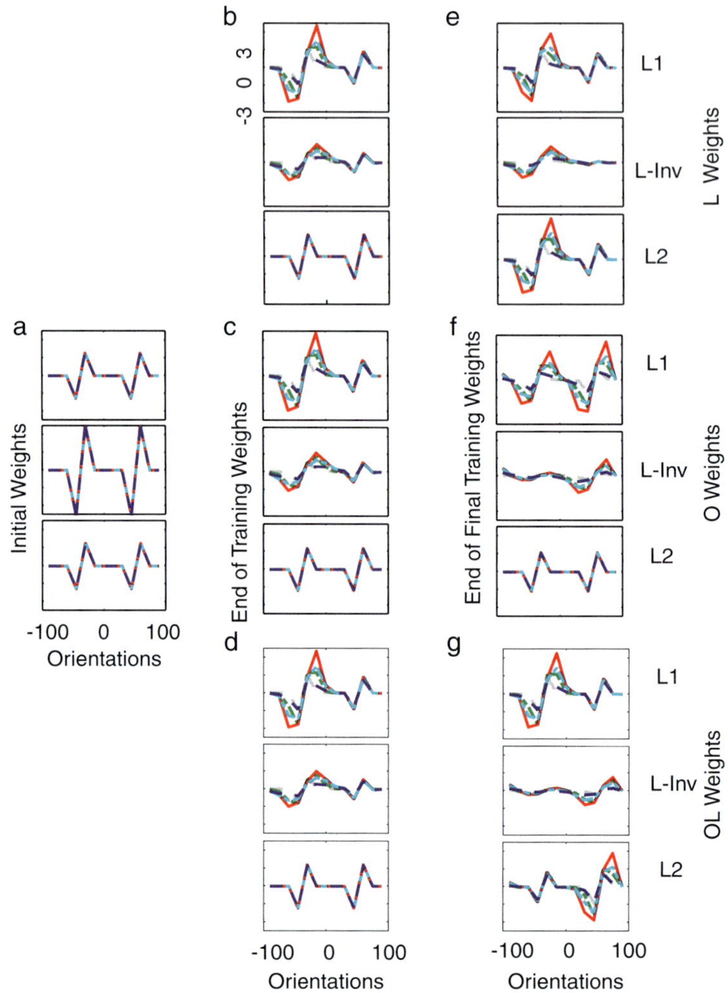

Plate 7

IRT weight structures expressing perceptual learning and transfer to new retinal locations and/or orientations in an orientation-discrimination task. Weight structures at the beginning of initial training for all three groups (a), at the end of initial training (b, c, d), and at the end of the training in the transfer phase (e, f, g), for the L, O, and OL groups (see the text). In each set, the middle represents the location-invariant weights and the top and bottom show the two location-specific weights. Redrawn from Dosher et al., figure S3.

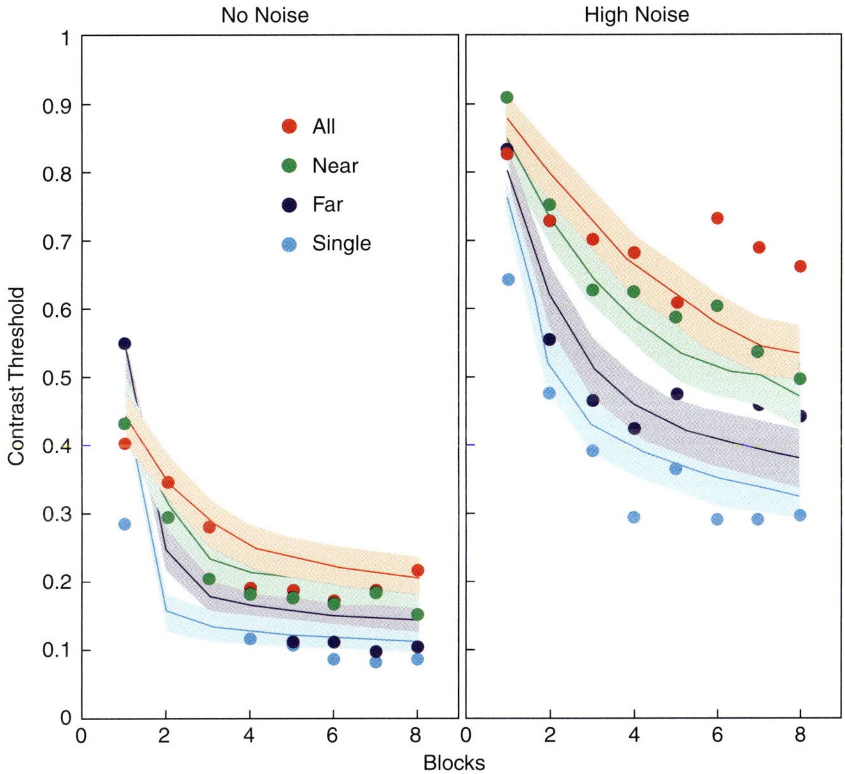

Plate 8
Intermixing training at four different locations shows interactions in learning, depending on the relationship between the orientation-discrimination tasks in those locations. Learning is fastest when the same reference angle is trained in all locations or for widely separated reference angles, slower for similar reference angles, and slowest for four reference angles, as seen in learning curves for the four groups. Lines with bands show the predictions of a best-fitting IRT model fit. From Dosher et al., with permission of the authors.

Plate 9

Simulations illustrate perceived shifts in color appearance following adaptation to the color distributions in lush or arid environments. After Webster, figure 2, with permission.

experimental findings. In particular, that learning sometimes still occurred in the absence of feedback focused our choice of learning rules for the AHRM[10, 11] on hybrid systems, such as augmented Hebbian learning. In this regimen, learning occurs in a purely unsupervised mode without feedback, yet feedback or supervision can be either necessary or just useful in other situations. The way that feedback operates in the AHRM, by shifting the (late) activation toward a correct response (after the response and feedback but before learning), requires trial-by-trial feedback; it cannot occur with block feedback.

In a related model of perceptual learning, Law and Gold focused on reinforcement learning, citing the associated neurophysiological concepts of reward and reward expectation (see chapter 9).[44] They applied a reinforcement model in an experiment testing coarse two-alternative forced-choice motion discrimination (i.e., left versus right) in monkeys. In this binary choice paradigm, unlike the more general situations in which reinforcement learning was developed, feedback does indeed provide information about the desired response and therefore the direction of the error. In Law and Gold's model, weights changed in this way (in the simple form): $\Delta W_i = \eta C(r - E[r])x_i$, where i is a single response unit, C is the choice on that trial (i.e., −1 or 1 for left or right, respectively), r is the reward (i.e., either 1 or 0 for correct or error responses, respectively), and $E[r]$ is the expected probability of reward. Although categorized as reinforcement learning, this rule is more fully supervised in the two-alternative context, given that it uses the sign and magnitude of the reward prediction error. As specified, there is no learning without feedback. In order to account for perceptual learning in the absence of feedback or reward, such reinforcement learning models require further elaboration—perhaps by specifying a system in which unsupervised learning rules are available but superseded by reinforcement learning in the presence of a reward.

Several other alternative learning modes for perceptual learning have recently been suggested. These include top-down recurrent inhibition[4] and attention-gated learning[45] (see chapter 6). Other learning rules, such as Kohonen learning,[12, 13, 46] often used to explain the development of unsupervised self-organizing maps based on similarity and grouping, might be relevant in cases where perceptual learning has been extended to more complex categorization or multiple-response tasks. A more intricate model elaborated from the AHRM, along with some variants, will be described in chapter 12. What is clear in any case is that manipulations of feedback and reward provide some of the strongest behavioral methods available to guide our understanding of the learning rule(s) actually used.

7.4 Feedback and the AHRM

The AHRM has been applied to a number of feedback experiments, where it makes several important predictions. It predicts that learning will occur in the absence of feedback in the unsupervised mode if the initial accuracy of performance is adequate; it makes new and specific predictions about interactions between feedback and the level of performance during training; it provides a possible explanation for the impact block feedback might have on learning (through bias control); and it makes predictions about the potentially damaging effects of false feedback. In the following sections, we detail some of the feedback phenomena predicted by the AHRM. Many of the studies featured here were carried out with Jiajuan Liu.

7.4.1 Feedback and Learning in Nonstationary External-Noise Contexts

The ability to learn in the absence of explicit feedback was one major reason for choosing a rule grounded in Hebbian learning. This rule was then augmented by the ability to use feedback in the development of the AHRM. The initial empirical tests compared perceptual learning with and without feedback in an alternating external-noise paradigm (for details, see subsection 6.4.1).[10, 11, 17] The experiments with and without feedback showed a striking similarity in the complex data patterns during learning (though, in the absence of feedback, responses were more biased in the direction of the oriented external noise). It turns out that close equivalence likely reflects choices made in the training protocols. In particular, the inclusion of high-contrast training stimuli is an important factor in learning without feedback. Adequate initial performance accuracy on at least some trials is necessary to support learning without feedback.

7.4.2 Target Training Accuracy and Trial-by-Trial Feedback

Learning has been shown to occur in the absence of feedback in tasks trained at relatively high levels of accuracy. In fact, feedback may be relatively unimportant in these circumstances. Furthermore, feedback can promote learning in tasks trained at low accuracy levels, when learning may not occur without it.

The AHRM model predicts an interaction between feedback and accuracy level during training in the success of learning. These predictions of the AHRM model were tested in an experiment that trained Gabor orientation discrimination (clockwise or counterclockwise orientations near oblique angles at the fovea and in high external noise).[47] Training accuracy was controlled using an adaptive procedure to set Gabor contrast, with both accuracy and trial-by-trial feedback manipulated in four groups by way of a factorial design (65% or 85% correct × with or without feedback). The data and fit of the AHRM are shown in figure 7.2.

As predicted, there was essentially no learning in the group trained at low accuracy without feedback (65%, no feedback), while perceptual learning was robust for groups trained at the higher accuracy, even without feedback (85%, no feedback). Feedback enabled learning in low-accuracy training (65%, with feedback), while adding feedback had little effect in high-accuracy training (85%, with feedback). The three groups where learning did occur showed statistically equivalent threshold improvements of 23%–33%, while the group with no feedback and low training accuracy showed essentially no learning. The AHRM, where training at higher accuracy capitalizes on the natural correlations between input and output (though feedback can still be used when necessary), predicted this interaction.

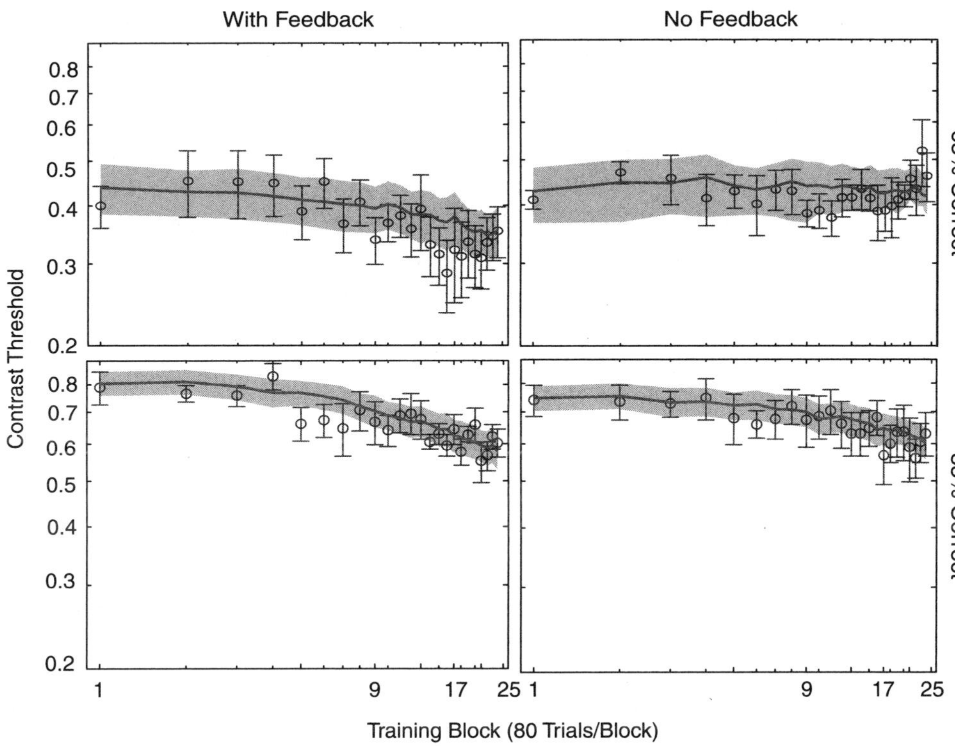

Figure 7.2
The AHRM predicted an interaction between feedback and training accuracy in learning, seen here in contrast-threshold learning curves in orientation discrimination in high external noise: 65% correct training with feedback, 65% correct training without feedback, 85% correct training with feedback, and 85% training without feedback; data (symbols) with AHRM predictions (line and gray bands). Feedback is important in learning only with 65% and not 85% training accuracy. Adapted from Liu, Lu, and Dosher,[32] figure 5.

The advantage of a quantitative model is that it predicts precisely which training accuracy conditions are likely to require feedback for learning. Although these predictions require estimating some parameter values of the model to account for the initial performance of the observer, this can be carried out using only a single initial performance measurement.

7.4.3 Mixtures Including High-Accuracy Trials

Another way to enable learning without feedback is to include high-performance trials in the same task, which can improve performance in low-accuracy trials that might otherwise require trial-by-trial feedback for learning. This has been examined in several studies that controlled performance by manipulating stimulus contrast.[10, 11] The idea is to induce initial high performance through easy first trials, a technique that takes its cue from the related phenomena of "insight" learning, or "Eureka" effects.[48–51]

The consequences of including high-accuracy training trials have been demonstrated in an experiment that intermixed training in two interleaved adaptive staircases in the orientation-discrimination task described in subsection 7.4.2.[52] There were six groups: 65%+65% with feedback, 65%+65% with no feedback, 85%+85% with feedback, 85%+85% with no feedback, 65%+85% with feedback, and 65%+85% with no feedback, with training accuracy controlled by target contrast (figure 7.3). The first four groups replicated those of the previous study, with equivalent results.[32] The critical new tests intermixed the two training accuracies, with and without feedback. Including 85% correct trials in the training protocol, even without feedback, produced learning for both 85% and 65% tests. The AHRM gives an excellent account of the data.

The changes in learned weights for the best-fitting model show patterns similar to those shown earlier (see chapter 6). Initial weights were set to incorporate some a priori knowledge of orientation and task instruction. Training in the 65%+65% with no feedback condition shows almost no change in weights, corresponding with the absence of learning. Where learning occurred, training increased the weights to decision on the most relevant orientation and spatial-frequency units and decreased the other weights. Although the 85%+85% with feedback condition led to the largest weight shifts, all conditions including either 85% trials or feedback yielded similar weights and predicted almost equivalent behavioral learning (see the source paper[52] for details).

Including high-accuracy stimuli may successfully promote learning in real-world situations in which trial-by-trial feedback would be impractical. Crucially, the AHRM model made it possible to predict how changing the proportion of 85% training trials might affect learning (see figure 7.4), with simulations generally predicting that increasing the proportion of high-performance training trials would also increase the size of threshold improvements from the same number of total training trials.

7.4.4 Modeling Trial-by-Trial, False, Random, and Reverse Feedback

If our goal is to understand how different kinds of feedback might influence the presence and/or rate of learning, it is only natural to compare learning outcomes across different feedback conditions. Although trial-by-trial feedback is the most commonly used in the literature (with no feedback the second most common), a few studies have investigated block feedback as well as manipulated, false, random, or exaggerated feedback.

In the experimental situation, feedback has been employed by using error messages, correct response messages, or both (e.g., a tone after errors, a tone after correct trials, or a different tone for each). In two-alternative forced-choice tasks, all three forms of feedback provide equivalent information (although they could potentially induce salience differences between errors and correct responses).

An important study by Herzog and Fahle compared many forms of feedback in the same task context (Vernier line judgments at the fovea with constant offsets and percentage correct as the dependent measure).[19] This study went well beyond many others that tested only

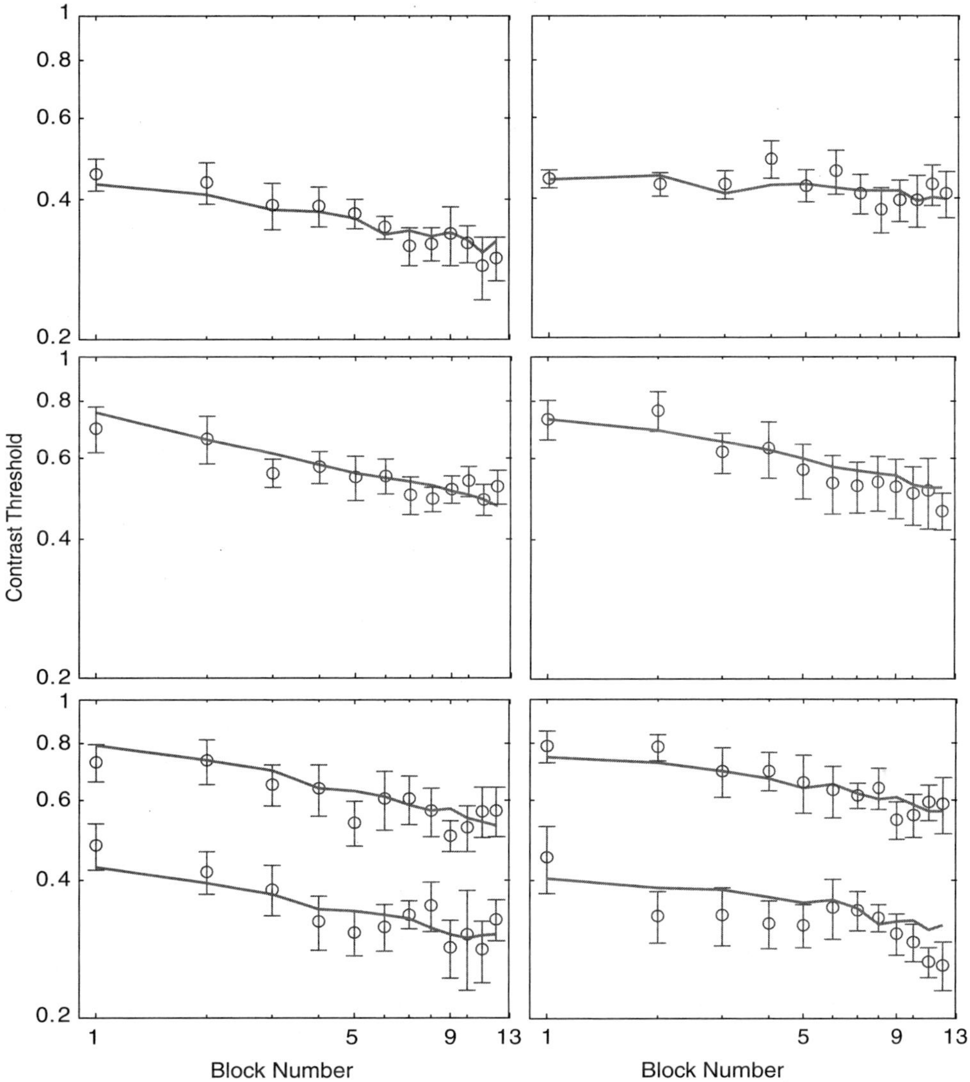

Figure 7.3
AHRM predicts the interactions of feedback and mixtures of high and low training accuracy during learning. Contrast thresholds for the six groups: 65%+65%, 85%+85%, or 65%+85%, with and without feedback. After Liu, Lu, and Dosher,[52] figure 5.

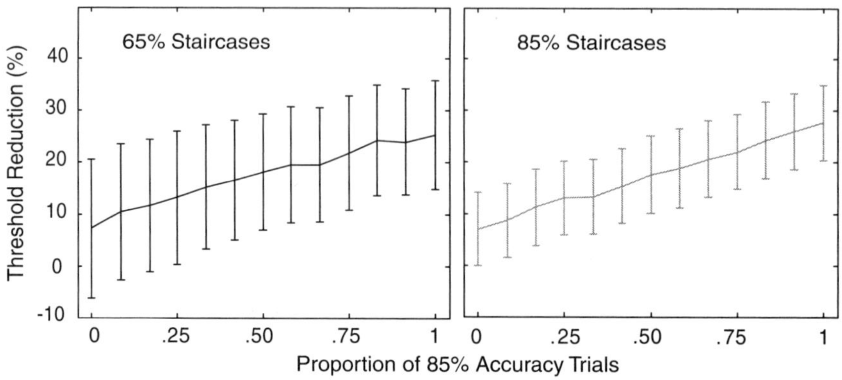

Figure 7.4
Percentage threshold reduction predicted by the AHRM model for the 65% and the 85% training staircases without feedback as a function of the proportion of 85% training accuracy trials mixed into training from simulations. After Liu, Lu, and Dosher,[52] figure 7.

one or two feedback conditions. Different groups of observers were trained using either no feedback, trial-by-trial feedback, block feedback, uncorrelated feedback, or several forms of manipulated feedback (see figure 7.5). The AHRM's ability to account for the learning in these different feedback conditions was similarly assessed in a simulation study.[53]

Despite setting initial Vernier offsets individually on the basis of pretests, initial performance accuracies of the different groups varied slightly (presumably reflecting variation between subjects assigned randomly to the groups). The theoretical focus was on learning *rates*. The two most commonly used forms of feedback showed the expected effects—essentially no learning in the absence of feedback (given initial accuracies near 60%) and robust learning with (accurate) trial-by-trial feedback. When trial-by-trial feedback occurred on only half the trials, learning still occurred. On the other hand, inaccurate trial-by-trial feedback was found to disrupt learning: feedback that was random or uncorrelated with the accuracy of the response was unhelpful, and like the no-feedback group, the uncorrelated-feedback group did not improve with practice. The reverse-feedback manipulation, which was ultimately tested in only one observer, produced very inconsistent performance over the course of training.

We modeled these results with the AHRM using the orientation and spatial-frequency representation module.[53] This followed the precedent of others, who accounted for Vernier line performance with orientation detectors.[36, 54] The AHRM predicted little or no learning in the no-feedback group, given the low initial accuracy before training. On the other hand, trial-by-trial feedback promoted strong learning. (These results paralleled those for the 65% accuracy training groups with and without feedback for the same reasons; see subsections 7.4.2 and 7.4.3.) The model predicted a slightly slower learning rate for partial (50%) trial-by-trial feedback, while the behavioral difference was insignificant. For random

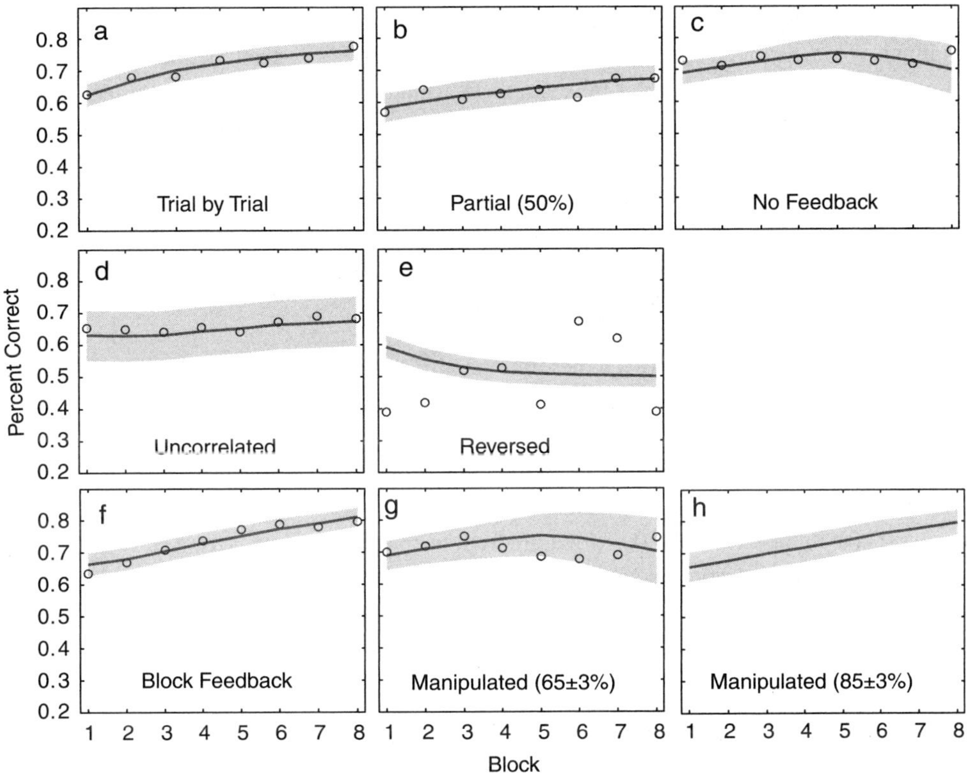

Figure 7.5
Different forms of feedback yield different learning rates in a Vernier line-offset task, as fitted by the AHRM model. Percentage correct as a function of training for (a) trial-by-trial feedback, (b) partial trial-by-trial feedback, (c) no feedback, (d) uncorrelated feedback, (e) reversed feedback, (f) block feedback, (g) manipulated block feedback (65%±3%), and (h) manipulated block feedback (85%±3%). Data are the symbols, from Herzog and Fahle,[19] the line and gray bands are optimized AHRM predictions, and n=6–10 in the data, except (e), where n=1. After Liu, Dosher, and Lu,[53] figures 4, 5, and 6.

feedback, it similarly predicted no learning or a very slight decrement in performance, again consistent with the behavioral observations. Finally, the model predicted a *decrease* in percentage correct classification, as reversed trial-by-trial feedback indicated the reverse response. (A real observer might have suspected inaccurate feedback and chosen to ignore it—a cognitive strategy that was outside the scope of the model.) In short, the AHRM predictions provided an excellent account of the differences in the rate and presence of learning in the various feedback conditions.

The patterns of weight changes estimated from the fits of the AHRM paralleled those seen in other applications of the model. Practice with accurate trial-by-trial feedback increased the weights on the units tuned closest to the very small orientations relevant for detecting the Vernier stimuli (e.g., the units tuned to ±15°) and reduced the weights on other channels. This occurred slightly more slowly for partial trial-by-trial feedback. The weight histories for uncorrelated and for false feedback groups differed from one another, but neither predicted any learning. Uncorrelated feedback tended to compress the weights toward zero, reflecting feedback unreliability, as well as exhibiting bias fluctuations over time in any individual simulation run. Anticorrelated or reverse feedback, as intuitively expected, pushed the weights of the most relevant channels first toward zero and then toward the reverse weights, while reducing the weights on other, less relevant channels, ultimately predicting a decline in performance accuracy.

Almost every simulation run, which represents the learning of one observer in trial-by-trail feedback conditions, followed the same reliable learning pattern. Although over many simulations the weight structures for the no-feedback group also showed a pattern of slightly increasing weights on relevant channels and decreasing weights on irrelevant ones, any single simulation was erratic and likely to develop biases (e.g., weights drifting above or below the balanced zero point). The average performance of random selections of ten observers (as in the experiment) yielded predictions that matched the group behavioral data. From this, a useful interpretation emerges: single simulation runs, taken together, can make important predictions about learning not only for one observer but also for the distribution of results one might expect across observers.

7.4.5 Modeling Block Feedback

Block feedback (e.g., percentage correct for every block of trials) can support learning in some cases, although in general it is not as useful as trial-by-trial feedback. With block feedback, learning is unsupervised within the block, but this nonetheless permits some learning. Herzog and Fahle examined several forms of block feedback (see figure 7.6).[19] One group received accurate block feedback every 100 trials, which somehow supported learning, although there was no learning in the absence of feedback. Meanwhile, another group received manipulated block feedback that conveyed (inaccurately) to the observer that accuracy hovered around 65% correct throughout training (65%±3%). This discouraging feedback led to no discernible learning.

Understanding block feedback may be especially useful, as this kind of intermittent feedback may be typical in many learning contexts. How can block feedback support learning within the rules of a quantitative learning model? Some researchers have suggested that observers discounted or rolled back weight changes from the prior block following poor block feedback, essentially returning to earlier weights from the beginning of the block.[4] By contrast, the AHRM presents an alternate picture. After investigating several alternatives, we proposed that block feedback changed the weight assigned to bias control in the subsequent block; in particular, our hypothesis was that the weight varied linearly between 0 and 1 in proportion to the value of the last block feedback, which ranged between 50% and 100% correct.[53] In this understanding, the bias-control unit is a corrective that counteracts biases observed in the recent response history. Higher block feedback was hypothesized to lead to

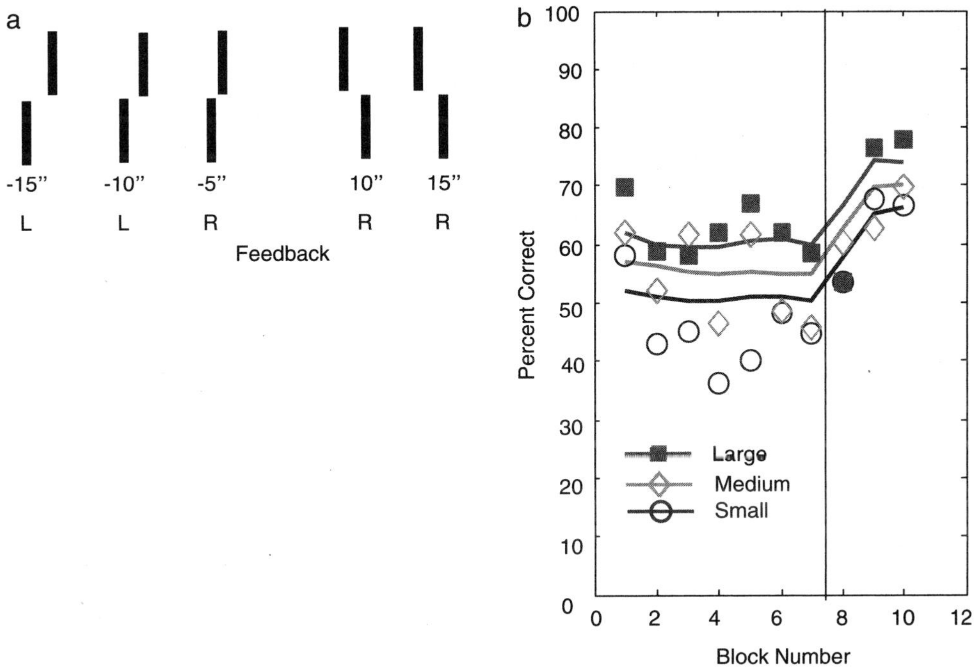

Figure 7.6
Bias induced by reverse feedback on subthreshold stimuli in asymmetric training sets, and the fits of the AHRM model. (a) Example stimuli and response feedback. (b) Percentage correct (symbols) for small, medium, and large left offsets and the fit (lines) of the AHRM model. Data from Herzog and Fahle,[28] figure 3. After Liu, Dosher, and Lu,[55] figure 1.

more bias correction, thus counteracting the tendency for biased responses in otherwise unsupervised learning within the block. The predictions of the AHRM for the two block-feedback conditions, and for a hypothetical condition in which feedback was artificially high (85°±3°), suggested that block feedback should move weights in the same direction as trial-by-trial feedback, though more slowly and with more variability. Correspondingly, the AHRM shows that exaggerated feedback will lead to better learning.[31]

7.4.6 Training Asymmetry and Induced Bias

False or reverse feedback has an effect on performance, and if false feedback occurs in one direction but not the other, the result is bias.[53] The AHRM model has been able to predict a number of bias-induction phenomena induced by selective false feedback. Asymmetric false feedback and the resulting induced response bias have been studied in a series of experiments by Herzog and his colleagues.[28–30] In one typical experiment, observers were trained with five Vernier offsets: large, medium, and small bottom-left offsets and medium and large bottom-right offsets (see figure 7.6). In experiment 3 of Herzog and Fahle,[28] the small (subthreshold) bottom-left offset received reverse feedback, indicating a bottom "right" response, and these stimuli sometimes represented fully one-third of all trials. A manipulation that favored "right" responses for many left stimuli reduced the proportion of "left" responses for all bottom-left stimuli (only data for left offsets were published). When accurate feedback was restored (at the vertical line), the biased percentage correct for left offsets quickly recovered.

The original interpretation of the induced bias results was that perceptual learning trained response bias (in the signal detection theoretic sense), thus lowering the criterion for the

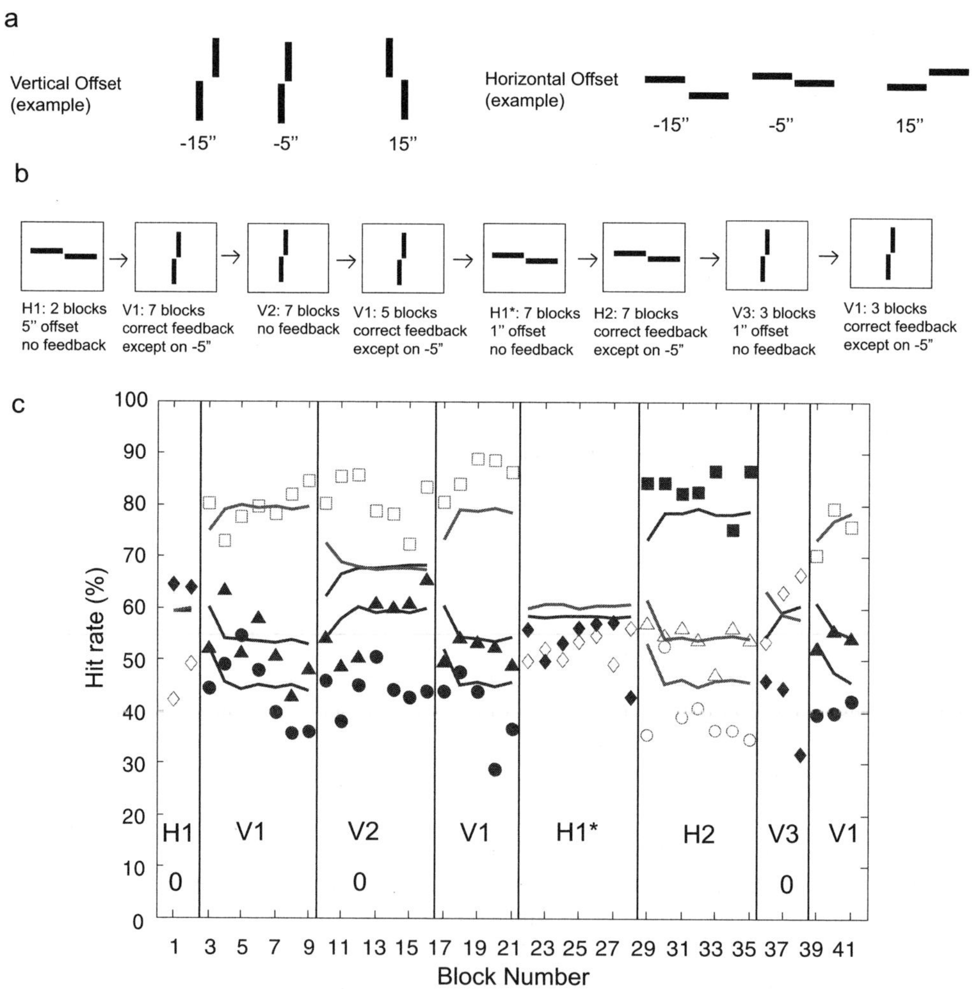

Figure 7.7
Bias induction in an experiment training horizontal and vertical Vernier offset judgments and the AHRM predictions. Stimuli and training protocol (a, b), and hit rate data (symbols) and the model fits (lines) (c). Data from Herzog et al.[29] Adapted from Liu, Dosher, and Lu,[55] figure 5d.

dominant feedback response, which was then reversed when accurate feedback was restored.[28] The AHRM model (lines in figure 7.7), by contrast, accounted for the data not by changing bias (where the bias unit opposes biases in the response history) but instead by changing weights to decision, which, over the course of training with false feedback, came to favor the false-feedback response. In essence, the weights on the relevant sensory representations shifted toward favoring the "right" response (analogous to shifting the evidence distributions in signal detection theory). This affected the performance on the small, medium, and large offsets because all these offsets predominantly activated the orientation channels just to the left and right of vertical, albeit to different degrees. Trials with false feedback indicating "right" (as opposed to "left") shifted the weights and changed responses to all the offset stimuli together.

These bias effects themselves can be specific to the training stimuli in various ways. Induced biases were found to be largely specific to the (widely separated) trained orientations, as shown in a complicated experiment that trained judgments for both horizontal and vertical lines (see figure 7.7).[29]

In this experiment, which used one smaller offset with false feedback and two larger offsets with accurate feedback, there were distinct phases of training: some used balanced sets of stimuli without feedback to assess response biases, while others used the standard bias-induction training (labeled V1 and H2). In short, biases were trained for vertical offset judgments (V1) that left horizontal offset judgments unaffected (H1*) and then independently trained for horizontal offset judgments (H2). This specificity of induced bias to orientation was a natural property of the account provided by the AHRM. The wide separation in orientation space between the horizontal and vertical Vernier stimuli meant the vertical and horizontal stimuli activated quite separate orientation- and spatial-frequency-tuned representations, such that the changes in the learned weights were naturally segregated (see the source paper[55] for details). Yet another study compared learning in groups with a systematic variation of forms of feedback. Groups either had no feedback, accurate trial-by-trial feedback, trial-by-trial feedback with the false (reversed) small offset, accurate block feedback, or block feedback that factored in the false feedback in two different block lengths.[30] In this experiment, biases emerged only with trial-by-trial false feedback, a necessary prediction of the AHRM model. In this case, block feedback was ineffective.

Overall, then, the AHRM model provides a compelling account of the bias phenomena reported by all the reverse feedback experiments conducted so far.[55] Indeed, it did so as a natural consequence of reweighting from sensory representations to decision; further elaborations were not necessary to make successful predictions. The trained changes in weights essentially shifted the evidence distributions at the decision unit. This suggests that, even though the criterion control unit might seem to be the natural locus for a "bias" effect, the effects of feedback on learned weights were in fact a more powerful and natural mechanism.[55]

7.5 Learning in Multistimulus Identification

With a few notable exceptions, perceptual learning has only been studied in simple two-alternative discrimination or detection tasks.[56, 57] These implicit procedural choices by the field may have unnecessarily limited the range of the research. At present, we are only beginning to understand the role and effectiveness of feedback in situations where observers are asked to classify stimuli into a larger number of categories, which has been labeled "n-alternative" identification.

How does feedback operate in these cases? When is it most useful? By what mechanisms does it influence learning? Future research in this seldom-studied domain has the potential to expand the research paradigm of perceptual learning significantly.

One of the most central questions to be asked here involves the presence and possible robustness of learning. Several studies, some dating back to the 1950s, in the related field of absolute identification, have historically led researchers to conclude that the identification of stimuli in a single dimension was limited to about four to seven categories and showed little potential for learning.[58–62] More recently, learning has been reported in certain cases, such as line-length discrimination, where new stimuli extended the range of the dimension.[63, 64] In this classic literature, different limits on identification were shown to occur for different stimulus dimensions.

The reweighting models turn out to provide a rather different approach for understanding multicategory identification, with the n-alternative paradigms providing a useful test bed for investigating the effect of feedback on learning. The limits on performance in tasks using different kinds of stimuli reflect how the front-end module represents these categories alongside the discriminability in weight space of the n-alternative identifications. If observers can improve with practice in some cases, this insight would shed new light on important questions regarding the kinds of supervision that might occur in learning.

Several different kinds of trial-by-trial feedback can be distinguished experimentally, going beyond simple comparisons of feedback versus no feedback. Response feedback, for example, is a form of full supervision: it tells the observer the response that they should have made, which can be compared with the response they did make. Accuracy feedback, however, is more analogous to supervision through reinforcement: it tells the observer whether their response was right or wrong but not *how* it was wrong or which response would have been the correct one. We have been able to find evidence of learning with full supervision (providing full feedback about the correct response) in n-alternative tasks in experiments involving both orientation and spatial-frequency judgments.[65, 66]

Perhaps surprisingly, performance in n-alternative identification can be modeled using only slightly more complicated decision and learning rules than the basic AHRM. An extension was recently developed to handle n-alternative tasks (and a corresponding version of this model, based on the integrated reweighting theory, or IRT, offers a multilocation, multilayer extension of the AHRM and is further discussed in chapter 8). In this extended model,

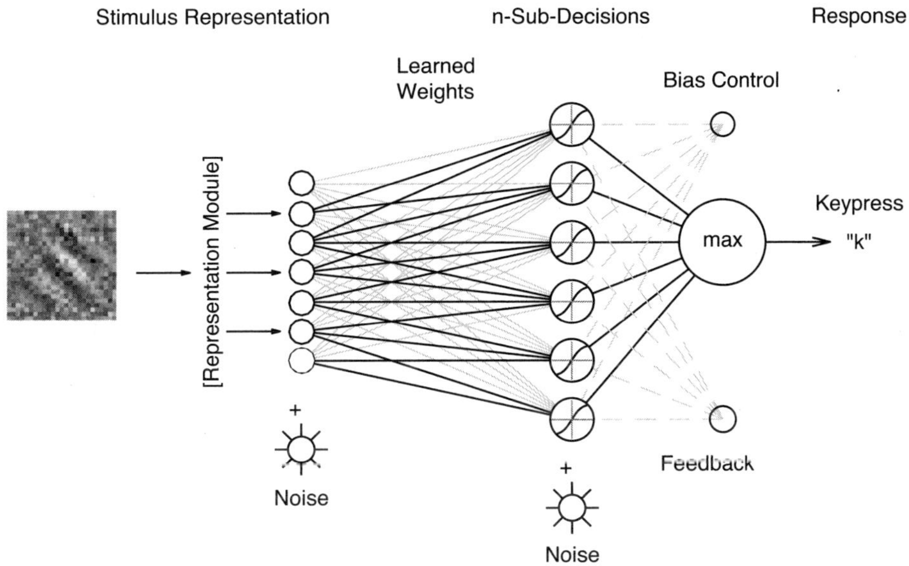

Figure 7.8
An outline of a reweighting model for n-alternative identification in which the evidence for each category of response is computed by a subdecision unit, and the response on the trial corresponds to the decision unit with the maximum response (a winner-take-all or max-rule decision). In this framework, variations in the nature of feedback, corresponding to different levels of supervision, result in different learning rates.

decision units—templates—are set up for each response category, with the final classification being made by choosing the decision unit or template with the strongest activation, a "max-rule" (figure 7.8).[65–67] In essence, each of the n decision units collects evidence for a binary decision regarding whether the stimulus matches that category. Learning then occurs as the weights defining the templates or decision units are improved.

For these extended models, feedback, when available, can be used to improve the weights to the decision units for each category. Response feedback on one trial drives all n of the category decision units toward either a match or mismatch decision, whichever is correct. With accuracy feedback, error responses provide less information, driving the decision unit for the incorrect response given by the observer toward a mismatch but providing no information for the other n–1 decision units. Without feedback, learning is likely to be less successful than in a two-alternative case unless prior knowledge and high visibility permit exceptionally high levels of accurate identification even before practice.

Researchers are just beginning to examine experimentally how visual perceptual learning operates in these n alternative task domains. So far, we have identified some eight-alternative tasks in which learning does occur, and we have found preliminary evidence supporting the predicted differences in learning rates between conditions of response feedback, accuracy feedback, and no feedback. A host of predictions follow from these n-AFC reweighting models, including predictions for learning at different levels of supervision. Such models furthermore promise to make revealing predictions about confusion data (i.e., the frequency pattern of responses given to each stimulus), how these change with training, and how discriminability is determined by the similarity of the patterns of activation to different stimuli in the representation module.

7.6 Conclusions

The manipulation of feedback in visual perceptual learning experiments has become the key method for researchers to investigate learning rules or algorithms. Although most experiments have used only trial-by-trial feedback (and therefore do not make comparisons), other experiments that manipulate the presence and/or form of feedback have yielded compelling findings. The progress in this nascent area of research has been remarkable. The patterns of learning resulting from different feedback conditions have led researchers to recognize that perceptual learning is neither purely supervised nor purely unsupervised but instead reflects a hybrid of the two that dominates in different and predictable contexts.

Learning rules and architectures first developed as artificial neural networks provided a theoretical structure to study analogous phenomena in visual perceptual learning. Existing behavioral data have so far been consistent with the augmented Hebbian reweighting model (AHRM), which uses an unsupervised learning rule that is also augmented by feedback and bias control. In this model, learning proceeds in an unsupervised mode in the absence of feedback as well as in a supervised mode when feedback is available.

The AHRM is especially useful, as it makes a wide-ranging set of testable predictions. It predicts that learning can still occur in the absence of feedback if the initial level of task performance was good enough. It also predicts that, when feedback is unavailable and the initial performance inadequate, even extensive practice would reap few benefits; in these conditions, the inclusion of feedback could release perceptual learning that would have failed without it.

One model of the effects of block feedback, developed in the context of the AHRM (that of increased weight on bias control with higher-accuracy block feedback), gave a good account of several feedback conditions. The model also predicted that random trial-by-trial feedback would damage or slow learning (so long as it is not discounted by the observer) and that false trial-by-trial feedback in favor of an objectively incorrect response for subsets of stimuli would induce systematic biases in stimulus classifications.

One of the most important principles emerging from model-inspired experiments involves the different circumstances in which feedback may be useful depending on the level of performance during training. Feedback is most necessary when the accuracy during training is sufficiently low that unsupervised learning would be unsuccessful, while feedback may be either unnecessary or even redundant when accuracy during training is high, especially for two-alternative tasks. As this predicted interaction was revealed in experiments that held performance at a controlled accuracy level throughout training with adaptive methods by adjusting stimulus contrast, it remains to be seen whether the analogous finding would occur when training occurs at different accuracy levels (e.g., by adjusting orientation differences for high-contrast stimuli). In this latter case, the observer would have to make increasingly fine discriminations between stimuli during learning, with the set of stimuli changing, especially during early learning. For tasks involving complex stimuli, keeping the stimuli constant and observing improvements in performance accuracy with learning may be the only practical experimental option; in such experiments, the effects of feedback are intermixed with the effects of training at gradually increasing performance levels.

It must be repeated that the benefits of a quantitative model are substantial. Only with such a model can researchers precisely study how feedback works in training and how this might depend on other factors within any given testing paradigm. As researchers work to expand the variable space to include multiple types of feedback, more complex tasks, and interactions with reward, it will surely be necessary to further refine existing models to account for the newly observed empirical phenomena of perceptual learning.

7.7 Future Directions

The most prominent current models of feedback in perceptual learning are reweighting models. In these models, reweighting operates within a relatively simple architecture, with only a few layers. Simple extensions of the architecture that add layers would lead to even richer possibilities.

Even in the relatively simplified models, information can be integrated over several layers, connecting sensory representations to decision units (through hidden layers) and integrating the top-down signals from feedback to augment learning. How these simplified learning models might be embedded within a more complex and biologically plausible network of modules and connections remains to be seen. The same can be said for how such models might be used to discover the biologically specific neural instantiations of perception, decision, interpretation, and feedback-specific learning. In this sense, there are several ways in which the experimental study of feedback relates to outstanding theoretical questions in the field and thus might guide future research.

The kinds of experiments used to study perceptual learning and feedback have been unusually simple and were concentrated in a few task domains. Experimentally investigating the range of tasks and variety of feedback, or even extending these investigations to the forms of feedback available in real-world settings, would provide new insights into perceptual learning systems and learning rules.

At this point, we know little about how learning rules may be affected by manipulations such as reward magnitude, probability, or timing—or, for that matter, by manipulations of attention. Discovering and refining our understanding of this relationship would meaningfully expand our experimental repertoire. Some possible ways to do this are considered in chapter 9.

Does the dependence of learning on feedback depend on the nature of the task? Comparisons of different forms of feedback have been focused on lower-level visual tasks. For such simple two-alternative low-level visual tasks, we already know that learning, which has been successfully modeled by selective reweighting of existing representations, can occur in the absence of feedback. What is less clear is whether higher-level visual tasks, possibly requiring the creation of representations for new feature combinations, also occur in the absence of feedback. Often these higher-level tasks have involved naming or identification of natural or synthetic objects from a larger set (*n*-alternative choice). One direction for future research could be to systematically compare feedback and learning in tasks associated with low-, middle-, and high-level visual processes.

With the exception of the work on induced bias, there has been little research about learning and/or feedback in situations in which the relative frequencies of the stimuli and responses are unbalanced or unequal. Yet, in the real world, we often encounter unequal stimulus probabilities, biased response environments, and situations in which feedback is intermittent, incomplete, or ambiguous. Extending the paradigms in which learning is assessed might demand a reconsideration of the learning rules, our understanding of how to model expectancy and bias, and other aspects of generalized task knowledge.

Investigations of learning in multialternative tasks ($n>3$) that provide many expanded opportunities for contrasting different levels of supervision for learning rules in these more complex situations might reveal the differential importance of feedback in learning them. Some new experiments related to this idea are described in chapter 8.

It is remarkable but true that abstract or symbolic feedback messages (beeps, buzzes, or words) influence learning in very basic visual tasks. These stimuli require translation into neural teaching signals delivered at just the right level, location, and time to affect learning. This might involve widely disseminating the top-down signal to many brain sites, with possible consequences for plasticity and stability in many regions.[68] How this is accomplished in a biological system remains a mystery, though it might be related to attention or the operations of reward and decision centers, issues that are taken up in chapter 9.

On the one hand, there has been substantial progress made over the last few decades. We now have a much clearer understanding of how perceptual learning works and how it often depends on training regimens for success. On the other hand, there is still a long way to go. An expansion of experimental and training paradigms, in tandem with the development of new, biologically plausible learning theories, offers the potential for even more fundamental discoveries.

References

1. Jordan MI, Rumelhart DE. Forward models: Supervised learning with a distal teacher. *Cognitive Science* 1992;16(3):307–354.

2. Reed RD, Marks RJ. *Neural smithing: Supervised learning in feedforward artificial neural networks.* MIT Press; 1998.

3. Barlow HB. Unsupervised learning. *Neural Computation* 1989;1(3):295–311.

4. Herzog MH, Fahle M. Modeling perceptual learning: Difficulties and how they can be overcome. *Biological Cybernetics* 1998;78(2):107–117.

5. Dosher BA, Lu Z-L. Hebbian reweighting on stable representations in perceptual learning. *Learning & Perception* 2009;1(1):37–58.

6. Mohri M, Rostamizadeh A, Talwalkar A. *Foundations of machine learning.* MIT Press; 2012.

7. Rumelhart DE, Durbin R, Golden R, Chauvin Y. Backpropagation: The basic theory. In Chauvin Y, Rumelhart DE, eds., *Backpropagation: Theory, architectures and applications.* Laurence Erlbaum Associates;1995:1–34.

8. Sutton RS, Barto AG. *Reinforcement learning: An introduction.* MIT Press; 1998.

9. Sejnowski TJ, Tesauro G. The Hebb rule for synaptic plasticity: Algorithms and implementations. In Byrne JH, Berry WO, eds., *Neural models of plasticity: Experimental and theoretical approaches.* Academic Press;1989:94–103.

10. Petrov AA, Dosher BA, Lu Z-L. The dynamics of perceptual learning: An incremental reweighting model. *Psychological Review* 2005;112(4):715–743.

11. Petrov AA, Dosher BA, Lu Z-L. Perceptual learning without feedback in non-stationary contexts: Data and model. *Vision Research* 2006;46(19):3177–3197.

12. Kohonen T. Self-organized formation of topologically correct feature maps. *Biological Cybernetics* 1982;43(1):59–69.

13. Kohonen T. The self-organizing map. *Proceedings of the IEEE* 1990;78(9):1464–1480.

14. Jain AK. Data clustering: 50 years beyond K-means. *Pattern Recognition Letters* 2010;31(8):651–666.

15. Mitchison GJ, Swindale NV. Can Hebbian volume learning explain discontinuities in cortical maps? *Neural Computation* 1999;11(7):1519–1526.

16. Oja E. Simplified neuron model as a principal component analyzer. *Journal of Mathematical Biology* 1982;15(3):267–273.

17. Sanger TD. Optimal unsupervised learning in a single-layer linear feedforward neural network. *Neural Networks* 1989;2(6):459–473.

18. Gerstner W, Kistler WM. *Spiking neuron models: Single neurons, populations, plasticity.* Cambridge University Press; 2002.

19. Herzog MH, Fahle M. The role of feedback in learning a Vernier discrimination task. *Vision Research* 1997;37(15):2133–2141.

20. Shiu L-P, Pashler H. Improvement in line orientation discrimination is retinally local but dependent on cognitive set. *Perception, & Psychophysics* 1992;52(5):582–588.

21. Shibata K, Watanabe T, Sasaki Y, Kawato M. Perceptual learning incepted by decoded fMRI neurofeedback without stimulus presentation. *Science* 2011;334(6061):1413–1415.

22. Choi H, Watanabe T. Perceptual learning solely induced by feedback. *Vision Research* 2012;61:77–82.

23. Ball K, Sekuler R. Direction-specific improvement in motion discrimination. *Vision Research* 1987;27(6):953–965.

24. Crist RE, Kapadia MK, Westheimer G, Gilbert CD. Perceptual learning of spatial localization: Specificity for orientation, position, and context. *Journal of Neurophysiology* 1997;78(6):2889–2894.

25. Fahle M, Edelman S. Long-term learning in Vernier acuity: Effects of stimulus orientation, range and of feedback. *Vision Research* 1993;33(3):397–412.

26. Karni A, Sagi D. Where practice makes perfect in texture discrimination: Evidence for primary visual cortex plasticity. *Proceedings of the National Academy of Sciences* 1991;88(11):4966–4970.

27. McKee SP, Westhe G. Improvement in Vernier acuity with practice. *Perception & Psychophysics* 1978;24(3):258–262.

28. Herzog MH, Fahle M. Effects of biased feedback on learning and deciding in a Vernier discrimination task. *Vision Research* 1999;39(25):4232–4243.

29. Herzog MH, Ewald KR, Hermens F, Fahle M. Reverse feedback induces position and orientation specific changes. *Vision Research* 2006;46(22):3761–3770.

30. Aberg KC, Herzog MH. Different types of feedback change decision criterion and sensitivity differently in perceptual learning. *Journal of Vision* 2012;12(3):3,1–11.

31. Shibata K, Yamagishi N, Ishii S, Kawato M. Boosting perceptual learning by fake feedback. *Vision Research* 2009;49(21):2574–2585.

32. Liu J, Lu Z, Dosher B. Augmented Hebbian learning accounts for the complex pattern of effects of feedback in perceptual learning. *Journal of Vision* 2010;10(7):1115.

33. Lu Z-L, Liu J, Dosher BA. Modeling mechanisms of perceptual learning with augmented Hebbian reweighting. *Vision Research* 2010;50(4):375–390.

34. Polat U, Sagi D. Spatial interactions in human vision: From near to far via experience-dependent cascades of connections. *Proceedings of the National Academy of Sciences* 1994;91(4):1206–1209.

35. Vaina LM, Sundareswaran V, Harris JG. Learning to ignore: Psychophysics and computational modeling of fast learning of direction in noisy motion stimuli. *Cognitive Brain Research* 1995;2(3):155–163.

36. Weiss Y, Edelman S, Fahle M. Models of perceptual learning in Vernier hyperacuity. *Neural Computation* 1993;5(5):695–718.

37. Rumelhart DE, McClelland JL, Group PR. *Parallel distributed processing*. Vol. 1. IEEE; 1988.

38. McClelland JL, Rumelhart DE, Group PR. *Parallel distributed processing*. Vol. 2. MIT Press; 1987.

39. Lillicrap TP, Cownden D, Tweed DB, Akerman CJ. Random synaptic feedback weights support error backpropagation for deep learning. *Nature Communications* 2016;7.

40. Neftci EO, Augustine C, Paul S, Detorakis G. Event-driven random back-propagation: Enabling neuromorphic deep learning machines. *Frontiers in Neuroscience* 2017;11:324.

41. Hebb DO. *The organization of behavior: A neuropsychological approach*. Wiley; 1949.

42. Sanger TD. Analysis of the two-dimensional receptive fields learned by the generalized Hebbian algorithm in response to random input. *Biological Cybernetics* 1990;63(3):221–228.

43. Gerstner W, Kistler WM. Mathematical formulations of Hebbian learning. *Biological Cybernetics* 2002;87(5–6):404–415.

44. Law C-T, Gold JI. Reinforcement learning can account for associative and perceptual learning on a visual-decision task. *Nature Neuroscience* 2009;12(5):655–663.

45. Roelfsema PR, van Ooyen A. Attention-gated reinforcement learning of internal representations for classification. *Neural Computation* 2005;17(10):2176–2214.

46. Kohonen T. The self-organizing map. *Neurocomputing* 1998;21(1):1–6.

47. Liu J, Lu Z-L, Dosher BA. Augmented Hebbian reweighting: Interactions between feedback and training accuracy in perceptual learning. *Journal of Vision* 2010;10(10):29,1–14.

48. Ahissar M, Hochstein S. Task difficulty and the specificity of perceptual learning. *Nature* 1997; 387(6631):401–406.

49. Papathomas TV, Gorea A, Feher A, Conway TE. Attention-based texture segregation. *Attention, Perception, & Psychophysics* 1999;61(7):1399–1410.

50. Rubin N, Nakayama K, Shapley R. Abrupt learning and retinal size specificity in illusory-contour perception. *Current Biology* 1997;7(7):461–467.

51. Goldhacker M, Rosengarth K, Plank T, Greenlee MW. The effect of feedback on performance and brain activation during perceptual learning. *Vision Research* 2014;99:99–110.

52. Liu J, Lu Z-L, Dosher BA. Mixed training at high and low accuracy levels leads to perceptual learning without feedback. *Vision Research* 2012;61:15–24.

53. Liu J, Dosher B, Lu Z-L. Modeling trial by trial and block feedback in perceptual learning. *Vision Research* 2014;99:46–56.

54. Sotiropoulos G, Seitz AR, Seriès P. Changing expectations about speed alters perceived motion direction. *Current Biology* 2011;21(21):R883–R884.

55. Liu J, Dosher BA, Lu Z-L. Augmented Hebbian reweighting accounts for accuracy and induced bias in perceptual learning with reverse feedback. *Journal of Vision* 2015;15(10):10,1–21.

56. Gold J, Bennett P, Sekuler A. Signal but not noise changes with perceptual learning. *Nature* 1999;402(6758):176–178.

57. Gold JM, Sekuler AB, Bennett PJ. Characterizing perceptual learning with external noise. *Cognitive Science* 2004;28(2):167–207.

58. Miller GA. The magical number seven, plus or minus two: Some limits on our capacity for processing information. *Psychological Review* 1956;63(2):81–97.

59. Garner WR, Hake HW. The amount of information in absolute judgments. *Psychological Review* 1951;58(6):446–459.

60. Hake HW, Garner W. The effect of presenting various numbers of discrete steps on scale reading accuracy. *Journal of Experimental Psychology* 1951;42(5):358–366.

61. Pollack I. The information of elementary auditory displays. *Journal of the Acoustical Society of America* 1952;24(6):745–749.

62. Shiffrin RM, Nosofsky RM. Seven plus or minus two: A commentary on capacity limitations. *Psychological Review,* 1994;101(2):357–361.

63. Rouder JN, Morey RD, Cowan N, Pealtz M. Learning in a unidimensional absolute identification task. *Psychonomic Bulletin & Review* 2004;11(5):938–944.

64. Dodds P, Donkin C, Brown SD, Heathcote A. Increasing capacity: Practice effects in absolute identification. *Journal of Experimental Psychology: Learning, Memory, and Cognition* 2011;37(2):477–492.

65. Dosher B, Liu J, Lu Z-L. Perceptual learning of spatial frequency identification through learned reweighting. *Journal of Vision* 2017;17(10):490.

66. Liu J, Dosher B, Lu Z-L. Perceptual learning in n-alternative forced choice with response and accuracy feedback, and a reweighting model. *Journal of Vision* 2017;17(10):1078.

67. Aberg KC, Herzog MH. About similar characteristics of visual perceptual learning and LTP. *Vision Research* 2012;61:100–106.

68. Seitz A, Watanabe T. A unified model for perceptual learning. *Trends in Cognitive Sciences* 2005;9(7):329–334.

8

Modeling Transfer and Specificity

Transfer and specificity represent two sides of a defining characteristic of perceptual learning. Specificity to stimulus attributes or to spatial location motivated speculations about the role of the early visual cortices in learning, but what then are the mechanisms of transfer when it does occur? In this chapter, we introduce an integrated reweighting theory (IRT) to account for transfer by way of improved reweighting and readout from higher-level representations. Transfer occurs when learning focuses on these higher-level representations, while specificity occurs when learning focuses on lower-level representations. This principle accounts for many challenging observations about the graded and variable nature of transfer during learning.

8.1 An Integrated Reweighting Theory (IRT)

One of the most discussed properties of perceptual learning is its specificity to the trained task. Learning is often specific to the task feature or even the trained retinal location (which historically became a central basis for the strong claims made about learning-induced neural retuning in the early visual cortex). Sometimes, however, learning does transfer. Why and when is this the case? What makes some kinds of training more transferable than others? When transfer does occur, how is the transfer accomplished?

We have already explored (in chapter 3) the experimental findings of specificity and transfer. In this chapter, we develop a theoretical framework and a corresponding computational model to account for when and how training transfers, with a preliminary focus on transfer across retinal locations.

Given that different low-level features and spatial locations have somewhat segregated representations in early cortical areas, it follows that specificity should be expected in tasks that rely on these early representations. According to this view, it is the ability to transfer, and not the specificity of learning, that requires an explanation. How does transfer occur? For those tasks trained to a specific feature or retinal location, how can training apply to other features or locations?

The bulk of this chapter develops one possible answer. As we will see, the integrated reweighting theory (IRT), an extension of the AHRM, explains transfer across task variants using a representational hierarchy in which higher-level invariant representations (relevant to many task variations) provide the scaffold for transfer. Though a number of competing hypotheses exist[1], our proposal is based on the idea that transfer occurs when the training and transfer tasks draw on common representations and processes.[2]

8.2 Everyday Analogies for Transfer

A colloquial analogy for specificity and transfer in perceptual learning can be found in language learning. An English speaker who has mastered Spanish may find it easier to learn Italian. On the other hand, knowing English and learning Spanish will likely provide little benefit for learning Chinese. In this case, we would say that learning Spanish transfers to learning Italian but shows mutual specificity for learning Chinese. Each language involves physical stimuli, phonemes, and tonality, as well as written characters; other levels of representation may also include those for words, syntax, and meaning. Sharing phonetic or symbolic inputs, cognate words, syntax, or other forms of compatibility may aid in the immediate understanding or subsequent learning of a new language. There may also be subtler ways in which the details of the two languages compete or interfere. This language analogy illustrates the broader point that it is the commonalities or differences between the representations and processes at different levels that create positive or negative interactions during learning.

Another useful analogy is found in the learning of multiple sports. Learning badminton from scratch would likely have positive transfer to learning another racquet sport, such as squash. Handling a racquet, judging distance to a target object, and managing aspects of the swing are all skills that apply to both games. Conversely, we might expect there to be negative transfer between some pairs, such as batting in baseball, where the ball often drops near the plate, and in softball, where it often rises near the plate (or so the experts say). Other sports may vary sufficiently from one another that there is neither positive nor negative transfer. (In the field of perceptual learning, such independence of learning is labeled full specificity.)

The empirical literature on specificity and transfer in perceptual learning (chapter 3) was sorted into four classes, based on the relationships between pairs of tasks between which learning might transfer (see figure 3.3). As in the intuitive analogies just mentioned, it was the overlap between representations and processes (or, in the language of neural networks, weighted connections and decision units) that determined both the likely results and the plausible interpretations of specificity and transfer. In simplified form, the relationship between two tasks depends on whether they share stimulus representations, decision judgments, both, or neither. These task relationships may of course be more complicated in multilayered networks, but the fundamental idea is the same. The relationships between the training and transfer tasks, whether they share representations and/or weights, determine whether training in the two tasks will be predicted to interact—whether learned weights can be shared or are separate and independent.

As we saw in chapter 3, this analysis, although simplified, provides a powerful theoretical framework that illuminates the ways different relationships between training and transfer tasks map onto a number of observed patterns in the behavioral data. For example, if two tasks use separate representations but the same judgment, the weights connecting representations to judgments must still be independent, and specificity is the default (this is true regardless of whether learning retunes the representations or reweights the sensory evidence to make the decision). After reviewing the literature, we concluded that the experimental data were either consistent with or actively supported the reweighting hypothesis, while representation change was ruled out in several key experiments. (Of course, as we indicated, the relationships could be more complex if the tasks required hierarchical architectures to account for the data, which might produce some hybrid explanations.) With these tentative conclusions in mind, we went on to develop the IRT framework for modeling transfer. This was based on the insight that network architectures with both *specific* representations and *invariant* representations could use learning through reweighting to explain many observed patterns of graded transfer or specificity.

In what follows, we focus on an IRT implementation that was designed to account for transfer over locations. We show how this model accounts for the data from a number of new experiments. The IRT implemented with invariant representations other than location can be used to account for other kinds of transfer. We consider alternative theories as well as possible developments for future models at the end of the chapter.

8.3 Hierarchical Representations and Transfer

In the context of perception and perceptual learning, the brain can be understood by analogy to a multilayer hierarchical network that connects stimulus to response, from early representations to decision. Each early representation is connected to some representations in higher layers, from the first analyses in the primary visual cortex, to many other representations in the secondary visual cortex, and then to higher cortical areas representing decision and action.

Within such a multilayer network model, reweighting could in principle alter the strength of connection weights from one representation layer to the next layer, within a representation layer, from lower sensory areas to higher decision areas, or from higher levels through feedback to lower ones. Any relevant representation could also be connected either directly or indirectly to a task-defined decision unit (or units). The weights at different levels might be learned either simultaneously (as in our models) or sequentially. In the network, the weights that connect high-quality information to a decision (and therefore to action) are what should be strengthened during learning.

These intuitions led to the integrated reweighting theory (IRT)[2]—a framework designed to account for transfer over location. The simpler augmented Hebbian reweighting model (AHRM) includes a layer of stimulus representations, with activations computed through front-end representations, and a layer with one decision unit (or sometimes multiple decision units). The IRT builds on this model by adding several sets of location-specific representations (e.g., one set per retinal location used in an experiment) and a layer of location-invariant representations with units that respond to stimuli in any location. (The location-invariant representations, for example, might code for the spatial-frequency and orientation content in the stimulus regardless of the location of its presentation.) All these representations at both the location-specific and the location-invariant levels are connected to a decision unit (or units). Positive transfer occurs when learned weight structures between training and transfer tasks overlap and are compatible, while negative transfer may occur when learned weight structures overlap and are incompatible. A third possibility is that training and transfer tasks are learned independently if the weight structures do not overlap, or overlap in a minimal way. In the IRT framework, transfer of learning from one location to another occurs because of learned weights on the location-invariant representations, while subsequent learning in the new task involves learning of weights on new location-specific representations and continued learning in weights on the location-invariant ones. Learning reweights the connections from the location-specific and the location-invariant representations to decision using the same augmented Hebbian learning rules as in the AHRM.[3, 4]

The same IRT framework, modified to include different kinds of invariant representations, could, in principle, be extended to other forms of transfer over invariant representations. In all these forms, however, the model makes a core claim: *perceptual learning transfers—either positively or negatively—from one task to another if and only if there is overlap in the weight structures connecting input representations, higher-level representations, and/or decision.* A schematic illustration of this can be found in figure 8.1, which shows overlapping or segregated task networks that mediate the interactions between training and transfer tasks. Furthermore, although these are all illustrated as feed-forward networks (and correspondingly all current IRT implementations use feed-forward networks), analogous concepts could similarly be developed in networks that include feedback or top-down weights or weights connecting units within layers.

In the original implementation of the IRT, perceptual learning was programmed to reweight all layers of the network to the decision unit simultaneously.[7] Weights from all the location-specific and location-invariant representations were updated on each trial with the same learning rate (a simplification that could easily be relaxed to allow different rates). Nevertheless, the weights on connections between location-specific or location-invariant representations and decision changed more or less quickly during learning in different applications. This occurred either because some representations carried information that was more useful (e.g., a better signal-to-noise ratio) and/or because connections from invariant representations experienced learning on more trials in the experimental design.

In essence, transfer operates through learning at different layers of a multilayer network. The core proposal, then, intuitively parallels the view that visual objects are represented at multiple levels with varying degrees of invariance. In this view, representations at early levels involve simple features (such as orientation and spatial frequency) separately

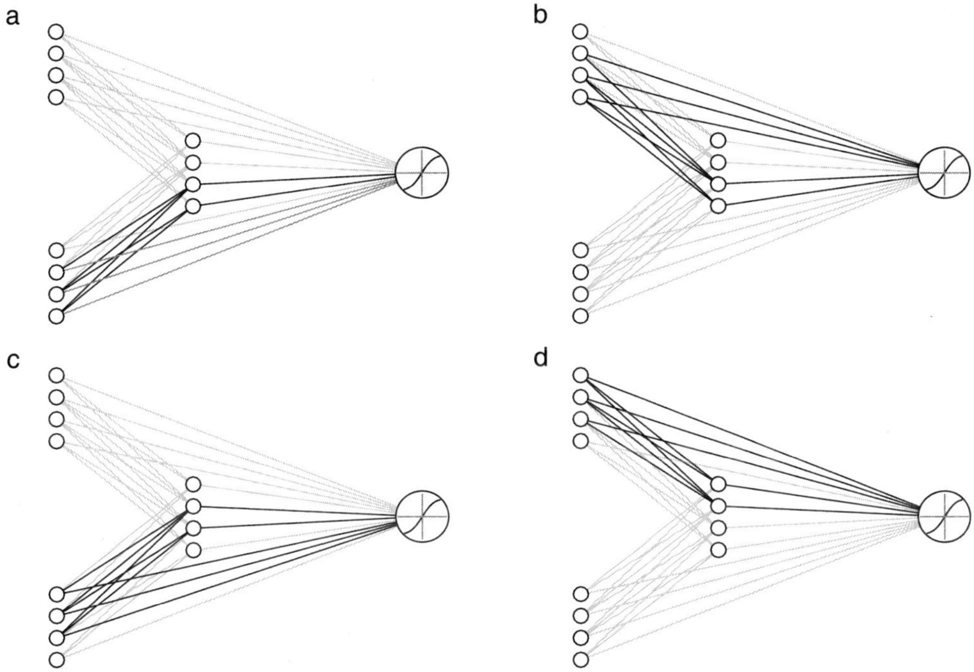

Figure 8.1
Hypothetical weight structures are shown for tasks with different kinds of overlap with that of an initial task. (a) Weights for an initial task (feed forward from left to right). (b) A task with different weights from low-level units (e.g., in different locations) but the same high-level weights to decision. (c) A task with partial overlap in weights from both low-level units and high-level units to decision. (d) A task with different weights at all levels from the initial task. Transfer is associated with overlap in weight structures.

represented in different locations in the visual field, while higher levels of representation combine or transform these simpler features to represent something more complicated. Higher-level representations are also often seen as more abstract. For example, they become invariant to other features, combine information from many locations (thus becoming location invariant), or combine inputs from different scales (thus becoming size invariant). These invariant features are then combined in unique ways to represent specific objects or patterns. Figure 8.2 shows a hierarchy of representations developed from common principles of computer vision[5] (see also Leonardis and Fidler[6]).

8.4 Preliminary Hierarchical Models

Before embarking on an analysis of the IRT, it is useful to note that there have already been a number of three-layer networks proposed to explain specificity and/or transfer in certain tasks. Essentially precursors to the IRT, these models figured in explanations of the patterns

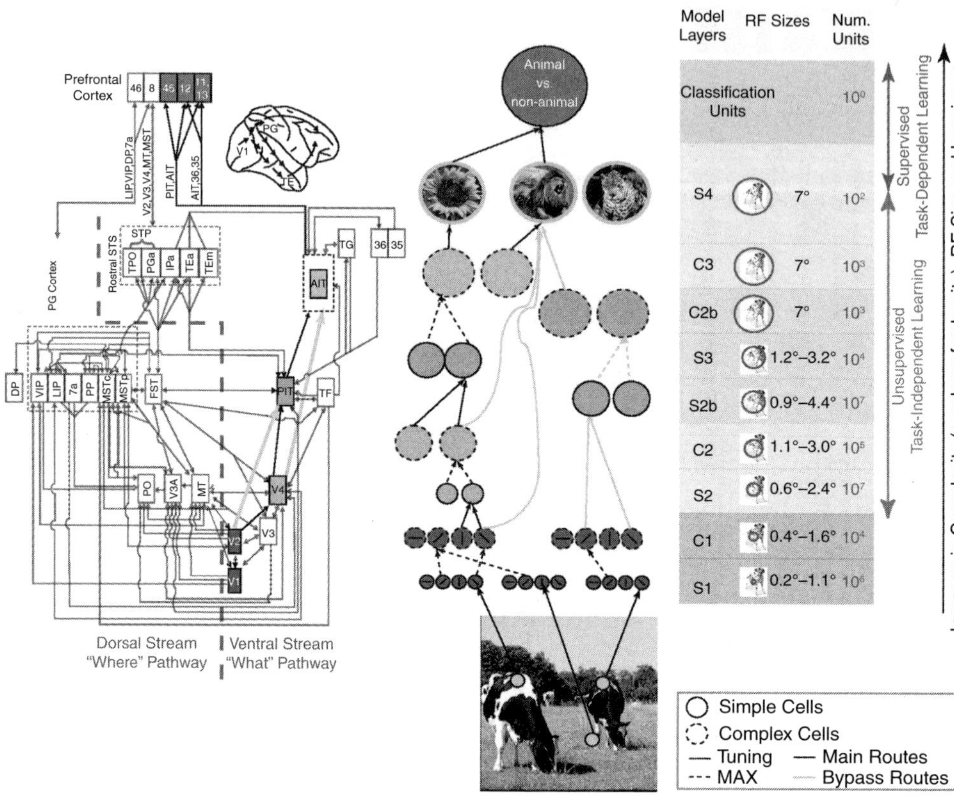

Figure 8.2
Illustration of a hierarchy of representations of a visual object, ranging from low-level orientation and spatial-frequency representations of the early visual cortex up to higher-level object representations. After Serre, Oliva, and Poggio,[5] figure 1. Copyright (2007) National Academy of Sciences. (See plate 5.)

of specificity and transfer involving eyes in one case (subsection 3.4.2)[7] and first- or second-order systems in another (subsection 3.4.4).[8–10] Explanations of eye specificity or transfer, for example, require at least three layers: left- and right-eye representations and a higher-level representation at or above binocular combination, plus a decision layer (see figure 8.3).[7, 11–13] In this (feed-forward) framework, learning transfers from one eye to the other if perceptual learning significantly improves the connection from the higher-level postbinocular representation to decision, and if task performance after training with one eye transfers essentially completely to the other eye, this implies that learning occurred predominantly at this higher level. To explain the pattern of transfer betwen first- and second-order textures[14, 15] or motion tasks[16] also requires a multilevel structure feeding into a decision to account for observed asymmetries in transfer in which learning a second-order task improved performance on a first-order task but not the reverse. First-order stimuli are thought to feed directly into a first-order representation that is connected to decision, while second-order stimuli must first be

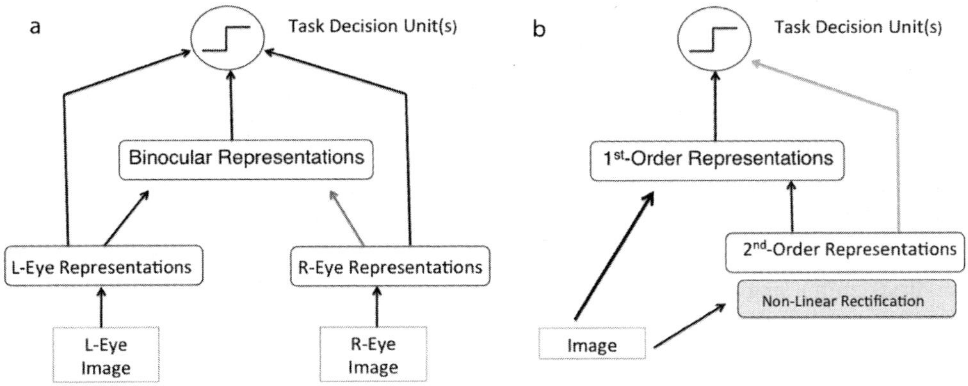

Figure 8.3
Schematic illustrations of IRT-type models are shown for different forms of asymmetric transfer. (a) Learning weights from individual eye representations to decision will not transfer to the other eye, while learning weights from a binocular representation to decision will transfer. (b) Improved weights from first-order representations to decision can be trained with either first- or second-order tasks, while only training with second-order stimuli can improve second-order tasks.

preprocessed by rectification (or other processes that "grab" second-order information) to activate the pattern analyzers, the output of which in turn passes through the first-order representation connected to decision.[15,17–19] Learning the second-order task requires training weights to process the noisy second-order information as well as the weights to decision, with these latter weights mediating transfer from the second-order task to the first-order task. Training the first-order task also trains these weights to the decision unit but not the preprocessing of noisy second-order stimuli.

Such three-layer network models can thus be seen as early precursors to the IRT, only with different content coded in the second layer of representations.

8.5 The AHRM as an IRT

The key step along the way to the IRT was, of course, the AHRM. It is worth noting that the AHRM itself makes many predictions about specificity or transfer within the same location. Although the AHRM can resemble a two-layer model in its learning characteristics, it is in fact more complicated than simple two-layer network models, because the front end, which incorporates a multilayered module designed to mimic the orientation and spatial-frequency responses of the early visual system, structures the stimulus space. It also differs from the simple two-layer neural network models in that it includes internal noise in the representations (and early nonlinearities).

The AHRM has made accurate and testable predictions for transfer and specificity for tasks trained in the same location. As described earlier, the model successfully predicts transfer when the stimulus conditions change from no external noise to high external noise[20] (see subsection 6.4.3); it makes systematic predictions about the consequences of pretraining in low or high external noise to tests involving both (see subsection 6.4.4);[21] and it predicts independence (specificity) of colearning two different judgments on the same stimuli (see subsection 6.4.5)[22] by assuming that very different judgments require separate decision unit(s) with independent weight structures. The AHRM also makes predictions for alternative stimuli using the same judgment. A related model that uses a simplified representation module similarly makes such predictions (see section 6.5).[23] In sum, the IRT, including the AHRM and related models by others, already has made many predictions about specificity and transfer in a single training location that have been verified in empirical studies. Nevertheless, the problem of how to *explain* transfer over retinal locations remains.

8.6 An IRT with Location-Invariant Representations

Our first computational implementation of a multilayer IRT was designed to investigate specificity of learning to trained locations and transfer over locations (figure 8.4).[2] As described earlier, it used an architecture with one layer consisting of sets of location-specific representations (one per retinal location of training and testing), a second layer of location-invariant representations, and a third layer for decision, all while relying on the learning rules of the AHRM.[3, 4] In addition, the representation activations were processed through the multilayer front end of the AHRM.

The first experimental tests of the IRT focused on orientation discrimination. Figure 8.4 illustrates the IRT model with two location-specific representations and one location-invariant representation, each set consisting of 35 units (7 orientations × 5 spatial frequencies), though several experimental applications required additional location-specific representations for additional testing locations or a wider range and sampling of orientations or spatial frequencies to cover all the stimuli. The location-invariant activations were thus computed from the stimuli with different bandwidth assumptions[2, 24,25] or, in a variant model, by using inputs from the location-specific representations.[23]

In applications of this model to the data, the parameters of the location-specific representations (orientation and spatial-frequency bandwidths and nonlinearities) were set from previous applications of the AHRM. The location-invariant representations were assumed to be noisier and more broadly tuned than the location-specific representations (the exact quantitative relationship varies slightly in different simulations, based on analyses of parameter sensitivity). This assumption was based on the intuition that the trade-off for representing inputs from many locations was likely to be a reduction in the precision and an increase in the noisiness of the representations. This scheme was generally consistent with an analogy of the invariant representations to orientation tuning in V4 (or higher visual areas) compared to V1.[26, 27] Broader bandwidths and higher internal noise levels also have been needed to fit behavioral data.[2] (So far, only a single model learning rate parameter has been assumed, although the location-specific and location-invariant learning rates logically could be

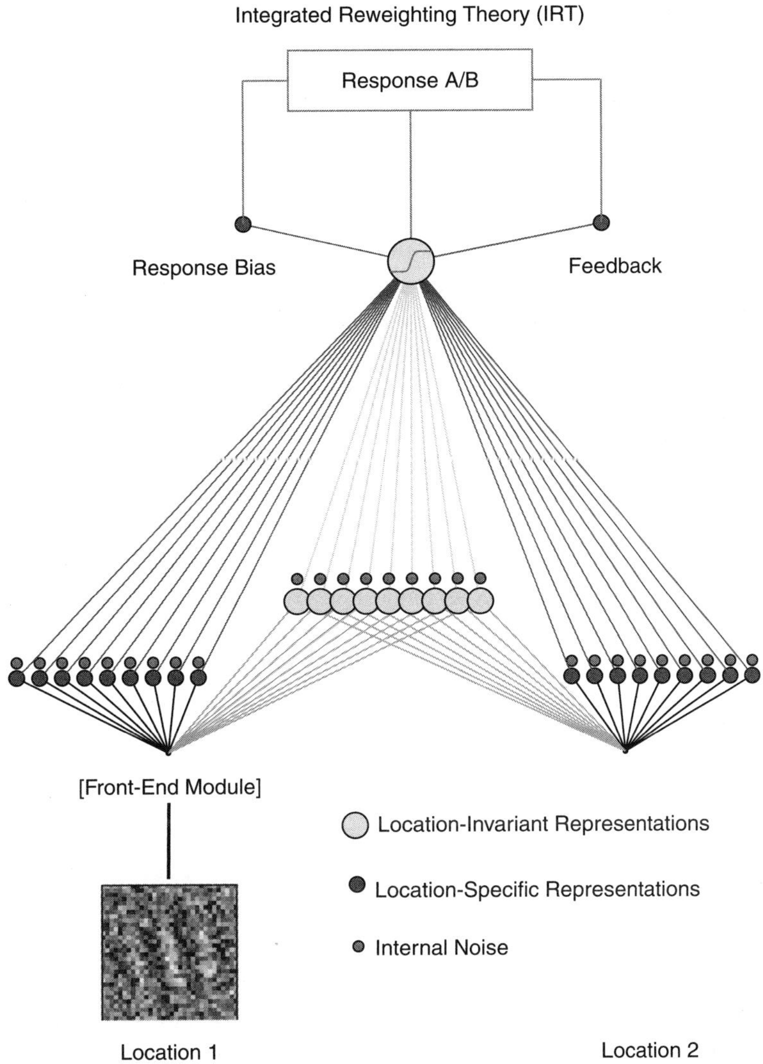

Figure 8.4
An integrated reweighting theory (IRT) designed to account for transfer over locations and to different stimuli. The architecture illustrated here includes two sets of location-specific representation units and one set of location-invariant representation units, each tuned for orientation and spatial frequency and computed by the front-end module. The weight structure connects each unit to the decision unit. A Hebbian learning rule, augmented with bias and feedback inputs, learns by reweighting the connections. After Dosher et al.,[2] figure 1. (See plate 6.)

different if this proved useful. Indeed, a modification of the IRT by other researchers has decoupled learning rates for weights connecting these two levels of representations to decision.)[28]

Simulating the IRT involves reprising the exact experimental trial sequence and stimuli in a behavioral experiment. The model generates simulated responses that are analyzed in the same way as the human data. It makes predictions about what will happen when the details of the training and/or transfer protocols are changed or interleaved in shorter or longer blocks.[29, 30] It can make predictions for learning with the method of constant stimuli, adaptive staircases with fixed stimuli for training at higher or lower accuracy levels,[31, 32] or with more or less precise judgments (more or less similar patterns). Notably, it also makes predictions about different feedback systems[33] and can be extended to incorporate attention or reward manipulations (see chapter 9). Additional front-end modules might be used to extend the IRT, depending on the nature of the stimuli used (e.g., motion, stereo, color), with other IRT variants potentially implementing different forms of invariance (e.g., orientation-invariance) to account for other types of transfer.

The possible applications of the IRT framework are only beginning to be explored.[2] Along with similar or competing models, it promises to generate new predictions about learning and transfer that in turn can be used to guide a program of empirical and theoretical investigations.[1, 28] Even in its initial stage, however, the IRT has yielded a host of compelling predictions that extend beyond location invariance to include the effects of task precision, the amount of training, and the extent of interaction in multiple tasks. Furthermore, an implemented IRT model could make predictions for different training protocols, while the generality of its basic architecture allows it to be adapted and altered depending on the task domain.

8.7 Applications of the IRT

In this section we examine a variety of applications of the IRT model to experimental data. These include location and feature specificity, the role of task precision in determining specificity and transfer, specificity of trained biases, double-training paradigms designed to improve generalization, and explanations of interactions between tasks trained together in task roving paradigms.

8.7.1 Location and Feature Specificity

Our first implementation of the IRT examined differential transfer related to changes in stimulus orientation and location. There was good reason to focus on these two variables. Many prior behavioral studies focused on changes in either retinal location or stimulus feature—but not both.[7, 34–38] The first IRT analysis, which tested three kinds of changes in the same context, made it possible to compare specificity in the three forms directly.[2] Contrary to strong early claims of specificity to location (sometimes even when separated by only a few degrees of visual angle[7]), a more extended review showed a mixture of transfer and specificity. The same was true for feature specificity, which sometimes showed very high levels of specificity and sometimes mixed effects. Unlike verbal claims, models are especially well suited to account for such effects, as they make graded quantitative predictions that can be tested and fit to data.

In this first application of the IRT, the initial training task was followed by a switch either to new locations, new orientations, or both, as tested in three separate groups of observers.[2] The experiment used relatively precise Gabor orientation judgments (tilted –35°±5° or 55°±5°) testing one or another diagonal of peripheral locations (NW/SE or SW/NE quadrants), with the target location precued on each trial. Tests occurred in both zero and high external noise, with staircases that estimated contrast threshold at 75% correct (the average of 3:1 and 2:1 staircases). Observers were trained in one orientation judgment on one diagonal in the initial phase (eight blocks), then transferred to the same orientations on the other diagonal (group L), to the other reference angle in the same location (group O), or to both the other reference angle and the other diagonal (group OL), and then trained again (another eight blocks). As expected, the contrast thresholds of the three groups were statistically equivalent during the initial learning phase, while they predictably diverged for the different transfer tasks (see figure 8.5). Judging the same orientations in new locations (e.g., NW/SE to SW/NE) led to the most transfer, while judging new orientations in the same locations (e.g., –35°±5° to 55°±5°) led to the least transfer. Changing both locations and orientations led to intermediate performance.

The IRT predicted this pattern of findings, as seen in figure 8.5. To explain in greater detail, the largest (positive) transfer occurred where the same orientations were tested in different locations (group L), in which learned weights from location-invariant representations to decision were the basis for initial transfer to the same task in the new locations. The least transfer occurred when the orientations were changed but the locations were not (group O), in which case new weights to decision must be learned for new orientations for both the location-specific and location-invariant representations. If both orientations and locations were switched (group OL), the results were intermediate—and the model suggested the partial transfer was caused by changes in weights throughout learning that favored representations better tuned to the spatial frequencies of the Gabor targets. Figure 8.6 shows representative weight structures from the best-fitting simulation of the behavioral data, including weights on the location-specific representations for the first training task, the location-invariant representations, and the location-specific representations for the transfer task (top to bottom) for initial weights, weights after initial training, and weights after practice on the transfer task (left to right).

The initial weights before training were set to include general knowledge based on task instructions about the orientations to be discriminated in the task. After practice in the initial training task, weights in all three groups had been changed through learning to upweight representation units closest to the Gabor orientations and spatial frequency. For this task, the magnitudes, either positive or negative, of the relevant weights in the

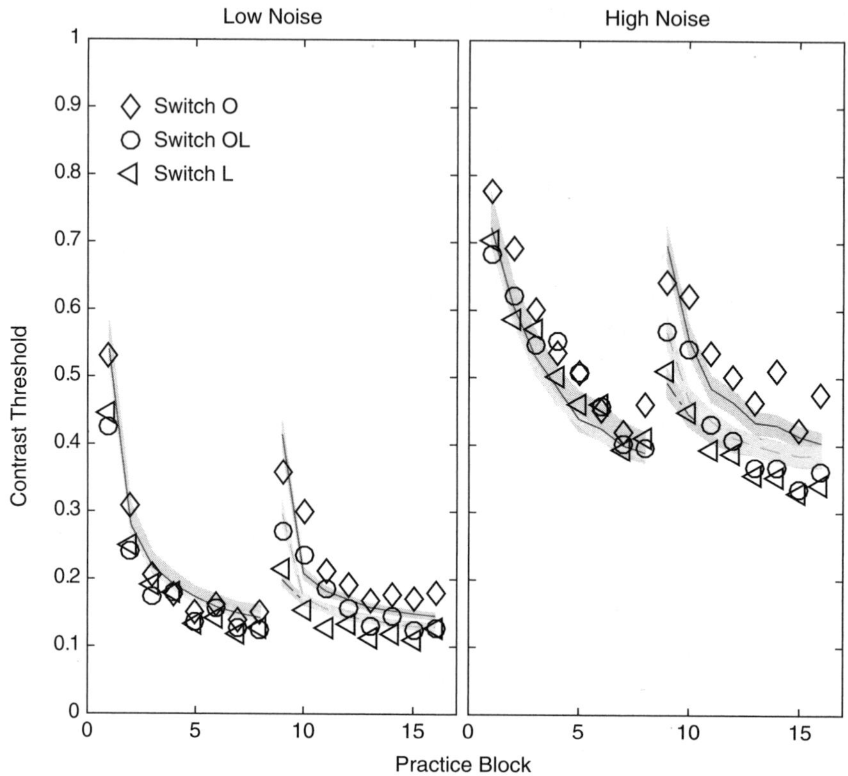

Figure 8.5
Perceptual learning improves contrast thresholds in an orientation task and transfer to new retinal locations and/or orientations for three groups of observers that changed orientation (O), location (L), or both (OL) (data points) with predictions of the IRT simulation (smooth curves). Redrawn from Dosher et al.,[2] figure 3.

location-specific representations increased, while those of the relevant location-invariant representations decreased (because the location-invariant representations were noisier and more broadly tuned, and the model upweights relevant representations based on their signal-to-noise ratios). After practice on the transfer task, the group that trained on the same orientation judgments in different locations (L) showed consistent orientation-tuned weights in all sets of units; the group that trained in the same locations on different orientations (O) showed some orientation tuning around both reference angles for relevant spatial frequencies, with larger weights for the more recently trained orientations (e.g., forgetting because of interference); and the group that switched both orientations and locations (OL) showed tuning around the respective reference orientations in the separate location-specific representations, with relatively low weights on the location-invariant representations that shared no experiences between the training and transfer tasks.

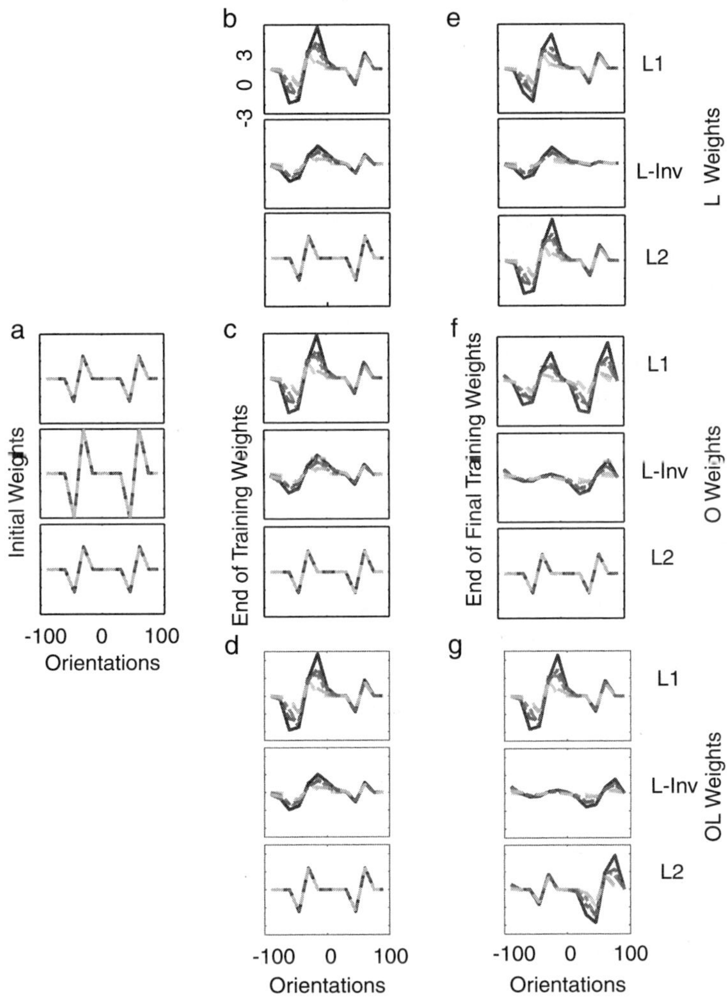

Figure 8.6
IRT weight structures expressing perceptual learning and transfer to new retinal locations and/or orientations in an orientation-discrimination task. Weight structures at the beginning of initial training for all three groups (a), at the end of initial training (b, c, d), and at the end of the training in the transfer phase (e, f, g), for the L, O, and OL groups (see the text). In each set, the middle represents the location-invariant weights and the top and bottom show the two location-specific weights. Redrawn from Dosher et al.,[2] figure S3. (See plate 7.)

As the preceding summary indicates, the IRT was able to explain the magnitude of transfer based on the compatibility of the learned weight structures between the initial training task and the transfer task. This is in stark contrast to the qualitative attributions of specificity to low-level retuning, which provide little basis on which to make predictions, especially quantitative ones.

8.7.2 Task Precision and Transfer

An early and influential claim in perceptual learning was that "the degree of specificity depends on the difficulty of the training conditions"[34] (p. 401). There is now evidence to suggest that specificity is in fact largely controlled by the demands of the transfer task and, in particular, the *precision* of the transfer task.[39]

The IRT provided an explanation for this phenomenon as well. The original claim that task difficulty at training produced specificity was based on findings in a texture-discrimination task. In that task, *difficulty* was manipulated by changing the angular difference between the target and background elements.[34] In the original study, training on an "easy" task (angular difference of 30° and targets in one of two locations) transferred to a similarly "easy" task with different orientations and locations (similar to OL here). On the other hand, training on a "difficult" task (angular difference of 16° with targets in one of two locations) failed to transfer to a correspondingly "difficult" task (see subsection 4.5.1 for a description). (We have suggested that such manipulations of angular difference should be called task *precision* because the word *difficulty* generally refers to how accurately a task can be performed, while these studies all use a 75% correct threshold that holds accuracy constant in all conditions.)[39]

In an experiment designed to more fully test the hypothesis, the manipulations of orientation judgments in the training and transfer tasks were decoupled. This made it possible to determine whether it was the nature of the training task, the nature of the transfer task, or some interaction that controlled specificity. The corresponding experiment crossed the precision of the training and transfer tasks in four training groups (low-low, high-low, high-low, and high-high).[39] In each group, orientation judgments (±5° or ±12°) were tested in zero and high external noise, and then both orientation and retinal location were changed between training and transfer (as in the original texture studies). In the high-low group, for example, an observer who first trained on the orientations –35°±5° in the NW/SE locations would have transferred to 55°±12° in the SW/NE locations. The IRT predicted the very surprising pattern of results observed in the behavioral data (see figure 8.7).

Remarkably, performance in the transfer task reflected *only* the precision of the transfer task: the two groups transferring to a high-precision task had essentially the same threshold functions regardless of whether initial training was on the low- or high-precision task (low-high and high-high). Likewise, the two groups transferring to a low-precision task also showed essentially the same threshold functions on the transfer task (low-low and high-low). That is, transferring to a high-precision task produced far more specificity, regardless of the precision of the original training.

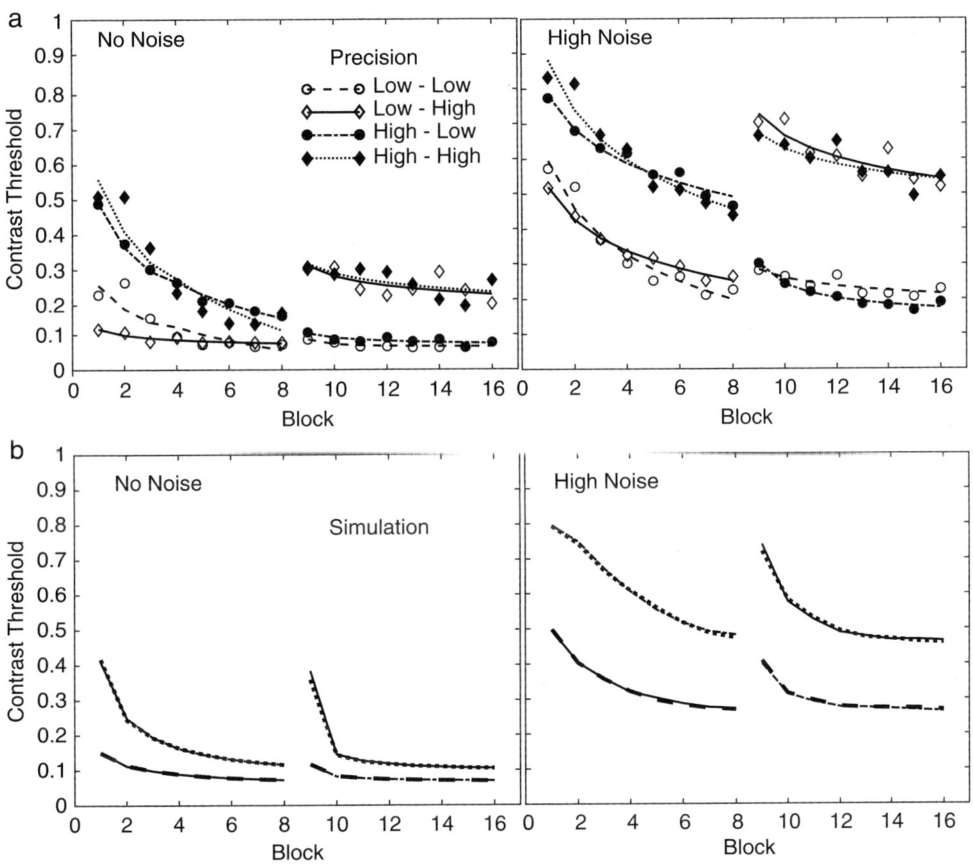

Figure 8.7
Perceptual learning and transfer to new orientations and retinal locations as a function of task precision in the training versus the transfer task. (a) Contrast thresholds for four groups of observers in experiments with no and high external noise and (b) predictions of the IRT simulation. Data in (a) redrawn from Jeter et al.,[39] figure 2c; model simulations from Liu, Dosher, and Lu,[40] with permission of the authors.

The intuitive reason for this is that precise judgments require finer tuning to relevant information. To distinguish two patterns differing by only a few degrees of rotation will require both better-tuned weights and higher contrasts to achieve the target accuracy level. On the other hand, distinguishing patterns differing by 20° or 30° can occur with weights that are less well tuned and can be successful with lower contrasts to achieve the target accuracy level. The initial training task tuned weights to favor orientations and spatial frequencies that matched the Gabor target stimuli, while also increasing the weights on the location-specific representations and decreasing the location-invariant representations. The switched task (different orientations and locations) required learning new weights. In the simulations, transfer in performance reflected a combination of improved spatial-frequency tuning and reductions of weights from the noisier and less precise location-invariant representations to decision (for the same reasons as described in subsection 8.7.1).

The dependence of transfer on the precision of the *transfer* task, rather than the training task, was predicted by the IRT.[39] What might appear to have been slight departures from these predictions in a few other studies almost surely reflect different procedural details in those experiments. For example, one experiment that did not control for accuracy in the training task (i.e., observers were trained on fixed stimuli to yield an improvement in percentage correct rather than being trained using staircases to control accuracy) reported a very slight influence of the training task on specificity in addition to the much more substantial effect of the precision of the transfer task.[41] One of the advantages of a framework such as the IRT (or competing quantitative models) is that such findings can be analyzed, predicted, and ultimately explained by the model.

8.7.3 Specificity and Transfer of Bias Training in Different Locations

The IRT has also been used to model a significant body of literature in perceptual learning on induced response biases. One set of experiments showed that false feedback can induce learned bias in responses and even opposite biases in separate locations (see subsection 7.4.6 for a discussion of the induced bias paradigm).[42] In a Vernier line task, a pair of larger offset stimuli that always received accurate feedback were mixed during training with one smaller singleton offset stimulus that received reversed, false feedback—and vice versa for another training location (i.e., a larger offset pair ±15″ with –5″ for the left location and +5″ for the right location). The training with false feedback shifted all the responses in the direction favored by that false feedback (graphed as increasing hit rate for the singleton and decreased hit rates for the offset stimuli in the other direction), followed by recovery as soon as the false feedback was removed. The data (see figure 8.8) showed opposite induced biases in two trained locations.

The IRT easily accounted for these results. Opposite biases were induced as the reversed feedback shifted the weights on the separate location-specific representations in the direction of the false feedback (opposite in the two locations). These opposite induced biases were contained in the location-specific weights, while the opposite false feedbacks in the two locations canceled each other in the learned weights on the location-invariant units (which nonetheless continued to support task learning by upweighting connections from useful orientation and spatial-frequency units to decision). A wide range of findings originally interpreted as changes in signal detection criteria were thus predicted quantitatively using the IRT model.[24]

8.7.4 Double Training, Paradigm Specificity, and Location Transfer

A great deal of recent research has focused on cross-training protocols designed to promote transfer to new retinal locations for training tasks that ordinarily demonstrate high levels of specificity. A number of different experimental cross-training protocols were

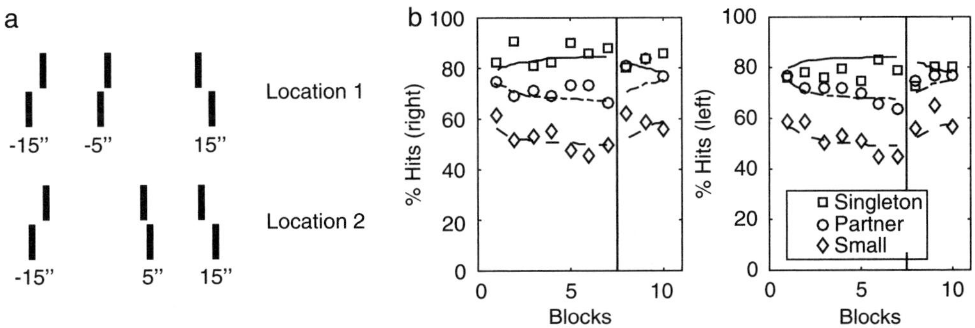

Figure 8.8
Inducing opposite biases in separate locations through false feedback, and predictions of the IRT model. (a) Vernier stimuli in the two locations, where the small offset stimuli receive false feedback. (b) Learning data (symbols) show increasing shifts in the direction of the false feedback that recover when false feedback is removed (at the vertical line) with fits of the IRT (lines), shown as opposing hit rates in the two locations. Redrawn from Liu, Dosher, and Lu,[40] figure 6.

discussed in subsection 3.5.4.[43–45] The precise causes and generality of the cross-training effects are still actively debated in the field.[30, 46–48]

Results for one such double-training experiment and a corresponding IRT simulation are shown in figure 8.9. In this experiment, contrast thresholds were pretested in two locations (contrast-L1 and L2), then contrast thresholds in L2 were assessed after contrast judgments for vertical Gabors were trained in L1 (contrast-L1) and then again after orientation judgments for Gabors around vertical were trained in L2 (orientation-L2).[49] The idea was that cross training with a different task in L2 should improve transfer of the contrast judgment (contrast-L2), which ordinarily shows specificity. The data showed that the cross training improved contrast thresholds in L2 to approximately the asymptotic levels in L1 (contrast-L1 after training). However, the independent impact of the two should be assessed.

The IRT approximately accounted for this pattern, suggesting that improved performance was caused by repeating the assessment of contrast-L2 as well as a general rebalancing of the relative weights on noisy location-invariant representations with successive learning.[50] A subsequent set of simulation studies, conducted by a different research group, developed a more flexible IRT-like model architecture. This explicitly added the ability to change the relative weights on location-specific (V1-like) and location-invariant (V4-like) representations to account for many of these double-training and cross-training paradigms.[28]

Once again, a quantitative model can provide unique insights that qualitative claims cannot. The degree of transfer in different cross-training paradigms may in principle depend on seemingly unimportant protocol choices, such as the nature of the training staircases.[46] In one example, training with a single long adaptive staircase (which tends to train on a narrower range of stimuli after initial settling) was shown to lead to more specificity, while training on a series of short staircases (which train stimuli in low-precision judgments

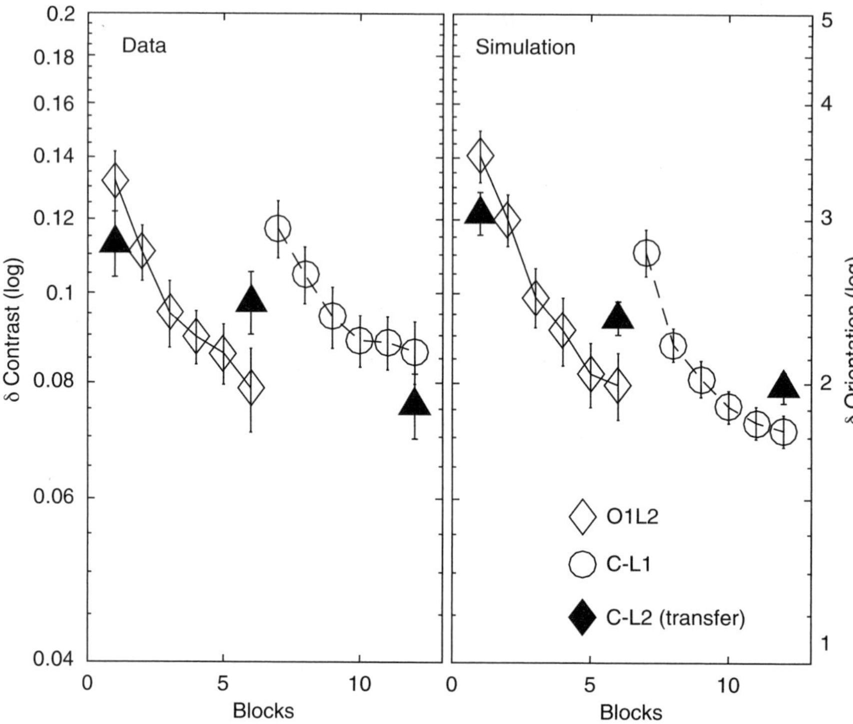

Figure 8.9
Results of a double-training experiment, and the corresponding predictions of the AHRM, in which training a horizontal Gabor orientation judgment on location 2 (O1-L2) completes transfer of a contrast judgment using vertical Gabors (C-L1) to a new location (C-L2). After Xiao et al.,[49] figure 2B (left), with permission; simulation from Liu, Lu, and Dosher,[50] with permission of the authors.

on more trials and use a wider range of stimuli) led to increased transfer.[46] Models like the IRT and its variants or contender models may be able to predict such dependencies (figure 8.9).[30]

8.7.5 Task Roving and Multiple Locations

Another interesting phenomenon for which the IRT makes predictions is task roving. Roving is the name given to a phenomenon whereby intermixed training of several tasks (or task variants) can seriously disrupt learning (see subsection 2.2.3), even when the same tasks can be easily learned when trained alone.[44, 51–53]

Such disruptions have been found in a variety of tasks, including auditory tasks.[54, 55] They make intuitive sense in the context of reweighting models. Furthermore, according to the IRT/AHRM, interference can be predicted to occur in certain circumstances but not in others.[56] Learning will be disrupted whenever the optimal weight structures of the intermixed task variants conflict with each other. When this occurs, changes that improve the weights for one task variant are likely to be erased over the next few trials as the other task variants are practiced. Indeed, the most disruptive examples of roving have been those with high stimulus overlap and the same spatial location. However, when the judgments required in the two tasks were quite different (requiring separate decision units and completely separate weight structures to perform), the two intermixed tasks were learned independently (e.g., the intermixed training of Vernier and bisection judgments, as in subsection 6.4.5).[22]

Even when interleaved tasks are similar, however, if the respective stimuli are far enough apart in the stimulus space, the weights that associate their representations to the decision will also be distinct. From this, it follows that it should be possible to learn both during mixed training. The ability to learn with intermixed training for sufficiently separated stimuli also occurs for the AHRM, which handles learning in a single location, as shown in the simulated predictions (see figure 8.10).

The IRT framework predicts many effects of roving on learning. All else being equal, for example, intermixed trainings on multiple tasks are expected to interact even when the tasks are trained in different locations. This is because all locations train the weights on the location-invariant representations.

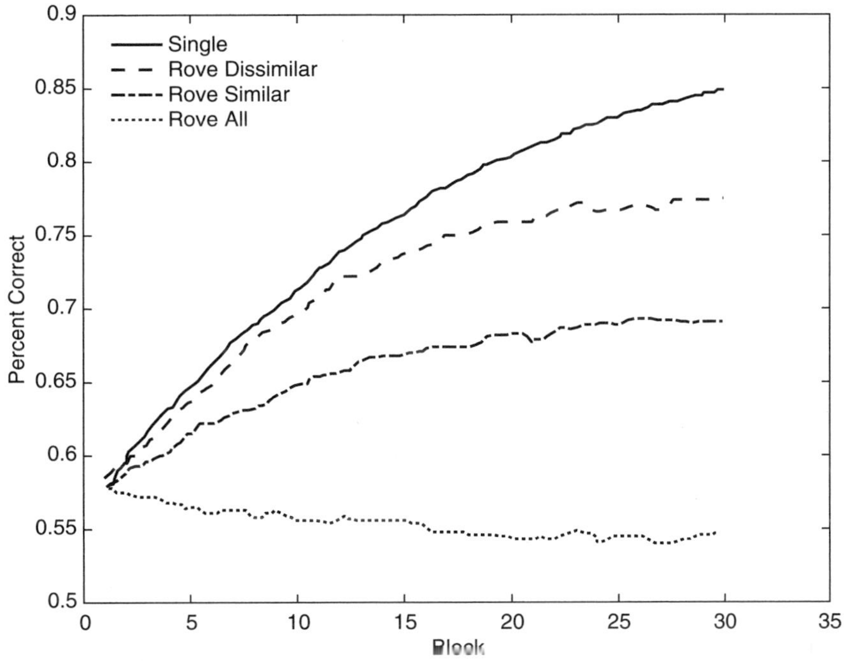

Figure 8.10
Predictions of an AHRM simulation for learning different roved mixtures of orientation judgments in one location. Learning is successively faster for training a single reference angle (no roving), two widely separated reference angles, two closer reference angles, and actual slight performance loss with intermixed training of four reference angles.

To test this prediction of the IRT model, we designed an experiment in which four different combinations of orientation-discrimination tasks were trained in different locations in four training groups. One group was trained for different reference angles (θ±12°, clockwise or counterclockwise) in each of four locations (e.g., −67.5°, −22.5°, 22.5°, or 67.5° from vertical), or maximum roving; a second was trained for two nearer reference angles (e.g., 22.5° in the NW and SE and 67.5° in the SW and NE locations); a third involved two widely spaced reference angles (e.g., −22.5° in the NW and SE and 67.5° in the SW and NE locations) trained in two locations each; and a fourth trained a single reference angle in all four locations (labeled all, near, far, and single). Training accuracy was held at a constant performance level of 75% correct, using adaptive methods.

The results of this experiment (figure 8.11) have profound implications for understanding the consequences of intermixed task training and also for reweighting theories of perceptual learning. The combination of tasks trained in the four locations strongly interacted, a result that demonstrates that learning cannot solely—or even predominantly—be the

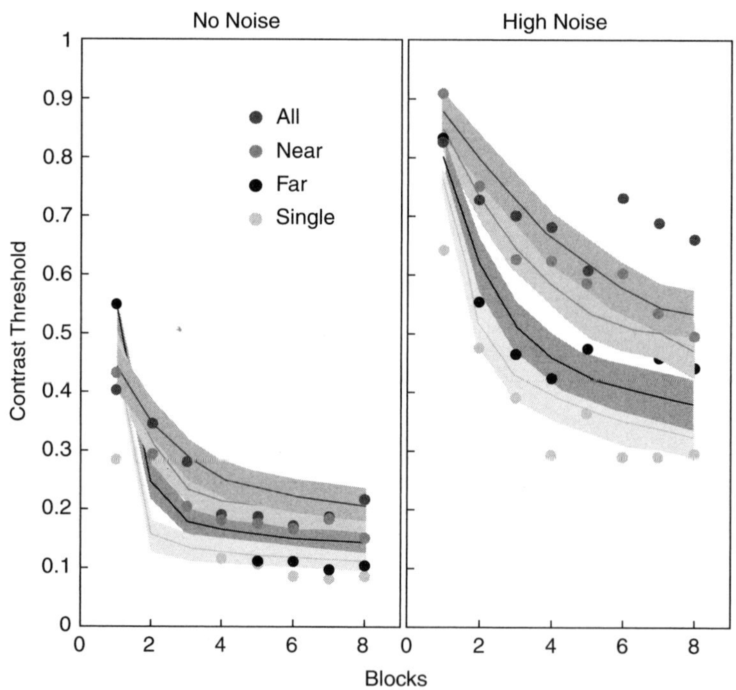

Figure 8.11
Intermixing training at four different locations shows interactions in learning, depending on the relationship between the orientation-discrimination tasks in those locations. Learning is fastest when the same reference angle is trained in all locations or for widely separated reference angles, slower for similar reference angles, and slowest for four reference angles, as seen in learning curves for the four groups. Lines with bands show the predictions of a best-fitting IRT model fit. From Dosher et al.,[25] with permission of the authors. (See plate 8.)

result of retuning (representation change) in the early retinotopic visual cortex. That combining two additional similar reference angles showed more disruption than two dissimilar reference angles further demonstrates the importance of stimulus dissimilarity in enabling learning in roving experiments.

Interference during learning was a direct consequence of the close representational overlap and incompatibility of optimal weights for cases in which similar orientation stimuli required opposite responses. All these results were predicted by the IRT, in which interference in the network weights occurs to different degrees. Our conclusion was that if the relevant weight spaces—here orientation tuning in the representation units—were separable, then the weight updates could be learned independently, so multiple tasks could be learned even when intermixed. Such a possibility, consistent with the IRT, further underscores the explanatory promise of the model.

8.8 Other Models

The IRT is not the only model to account for transfer in perceptual learning. Other theories have also contributed to an understanding of the phenomenon. Some of these are computational models, while most have relied on qualitative statements about the transfer processes. Furthermore, most of the early computational models (as discussed in chapter 6) were developed with only learning in mind; transfer as such was not their focus. A few, such as models for Vernier offset judgments, considered specificity to stimulus variants but did not explicitly consider how learning might transfer either to different retinal locations or to substantially different stimuli.

Perhaps the most famous claim about transfer, including transfer over location, derives from the reverse hierarchy theory (RHT).[57] This theory proposes that top layers of the visual hierarchy are trained first and that learning in the top layers is transferable for easy tasks; learning in the lower layers of the visual hierarchy (closer to V1), which might exhibit more specificity, is predicted to occur only later in training and only as required to perform the task. These broad verbal claims are among the more prominent in the field. Because the RHT follows intuitions about the physiological hierarchy of visual representations in the cortex, it has a strong intuitive pull, but it is noncomputational, and makes no specific quantitative predictions. (Similarly, it makes verbal claims stating that training in easy tasks should lead to more transfer were experimentally challenged, as described in subsection 8.7.2.)[39] Few additional specific predictions have been proposed that might test the theory.

Other ideas about transfer have focused on the notion that it follows from learning abstract rules. These claims, made largely in the context of cross-training papers, inferred that transfer to a new location after cross training must reflect some form of cognitive rule induction or else be based on other general learning, such as temporal patterning,[58, 59] but the exact nature of the abstract rule and its relationship to generalization have not been specified. Of course, the lack of implemented computational models need not imply that the ideas are incorrect. Indeed, instantiating these ideas in computational form (and then using them in competitive tests with the IRT and its variants) has the potential to bring new and important features to existing models.

The key idea of the IRT was that transfer was scaffolded by learning weights for higher-level representations that embodied some form of invariance. This chapter reviewed how the original IRT, or the simpler AHRM in some cases, accounted for patterns of transfer over a wide range of experimental manipulations. The successful predictions included transfer with switched orientation and/or location, the role of task precision in transfer, some forms of cross training, and learning in multitask roving designs.

In many ways, the reach of the IRT has gone far beyond what was initially expected of it. In addition, related approaches have been developed to account for other transfer phenomena. Starting with the IRT architecture, several related models have been explored. These include a model with a simplified front end that examined transfer over different stimuli[23] and a variant of the IRT that derived activations in the location-invariant representations somewhat differently.[23] In yet another case, researchers modified the learning rules and made it possible to use confidence calculations to change the learning rates at the "V1" and "V4" levels.[30] Each of these models differs in its details from the original IRT, though all broadly follow the same framework. In one case, the new model introduced more flexibility (e.g., several learning rates) and thus should be able to account for datasets that are more complex. Another model used both feed-forward and feedback (bottom-up and top-down) connections in order to account for transfer as a form of rapid self-organized learning.[1, 60] In these various models, the broad structure of the IRT architecture has proven to be flexible and expandable. Future modifications may permit new predictions that account for transfer in other task domains.

One important feature of the IRT/AHRM framework is that the front-end representation module was designed to be consistent with the signal-to-noise properties of visual responses revealed by external-noise studies and the perceptual template model (PTM), as described in chapter 4. The IRT explicitly incorporates internal noise and nonlinearity in the front end, which converts stimulus images into activations over a set of stimulus representation units; it likewise incorporates internal noise at every stage from representation to decision, and nonlinearity in the responses of stimulus representations and in the decision rules. All these ingredients contribute to the generality and the robustness of the model.

Another recent development has been the interest in using convolution neural networks (CNNs), and in some cases deep (many layered) CNNs (DCNNs), to account for perceptual learning. Developed originally for object recognition in image processing, these networks, when trained with large sets of images and object labels, can be relatively successful in their object classifications.

As discussed previously (see subsection 5.2.2), lower layers of a many-layered CNN (usually the first three to five layers) have responses that have been claimed to approximate those in the early visual cortex.[61–63] Indeed, some researchers have argued that the deep CNNs may provide better explanations of the behavior of cells in visual areas such as IT than those inferred from single-cell recording studies.[61, 64] Extensively studied in the realm of object recognition and increasingly thought to be useful in modeling responses of the early visual system in fMRI and other brain imaging measures, these deep learning networks, or shallow two- or three-layer variants of them, have recently been applied to perceptual learning.

One such example used a shallow two-layer CNN based on the "neurocognitron" model[65, 66] and a two-unit output layer corresponding with the two responses in a two-alternative forced-choice task.[67] These are described as follows: "Each layer in the [CNN] network computes a number of feature maps (or channels), where each channel corresponds to a certain filter which convolves with patches in the image.... In addition to a convolution sub-layer, each layer includes additional operations (sub-layers). Some correspond with known operations in the visual system"[67] (p. 2).

The shallow CNN-based network model was used to generate predictions corresponding to the ordinal properties of data on orientation and location transfer and precision (figures 8.5 and 8.6). These simulations made general qualitative predictions about improvements in generic error rates as a function of relatively long "epochs" of training (see figure 8.12). (This simulated model was not fit to the contrast threshold data of the behavioral experiments, nor did the training history duplicate the number of training trials or the trial structure of the experiments.) In this modeling exercise, the network had 8,846 weight and bias parameters, was initialized using a randomly generated set of weights, and was trained to convergence in an initial training task—the point at which predictions were generated. The focus was on showing that the extensive early training derived "edge-like features matching the displayed stimuli" with the training location "highlighted" against the background[67] (p. 6). That is, the goal of this simulation exercise was primarily to understand development of the feature channels.

Another application of deep CNNs to visual perceptual learning aimed to provide a linking function between patterns of responses in the model before and after training to some key findings about perceptual learning from physiology.[68] This project examined changes in multiple early layers of the CNN that have been claimed to correspond with regions of the early visual cortex (V1, V2, V4,...) as described in the physiological literature. Here the focus was on linking changes in response patterns of pseudoneurons in the model to similar patterns reported in certain single-cell recording studies (chapter 5). Here, too, the training of the CNN was somewhat general; it did not explicitly model the specific tasks and/or training protocols as found in the corresponding single-cell studies.

Such CNN models, especially the deep-learning counterparts, are tremendously powerful learning machines. When properly constrained by system principles, they promise to provide a broad and powerful framework for incorporating changes in representation as

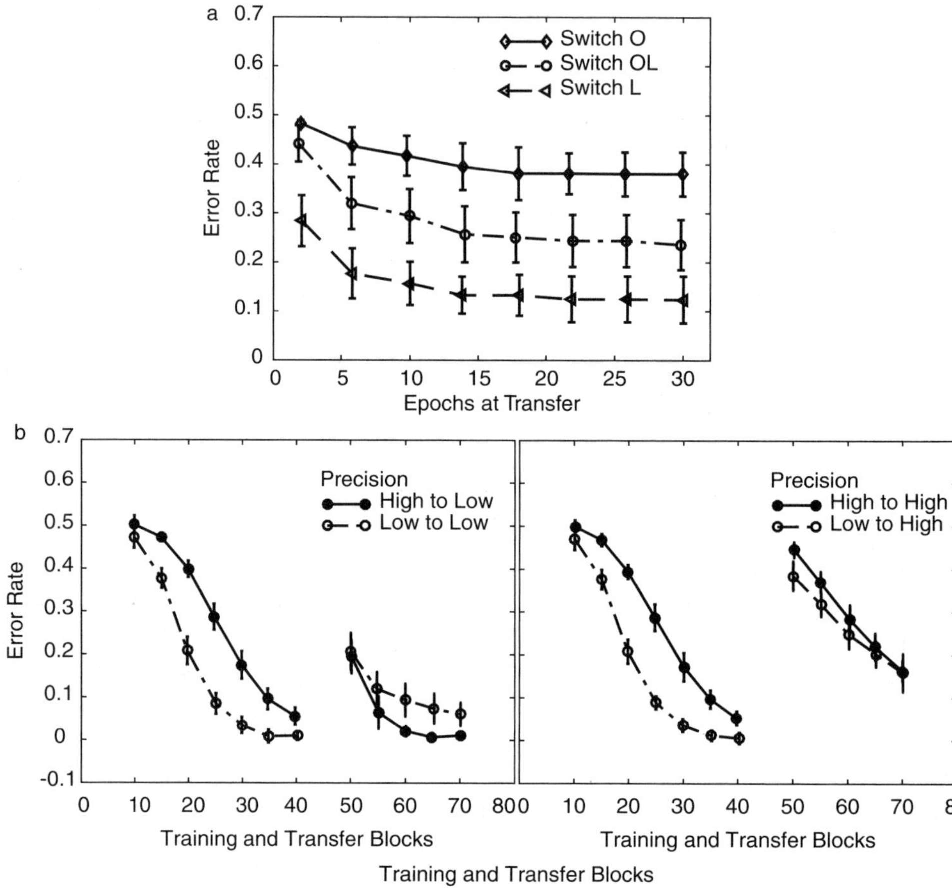

Figure 8.12
Simulated predictions of a network with two convolution neural network layers and an output layer, in the form of generic error rates as a function of training epoch, for transfers involving location and orientation switches (a), or differential task precision for the training and transfer tasks (b). Corresponding fits of the IRT model to the actual contrast-threshold data of the target experiments are shown in figures 8.5 and 8.6, respectively. After Cohen and Weinshall,[67] figures 7 and 6 (open access via the Computer Vision Foundation, 2017).

well as reweighting of evidence to decision. Indeed, one interpretation is that the early convolution layers of the deep-learning model carry out operations that seek to serve the function of the front end of the IRT/AHRM. In particular, learning weights in the IRT may be similar to learning weights in the top few layers of a CNN for training a specific task (while the early layers of the CNN have been massively trained by prior experience). One interesting difference between these models and the IRT, however, is that they often "yoke" changes in weights at early levels across other similar representations—assuming de facto that training in one location propagates to other analogous parts of the network. This occurs so that an object trained in one location can be recognized in another, thus building location transfer into the system by assumption.

This is just one example suggesting a more general point. While these powerful CNN models hold great promise, they also present many technical challenges. Furthermore, as we have argued earlier, the DNN or CNN models so far have not incorporated meaningful treatments of internal noise, a fundamental property of human systems. Meanwhile, the IRT—a far simpler computational model—has provided a quantitative account of the actual behavioral data.

Yet another model based on very different principles has been proposed to explain visual learning. The so-called where-what network (WWN) model uses a "brain-inspired neuromorphic computational model of the Where-What visuomotor pathways" to model learning and transfer[1] (p. 1). This WWN model involves multilayered laminar cortical structures, loosely inspired by the cortical layers and columns in the visual cortex: feature neurons combine bottom-up sensory inputs with top-down motor inputs and develop their tuning using Hebbian learning and "k-winners take all" competition. The model proposes that "gated self-organization of the connections during the off-task processes" accounts for transfer—essentially that top-down implicit rehearsal processes precondition rapid learning during transfer[1] (p. 1). The paper just quoted tested the model by simulating predictions for a double-training result, with the authors arguing that feed-forward reweighting models (such as the IRT/AHRM or, for that matter, the deep CNNs) were categorically inconsistent with known top-down recurrent inputs into lower-level cortical representations. One response to this criticism might be to consider versions of an IRT with feedback as well as feed-forward connections.

The WWN model also made an interesting comparison of representation change and readout. The model was based on the idea that learning causes changes in both lower cortical representations and in readout. As a counterargument, it is interesting to focus on one aspect of the authors' modeling exercise.[1] The reported computations estimated the learned change in d′ caused by sensory retuning as 0.0098 and that from reweighting as 0.247, suggesting that behavioral improvements resulting from sensory retuning are about 5% of the overall improvement. These estimates are remarkably similar to our own estimates of the maximum size of improvements resulting from retuning in the AHRM (as less than 10%). In other words, taken together, these results might suggest that changes in readout, such as those modeled in the IRT/AHRM, still account for the lion's share of learning. Additionally, as with the early learning models and the DNN/CNN models, the WWN model does not explicitly include internal noise.

In sum, simplified feed-forward reweighting models have provided strong quantitative accounts of many phenomena of perceptual learning, feedback, and transfer. Even so, these simplified models could be expanded to include reweighting of feedback and recurrent connections within a module in addition to feed-forward connections. One proposed model, for example, used fixed feed-forward connections (all weights set to 1) and differently weighted inhibitory top-down connections, together with anti-Hebbian rules, to account for perceptual learning.[69] (Such models are closely equivalent to corresponding feed-forward Hebbian networks.) Expanding the nature of network connections in these ways could easily generalize the IRT framework, making it more flexible and perhaps more consistent with physiology. In addition, future models might seek to take into account aspects of brain microstructure and function more directly. The general issue of transfer is so important to both the theory of perceptual learning and to its usefulness in practical applications that it deserves further development and testing.

8.9 Future Directions

In this chapter, we examined the predictions made by the integrated reweighting theory (IRT) about transfer and specificity. The key insight was to use higher invariant levels of representations as the scaffold for transfer. Our initial implementation focused on transfer over spatial locations. In this framework, location transfer occurs when weights from higher-level location-invariant representations to decision are both useful and consistent between the initial training task and the transfer task. As the applications of the model to data illustrated, the framework provided interesting new predictions that ranged from cross-location interactions in training to the role of task precision. This framework also seems to predict a number of other phenomena. These include the effects of longer training (e.g., increased specificity), the consequences of using different training paradigms (e.g., one long staircase as opposed to several short staircases including easy trials), and others.

Taking further inspiration from the multilevel hierarchies used in computer science for object recognition suggests other forms of higher-order representational invariance. There are certain visual perceptual tasks, for example, that show high levels of scale invariance (although others do not). Likewise, object recognition has been shown to be partly and locally rotation invariant (at least in some cases and with smaller rotations). Many tasks show color invariance in pattern judgments. There are other examples as well. In at least some of these cases, it may be possible to create other kinds of invariant representations to make new predictions about other forms of transfer or specificity. Indeed, the development of new invariant representations may occur through recruiting or creating new representation units, possibly through pooling over separate lower-level representations by reweighting.

Additional ways to expand the model might include schemes for programming distinct learning rates at different levels of representation, the introduction of different computational forms for the representational front ends, the use of more complex deep-learning networks, or the integration of so-called neuromorphic learning systems.

Each of these theoretical innovations could potentially lead to a series of experimental investigations, motivated by computational predictions of the new extended models. Hypotheses about each kind of invariance could easily generate an entire series of experiments and model studies in different task domains. For example, is spatial-frequency invariance more powerful than orientation invariance? When does phase invariance (or phase quadrature pooling, as in the current front-end implementations) characterize performance? Are there tasks in which phase specificity is natural or can be developed with practice? Such questions only hint at the many directions future explorations could take.

The dominant models for perceptual learning might one day become as complex as the deep convolutional neural networks (deep CNNs) of recent interest in computer science and image processing. At present, however, the current implementation of the IRT framework and the shallow CNN models use only a few layers for learning the perceptual task itself (massive training may be used to develop the early layers). Since intralayer reweighting or reweighting to a higher level can appear as a representation change to subsequent layers, these hierarchical forms may also be able to integrate representation change as a special case of reweighting.[2–4]

Models that are more complex are likely to emerge in the years to come. Yet even as models come to more closely mimic brain anatomy and physiology, the simplified models such as the IRT may retain an advantage in certain contexts in which efficient prediction is the primary concern. This may be especially true if we can show that simpler approximating models provide a sufficiently good account of the relevant behavioral observations of specificity and transfer.

References

1. Solgi M, Liu T, Weng J. A computational developmental model for specificity and transfer in perceptual learning. *Journal of Vision* 2013;13(1):7,1–23.

2. Dosher BA, Jeter P, Liu J, Lu Z-L. An integrated reweighting theory of perceptual learning. *Proceedings of the National Academy of Sciences* 2013;110(33):13678–13683.

3. Petrov AA, Dosher BA, Lu Z-L. The dynamics of perceptual learning: An incremental reweighting model. *Psychological Review* 2005;112(4):715–782.

4. Petrov AA, Dosher BA, Lu Z-L. Perceptual learning without feedback in non-stationary contexts: Data and model. *Vision Research* 2006;46(19):3177–3197.

5. Serre T, Oliva A, Poggio T. A feedforward architecture accounts for rapid categorization. *Proceedings of the National Academy of Sciences* 2007;104(15):6424–6429.

6. Leonardis A, Fidler S. Learning hierarchical representations of object categories for robot vision. In Kaneko M, Nakamura Y, eds., *Robotics Research*. Springer; 2010:66:99–110.

7. Schoups AA, Vogels R, Orban GA. Human perceptual learning in identifying the oblique orientation: Retinotopy, orientation specificity and monocularity. *Journal of Physiology* 1995;483(3):797–810.

8. Zanker JM. Perceptual learning in primary and secondary motion vision. *Vision Research* 1999;39(7):1293–1304.

9. Chen R, Qiu Z-P, Zhang Y, Zhou Y-F. Perceptual learning and transfer study of first- and second-order motion direction discrimination. *Progress in Biochemistry and Biophysics* 2009;36:1442–1450.

10. Petrov AA, Hayes TR. Asymmetric transfer of perceptual learning of luminance- and contrast-modulated motion. *Journal of Vision* 2010;10(14):11,1–22.

11. Karni A, Sagi D. Where practice makes perfect in texture discrimination: Evidence for primary visual cortex plasticity. *Proceedings of the National Academy of Sciences* 1991;88(11):4966–4970.

12. Zhang J-Y, Zhang G-L, Xiao L-Q, Klein SA, Levi DM, Yu C. Rule-based learning explains visual perceptual learning and its specificity and transfer. *Journal of Neuroscience* 2010;30(37):12323–12328.

13. Sowden PT, Rose D, Davies IR. Perceptual learning of luminance contrast detection: Specific for spatial frequency and retinal location but not orientation. *Vision Research* 2002;42(10):1249–1258.

14. Chubb C, Sperling G, Solomon JA. Texture interactions determine perceived contrast. *Proceedings of the National Academy of Sciences* 1989;86(23):9631–9635.

15. Chubb C, Sperling G. Drift-balanced random stimuli: A general basis for studying non-Fourier motion perception. *Journal of the Optical Society of America A* 1988;5(11):1986–2007.

16. Baker CL. Central neural mechanisms for detecting second-order motion. *Current Opinion in Neurobiology* 1999;9(4):461–466.

17. Graham N, Sutter A. Normalization: Contrast-gain control in simple (Fourier) and complex (non-Fourier) pathways of pattern vision. *Vision Research* 2000;40(20):2737–2761.

18. Wilson HR. Non-Fourier cortical processes in texture, form, and motion perception. *Cerebral Cortex* 1999;13:445–477.

19. Dosher BA, Lu Z-L. Level and mechanisms of perceptual learning: Learning first-order luminance and second-order texture objects. *Vision Research* 2006;46(12):1996–2007.

20. Dosher BA, Lu Z-L. Perceptual learning in clear displays optimizes perceptual expertise: Learning the limiting process. *Proceedings of the National Academy of Sciences* 2005;102(14):5286–5290.

21. Lu Z-L, Chu W, Dosher BA. Perceptual learning of motion direction discrimination in fovea: Separable mechanisms. *Vision Research* 2006;46(15):2315–2327.

22. Huang C-B, Lu Z-L, Dosher BA. Co-learning analysis of two perceptual learning tasks with identical input stimuli supports the reweighting hypothesis. *Vision Research* 2012;61:25–32.

23. Sotiropoulos G, Seitz AR, Seriès P. Perceptual learning in visual hyperacuity: A reweighting model. *Vision Research* 2011;51(6):585–599.

24. Liu J, Dosher BA, Lu Z-L. Augmented Hebbian reweighting accounts for accuracy and induced bias in perceptual learning with reverse feedback. *Journal of Vision* 2015;15(10):10,1–21.

25. Dosher B, Liu J, Chu W, Lu Z-L. Roving: The causes of interference and re-enabled learning in multi-task visual training. *Journal of Vision* 2020; in press.

26. Ringach DL, Hawken MJ, Shapley R. Dynamics of orientation tuning in macaque primary visual cortex. *Nature* 1997;387(6630):281–284.

27. McAdams CJ, Maunsell JH. Effects of attention on orientation-tuning functions of single neurons in macaque cortical area V4. *Journal of Neuroscience* 1999;19(1):431–441.

28. Sotiropoulos G, Seitz AR, Seriès P. Performance-monitoring integrated reweighting model of perceptual learning. *Vision Research* 2018;152:17–39.

29. Aberg KC, Herzog MH. About similar characteristics of visual perceptual learning and LTP. *Vision Research* 2012;61:100–106.

30. Talluri BC, Hung S-C, Seitz AR, Seriès P. Confidence-based integrated reweighting model of task-difficulty explains location-based specificity in perceptual learning. *Journal of Vision* 2015;15(10):17,1–12.

31. Liu J, Lu Z, Dosher B. Augmented Hebbian learning accounts for the complex pattern of effects of feedback in perceptual learning. *Journal of Vision* 2010;10(7):1115.

32. Liu J, Lu Z-L, Dosher BA. Mixed training at high and low accuracy levels leads to perceptual learning without feedback. *Vision Research* 2012;61:15–24.

33. Liu J, Dosher B, Lu Z-L. Modeling trial by trial and block feedback in perceptual learning. *Vision Research* 2014;99:46–56.

34. Ahissar M, Hochstein S. Task difficulty and the specificity of perceptual learning. *Nature* 1997;387(6631): 401–406.

35. Shiu L-P, Pashler H. Improvement in line orientation discrimination is retinally local but dependent on cognitive set. *Perception, & Psychophysics* 1992;52(5):582–588.

36. Dosher BA, Lu Z-L. Perceptual learning reflects external noise filtering and internal noise reduction through channel reweighting. *Proceedings of the National Academy of Sciences* 1998;95(23):13988–13993.

37. Dosher BA, Lu Z-L. Mechanisms of perceptual learning. *Vision Research* 1999;39(19):3197–3221.

38. Crist RE, Kapadia MK, Westheimer G, Gilbert CD. Perceptual learning of spatial localization: Specificity for orientation, position, and context. *Journal of Neurophysiology* 1997;78(6):2889–2894.

39. Jeter PE, Dosher BA, Petrov A, Lu Z-L. Task precision at transfer determines specificity of perceptual learning. *Journal of Vision* 2009;9(3):1,1–13.

40. Liu J, Dosher B, Lu Z-L. An integrated reweighting theory accounts for the role of task precision in transfer of perceptual learning for similar orientation tasks. *Journal of Vision* 2015;15(12):34.

41. Wang X, Zhou Y, Liu Z. Transfer in motion perceptual learning depends on the difficulty of the training task. *Journal of Vision* 2013;13(7):5,1–9.

42. Herzog MH, Ewald KR, Hermens F, Fahle M. Reverse feedback induces position and orientation specific changes. *Vision Research* 2006;46(22):3761–3770.

43. Zhang J-Y, Wang R, Klein S, Levi D, Yu C. Perceptual learning transfers to untrained retinal locations after double training: A piggyback effect. *Journal of Vision* 2011;11(11):1026.

44. Zhang J-Y, Kuai S-G, Xiao L-Q, Klein SA, Levi DM, Yu C. Stimulus coding rules for perceptual learning. *PLoS Biology* 2008;6(8):e197.

45. Zhang J-Y, Yang Y-X. Perceptual learning of motion direction discrimination transfers to an opposite direction with TPE training. *Vision Research* 2014;99:93–98.

46. Hung S-C, Seitz AR. Prolonged training at threshold promotes robust retinotopic specificity in perceptual learning. *Journal of Neuroscience* 2014;34(25):8423–8431.

47. Liang J, Zhou Y, Fahle M, Liu Z. Specificity of motion discrimination learning even with double training and staircase. *Journal of Vision* 2015;15(10):3,1–10.

48. Liang J, Zhou Y, Fahle M, Liu Z. Limited transfer of long-term motion perceptual learning with double training. *Journal of Vision* 2015;15(10):1,1–9.

49. Xiao L-Q, Zhang J-Y, Wang R, Klein SA, Levi DM, Yu C. Complete transfer of perceptual learning across retinal locations enabled by double training. *Current Biology* 2008;18(24):1922–1926.

50. Liu J, Lu Z-L, Dosher B. Multi-location augmented Hebbian re-weighting accounts for transfer of perceptual learning following double training. *Journal of Vision* 2011;11(11):992.

51. Yu C, Klein SA, Levi DM. Perceptual learning in contrast discrimination and the (minimal) role of context. *Journal of Vision* 2004;4(3):4,169–182.

52. Kuai S-G, Zhang J-Y, Klein SA, Levi DM, Yu C. The essential role of stimulus temporal patterning in enabling perceptual learning. *Nature Neuroscience* 2005;8(11):1497–1499.

53. Parkosadze K, Otto TU, Malania M, Kezeli A, Herzog MH. Perceptual learning of bisection stimuli under roving: Slow and largely specific. *Journal of Vision* 2008;8(1):5,1–8.

54. Amitay S, Hawkey DJ, Moore DR. Auditory frequency discrimination learning is affected by stimulus variability. *Perception & Psychophysics* 2005;67(4):691–698.

55. Banai K, Amitay S. Stimulus uncertainty in auditory perceptual learning. *Vision Research* 2012;61:83–88.

56. Dosher B, Chu W, Liu J, Lu Z-L. Perceptual learning of task mixtures. *Journal of Vision* 2012;12(9):767.

57. Ahissar M, Hochstein S. The reverse hierarchy theory of visual perceptual learning. *Trends in Cognitive Sciences* 2004;8(10):457–464.

58. Wang R, Cong L-J, Yu C. The classical TDT perceptual learning is mostly temporal learning. *Journal of Vision* 2013;13(5):9,1–9.

59. Wang R, Zhang J-Y, Klein SA, Levi DM, Yu C. Task relevancy and demand modulate double-training enabled transfer of perceptual learning. *Vision Research* 2012;61:33–38.

60. Ji Z, Weng J, Prokhorov D. Where-what network 1: "Where" and "What" assist each other through top-down connections. Paper presented at IEEE International Conference on Development and Learning; 2008.

61. Yamins DL, Hong H, Cadieu CF, Solomon EA, Seibert D, DiCarlo JJ. Performance-optimized hierarchical models predict neural responses in higher visual cortex. *Proceedings of the National Academy of Sciences* 2014;111(23):8619–8624.

62. Yamins DL, DiCarlo JJ. Using goal-driven deep learning models to understand sensory cortex. *Nature Neuroscience* 2016;19(3):356–365.

63. Yamins DL, DiCarlo JJ. Eight open questions in the computational modeling of higher sensory cortex. *Current Opinion in Neurobiology* 2016;37:114–120.

64. Cadieu CF, Hong H, Yamins DL, Pinto N, Ardila D, Solomon EA, Majaj NJ, DiCarlo JJ. Deep neural networks rival the representation of primate IT cortex for core visual object recognition. *PLoS Computational Biology* 2014;10(12):e1003963.

65. Fukushima K. Neocognitron: A hierarchical neural network capable of visual pattern recognition. *Neural Networks* 1988;1(2):119–130.

66. Fukushima K. Artificial vision by multi-layered neural networks: Neocognitron and its advances. *Neural Networks* 2013;37:103–119.

67. Cohen G, Weinshall D. Hidden layers in perceptual learning. *Proceedings of the IEEE Conference on Computer Vision & Pattern Recognition (CVPR)*;2017:4554–4562.

68. Wenliang LK, Seitz AR. Deep neural networks for modeling visual perceptual learning. *Journal of Neuroscience* 2018;38(27):6028–6044.

69. Herzog MH, Fahle M. Modeling perceptual learning: Difficulties and how they can be overcome. *Biological Cybernetics* 1998;78(2):107–117.

9

Top-Down Influences of Task, Attention, and Reward

Perceptual learning very often occurs in the context of goal-directed tasks. Attention, reward, and task requirements all provide ways to select the relevant stimulus features, and so have the potential to influence both immediate performance and learning. In this chapter, we show how task demands influence what is learned and how attention can influence perceptual learning through improved perceptual coding, and conversely how perceptual learning can reduce the requirement for attention in challenging tasks. Reward can also influence perceptual learning either through direct modulation of the learning process or through enhancement of stimulus coding. These top-down factors delineate a broader network of brain functions and may be critical to determining the rate and efficiency of learning. They can furthermore be integrated into learning rules that are more elaborate.

9.1 Perceptual Learning and Selectivity

Learning never occurs in a vacuum. And while some learning may be implicit and based on exposure alone, most learning occurs in the course of performing goal-directed tasks, with the relevant stimuli and judgments determined by circumstance or instruction. These goal-directed activities must be embodied in brain processes and must involve some form of top-down guidance.

Prominent among the top-down influencers of learning—and even of task performance itself—are task structure, attention, and reward. Investigating these possible influences leads to a number of questions: Does learning go beyond the required task? Does learning require attention? Is there a relationship between learning and reward? If so, how does reward differ from the information delivered by feedback?

It seems natural that learning might be specific to inputs directly relevant to the task and that attention might also play a role. It is similarly taken as a truism that reward incentives improve learning. But how much evidence is there in the existing literature that empirically supports these claims and, furthermore, answers these questions? This chapter explores what we know about the complex influence of all these top-down factors on the learning process.

To illustrate the many issues involved, consider an experiment in which the observer judges the orientation of a sine-wave pattern on a screen in the laboratory. The relevant pattern must be selected from other aspects of the visual, auditory, and tactile environment, such as the surround of the computer screen or the ticking of a clock on the wall. The observer's performance, which almost surely improves with training, is measured only for orientation judgments and might either be informed by feedback or influenced by reward. The judgment task specifies the relevant features of the stimuli (orientation), the required decision (clockwise or counterclockwise), and the overt behavioral response (press the right or left key).

Yet this scheme leaves many questions open. Are other aspects of the target stimuli coded (such as their spatial frequency, size, or contrast)? Are only the instructed and attended stimuli involved in learning, or does learning extend to unattended features or perhaps even other stimuli as well? Do the details of feedback or reward influence how quickly a task is learned?

Similar questions occur when considering brain processes. The visual hierarchy has many modules representing the stimulus that are active nearly simultaneously, and learning requires the observer to focus on those that most efficiently code the targeted feature and connect them to decision. Perceptual learning may involve all these potential influences.

In considering current theories about top-down influence, it is important to distinguish between what has been assumed or inferred and what has been based on empirical results. If researchers demonstrate learning in a particular task and infer a role that task relevance, attention, or reward might play, this inference may open as many questions as it answers. To investigate the role of top-down influences more precisely, experiments must involve comparisons of learning in conditions in which either the task, attention, or reward has been explicitly manipulated.

A few experiments—but not as many as one might think—have explored the roles of these factors via explicit manipulation. Other investigations merely *suggested* influences or mechanisms. By explicitly focusing on these top-down factors, future experiments promise to specify the more general principles of selectivity active in learning while also suggesting possible avenues for other forms of intervention.

9.2 Task-Relevant and Task-Irrelevant Learning

Any visual stimulus is composed of multiple features. In the example at the start of the chapter, an oriented pattern stimulus by definition will also have a spatial extent, a spatial frequency or texture, contrast, location on the screen, and other characteristics. The question then arises: is perceptual learning focused solely on the feature(s) relevant to the task, or does learning incorporate other features of the stimulus?

A series of studies have investigated this distinction between task-relevant learning[1] and so-called task-irrelevant learning. In the former, research has centered on whether only those features or stimuli most relevant to the goal-directed task are learned or whether additional features of the task-relevant stimuli are learned incidentally. The latter has explored whether and in what circumstances aspects of task-*irrelevant* stimuli (those that are extraneous to the goal-directed task) are learned implicitly.

9.2.1 Learning Task-Relevant Judgments

There is some evidence that perceptual learning can focus on a single task-relevant *feature* or dimension of a complex display.[2] In one study, two separate judgments could be made on the same line texture stimuli that varied in layout (7×5 or 5×7) and either did or did not contain a differently oriented line target (see figure 9.1).[2] Learning of the two judgments was found to be essentially independent. Although this study varied only task instructions, it has been widely cited as the definitive demonstration that *attention* will gate only task-relevant features for perceptual learning (see also subsection 9.3.2), a conclusion that follows from an inference that the two tasks had guided attention differently. Other studies have similarly investigated the role of instructed task relevance on learned perceptual judgments. These have included demonstrations of independent perceptual learning related to the contrast and orientation of line stimuli[3] as well as independence of horizontal or vertical offset judgments in compound Vernier stimuli that included stimuli for both.[4–6] (An alternative interpretation of these studies is that learning occurs for the practiced judgments.)

In addition to these studies, the selective effects of task relevance have also been investigated in more nuanced experiments, where they have been shown to interact with other factors. In one study, observers practiced motion judgments in displays that included both task-relevant rightward-moving dots and task-irrelevant downward-moving dots. Learning failed to improve the practiced speed discriminations, even as coherence thresholds in the task-relevant direction and binocular rivalry dominance favoring task-relevant motion were improved by the training.[7] Similar effects were reported for tasks involving directional motion aftereffects.[8] These examples show something surprising: practicing a

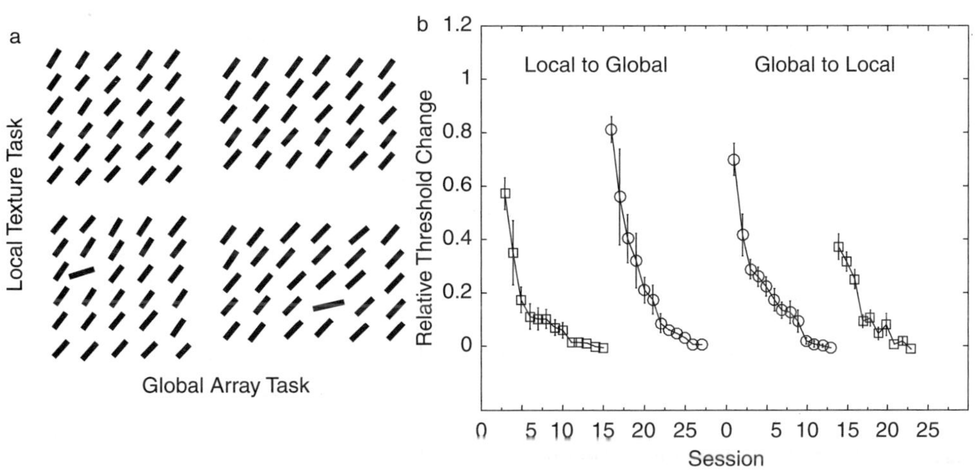

Figure 9.1
Practice trains only the task-relevant stimulus features. (a) Stimuli varying in the presence of a local target and global shape layout. (b) Global shape and local texture orientation detection are learned independently, as seen in shorter threshold stimulus onset asynchrony with practice. After Ahissar and Hochstein,[2] figures 1 and 4. Copyright 1993 National Academy of Sciences.

particular task may actually influence other related judgments even when the learning is not sufficient to improve the task-relevant judgments themselves.

In other cases, meanwhile, learning to make judgments about task-relevant stimuli had the side effect of learning to suppress task-irrelevant features. In one study, observers who were trained to make motion-direction judgments also showed reduced coherence thresholds in the task-relevant dot-motion direction while elevating coherence thresholds in the direction of a distractor motion.[9] It is now believed that suppression of irrelevant aspects of a stimulus occurs only for suprathreshold features that compete for processing with the primary task; furthermore, task-relevant learning and learned suppression of task-irrelevant features can occur together. (Note that this conclusion holds for a different experimental regime than so-called exposure-based task-irrelevant learning, which requires task-irrelevant stimuli to be near threshold; see subsection 9.2.2). Another example of secondary learning sometimes occurred when a lower-level task inherited the benefits of training a higher-level task. For example, training random dot-motion-direction judgments was shown to improve detection as well as discrimination in the trained direction, while training detection failed to improve discrimination.[10]

Taken together, the preceding results demonstrate that the explicitly practiced task judgment is a key selection mechanism and that plasticity may be focused primarily on those features of complex stimuli that are directly related to the training task or judgment. There are exceptions, however. When an irrelevant feature or stimulus presents strong competition, learned suppression may also sometimes occur. Task-relevant selection along with the occasional deselection of task-irrelevant competitors presents a powerful compound principle for selecting the sensory representations involved in learning. Both upweighting the task-relevant sensory representations and, if necessary, downweighting competitive task-irrelevant sensory representations are consistent with the selective reweighting framework for perceptual learning.

In considering these theoretical positions, notice that while the conclusions about task-relevant learning are often stated in strong form, there is only one study (to our knowledge) to have explicitly tried to evaluate what has been learned about the incidental features of stimuli.[11] Many questions thus remain for researchers to investigate. To take the example of learning in random-dot motion displays: Is motion direction the only feature that has been incorporated into learning? What would happen, for instance, if the dots were switched from dark to light (as with flankers with the same or a different color[12])? Or, what would be the effect of switching to a different speed, number of dots, or region of motion? Would there be some specificity of learning to these incidental features?

9.2.2 Task-Irrelevant Perceptual Learning

Though there are many cases where learning causes the suppression of suprathreshold task-irrelevant stimuli, there are also cases where (positive) perceptual learning occurs for *subliminal* (subthreshold) task-irrelevant stimuli. Labeled task-irrelevant perceptual learning (TIPL) or passive perceptual learning of task-irrelevant features,[13, 14] this phenomenon has been explored in a series of studies. These studies found that task-irrelevant stimuli that appear in close temporal association with targets in a main task can experience perceptual learning.[1, 7, 13, 15–21]

The classic demonstration of TIPL paired the primary task of detecting and reporting target letters in a ten-letter rapid serial visual presentation (RSVP) stream at fixation with weak random-dot motion in an annulus around the letters.[13] During the exposure phase, observers reported two lighter letters among darker letters at the end of the RSVP stream. Meanwhile, stimuli with very low coherent motion (5%) that were paired in time with the target letters all moved in the same motion direction, sometimes called the exposed direction (see figure 9.2). Task-irrelevant perceptual learning was measured by comparing pretests and posttests in which observers identified which of eight motion directions they perceived for either 5% or 10% coherence motion stimuli. While the direction reported for 5% coherence motion stimuli was at chance both before and after exposure training, for 10% coherence stimuli the exposed direction and to a lesser degree the two adjacent directions were identified more frequently after training. Temporal pairing of a task-relevant target and a subliminal task-irrelevant stimulus has been proposed as the critical ingredient of task-irrelevant perceptual learning.

Researchers argued that the task-irrelevant learning in this study was driven by an internally generated reward signal in the brain that was triggered by target detection (see the discussion in section 9.4 on reward and perceptual learning). A simplified study using only four motion directions during the learning phase found that only the direction temporally paired with (overlapped and slightly preceding) target letters showed learning.[20]

In another study, task-irrelevant learning failed when target letter detection was suppressed by an attention blink (a phenomenon in which detection of a second target very shortly following a first target is reduced).[18] Task-irrelevant learning was subsequently shown to occur only when the task-irrelevant stimulus was hard to see.[1] Groups in which the motion stimuli were subliminal or nearly subliminal showed evidence of task-irrelevant perceptual learning, while groups receiving suprathreshold motion stimuli did not. Furthermore, when two task-irrelevant local random-dot-motion distributions were integrated to form a global motion direction, the local motion components were learned first,[21] a result seen in several other papers.[22, 23] In most experiments, task-irrelevant dots were high contrast and the weak motion signals resulted from low motion coherence; an alternative demonstration used high-coherence motion of low-contrast dots and found a learned bias toward reporting the task-irrelevant motion directions that were paired with letter targets, even when there was no motion stimulus at all.[18]

Task-irrelevant perceptual learning has been found in several domains other than dot motion, including the orientation of dim Gabor patches temporally paired with RSVP letter targets[17] as well as subliminal global contours created from local orientations paired with high-contrast shape targets in the foreground.[26] Another example trained arbitrary

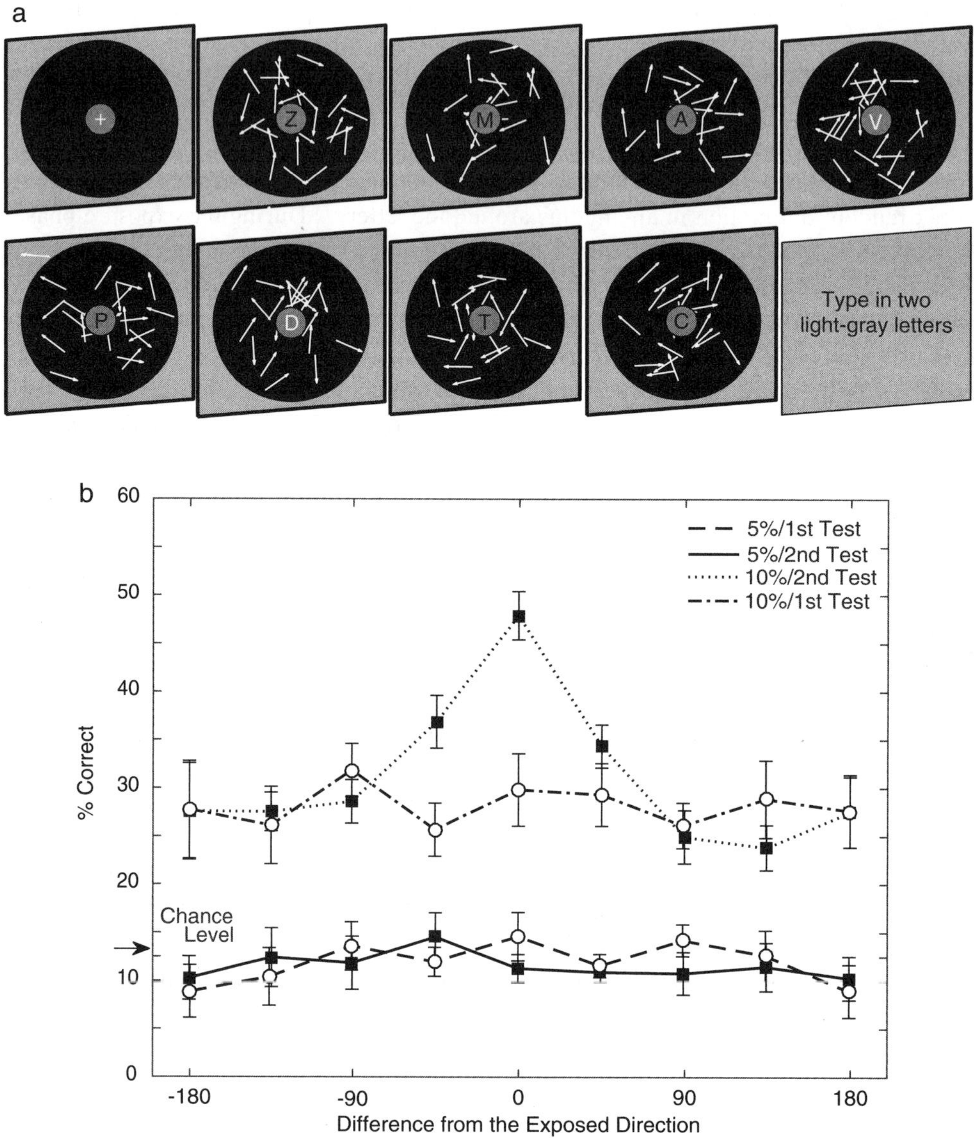

Figure 9.2
Task-irrelevant learning of motion direction in random dot displays. (a) Illustration of the training task during the exposure stage. (b) Task-irrelevant learning of exposed direction. After Tsushima and Watanabe,[1] figure 1, with permission.

complex line shapes in the context of object naming or visual search, similarly reporting the development of some task-irrelevant perceptual expertise.[27]

9.2.3 Summary

Task relevance is considered a very powerful principle for determining what is learned in complex visual environments. The principle of task relevance dictates that the task-relevant features will be learned, while learned suppression of competing suprathreshold task-irrelevant stimuli may also occur in some cases. Nevertheless, perceptual learning of task-irrelevant stimuli has also been demonstrated in some circumstances, namely with subliminal or nearly subliminal stimuli. This led Watanabe, Seitz, and their colleagues to propose a key role for temporally paired internal reward responses in visual perceptual learning. The role of reward in perceptual learning and its theories is considered again in section 9.4.

Considerable work could still be done in further testing these phenomena. While existing experimental tests of specificity and transfer have largely focused on changing the primary feature of judgment (e.g., horizontal versus vertical Vernier offsets, up-down motion versus left-right motion), evaluations of whether and to what extent irrelevant features of the task-relevant stimuli are integrated into learning remain relatively underexplored.

9.3 Attention and Perceptual Learning

Task relevance—itself driven by the task judgment—strongly influences what is learned. This is the case almost by definition. Several researchers, however, have proposed that attention is actually the primary gating mechanism—some even claiming that attention is a necessary precondition to learning.[26]

Although many researchers have argued for a tight coupling between attention and learning, the explanations for this coupling have varied.[2, 28–31] Furthermore, the fact that learning can occur for subliminal (and so unattended) task-irrelevant stimuli suggests that the connection is not absolute.[13, 19] In fact, as we will see, the relationship between perceptual learning and attention can actually be a two-way street. Just as attention may influence learning, perceptual learning can change the reliance on or deployment of attention during task performance.[27]

The following quotations illustrate the different explanations that have been given for why attention is important to learning:

> Learning is therefore attention driven, where attention is the mechanism for choosing the relevant neuronal population, by increasing its functional weight[29] (p. 460).

> Perceptual learning involves direct interactions between areas involved in face [and object] recognition and those involved in spatial attention, feature binding, and memory recall.[30] (p. 596).

Perceptual learning shows strong interaction with attention, indicating that it is under top-down control. Attention is necessary for consolidation[31].

We hypothesize that location learning improves spatial attention, which is stimulus non-specific, to a peripheral location[32] (p. 1924).

In these quotations, attention is thought to operate by selecting the relevant neural populations, by coordinating feature and memory, by improving consolidation, or, in an obverse explanation, by guiding attention to the right location for learning. Yet, in the same papers from which these quotations have been drawn, attention was rarely explicitly manipulated.

The conflation of perceptual learning with attention makes sense on an intuitive level. Both tend to improve performance, and both can have similar physiological manifestations. Having said this, the connection between attention and learning is almost surely more nuanced than straightforward. First, attention is not unitary: spatial attention, feature attention, or object attention could, in principle, each operate differently in learning. Second, interactions between learning and attention need not imply that attention gates learning. The interaction may also work the other way around: a difficult task that initially requires attention to perform may become increasingly automated through perceptual learning, such that the progression of learning may obviate the need for attention rather than vice versa (see subsection 9.3.3).

In 9.3.1–9.3.5, we consider the literature that explores the connection between attention and perceptual learning, with an eye toward distinguishing cases in which attention was explicitly manipulated and those in which its role was simply inferred.

9.3.1 The Attention-Control System

The brain circuits of attention have been the topic of significant research since the 1990s. Most of these studies have involved brain imaging in humans[33–35] and have been summarized in several integrative reviews.[36–38] Two partially intertwined systems are identified with the control of attention: a *dorsal* frontoparietal system associated with top-down guided voluntary attention to features or to space; and a *ventral* frontoparietal system, believed to be engaged when events in the outside world are detected and trigger a shift of attention.[34] These two systems roughly correspond with the distinction between endogenous (voluntary or goal-directed) and exogenous (involuntary or alerted) attention. Several connected brain areas participate in these attention networks: the visual cortex, the frontal eye fields (FEFs), and the intraparietal sulcus (IPS) in the dorsal system; and the visual cortex, the temporoparietal junction (TPJ), and the ventral frontal cortex (VFC) in the ventral system. The possible hemispheric lateralization of these networks and the degree of cooperation between them, as well as details of the networks themselves, remain open topics of study.

Experimental studies of these systems have relied on different kinds of experiments. Investigations of the dorsal network have primarily focused on precued visuospatial attention, although the network has also been associated with feature attention[39] (see subsection 9.3.2). Measures of effective functional connectivity (e.g., by Granger causal analysis of fMRI) suggest top-down influences from the dorsal system into the visual cortex, as well as bottom-up connections, while transcranial magnetic stimulation of FEF/IPS has been shown to affect responses in the visual cortex,[40–43] in a complex network.[38]

The function of the ventral system, and especially the TPJ, is somewhat more controversial, though it is believed that it is suppressed while top-down attention is engaged, reactivates when a salient but unexpected stimulus is processed through bottom-up systems, and is involved in switching attention to the new location.[44] (Other functions, such as social cognition and theory of mind, have also been associated with TPJ.)[38] From these hypotheses, it seems that the dorsal system largely manages voluntary deployment of attention, while the ventral system handles alerting and switching. This division of labor implies some coordination or handoff of information from the alerting system to the voluntary attention system, and some ability of the voluntary system to partially suppress or modulate the alerting system.[34, 44, 45]

The dorsal and ventral attention networks of the brain are the control systems of attention, but they also modulate how stimuli are processed and analyzed bottom-up through their connections to visual cortical areas. Because of this, research has also focused on the impact of attention on the neural responses in the visual cortex. Research into this top-down effect of attention (using single-cell recording methods in monkeys and sometimes by fMRI in humans) has reported that attention and perceptual learning seem to induce similar patterns of change in the visual cortex, though to better understand one, we may also need to understand the other, as well as their potential interrelation.

As with the subsequent studies investigating the effects of perceptual learning on visual cortical neurons, a relatively large body of single-cell recording and fMRI literature has correlated spatial attention in particular (but also feature attention in a few cases) with small to moderate changes in the responses in the early visual cortex, such as V4, or possibly V1 (although the latter may reflect feedback from higher visual areas).[9, 35, 46–66] As we will see, the apparent parallels between the two phenomena cover a range of possibilities. Attention may influence immediate behavior, and deployment of attention may also offer alternative explanations for some physiological changes observed in perceptual learning studies, especially those measured during the active performance of a task and those that occur early in training (see chapter 5).

9.3.2 Types of Attention and Basic Attention Paradigms

The influence of attention on perceptual learning has generally been treated as a unitary phenomenon. Both theoretically and experimentally, however, the literature distinguishes three widely studied forms of attention: spatial, feature, and object (not to mention attention associated with vigilance). Each form operates somewhat differently and has been tested in different characteristic behavioral paradigms, although they all share certain attributes. In all three, attention often (though not always) improves the accuracy of detection or discrimination, or the response time, and is especially important in the presence of external noise or in cluttered displays.[67, 68] When measured with the perceptual template model (PTM) and external noise manipulations, attention has been shown to exclude external noise/distractors and/or enhance stimulus representations, with external-noise exclusion often dominating the results[69] (see chapter 4). Nevertheless, each form of attention also seems to have its own properties and is thought to operate differently in different situations.

Spatial attention enhances processing in a region of space. It may coincide with the point at which the eye is fixated, or it can be shifted away from that point by an external cue or internal goals.[70, 71] Whether or not spatial attention is deployed to a stimulus that draws attention to itself (exogenous attention) or based on top-down selection that may be oriented by a more symbolic external cue (endogenous attention), it favors processing in the attended spatial regions, while withdrawing processing resources from elsewhere. Spatial attention is generally manipulated in the laboratory by cuing the observer to attend to one or several locations as they perform a visual task, usually a task involving a single stimulus and response.

Feature attention selects inputs based on a feature value, such as attending to a color or an orientation.[72, 73] This seems a natural facet of perceptual behavior, one that might be engaged when looking for something particular, such as a friend in a crowd who is wearing a red jacket. It is generally believed that attending to a feature in one location also promotes attention to that same feature across the visual field.[74–78] In the laboratory, feature attention is generally manipulated by providing a cue, well in advance of the trial, instructing the observer to focus on an attended feature value; it also typically uses tasks requiring a single response.

Object attention selects an object and has been claimed to simultaneously process and bind together several features of the object without loss.[79] In the original studies, object attention was indexed by the ability to report several features from an object just as well as it is possible to report one feature. Object attention thus often uses laboratory paradigms contrasting multiple judgments within a single object to the same judgments across objects (e.g., report the color of one object and the orientation of another object).[80] Dividing attention over objects is especially challenging when different features are being judged, such as the color of one and the orientation of another.[80, 81]

Each of the three commonly identified forms of attention could, in principle, operate somewhat differently during perceptual learning. Demonstrating that attention affects perceptual learning requires an experiment that *manipulates* attention in an otherwise equivalent task. This manipulation could compare learning with and without attention, or it could manipulate the degree of attention in a graded way. If minimal attention is sufficient to permit perceptual learning, does more attention promote more or faster learning? If attention filters out a location, feature, or object, is learning prevented, or does it sometimes occur anyway? There are some recent studies investigating these questions, yet many kinds of attention manipulations and their influence on perceptual learning remain to be explored.

Most assertions about attention and learning, as illustrated by the quotations listed previously, claim that attention is a precondition for or increases the amount of perceptual learning. A separate question, however, focused on the opposite direction of influence, is whether the state of learning has any effect on the need for attention deployed during the task and, if so, by what amount.[27] In what follows, we address both questions.

9.3.3 Effects of Attention on Task-Relevant Perceptual Learning

As we have seen, one primary observation taken to support the central role of attention in perceptual learning is that learning is largely restricted to the feature that drives the task response (see subsection 9.2.1).[2, 3, 5, 6–8] A typical statement of this claim is that "perceptual learning cannot occur without persistent and intensive attention to the feature to be learned."[19] That attention to the task-relevant feature is causal is often an assumption, however, not an inference. Only a few experiments have directly manipulated attention to compare task-relevant perceptual learning in attended and unattended conditions.

One important experiment that did explicitly manipulate attention looked at perceptual learning in locations assigned either to a focal attention, a divided attention, or an unattended condition, intermixed over training trials (see figure 9.3).[82] In the focal attention

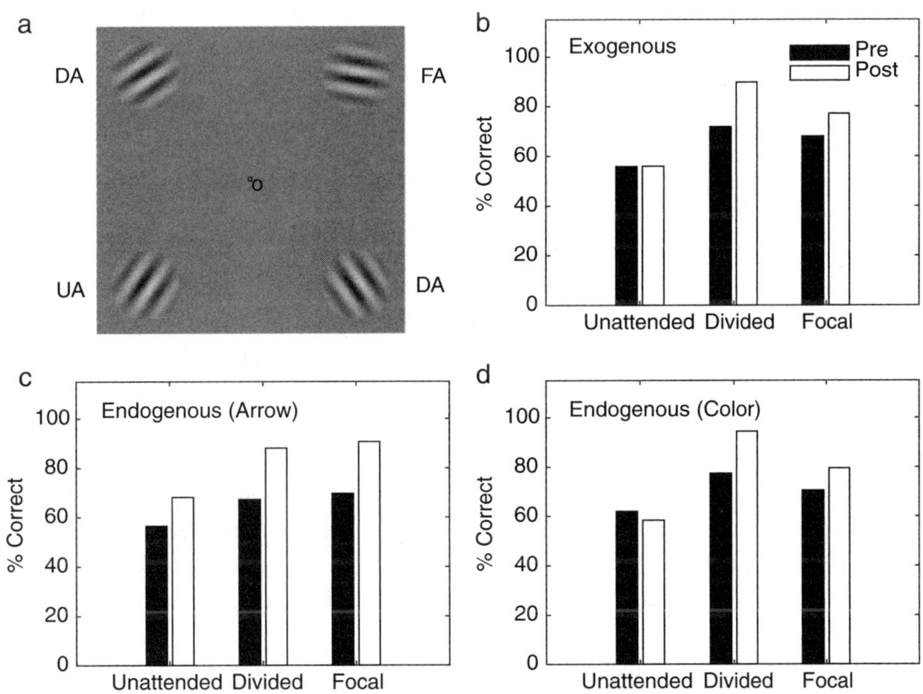

Figure 9.3
Spatial attention affects perceptual learning of orientation discrimination. (a) Focal attention (FA), divided attention (DA), and unattended (U) locations defined by cuing. Percentage correct before and after training for the exogenous (b), endogenous arrow (c), and endogenous color (d) cue groups. Redrawn from selected data in Mukai et al.,[82] figure 5.

condition, one location was cued; in the divided attention condition, two were cued together; and a fourth location never received attention cuing. (These were manipulated by precues that were either exogenous or endogenous in different groups of observers.) Later, after stimulus offset, one of the four locations was postcued, and the observer made a coarse orientation-discrimination judgment.

Even before learning, attention affected performance accuracy, and depended on the precued condition: unattended and divided attention locations exhibited poorer accuracy than in the focal attention location. These are standard attention effects. Following practice, locations receiving some attention showed improvements in orientation judgments, with learning in the focal attention location occurring slightly, but nonsignificantly, faster, while the unattended location showed no learning, even in the presence of feedback (though it is unclear whether this might simply reflect the low accuracy in the unattended location together with the delayed cue and feedback; see chapter 7).

The interpretation of this study has been questioned on the grounds that it used a within-subject design.[11, 83] It is our view, however, that the clear demonstration of attention and differential learning in different locations within the same display provided the first strong evidence for the idea that attention controls the rate and presence of learning.

Another finding that is often cited to support the idea that spatial attention gates learning trained several groups of observers in a texture task with different spatial distributions of the target.[26] In different groups, the target could appear either in one of two locations to the left and right of the fixation (two-location horizontal), diagonally from fixation (two-location diagonal), or anywhere in a central region (20-location center). Learning, as shown by reductions in threshold (stimulus onset asynchrony, or SOA), and the detection maps measured at all display locations after training are shown in figure 9.4. The detection maps of the different groups favored a horizontal patch across fixation following horizontal training, a diagonal patch across fixation following diagonal training, and the central region following central patch training. The authors concluded that attention spreads out from fixation to encompass the target regions and that "attention is both necessary and sufficient for learning" (p. 1360) and that "attention suffices to improve performance even at positions where the target never appears" (p. 1357).[26] Still, however intuitive, the attribution of attention as causal for learning is an assertion here. Other alternative interpretations are possible: for example, that perceptual training guides attention to focus on certain locations (the opposite direction of causation) or that training created more accurate target-response associations.

In one counterexample to the strong coupling hypothesis, attention had only very subtle effects in an experiment training eye dominance.[84] Eye dominance, defined as the likelihood that an observer will report the content displayed to the strong eye over the weak eye when the eyes see different images, has been measured experimentally by how much extra contrast must be placed in the weak eye to equate report probability. In this study, dominance of the strong eye over the weak eye for a trained orientation was rebalanced through a "push-pull"

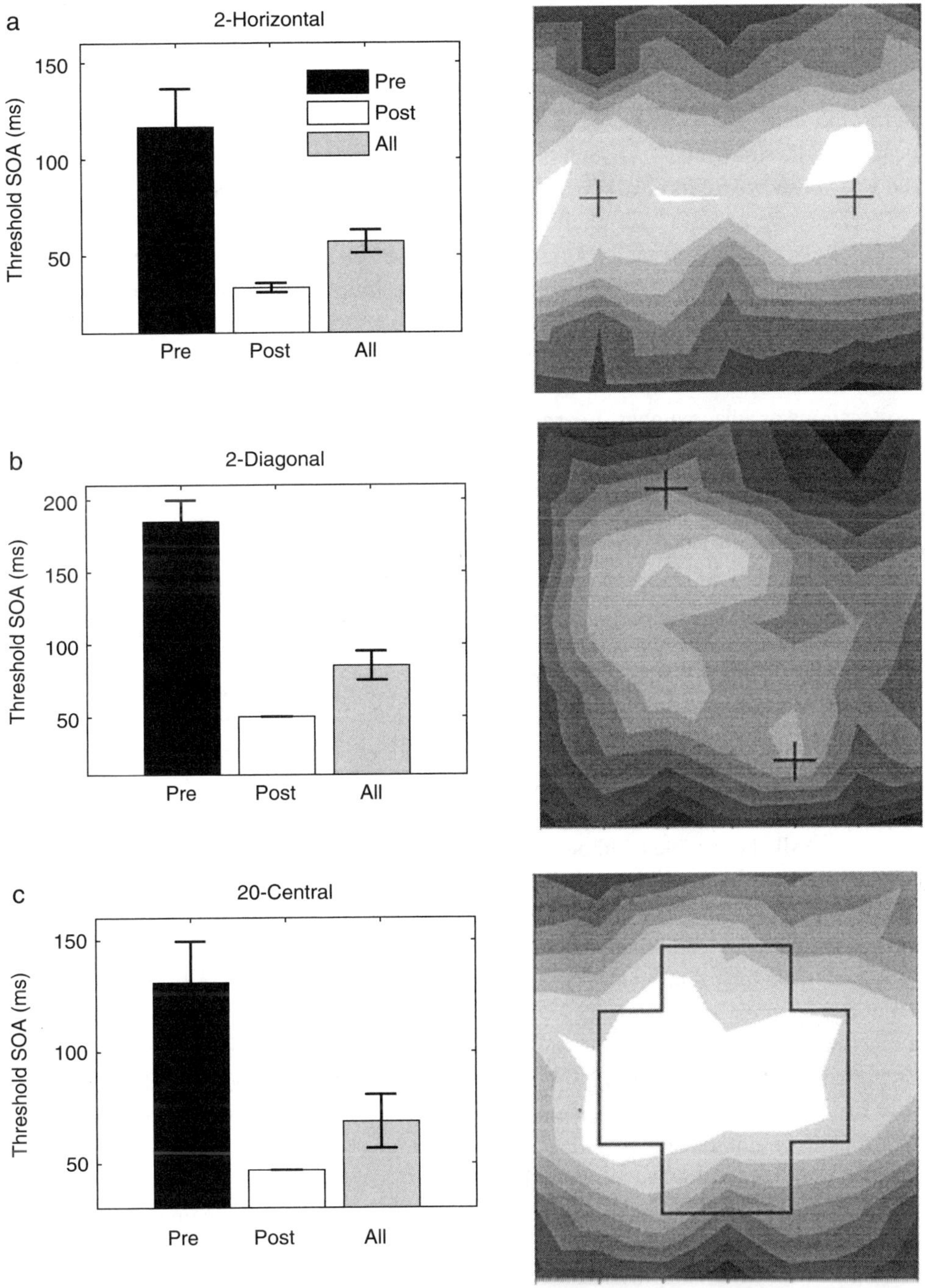

Figure 9.4
Texture-discrimination training with different target distributions differentially affects detection thresholds measured after training. Threshold reductions during learning (left) and posttest detection distributions (right) are shown for (a) two-location horizontal, (b) two-location diagonal, and (c) 20-central-position training, indicated by the + signs or outline; higher accuracies are shown as lighter values. After Ahissar and Hochstein,[26] figure 6, with permission, and redrawn from data in figure 4.

procedure in which high-contrast monocular cues (outline squares) favored an attended location in the weak eye while the observer ignored a highly visible grating in an unattended location in the strong eye. Although there were some subtle effects in auxiliary tests, attention was *not* shown to improve perceptual learning of rebalanced eye dominance.

Several recent studies have explicitly manipulated spatial attention in between-group designs, with a focus on transfer. One of these focused on differential location transfer following learning in a high-precision orientation-discrimination task with neutral cues (dot at fixation) in one group and with valid exogenous cues (dots right above the relevant display) in another group.[83] Initial training displays had objects to the left and right of fixation, one of which was postcued for response, while a second phase used either the original two locations or adjacent ones. Learning occurred both with and without attention (exogenous cue and neutral cue) at approximately the same rate in the first phase—yet another counterexample to the effects of attention on the learning rate of the primary task.

In another study, training conditions were chosen to yield no learning in an unattended group, whereas valid exogenous precuing enabled both learning and transfer.[11] This pattern can be explained within the framework of an IRT model: informative precues permit location-specific stimuli in uncued locations to be gated out early, thereby permitting learning involving the location-invariant representations that are then the basis for transfer, whereas with neutral cues, the stimuli from all locations would be conflated in the location-invariant representations, so learning could only involve location-specific representations. Yet another study frequently cited as evidence that attention gates learning may not be truly diagnostic.[85] In this study, attention-gated learning was claimed to explain specificity of learning for horizontal and vertical Vernier judgments, even as specificity of these judgments can easily be modeled without assuming a role for attention (see chapter 6).

Several reasonable conclusions can be drawn from this collection of studies. First, spatial attention supports or even enables perceptual learning, though spatially divided attention may be sufficient, and focal attention may not be required. In other cases, evidence of a necessary role for attention is less compelling and may suggest the opposite causal relation: that perceptual learning trained the distribution of attention, not the other way around. In cases such as implicit eye dominance, when attention was only indirectly related to the perceptual task, it seemed to have little effect on learning. It should be noted that the majority of the experiments seeking evidence for a connection between attention and learning, at least at this point, have involved spatial attention, paralleling the dominance of spatial attention in the physiological and fMRI literature. The potential roles of feature or object attention in perceptual learning may or may not be equivalent, and remain open topics for future investigation.

9.3.4 Effects of Attention on Task-Irrelevant Learning

Attention has also been theorized to play a gating function that determines when task-irrelevant learning occurs.[1, 86, 87] In this case, attention is seen as central in *preventing* learning. Learning from a task-irrelevant motion stimulus paired in time with RSVP letter targets[13] seems to occur only so long as the motion signal is subliminal, or nearly so, but not when it is more obvious.[1] The proposed explanation is that suprathreshold stimuli compete with the central RSVP task, thus triggering the attention system to actively suppress the competing task-irrelevant signals and thus eliminate task-irrelevant learning.[87, 88] From this observation, some have suggested that learning signals and attention signals jointly gate task-irrelevant learning. This idea was recently codified in a conceptual model whose primary distinction is between task-relevant and task-irrelevant stimuli, such that attention sometimes suppresses task-irrelevant stimuli, while reward enhances selected or subliminal task-relevant and task-irrelevant stimuli.[89] In support of this view, higher levels of fMRI BOLD activity have been reported in the lateral prefrontal cortex (LPFC), which is associated with attention control, for easily perceivable higher-coherence task-irrelevant motion stimuli.[87]

9.3.5 Perceptual Learning Alters the Need for Attention

Although the primary focus has been on attention as a gateway to learning, there is clear empirical evidence for the reverse: that perceptual learning can alter the *need* for attention while performing a task. The role of learning in automating previously attention-demanding processes goes back to, among other studies, the seminal work of Shiffrin and Schneider,[90] who showed that finding a target letter among letter arrays was transformed over thousands of trials of consistent practice from a slow and attention-demanding process limited by the number of targets and the size of the display to an automatic one that depended on neither. This direction of influence—in which learning changes the need for attention rather than attention gating learning—is one that has a longer and more substantial body of literature.

In one example, the limits of object attention were reduced through learning. As originally defined, object attention selects an object, allowing its multiple features to be more easily reported than if the same two features appeared on different objects—this difference being called the dual-object report deficit.[79] Dual-object report deficits were systematically reduced through training for a task in which two Gabor objects appeared in diagonal quadrants from fixation, as shown in figure 9.5.[27] Each object had two features: orientation (top tilted left or right) and phase (center light or dark). At the beginning of learning, performance was worse for reporting the orientation of one object and the phase of another (two objects, two responses, 2O2R) relative to reporting the same two features of one object (1O2R) or either feature by itself (1O1R)—the classic dual-object deficit. Subsequent practice improved performance in all conditions but especially in the critical dual-object (2O2R) condition, such that it approached the performance of the single-object or even single-response conditions after some 12 sessions of practice. These improvements were partially location specific: the dual-object deficit reappeared after a switch to the alternate diagonal locations (marked by the vertical dashed line in figure 9.5). In another study that used a temporal-sequence task, the negative consequences of an attention blink were also

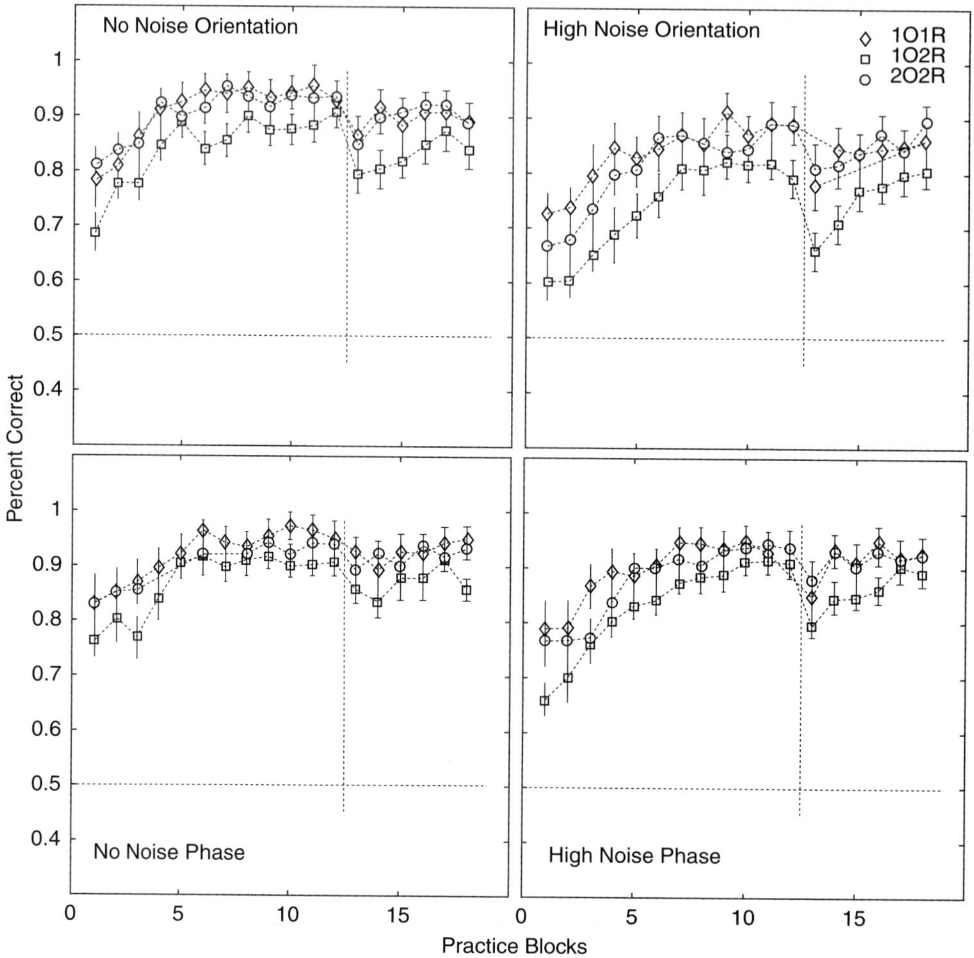

Figure 9.5
Perceptual learning reduces limitations of dual-object attention deficits for orientation judgments (top) and phase judgments (bottom) tested without (left) and with (right) external noise. Observers either report a Gabor orientation (top rotated right or left) and phase (center dark or light) of a single object (1O2R), the orientation of one and the phase of another (2O2R), or just one feature of one object (1O1R). The trained locations of the two objects switched from one diagonal to the other at the vertical dashed line. Insets show the changes in the dual-object deficit (2O2R-1O1R). Redrawn from Dosher, Han, and Lu,[27] figures 1 and 2.

reduced with practice. The attention blink is the reduced ability to detect a second target appearing very shortly after the first target in rapid serial visual presentation.[91–93] Even moderate training largely restored reporting of the second target.[94] In both cases, practice reduced or eliminated attention bottlenecks in performance.

Practice can reduce the need for attention even in basic perceptual tasks such as the discrimination of brightness.[95] In these experiments, prior to training, brightness discrimination thresholds were very much higher when the target location was unknown and attention was distributed compared with brightness discrimination in a single focal location. After training, the costs of distributed attention were essentially eliminated, while focal attention performance was largely unchanged (see figure 9.6).

It is also possible to train attention set. The texture-discrimination study (see figure 9.4), originally interpreted as evidence for attention gating of perceptual learning, may just as plausibly reflect learning as a distributive mechanism for attention.[26] Similarly, training observers to identify targets by color rather than shape has been claimed to improve the suppression of shape by attention.[96] Long-term training on visual search for an attended color (feature) sensitized that color in an independent third-order motion-direction task.[97, 98] Another facet of the interplay between attention and learning is that, after a task is learned—regardless of whether learning requires attention—the new expertise may be successfully expressed even in the absence of attention.[99]

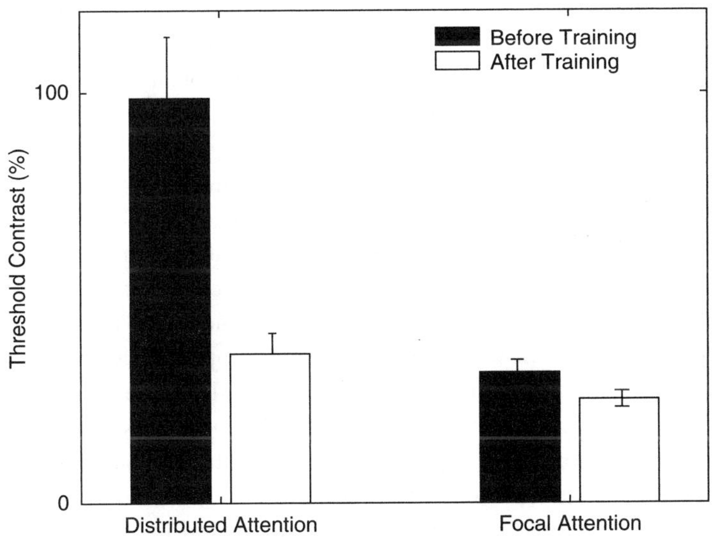

Figure 9.6
Training brightness discrimination in four-location displays with known or unknown target location eliminated the costs of distributing attention over space, while leaving performance with focal attention largely unchanged. Redrawn from data in Ito, Westheimer, and Gilbert,[95] figure 5B.

In terms of physiological evidence, responses normally associated with attention often change over the course of training. In fact, some researchers have even suggested that "the [different] stimulus-evoked responses seen during learning [could] reflect changes in attention modulation alone, rather than changes in bottom-up processing of the stimulus"[100] (p. 3899). Related fMRI findings were likewise reported for orientation discrimination under attended and unattended conditions,[57] while attention signatures and evoked responses also reportedly change over the course of learning in electrophysiological (EEG) studies of visual search.[101] Furthermore, the consequences of learning, such as changed amplitudes of early visual cortical responses, have sometimes been observed in anesthetized animals and thus by definition depend neither on attention nor on consciousness.

9.3.6 Summary

The proposal that attention plays an essential role in enabling perceptual learning has been a central explanatory hypothesis. The literature, however, suggests that the strongest claims were in fact based on manipulations of task relevance rather than attention as such. An alternative conclusion is that practice or training improves decision or that task relevance is mediated both by attention to improve perceptibility *and* by practice with the decision that fine-tunes the decision boundaries.

The existence of task-irrelevant perceptual learning, which occurs more reliably for subliminal stimuli than others, led to theories that attention-based suppression of task-irrelevant stimuli was triggered when those stimuli were supraliminal. Although it should be possible to manipulate attention in task-irrelevant learning, this has not been the focus of these studies. One possible prediction is that task-irrelevant learning should be more likely to occur for supraliminal stimuli under diverted attention conditions, which might eliminate attention-based suppression. Similar arguments could be made for attention processes assumed to track the distribution of target locations, which almost always also create differences in the number of targets practiced in each location.

Perceptual learning may indeed be influenced by attention in many cases. In the most commonly cited studies, however, its role has been inferred rather than measured. Only a few studies have explicitly manipulated attention and measured perceptual learning in otherwise equivalent task situations. One such study compared learning in focal, distributed, and no attention conditions and found that even divided attention was sufficient for learning, while the data on learning in the unattended location was ambiguous.[82] Other studies compared exogenously cued attention to a neutral attention condition, with a primary focus on the consequences for transfer.[11, 83] So far, then, only a few studies have carried out explicit attention manipulations. It is possible to imagine future studies that would manipulate the degrees of feature attention (none, singular, multiple) or object attention and measure both learning rates and transfer.

In contrast, demonstrations that practice can have profound effects on attention—or on the need for attention—have a longer and more robust history. Perceptual learning can train particular spatial distributions of attention[26] or feature selection.[96] Perhaps more commonly, conditions in which task performance is initially challenging and therefore require deployment of attention can be performed well without attention, given enough practice. In the language of Shiffrin and Schneider,[90] these tasks have become automatic and no longer require attention.

The interactions of attention with perceptual learning, as we have emphasized, move along a two-way street. Intuition suggests that both directions of influence may occur: attention may be required for performing the task early and for early learning, after which the need for attention may disappear as the trained task becomes automatic. There are many ways that future researchers might explore this relationship more fully and precisely.

9.4 Reward and other Interventions in Perceptual Learning

Another potentially powerful form of top-down control that could modulate perceptual learning is reward. Here as well, there have been relatively few empirical investigations of perceptual learning that explicitly manipulate reward. Of course, reward has a central role in the long conceptual history of learning going back to the early investigations of reinforcement conditioning, resulting in the impression that the importance of reward is almost a truism.[102] Rewards with positive valence typically lead to an increase in the rewarded behavior, while those with negative valence lead to a decrease in the target behavior (or avoidance). As we know from years of research in this broader context, rewards come in several types. Primary rewards are direct and physical: examples include water or food for deprived animals or shock or immersion of a limb in ice water. Secondary rewards are often symbolic. For humans, examples include money or pictures of money and coupons for free goods, among others. Some secondary rewards can be indirect, such as verbal praise from a supervisor or critiques from a colleague, or the implicit value of detecting a target in a rapid serial visual display.

Analogous to its role in reinforcement learning or operant conditioning, reward could theoretically play a major role in perceptual learning. However, while many perceptual learning studies in humans have used some form of feedback, very few studies to date have used either primary or secondary rewards that denote something tangible. Even in studies of perceptual learning in monkeys (which use primary rewards such as water or juice), the effects of *differential* rewards have been examined less frequently than one would expect.

The next subsection very briefly considers the brain circuitry for reward and possible ways reward might affect learning through connections to sensory and decision systems (see also chapter 5). The concepts from the physiology of reward expectation and reward-prediction error are relevant, as is their use in reinforcement learning algorithms. We then review the relatively small amount of literature on human visual perceptual learning that uses explicit reward and end with a brief mention of some related pharmaceutical interventions that may affect perceptual learning.

9.4.1 The Reward System

The reward circuits in the brain associated with goal-directed behavior interact with cognitive brain centers and motor-control regions. We know from the physiology that activity in the reward system responds not only to a delivered reward but also in anticipation or expectation of the reward. Reaction to delivered reward outcomes is thought to recruit the medial caudate, putamen, and the dorsal caudate with supplementary motor area, while anticipation of reward has been associated with activity in midbrain and basal forebrain regions such as the nucleus accumbens (NAcc). One conceptualization is that processing rewards involves the dopamine pathways and a convergence of several corticostriatal projections. These circuits and their interactions have been extensively investigated in animals[103] and continue to be studied in monkeys and humans.[104]

Sequential activation in these cortico–basal ganglia circuits occurs during (general) learning.[105] A single episode in which a stimulus is processed, a choice is made, and a behavior is executed likely involves an anticipatory phase and a subsequent reward-processing phase. Indeed, a number of models focus on *reward-prediction error*, in which learning occurs in response to the deviation between the reward outcome and reward expectation.[106–110]

Reward-prediction error is a typical ingredient in reinforcement-learning algorithms.[111] In these algorithms, the predicted reward is updated on a trial-by-trial basis, controlled by the reward-prediction error and a learning-rate parameter. If the reward contingencies are stable, the predicted reward ultimately converges to the expected value of the actual reward, which provides a nice way to stabilize learning. Signals related to reward-prediction errors have been found in several brain areas. The ventral tegmental area (VTA) in monkeys, for example, has been implicated in reinforcement learning by broadcasting prediction error signals throughout the reward system, a theoretical conclusion also supported by electrical microstimulation of the VTA in rodents.[112–114] Reward-prediction error signals have also been found in other areas, depending on the task.[114–117]

There is another way in which reward might affect visual perceptual learning—by directly affecting the responses in the *primary visual cortex*. Reward signals are thought to modulate contrast sensitivity in the primary visual cortex through connections from the basal forebrain to the cortex either directly or indirectly via the lateral geniculate nucleus (LGN), or they may connect to reward-processing centers in the basal forebrain, as suggested in some studies (see figure 9.7).[118] Whether the influences are excitatory or disinhibitory, the reward system may directly affect responses in several visual areas, such as V1 and V4, through these circuits.[119–122] If reward were to influence visual learning by altering the activity in the early sensory representations in this way, such impacts of reward would operate similarly to attention, though the effects may be larger. Analogous effects attributed to reward or to reward-prediction error on early visual responses or in higher visual association areas of the cortex have been seen in fMRI in humans, where in some studies the reward on one trial was shown to alter anticipatory baselines in the next trial.[58, 123, 124]

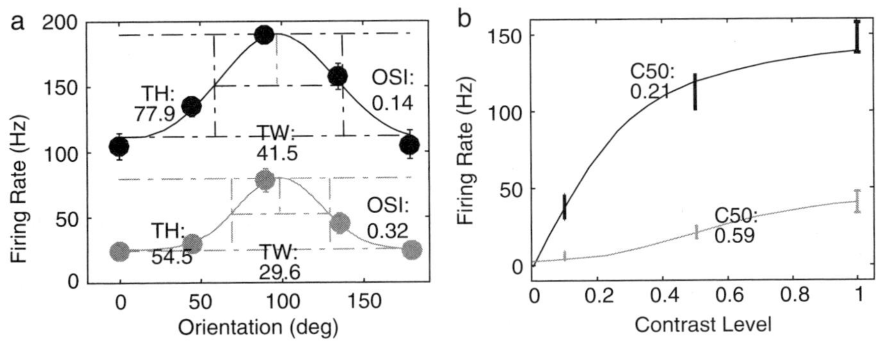

Figure 9.7
Deep brain stimulation of the basal forebrain affects the orientation sensitivity (a) and contrast sensitivity (b) of neurons in V1, illustrating a possible effect of activation in reward circuitry on visual cortex activity, as seen in functions with (top curves) and without (bottom curves) stimulation. After Bhattacharyya et al.,[118] parts of figure 2 (open access).

Reward-induced changes in the responses to visual stimulation in the early visual cortex might have several different effects, and these have been interpreted variously in the literature. Such reward effects on the visual cortex could drive anticipatory changes in baseline firing, as well as modulating contrast sensitivity. One theoretical interpretation is that these modulations are actually mediated by changes in attention; the notion here is that reward affects attention and then attention alters the response of early visual cortices to incoming stimuli.[125] Other researchers have claimed instead that the reward system and the attention system simply have high functional overlap and similar intertwined effects on early visual areas.[127] Still others have observed that reward and attention may operate either in concert or independently to modulate the responses of the early visual cortex.[128] The relationship of attention and reward in learning and performance of visual tasks remains an open question.

In the *decision system*, reward seems to be integrated with sensory information to select a response. In addition to potential direct effects of reward on sensory responses or effects mediated through changes in attention, reward information has been reported to alter the behavior of decision neurons that integrate sensory information toward a decision and response. So-called decision neurons have been shown to integrate prior probabilities and reward information as well as sensory information into a decision that then becomes the basis of an action. These hypothetical effects of reward on sensory information and on decision could potentially be further estimated separately within a signal detection theoretical framework and, by extension, through evidence accumulation in random walk models of response, as explored in some example studies.[129, 130]

Evidence that reward influences the responses of decision neurons has been reported for several brain areas. Accumulation of sensory evidence toward a decision has been identified in neural responses in the LIP in monkeys, where the neural firing patterns were correlated with the choice and timing of behavioral responses.[131–134] These neurons responded to stimulus information but also to other factors.[135] For example, one study[136] found monkey LIP neurons that coded for both absolute and relative rewards as well as responding to sensory inputs. The activity of these neurons was driven by (expected) relative reward value, then later in the trial by the sensory evidence, and finally by the actual behavioral choice in that trial, as shown in figure 9.8.[137] Similar or related effects have been reported in the superior colliculus.[138, 139] On this basis, researchers have proposed that there may be a network of such reward-influenced decision neurons in different brain regions.

In summary, there is growing evidence for the effects of reward on the responses of early visual cortical areas. The modulations of these sensory responses may occur either through attention circuitry or independently and directly through interactions between the reward system and relevant early visual cortical areas. In either case, changes in early visual

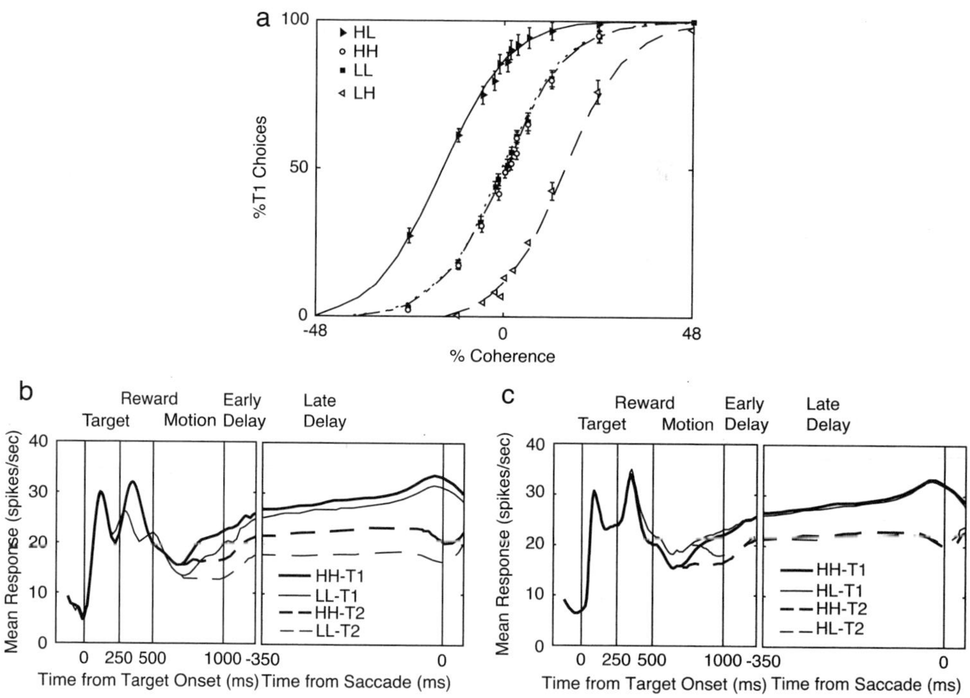

Figure 9.8
Neural activity in the macaque LIP depends on integrated decision variables, including reward condition in a motion-coherence task. A precue indicates the relative reward size (low or high) in left and right locations (LL, HH, LH, and HL). (a) Behavioral choice probabilities for monkey A as a function of motion coherence are shifted left when the reward favors response 1 (HL), shifted right when the reward favors response 2 (LH), and are intermediate for balanced rewards (LL or HH). Mean LIP firing rates during different trial phases depend on (b) absolute reward (HH versus LL) and (c) relative reward (HH versus HL) for monkey A. After Rorie et al.,[136] figures 2c, 3a, and 4a. Copyright 2010 Rorie et al.

cortical responses could very easily have consequences for learning in perceptual tasks, either through baseline elevation or through an enhanced contrast response to the stimulus. There is also evidence for an influence of relative reward or reward expectation on imputed accumulator or decision neurons in higher areas such as the LIP and others. Altering decision computations should also have consequences for perceptual learning, one rationale being that changes in either early sensory representation or decision have implications for the evidence available to even unsupervised learning rules such as Hebbian reweighting, beyond the direct informational value of reward. Potential ways to incorporate reward or reward-prediction error in learning algorithms are considered in section 9.7.

9.4.2 Reward in Human Perceptual Learning

In general learning theory, reward is known to alter the relative frequency of potential responses or behaviors. In perceptual learning, the question is different: can reward improve either the performance or the learning of a visual discrimination judgment?

This question will not be answered by experiments designed to show that an animal (or human) learns to produce a rewarded response more often or to move their eyes to a location where more reward is experienced. Instead, the relevant question is whether reward improves the ability to *discriminate* visual stimuli or changes the rate of learning to do so. Furthermore, we can also ask whether reward has any influence beyond the information it intrinsically provides (e.g., regarding the accuracy of a response).

While there have been few reward manipulations used in human perceptual learning studies, they cover a curious range. This has included physical rewards such as juice or water given to a deprived person, as well as symbolic secondary rewards, such as pictures of coins or monetary rewards to be delivered later. Some researchers have even proposed that detecting a target by itself triggers endogenous, or self-generated, reinforcement or reward signals.[19, 140] This latter observation led to one of the first proposals about reward in perceptual learning, in the context of task-irrelevant learning. These authors proposed that detecting a target in the main task generates an endogenous reward signal that in turn affects learning about other stimuli occurring at nearly the same time.[19] This idea was further supported by a study in which participants who were told that detection of a central target would lead to a larger end-of-session reward showed more robust task-irrelevant learning effects.[140] The corresponding endogenous reward signals, if we were able to measure them, could, in principle, be either anticipatory or responsive, and might operate through the same pathways as a physical reward outcome.

A more direct exploration of reward manipulated the probability of an explicit reward and observed effects on perceptual learning.[141] In this study, one of three noisy oriented stimuli, each assigned a different reward probability (80%, 50%, or 20%), was presented on each trial. The differential rewards altered the response probabilities. Learning was claimed only for the stimulus with high reward probability (80%), as this was the only stimulus for which the corresponding response increased (in a go or no-go paradigm), while responses to stimuli receiving a 50% or 20% reward declined slightly (figure 9.9). An alternative interpretation, however, is that increasing the reward probability influenced response choice, operating effectively as a stimulus-contingent bias to respond.

In one of the first studies to demonstrate the effects of reward on discrimination, drops of water were paired with nearly subliminal sine-wave patches of one orientation, but not a second control orientation—all in the absence of an explicit task for the observer. In separate experiments, the orientation stimuli were made subliminal by low contrast or were made unconscious through continuous flash suppression (e.g., by presenting them briefly in one eye while the other eye was continuously stimulated with flashes of complex contour stimuli).[142] Discrimination of the reward-paired orientation, but not the control orientation, improved after reward exposure compared with a pre-exposure baseline.[14] The authors concluded that human adults learned through stimulus-reward pairing "in the absence of a task and without awareness of the stimulus presentation or reward contingencies"[14] (p. 700). This study went beyond other demonstrations of conditioning by subliminal stimuli to suggest changes in perceptual processing.[143, 144] It appears from the data that reward can induce perceptual learning and that this may occur even when the stimuli were essentially subliminal.[145]

While compelling, these studies left several outstanding questions. Does the relative magnitude or timing of reward affect the speed or generalizability of learning? Does reward

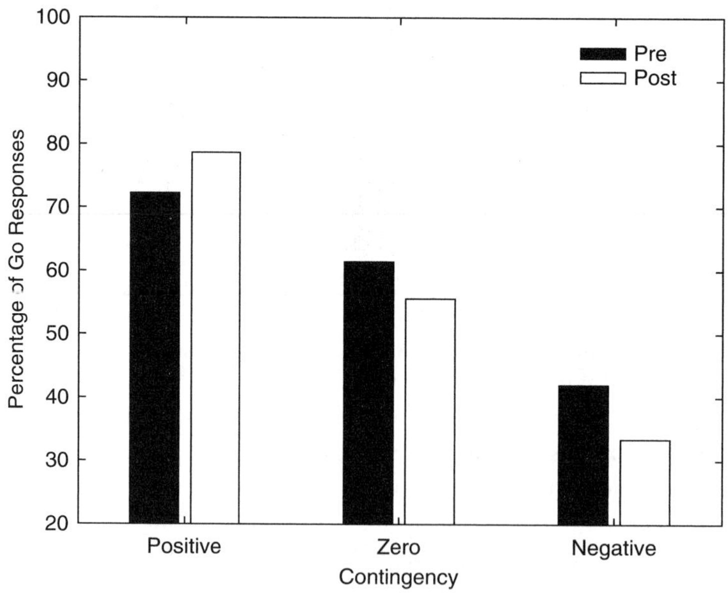

Figure 9.9
Training affects the inclination to respond to stimuli with different reward probabilities in a go or no-go paradigm. Redrawn from data in Kim, Seitz, and Watanabe,[141] figure 4a.

operate differently than feedback? Is the reward itself important, or is the information that the reward conveys what is really significant? Finding answers to these questions could improve our ability to exploit reward systems in optimizing perceptual learning.

One set of studies tried to answer some of these outstanding questions by manipulating the magnitude and type of reward; these studies also examined several measures of transfer.[146] The rate of perceptual learning was higher for high-magnitude trial-by-trial reward conditions, even when accuracy feedback was provided in all conditions, thus equating information about response accuracy on each trial. In the main experiment, contrast sensitivity (tested with Gabor patches) was practiced with five forms of reward: high, subliminal, and low trial-by-trial reward, block reward, and no reward. Pre- and posttraining assessments of the contrast-sensitivity function were also measured in the trained and untrained eyes (see figure 9.10). The rewards were combinations of images of (Chinese) currency, point counters shown trial by trial, and various measures of block performance (where relevant to the reward condition). The monetary value of compensation also depended on how points were translated to monetary compensation: high, subliminal, and block reward conditions received low base pay, such that total compensation depended heavily on performance-based reward points, while the low and no reward conditions used a low conversion rate and high base pay. Other experiments compared trial-by-trial reward to no reward in Gabor Vernier offset and global motion-direction tasks. In all these experiments, reward increased the rate of learning. In addition, the amount of transfer (to another eye, location, or stimulus) depended on the amount of learning in the trained task. That is, significant learning must occur in order to have anything to transfer.

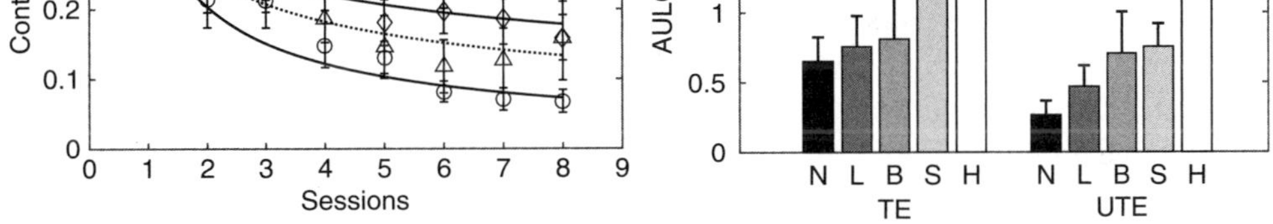

Figure 9.10
Five different reward schedules produce different rates of learning in sine wave grating detection (a), with corresponding different amounts of improvement in the contrast-sensitivity function in the trained and untrained eyes (TE and UTE) (b). In order of effectiveness, these are high trial-by-trial reward, subliminal trial-by-trial reward, block reward, low trial-by-trial reward, and no reward (H, S, B, L, N), added to trial-by-trial feedback about response accuracy. After Zhang et al.,[146] parts of figure 1.

To summarize, reward in perceptual learning has been studied far less than feedback (or even learning in the absence of any feedback). Some studies simply documented the effects of reward probability or reward magnitude on performance, implicating a role for reward contingency in learning. Another set of studies showed that learning could occur in response to reward signals alone, even when stimuli were subliminal or nearly so. A final set of studies demonstrated the role of reward on the rate of visual perceptual learning, even when information about response accuracy was provided by feedback.[147] At the same time, open questions remain, with ample room for further research on the role of reward in visual perceptual learning, and its possible interactions with attention and task relevance.

9.4.3 Pharmaceutical Interventions in Perceptual Learning

Pharmaceutical interventions are another potentially powerful means of altering perceptual learning, with a few existing reports of experiments in humans. This area of research is considered here because some (though not all) chemical agents may achieve their effects through mechanisms similar to attention or reward. However, an alternative, and quite different, possibility is that some pharmaceutical interventions might alter the consolidation of perceptual learning. To quote one review: "Learning… might be regulated through the release of neuromodulators, such as acetylcholine and dopamine, which gate learning and thus restrict sensory plasticity and protect sensory systems from undesirable plasticity"[148] (p. 149). A systematic research program in humans would involve finding agents with few side effects or unintended consequences that influence each of these hypothesized components. The hope would be to improve perceptual task performance and/or learning, either separately or together.

The influence of pharmacological agents or drugs on perceptual learning has been examined for a few possible modes of influence on learning. The neurotransmitter acetylcholine (ACh), for instance, modulates a number of cognitive functions, including attention and memory. One idea has been that acetylcholine affects neural plasticity by selectively enhancing the sensory responses to behaviorally relevant or attended stimuli.[149, 150] When acetylcholine is released under conditions of sustained attention, it is known to affect sensory responses.[151–154] In one study, cholinergic enhancement (by dosing with donepezil, a cholinesterase inhibitor) was shown to affect perceptual learning in a random dot motion-direction discrimination task.[150] Training improved motion-direction difference thresholds faster with the drug than with a placebo, with the learning effects reported to be more specific (see figure 9.11). As is typical for perceptual learning, these training effects were relatively long lasting and were still present 5 to 15 months after training.[155]

Another potential influence of pharmacological agents involves changes in memory consolidation, analogous to cholinergic effects on declarative learning.[156] At least one study suggested cholinergic modulation of consolidation in visual perceptual learning. Chewing nicotine gum compared to a placebo immediately after the end of the practice session enhanced the next-day expression of learning in a visual texture-discrimination task (such

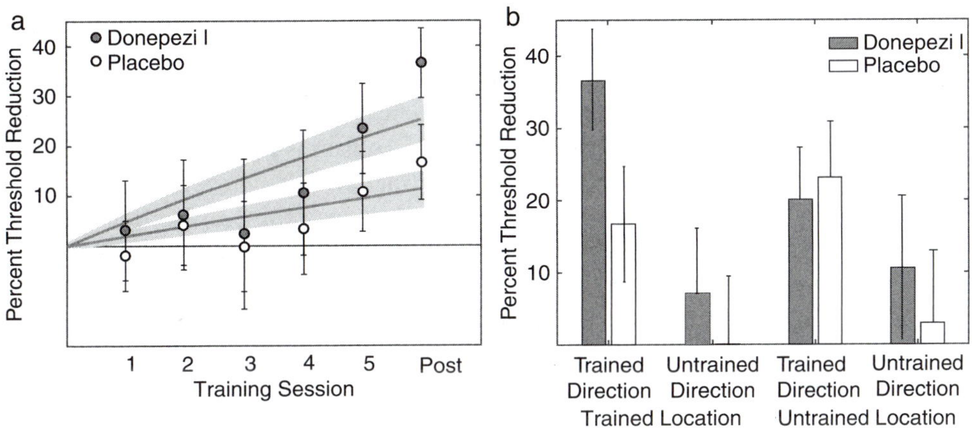

Figure 9.11
Cholinergic enhancement increases the magnitude and specificity of perceptual learning in humans. (a) Threshold reduction as a function of training with donepezil or a placebo, and (b) improved thresholds in trained and untrained motion directions in the trained and in an untrained location. After Rokem and Silver,[150] figures 3 and 4, with permission.

nicotine state changes were validated by EEG measures).[157] In this study, higher levels of ACh were thought to promote consolidation in perceptual learning, even though lower levels have typically been shown to promote consolidation in declarative memory. (The authors were aware of this difference, suggesting it likely indicated either different learning mechanisms or complicated dose-response effects.)

Among a number of possible mechanisms involved in pharmacological influence, dopamine is also likely to be central. This reflects the recognized centrality of the relationship between reward and the dopaminergic system, as well as widespread evidence for the role of dopamine in reinforcement learning.[158] It has also been reported that changes to dopamine in the prefrontal cortex can alter the tuning of V4 neurons in monkeys in ways that parallel those observed for attention.[159] Other observations support this general view. For example, some preliminary evidence exists that dopamine dosing influences perceptual learning in amblyopia. Dosing with levodopa-carbidopa combined with part-time occlusion of the good eye reportedly improved performance in the amblyopic eye and improved binocular fusion in children past the typical age of eye patching (although the interaction with practice under occlusion conditions was inferred rather than manipulated in these studies).[160, 161] In a possibly related finding, reduced learning was demonstrated in Parkinson's disease for sequence and pattern learning,[162] and levodopa was shown to improve motor learning in chronic stroke patients.[163]

It should be noted that the few reports of pharmacological interventions detailed here have studied their effects on humans. Alongside this, there exist many other examples of pharmacological effects on perceptual learning in audition and tactile learning in rodents or monkeys, where the research has emphasized the modulation of changes in the early auditory and tactile areas.[164, 165] These may all be related to the focus on neuromodulatory systems in certain theories of reinforcement learning[158, 166] and category learning.[167–170]

To summarize, the potential for using pharmaceutical agents to enhance or inhibit perceptual learning seems tantalizing, but more work in humans is still needed to understand the mechanisms of action in each of the potential drug interventions. The point of action could in principle be on sensory responses, the attention system, decision, sensitivity to reward, consolidation, or some combination of these. Such investigations present exciting possibilities for enhancing perceptual learning generally and in providing information that may be especially relevant in the context of rehabilitation (see chapter 12).

9.4.4 Summary

A full analysis of the influence of reward on perceptual learning—beyond whatever pure information value it carries—is just under way. Reward may shift the inclination to seek information in certain locations or in stimuli with certain features and may likewise lead to experience-induced shifts in the selection of different responses. While there is information to suggest that reward may enhance learning in discrimination tasks, the ways in which this may be mediated by shifts in attention still require study. Similarly, the relative value of endogenous signals or subliminal reward messages when compared to external rewards remains largely unexplored. It seems plausible that internal reward signals play an important role in maintaining behavior in situations in which rewards are intermittent, especially in light of related findings in rodents. There are many potential directions that future studies could take.

Reward could influence perceptual learning in any number of ways. It may directly affect early sensory processing, leading to increased amplitude or altered selection in sensory encoding of the stimulus. Alternatively, it may engage the attention system, which could also lead to these changes in sensory encoding. Another alternative is that reward changes the dynamics and outcome of a decision and response through reward-sensitive shifts in biases favoring rewarded stimuli or behaviors. Another possibility is that reward or reward magnitude might alter the rate of learning. Yet another possibility is that reward might affect consolidation. Although the empirical evidence exploring these different influences is relatively sparse, each mechanism has been related to known physiology of reward coding in the brain.

There are important theoretical distinctions to be drawn between the effects of reward outcome, reward expectation, and reward-prediction error, and how they affect learning. Perceptual performance and perceptual learning could be sensitive to any one of these theoretical constructs of reward. Indeed, one of the fascinating questions in some of the studies already carried out is the potentially powerful role of endogenous rewards, and whether this might contribute to apparently unsupervised learning in no-feedback conditions. Systematic study of the role of reward in perceptual learning could further disambiguate these mechanisms, leading not only to a better theoretical understanding of the role of reward in learning but also more effective training protocols and, potentially, pharmaceutical interventions that enhance learning.

9.5 Top-Down Influences, Reweighting, and Selection versus Creation

Many top-down influences, supported by a collection of brain functions, are required to organize performance—and learning—in the context of any goal-directed activity, including perceptual tasks. Carrying out such a task involves identifying the relevant stimuli and selecting appropriate decision computations. It may also engage attention to enhance the representation of certain stimuli or filter out the impact of others. Finally, it may be influenced by the nature of the reward.

In the case of attention and reward, specific brain systems for control have been identified by independent research, and there is growing evidence connecting these systems to the changes in the responses of early visual cortical areas. As we see it, potential effects of task, attention, and reward should be incorporated into the reweighting framework that explains perceptual learning. We briefly point to one overall approach for this here, while developing the framework in some detail in section 9.7 (appendix).

Our analysis here focuses on the basic Hebbian learning rule

$$\delta_i = \eta a_i (o - \bar{o}), \tag{9.1}$$

where a_i is the activity in representation unit i, η is the learning rate, $(o - \bar{o})$ is the output (compared to its long-term average), and δ_i is the change in the weight connecting the representation to decision. A potential role for reward may be incorporated into this basic equation as

$$\delta_i = \eta a_i (o - \bar{o})(r - \hat{r}), \tag{9.2}$$

where r is the reward on the trial and $\hat{r}$ is the reward expectation. This may be taken a step further by recognizing (as demonstrated empirically in a number of studies discussed earlier) that activation of the representations a_i and the learning rate η itself may be influenced by task, attention, and reward.

In short, all three top-down influences can be incorporated into the learning rules driving learning. It is relatively straightforward to incorporate these effects into the learning rules of the augmented Hebbian reweighting model (AHRM) and the integrated reweighting theory (IRT).[171–173] The changes in the learning rules of the AHRM/IRT, as well as several alternative frameworks that have been proposed to integrate reward, are the topic of section 9.7.[166, 174, 175]

Before delving further into the complex mechanisms involved in each of these top-down influences, there are two points to consider. First, there are multiple ways that task, attention, and reward might change the learning rule or the inputs attached to it, and many of these could be mapped equivalently onto changes in the rate of learning. Second, to further pursue and test specific predictions will require a generative model strong enough to make predictions about different experimental conditions. This is because the influence of top-down factors on the rate of learning may depend on the hierarchical architecture of the representation, the required decision, and the model of performance in that task.

Another question to consider is how these top-down factors on learning may complicate or interact with the functional difference between the selection of existing representations and the creation of new combinations of representations to define unique objects. We argued in chapter 2 that the first process was more consistent with learning in low-level visual tasks, whereas the second process was consistent with tasks at higher visual levels, including those related to the processing of objects. One hypothesis is that the task's structuring of the decision is more critical for the low-level tasks in which learning involves selecting the best representation inputs to decision, while attention may be more critical when learning requires the creation of representations that integrate the feature combinations defining the objects in the task. The second case may involve finding partial representations that are then elaborated over learning. This hypothesis provides a rationale for the future experimental investigation of the impact of attention and reward in tasks associated with different levels of the visual hierarchy.

9.6 Conclusions and Future Directions

We began this chapter by observing that learning rarely occurs in a vacuum. It almost always occurs as part of goal-directed tasks within specific contexts (although learning from mere exposure cannot be ruled out completely). Carrying out goal-directed behaviors requires the identification and selection of relevant sensory inputs, the construction of a behaviorally relevant decision, and the determination of where to engage plasticity during learning. Three major top-down factors related to selectivity were detailed: task structure, attention, and reward. Any or all of these factors could be in play in learning.

This review of the literature suggested that learning primarily occurs for the task-relevant features or stimuli in service of a particular task judgment, although there do seem to be exceptions. Near-threshold task-irrelevant stimuli co-occurring with targets (and therefore with rewards) may also sometimes be learned. The literature on attention in visual perceptual learning, at least so far, presents a more ambiguous picture. One interpretation is that attention mediates task-relevant learning while also suppressing task-irrelevant learning for suprathreshold stimuli. Too few studies, however, have explicitly manipulated attention and measured learning; from those few that have, there is modest evidence that more attention can lead to more learning. The evidence for a relationship in the other direction, that perceptual learning can reduce the need for attention in an initially attention-demanding behavioral task, is more robust. Finally, reward and reinforcement have been important theoretical concepts in the field, largely inherited from broader theories of reinforcement and learning. Here, too, however, research that disambiguates feedback information from the influences of reward, while suggestive, has only begun. There is some preliminary evidence to suggest that higher magnitudes of reward produce higher rates of learning. This work should be replicated and extended.

As part of our analysis of top-down influences, it has remained our view that selective reweighting provides the strongest theoretical structure for understanding visual perceptual learning. Although task structure, attention, and reward will almost surely have immediate effects on activation and response during each trial, the consequences of these immediate trial-by-trial differences will also be incorporated into *learned weight structures*. Exploring the importance of these factors requires new experiments with explicit manipulations in otherwise similar or equivalent learning tasks. In concert with these empirical investigations, a new framework marking the influence of task, attention, and reward in expanded augmented Hebbian learning rules—or alternative models—should be modeled computationally and tested experimentally.

9.7 Appendix: Expanding Models of Perceptual Learning

This appendix considers how task, attention, and reward might be integrated into existing models of perceptual learning, with potential consequences for understanding relevant brain functions. Performing a task implies both a selection of relevant stimuli and setting up a decision structure to organize choice behavior. Carrying out the task may engage attention to enhance the representations of stimuli or stimulus features. The effects of reward and/or reward expectation could influence choice bias, decision, or learning. The experimental evidence reviewed in this chapter suggests how each of these factors could operate and provides some general principles concerning the function of interacting brain systems in perceptual learning. Working either individually or collectively, modulating the stimulus representation, decision and bias, interpretation of feedback, and/or rate of learning could all impact perceptual learning. Models, whether quantitative or process models or brain system models, would need to specify how each of these potential processes is in play. In this appendix, we propose an expanded framework of learning equations that includes task, attention, and reward within the context of an extended model based on the augmented Hebbian reweighting model (AHRM).[172, 173] The same developments could be applied within the integrated reweighting model (IRT).[171]

In the original AHRM (see chapter 6), the representation module computes activations in the representation units from visual input images. These noisy activations, weighted by the learned weights (w_i), together with a weighted measure of bias in recent responses ($w_b b$), drive the output of the decision unit (or units). Then, after the response on each trial, the weights are updated, incorporating feedback ($w_f f$) within the Hebbian learning rule. Implicitly, the model selects task-relevant stimuli and the initial weights connecting the corresponding feature representations,(e.g., spatial frequency and orientation, motion, spatial location) to the decision unit so as to embody task instructions and prior knowledge of the domain. New versions of the learning equations explicitly mark potential dependencies on the task (T), attention (A), and reward (R). For example, the strength of representation activations might depend jointly on attention, reward, and task; the initial weights should depend on the task and attention; and aspects of bias or feedback could depend on either reward or reward expectation. We have used prior research to guide intuitions about which factors could influence task, attention, and reward.

In the AHRM, the weighted sum of the activation of representation units feeding into the decision unit (the same as equation 6.6) is

$$u = \sum_{i=1}^{n\ channels} w_i a_i - w_b b + \varepsilon. \tag{9.3}$$

A corresponding equation marking potential dependencies on task, attention, and reward is

$$u = \sum_{i=1}^{n\ channels} w_i(A,T) a_i(T,A,R) - w_b(R) b(R) + \varepsilon. \tag{9.3'}$$

The T, A, and R make explicit the dependencies of each parameter on the three factors. (Note that we have chosen the form $w_i(A,T)$ to denote that the values of the weights may depend on attention and task. An alternate notation could have been $w_i^{A,T}$.) The dependence of any component of the equation on T, A, and/or R reflects the influence of these factors. The aggregate input u is then passed through a nonlinearity to produce the early output at the decision unit o', which determines the response on that trial. If feedback is available, then this early output is shifted to a new and more accurate late output o before the learning cycle; if there is no feedback, $o = o'$. The original augmented Hebbian learning rule (equation 6.10) is

$$\delta_i = \eta a_i (o - \bar{o}). \tag{9.4}$$

To reprise the formula, the change of weight for representation unit i is proportional to δ_i, which depends on the learning rate η, the activation in that representation unit a_i, and the difference between the output at the decision unit and its long-term recency-weighted running average, or $(o - \bar{o})$ (a normalized form of the output activity).

In the expanded framework, the learning rate η could depend on either attention, reward, or both; that is, $\eta(A,R)$. The physiological and empirical literature also suggests that perceptual learning could be modulated by reward or *reward-prediction error.* The potential impact of reward-prediction error is entered into the learning equations in a reward term $(r - \hat{r})$ that modulates the size of the change signal δ_i, leading to a new expanded δ-rule:

$$\delta_i = \eta(A,R) a_i(T,A,R)(o - \bar{o})(r - \hat{r}). \tag{9.4'}$$

In this extension to the Hebbian rule, the magnitude of the learning signal depends on the deviation of the actual reward r and the expected reward $\hat{r}$. This expanded rule represents one idea about reward, task and attention in the learning rule. Learning with these rules, as with the AHRM/IRT, remains a hybrid system that can operate with or without feedback or without reward (by setting the reward term to 1).

Considering the expanded equation by itself, changes in learning rate, activation, and reward might map equivalently into changes in the learning rate η (while incorporating the reward-expectation error term could provide a way in which the effective learning rate could vary from trial to trial). Understanding in more detail how this extended learning rule would operate in the context of a specific architecture of representations and decision units in a particular task would, almost surely, require simulation modeling. This is because of the requirements to include internal noise and nonlinearity in order to predict actual behavioral performance and how learning is influenced by the distribution of useful signal and distracting noise in the different stimuli used during training in the task. It seems possible or even likely that the reward terms may serve essentially to modify the learning rate, while the impact of task and attention may be mediated in more complex ways that depend on the stimuli.

We note that incorporating reward into learning rules was proposed previously by other researchers, and these alternative forms could be compared to those developed here for augmented Hebbian reweighting. The reward prediction error term in the δ-rule here is similar to a proposal by Herzog and colleagues, who distinguished between Hebbian rules (equation 9.5) and Hebbian rules augmented by either fully supervised error correction (equation 9.6), reward-based learning (equation 9.7), or reward-prediction error (equation 9.8):[174]

$$\Delta w_{ij} = pre_i \times post_j \text{ (Hebbian)}, \tag{9.5}$$

$$\Delta w_{ij} = pre_i \times E_{ij} \text{ (supervised Hebbian)}, \tag{9.6}$$

$$\Delta w_{ij} = pre_i \times (post_j - \overline{post_j}) \times R \text{ (reward-based Hebbian)}, \tag{9.7}$$

$$\Delta w_{ij} = pre_i \times post_j \times (R - \hat{R}) \text{ (reward-prediction error Hebbian)}. \tag{9.8}$$

In their equations, the postsynaptic activation $post_j$ is equivalent to the o in equation (9.4′). The E_{ij} in equation (9.6) is a fully supervised error term that compares the postsynaptic output to a target value provided by a teacher. Technically, this form of reward-based Hebbian learning rule in equation (9.7) corresponds to the so-called R_{max} form.[107] It has been argued that the R_{max} reward rule is too powerful, while the reward prediction error relies on estimations of the predicted reward.[174] Finally, some have argued that fully supervised rules may be too powerful to be consistent with observed learning, and their neural plausibility has been questioned, although more plausible new supervised forms of learning are being developed.[176]

The expanded δ-rule of the AHRM (equation 9.4′) also bears a similarity to the attention-gated reinforcement learning (AGREL) model by Roelfsma and colleagues.[166, 175] Attention-gated reinforcement learning combines a broad reward signal with an attention function that limits weight changes to those units that are deemed to be the primary drivers of the response—by "assigning credit" to individual sensory or input units. Roelfsma et al. propose this equation for changing weights during learning:

$$\Delta w_{ij} = \eta \cdot a_i \cdot f(o_j) \cdot g(R) \cdot FB_{sj} \text{ (attention-gated reinforcement)}. \tag{9.9}$$

In this equation, $f(o_j)$ is some function of the postsynaptic activity in response (decision) unit j, $g(R)$ is some function of the reward outcome on that trial, and FB_{sj} is the attention feedback signal from the winning response unit s. This last factor is the critical departure from reinforcement or reward-based Hebbian learning. It limits weight changes to only those few connections that most strongly supported the selected response (similar to network penalties that encourage sparse representations). This attention-weighted learning model is also functionally similar to the fully supervised back-propagation models of learned weight change.[166]

We chose the particular form of the δ-rule for changing the weights in equation (9.4′) for backward compatibility with the AHRM learning rule—which has been extensively tested quantitatively against many datasets in perceptual learning. The architecture and learning rule(s) of the proposed extensions of the augmented Hebbian learning rules could easily be directly incorporated into integrated reweighting theory (IRT) (chapter 8).

This theoretical framework provides a structure within which to develop tests of the roles of task, attention, and reward in perceptual learning. Model simulations or derivations may generate new predictions about how these top-down influences could be empirically tested in specific tasks and stimuli. Situations that require n-alternative decisions rather than two-alternative decisions (described in chapters 6 and 8) may be one approach to distinguishing augmented Hebbian learning from other forms of reinforcement learning. Together, these approaches provide a framework for considering task-selective reweighting in perceptual learning.

References

1. Tsushima Y, Seitz AR, Watanabe T. Task-irrelevant learning occurs only when the irrelevant feature is weak. *Current Biology* 2008;18(12):R516–R517.

2. Ahissar M, Hochstein S. Attentional control of early perceptual learning. *Proceedings of the National Academy of Sciences* 1993;90(12):5718–5722.

26. Ahissar M, Hochstein S. The spread of attention and learning in feature search: Effects of target distribution and task difficulty. *Vision Research* 2000;40(10):1349–1364.

27. Dosher BA, Han S, Lu Z-L. Perceptual learning and attention: Reduction of object attention limitations with practice. *Vision Research* 2010;50(4):402–415.

3. Shiu L-P, Pashler H. Improvement in line orientation discrimination is retinally local but dependent on cognitive set. *Attention, Perception, & Psychophysics* 1992;52(5):582–588.

4. Herzog M, Fahle M. Learning without attention. Paper presented at the 22nd Gottingen Neurobiology Conference; 1994.

5. Hung S-C, Seitz AR. Prolonged training at threshold promotes robust retinotopic specificity in perceptual learning. *Journal of Neuroscience* 2014;34(25):8423–8431.

6. Fahle M, Morgan M. No transfer of perceptual learning between similar stimuli in the same retinal position. *Current Biology* 1996;6(3):292–297.

7. Paffen CL, Verstraten FA, Vidnyánszky Z. Attention-based perceptual learning increases binocular rivalry suppression of irrelevant visual features. *Journal of Vision* 2008;8(4):25,1–11.

8. Vidnyánszky Z, Sohn W. Learning to suppress task-irrelevant visual stimuli with attention. *Vision Research* 2005;45(6):677–685.

9. Gal V, Kozák LR, Kóbor I, Bankó EM, Serences JT, Vidnyánszky Z. Learning to filter out visual distractors. *European Journal of Neuroscience* 2009;29(8):1723–1731.

10. Huang X, Lu H, Tjan BS, Zhou Y, Liu Z. Motion perceptual learning: When only task-relevant information is learned. *Journal of Vision* 2007;7(10):14,1–10.

11. Szpiro SF, Carrasco M. Exogenous attention enables perceptual learning. *Psychological Science* 2015;26(12):1854–1862.

12. Sayim B, Westheimer G, Herzog MH. Contrast polarity, chromaticity, and stereoscopic depth modulate contextual interactions in Vernier acuity. *Journal of Vision* 2008;8(8):12,1–9.

13. Watanabe T, Náñez JE, Sasaki Y. Perceptual learning without perception. *Nature* 2001;413(6858):844–848.

14. Seitz AR, Kim D, Watanabe T. Rewards evoke learning of unconsciously processed visual stimuli in adult humans. *Neuron* 2009;61(5):700–707.

15. Gutnisky DA, Hansen BJ, Iliescu BF, Dragoi V. Attention alters visual plasticity during exposure-based learning. *Current Biology* 2009;19(7):555–560.

16. Ludwig I, Skrandies W. Human perceptual learning in the peripheral visual field: Sensory thresholds and neurophysiological correlates. *Biological Psychology* 2002;59(3):187–206.

17. Nishina S, Seitz AR, Kawato M, Watanabe T. Effect of spatial distance to the task stimulus on task-irrelevant perceptual learning of static Gabors. *Journal of Vision* 2007;7(13):2,1–10.

18. Seitz AR, Nanez JE, Holloway SR, Koyama S, Watanabe T. Seeing what is not there shows the costs of perceptual learning. *Proceedings of the National Academy of Sciences* 2005;102(25):9080–9085.

19. Seitz A, Watanabe T. A unified model for perceptual learning. *Trends in Cognitive Sciences* 2005;9(7):329–334.

20. Seitz AR, Watanabe T. Psychophysics: Is subliminal learning really passive? *Nature* 2003;422(6927):36.

21. Watanabe T, Náñez JE, Koyama S, Mukai I, Liederman J, Sasaki Y. Greater plasticity in lower-level than higher-level visual motion processing in a passive perceptual learning task. *Nature Neuroscience* 2002;5(10):1003–1009.

22. Pilly PK, Grossberg S, Seitz AR. Low-level sensory plasticity during task-irrelevant perceptual learning: Evidence from conventional and double training procedures. *Vision Research* 2010;50(4):424–432.

23. Wehrhahn C, Rapf D. Perceptual learning of apparent motion mediated through ON- and OFF-pathways in human vision. *Vision Research* 2001;41(3):353–358.

24. Rosenthal O, Humphreys GW. Perceptual organization without perception: The subliminal learning of global contour. *Psychological Science* 2010;21(12):1751–1758.

25. Wong YK, Folstein JR, Gauthier I. Task-irrelevant perceptual expertise. *Journal of Vision* 2011;11(14):3,1–15.

29. Ahissar M, Hochstein S. The reverse hierarchy theory of visual perceptual learning. *Trends in Cognitive Sciences* 2004;8(10):457–464.

31. Crist RE, Li W, Gilbert CD. Learning to see: Experience and attention in primary visual cortex. *Nature Neuroscience* 2001;4(5):519–525.

30. Dolan R, Fink G, Rolls E, Booth M, Holmes A, Frackowiak RSJ, Friston KJ. How the brain learns to see objects and faces in an impoverished context. *Nature* 1997;389(6651):596–599.

28. Goldstone RL. Perceptual learning. *Annual Review of Psychology* 1998;49(1):585–612.

32. Xiao L-Q, Zhang J-Y, Wang R, Klein SA, Levi DM, Yu C. Complete transfer of perceptual learning across retinal locations enabled by double training. *Current Biology* 2008;18(24):1922–1926.

33. Posner MI, Petersen SE. The attention system of the human brain. *Annual Review of Neuroscience* 1990;13(1):25–42.

34. Corbetta M, Shulman GL. Control of goal-directed and stimulus-driven attention in the brain. *Nature Reviews Neuroscience* 2002;3(3):201–215.

35. Desimone R, Duncan J. Neural mechanisms of selective visual attention. *Annual Review of Neuroscience* 1995;18(1):193–222.

36. Petersen SE, Posner MI. The attention system of the human brain: 20 years after. *Annual Review of Neuroscience* 2012;35:73–89.

37. Posner MI. *Cognitive neuroscience of attention.* Guilford Press; 2011.

38. Vossel S, Geng JJ, Fink GR. Dorsal and ventral attention systems: Distinct neural circuits but collaborative roles. *The Neuroscientist* 2014;20(2):150–159.

39. Liu T, Hospadaruk L, Zhu DC, Gardner JL. Feature-specific attentional priority signals in human cortex. *Journal of Neuroscience* 2011;31(12):4484–4495.

40. Vossel S, Weidner R, Driver J, Friston KJ, Fink GR. Deconstructing the architecture of dorsal and ventral attention systems with dynamic causal modeling. *Journal of Neuroscience* 2012;32(31):10637–10648.

41. Ruff CC, Bestmann S, Blankenburg F, Bjoertormt O, Josephs O, Weiskopf N, Deichmann R, Driver J. Distinct causal influences of parietal versus frontal areas on human visual cortex: Evidence from concurrent TMS–fMRI. *Cerebral Cortex* 2007;18(4):817–827.

42. Ruff CC, Blankenburg F, Bjoertomt O, Bestmann S, Freeman E, Haynes J-D, Rees G, Josephs O, Deichmann R, Driver J. Concurrent TMS–fMRI and psychophysics reveal frontal influences on human retinotopic visual cortex. *Current Biology* 2006;16(15):1479–1488.

43. Blankenburg F, Ruff CC, Bestmann S, Bjoertomt O, Josephs O, Deichmann R, Driver J. Studying the role of human parietal cortex in visuospatial attention with concurrent TMS–fMRI. *Cerebral Cortex* 2010;20(11):2702–2711.

44. Corbetta M, Patel G, Shulman GL. The reorienting system of the human brain: From environment to theory of mind. *Neuron* 2008;58(3):306–324.

45. Bressler SL, Tang W, Sylvester CM, Shulman GL, Corbetta M. Top-down control of human visual cortex by frontal and parietal cortex in anticipatory visual spatial attention. *Journal of Neuroscience* 2008;28(40):10056–10061.

46. Luck SJ, Chelazzi L, Hillyard SA, Desimone R. Neural mechanisms of spatial selective attention in areas V1, V2, and V4 of macaque visual cortex. *Journal of Neurophysiology* 1997;77(1):24–42.

47. Motter BC. Neural correlates of feature selective memory and pop-out in extrastriate area V4. *Journal of Neuroscience* 1994;14(4):2190–2199.

48. Motter BC. Focal attention produces spatially selective processing in visual cortical areas V1, V2, and V4 in the presence of competing stimuli. *Journal of Neurophysiology* 1993;70(3):909–919.

49. Kastner S, Pinsk MA, De Weerd P, Desimone R, Ungerleider LG. Increased activity in human visual cortex during directed attention in the absence of visual stimulation. *Neuron* 1999;22(4):751–761.

50. Kastner S, De Weerd P, Desimone R, Ungerleider LG. Mechanisms of directed attention in the human extrastriate cortex as revealed by functional MRI. *Science* 1998;282(5386):108–111.

51. Connor CE, Preddie DC, Gallant JL, Van Essen DC. Spatial attention effects in macaque area V4. *Journal of Neuroscience* 1997;17(9):3201–3214.

52. McAdams CJ, Maunsell JH. Attention to both space and feature modulates neuronal responses in macaque area V4. *Journal of Neurophysiology* 2000;83(3):1751–1755.

53. Williford T, Maunsell JH. Effects of spatial attention on contrast response functions in macaque area V4. *Journal of Neurophysiology* 2006;96(1):40–54.

54. Mitchell JF, Sundberg KA, Reynolds JH. Differential attention-dependent response modulation across cell classes in macaque visual area V4. *Neuron* 2007;55(1):131–141.

55. Mitchell JF, Sundberg KA, Reynolds JH. Spatial attention decorrelates intrinsic activity fluctuations in macaque area V4. *Neuron* 2009;63(6):879–888.

56. Byers A, Serences JT. Exploring the relationship between perceptual learning and top-down attentional control. *Vision Research* 2012;74:30–39.

57. Byers A, Serences JT. Enhanced attentional gain as a mechanism for generalized perceptual learning in human visual cortex. *Journal of Neurophysiology* 2014;112(5):1217–1227.

58. Serences JT. Value-based modulations in human visual cortex. *Neuron* 2008;60(6):1169–1181.

59. Serences JT, Yantis S. Selective visual attention and perceptual coherence. *Trends in Cognitive Sciences* 2006;10(1):38–45.

60. Martinez A, Anllo-Vento L, Sereno MI, Frank LR, Buxton RB, Dubowitz DJ, Wong EC, Hinrichs H, Heinze HJ, Hillyard SA. Involvement of striate and extrastriate visual cortical areas in spatial attention. *Nature Neuroscience* 1999;2(4):364–369.

61. Duncan J, Humphreys G, Ward R. Competitive brain activity in visual attention. *Current Opinion in Neurobiology* 1997;7(2):255–261.

62. Reynolds JH, Pasternak T, Desimone R. Attention increases sensitivity of V4 neurons. *Neuron* 2000;26(3):703–714.

63. Sundberg KA, Mitchell JF, Reynolds JH. Spatial attention modulates center-surround interactions in macaque visual area v4. *Neuron* 2009;61(6):952–963.

64. Buracas GT, Boynton GM. The effect of spatial attention on contrast response functions in human visual cortex. *Journal of Neuroscience* 2007;27(1):93–97.

65. Li X, Lu Z-L, Tjan BS, Dosher BA, Chu W. Blood oxygenation level-dependent contrast response functions identify mechanisms of covert attention in early visual areas. *Proceedings of the National Academy of Sciences* 2008;105(16):6202–6207.

66. Lu Z-L, Li X, Tjan BS, Dosher BA, Chu W. Attention extracts signal in external noise: A BOLD fMRI study. *Journal of Cognitive Neuroscience* 2011;23(5):1148–1159.

67. Dosher BA, Lu Z-L. Noise exclusion in spatial attention. *Psychological Science* 2000;11(2):139–146.

68. Lu Z-L, Dosher BA. Spatial attention: Different mechanisms for central and peripheral temporal precues? *Journal of Experimental Psychology: Human Perception and Performance* 2000;26(5):1534–1548.

69. Hetley R, Dosher BA, Lu Z-L. Generating a taxonomy of spatially cued attention for visual discrimination: Effects of judgment precision and set size on attention. *Attention, Perception, & Psychophysics* 2014;76(8):2286–2304.

70. Posner MI. Orienting of attention. *Quarterly Journal of Experimental Psychology* 1980;32(1):3–25.

71. Wundt W. *Elemente der völkerpsychologie*. A. Kröner; 1912.

72. Egeth HE, Virzi RA, Garbart H. Searching for conjunctively defined targets. *Journal of Experimental Psychology: Human Perception and Performance* 1984;10(1):32–39.

73. Dosher BA. Models of visual search: Finding a face in the crowd. In Scarborough D, Sternberg S, eds., *An invitation to cognitive science: Methods, models, and conceptual issues*. Vol. 4. 1998;455–521.

74. Maunsell JH, Treue S. Feature-based attention in visual cortex. *Trends in Neurosciences* 2006;29(6):317–322.

75. Cohen MR, Maunsell JH. Using neuronal populations to study the mechanisms underlying spatial and feature attention. *Neuron* 2011;70(6):1192–1204.

76. Saenz M, Buracas GT, Boynton GM. Global effects of feature-based attention in human visual cortex. *Nature Neuroscience* 2002;5(7):631–632.

77. Tse C-H, Lu Z-L, Sperling G. Attending to red and green concurrently in different areas reduces attentional capacity. *Investigative Ophthalmology and Visual Research* 2000;41:S42.

78. Treue S. Neural correlates of attention in primate visual cortex. *Trends in Neurosciences* 2001;24(5):295–300.

79. Duncan J. Selective attention and the organization of visual information. *Journal of Experimental Psychology: General* 1984;113(4):501–517.

80. Dosher B, Lu Z-L. Object attention. In Raaijmakers JGW, Criss AH, Goldstone RL, Nosofsky RM, Steyvers M, eds., *Cognitive modeling in perception and memory: A festschrift for Richard M. Shiffrin*. 2015:35–62.

81. Han S, Dosher BA, Lu Z-L. Object attention revisited: Identifying mechanisms and boundary conditions. *Psychological Science* 2003;14(6):598–604.

82. Mukai I, Bahadur K, Kesavabhotla K, Ungerleider LG. Exogenous and endogenous attention during perceptual learning differentially affect post-training target thresholds. *Journal of Vision* 2011;11(1):25,1–15.

83. Donovan I, Szpiro S, Carrasco M. Exogenous attention facilitates location transfer of perceptual learning. *Journal of Vision* 2015;15(10):11,1–16.

84. Xu JP, He ZJ, Ooi TL. Perceptual learning to reduce sensory eye dominance beyond the focus of top-down visual attention. *Vision Research* 2012;61:39–47.

85. Fahle M. Perceptual learning: A case for early selection. *Journal of Vision* 2004;4(10):4,879–890.

86. Tsushima Y, Sasaki Y, Watanabe T. Greater disruption due to failure of inhibitory control on an ambiguous distractor. *Science* 2006;314(5806):1786–1788.

87. Tsushima Y, Watanabe T. Roles of attention in perceptual learning from perspectives of psychophysics and animal learning. *Learning & Behavior* 2009;37(2):126–132.

88. Choi H, Seitz AR, Watanabe T. When attention interrupts learning: Inhibitory effects of attention on TIPL. *Vision Research* 2009;49(21):2586–2590.

89. Seitz AR, Watanabe T. The phenomenon of task-irrelevant perceptual learning. *Vision Research* 2009;49(21):2604–2610.

90. Shiffrin RM, Schneider W. Controlled and automatic human information processing: II. Perceptual learning, automatic attending and a general theory. *Psychological Review* 1977;84(2):127–190.

91. Weichselgartner E, Sperling G. Dynamics of automatic and controlled visual attention. *Science* 1987;238(4828):778–781.

92. Duncan J, Ward R, Shapiro K. Direct measurement of attentional dwell time in human vision. *Nature* 1994;369(6478):313–315.

93. Shapiro KL, Raymond JE, Arnell KM. Attention to visual pattern information produces the attentional blink in rapid serial visual presentation. *Journal of Experimental Psychology: Human Perception and Performance* 1994;20(2):357–371.

94. Choi H, Chang L-H, Shibata K, Sasaki Y, Watanabe T. Resetting capacity limitations revealed by long-lasting elimination of attentional blink through training. *Proceedings of the National Academy of Sciences* 2012;109(30):12242–12247.

95. Ito M, Westheimer G, Gilbert CD. Attention and perceptual learning modulate contextual influences on visual perception. *Neuron* 1998;20(6):1191–1197.

96. Dixon ML, Ruppel J, Pratt J, De Rosa E. Learning to ignore: Acquisition of sustained attentional suppression. *Psychonomic Bulletin & Review* 2009;16(2):418–423.

97. Tseng C-H, Gobell JL, Sperling G. Long-lasting sensitization to a given colour after visual search. *Nature* 2004;428(6983):657–660.

98. Tseng C-H, Vidnyanszky Z, Papathomas T, Sperling G. Attention-based long-lasting sensitization and suppression of colors. *Vision Research* 2010;50(4):416–423.

99. Bao M, Yang L, Rios C, He B, Engel SA. Perceptual learning increases the strength of the earliest signals in visual cortex. *Journal of Neuroscience* 2010;30(45):15080–15084.

100. Bartolucci M, Smith AT. Attentional modulation in visual cortex is modified during perceptual learning. *Neuropsychologia* 2011;49(14):3898–3907.

101. Hamamé CM, Cosmelli D, Henriquez R, Aboitiz F. Neural mechanisms of human perceptual learning: Electrophysiological evidence for a two-stage process. *PLoS One* 2011;6(4):e19221.

102. Skinner BF. *The behaviour of organisms: An experimental analysis*. D. Appleton-Century; 1938.

103. Belin D, Everitt BJ. Cocaine seeking habits depend upon dopamine-dependent serial connectivity linking the ventral with the dorsal striatum. *Neuron* 2008;57(3):432–441.

104. Haber SN, Knutson B. The reward circuit: Linking primate anatomy and human imaging. *Neuropsychopharmacology* 2010;35(1):4–26.

105. Tanaka SC, Doya K, Okada G, Ueda K, Okamoto Y, Yamawaki S. Prediction of immediate and future rewards differentially recruits cortico–basal ganglia loops. *Nature Neuroscience* 2004;7(8):887–893.

106. Farries MA, Fairhall AL. Reinforcement learning with modulated spike timing-dependent synaptic plasticity. *Journal of Neurophysiology* 2007;98(6):3648–3665.

107. Frémaux N, Sprekeler H, Gerstner W. Functional requirements for reward-modulated spike-timing-dependent plasticity. *Journal of Neuroscience* 2010;30(40):13326–13337.

108. Izhikevich EM. Solving the distal reward problem through linkage of STDP and dopamine signaling. *Cerebral Cortex* 2007;17(10):2443–2452.

109. Legenstein R, Pecevski D, Maass W. A learning theory for reward-modulated spike-timing-dependent plasticity with application to biofeedback. *PLoS Computational Biology* 2008;4(10):e1000180.

110. Loewenstein Y. Robustness of learning that is based on covariance-driven synaptic plasticity. *PLoS Computational Biology* 2008;4(3):e1000007.

111. Sutton RS, Barto AG. *Reinforcement learning: An introduction.* MIT Press; 1998.

112. Arsenault JT, Rima S, Stemmann H, Vanduffel W. Role of the primate ventral tegmental area in reinforcement and motivation. *Current Biology* 2014;24(12):1347–1353.

113. Bromberg-Martin ES, Matsumoto M, Hikosaka O. Dopamine in motivational control: Rewarding, aversive, and alerting. *Neuron* 2010;68(5):815–834.

114. Schultz W, Dayan P, Montague PR. A neural substrate of prediction and reward. *Science* 1997;275(5306):1593–1599.

115. McClure SM, Berns GS, Montague PR. Temporal prediction errors in a passive learning task activate human striatum. *Neuron* 2003;38(2):339–346.

116. Nomoto K, Schultz W, Watanabe T, Sakagami M. Temporally extended dopamine responses to perceptually demanding reward-predictive stimuli. *Journal of Neuroscience* 2010;30(32):10692–10702.

117. O'Doherty JP. Reward representations and reward-related learning in the human brain: Insights from neuroimaging. *Current Opinion in Neurobiology* 2004;14(6):769–776.

118. Bhattacharyya A, Veit J, Kretz R, Bondar I, Rainer G. Basal forebrain activation controls contrast sensitivity in primary visual cortex. *BMC Neuroscience* 2013;14(1):55,1–13.

119. Frankó E, Seitz AR, Vogels R. Dissociable neural effects of long-term stimulus–reward pairing in macaque visual cortex. *Journal of Cognitive Neuroscience* 2010;22(7):1425–1439.

120. Goard M, Dan Y. Basal forebrain activation enhances cortical coding of natural scenes. *Nature Neuroscience* 2009;12(11):1444–1449.

121. Pinto L, Goard MJ, Estandian D, Xu M, Kwan AC, Lee S-H, Harrison TC, Feng G, Dan Y. Fast modulation of visual perception by basal forebrain cholinergic neurons. *Nature Neuroscience* 2013;16(12):1857–1863.

122. Shuler MG, Bear MF. Reward timing in the primary visual cortex. *Science* 2006;311(5767):1606–1609.

123. Krawczyk DC, Gazzaley A, D'Esposito M. Reward modulation of prefrontal and visual association cortex during an incentive working memory task. *Brain Research* 2007;1141:168–177.

124. Weil RS, Furl N, Ruff CC, Symmonds M, Flanding G, Dolan RJ, Driver J, Rees G. Rewarding feedback after correct visual discriminations has both general and specific influences on visual cortex. *Journal of Neurophysiology* 2010;104(3):1746–1757.

125. Della Libera C, Chelazzi L. Visual selective attention and the effects of monetary rewards. *Psychological Science* 2006;17(3):222–227.

126. Stănişor L, van der Togt C, Pennartz CM, Roelfsema PR. A unified selection signal for attention and reward in primary visual cortex. *Proceedings of the National Academy of Sciences* 2013;110(22):9136–9141.

127. Engelmann JB, Damaraju E, Padmala S, Pessoa L. Combined effects of attention and motivation on visual task performance: Transient and sustained motivational effects. *Frontiers in Human Neuroscience* 2009;3:4,1–17.

128. Baldassi S, Simoncini C. Reward sharpens orientation coding independently of attention. *Frontiers in Neuroscience* 2011;5:13,1–11.

129. Green D, Swets J. *Signal detection theory and psychophysics.* Wiley;1966.

130. Ratcliff R, McKoon G. A retrieval theory of priming in memory. *Psychological Review* 1988;95(3):385–408.

131. Gold JI, Shadlen MN. Neural computations that underlie decisions about sensory stimuli. *Trends in Cognitive Sciences* 2001;5(1):10–16.

132. Roitman JD, Shadlen MN. Response of neurons in the lateral intraparietal area during a combined visual discrimination reaction time task. *Journal of Neuroscience* 2002;22(21):9475–9489.

133. Mazurek ME, Roitman JD, Ditterich J, Shadlen MN. A role for neural integrators in perceptual decision making. *Cerebral Cortex* 2003;13(11):1257–1269.

134. Huk AC, Shadlen MN. Neural activity in macaque parietal cortex reflects temporal integration of visual motion signals during perceptual decision making. *Journal of Neuroscience* 2005;25(45):10420–10436.

135. Gold JI, Shadlen MN. The neural basis of decision making. *Annual Review of Neuroscience* 2007;30:535–574.

136. Rorie AE, Gao J, McClelland JL, Newsome WT. Integration of sensory and reward information during perceptual decision-making in lateral intraparietal cortex (LIP) of the macaque monkey. *PloS One* 2010;5(2):e9308.

137. Bendiksby MS, Platt ML. Neural correlates of reward and attention in macaque area LIP. *Neuropsychologia* 2006;44(12):2411–2420.

138. Ratcliff R, Cherian A, Segraves M. A comparison of macaque behavior and superior colliculus neuronal activity to predictions from models of two-choice decisions. *Journal of Neurophysiology* 2003;90(3):1392–1407.

139. Ratcliff R, Hasegawa YT, Hasegawa RP, Smith PL, Segraves MA. Dual diffusion model for single-cell recording data from the superior colliculus in a brightness-discrimination task. *Journal of Neurophysiology* 2007;97(2):1756–1774.

140. Pascucci D, Mastropasqua T, Turatto M. Monetary reward modulates task-irrelevant perceptual learning for invisible stimuli. *PloS One* 2015;10(5):e0124009.

141. Kim D, Seitz AR, Watanabe T. Visual perceptual learning by operant conditioning training follows rules of contingency. *Visual Cognition* 2015;23(1–2):147–160.

142. Tsuchiya N, Koch C. Continuous flash suppression reduces negative afterimages. *Nature Neuroscience* 2005;8(8):1096–1101.

143. Öhman A, Mineka S. Fears, phobias, and preparedness: Toward an evolved module of fear and fear learning. *Psychological Review* 2001;108(3):483–522.

144. Jessup RK, O'Doherty JP. It was nice not seeing you: Perceptual learning with rewards in the absence of awareness. *Neuron* 2009;61(5):649–650.

145. Xue X, Zhou X, Li S. Unconscious reward facilitates motion perceptual learning. *Visual Cognition* 2015;23(1–2):161–178.

146. Zhang P, Hou F, Yan F-F, Xi J, Lin B-R, Zhao J, Yang J, Chen G, Zhang M-Y, He Q, Dosher BA, Lu Z-L, and Huang C-B. High reward enhances perceptual learning. *Journal of Vision* 2018;18(8):11,1–21.

147. Anderson BA. The attention habit: How reward learning shapes attentional selection. *Annals of the New York Academy of Sciences* 2016;1369(1):24–39.

148. Seitz AR, Dinse HR. A common framework for perceptual learning. *Current Opinion in Neurobiology* 2007;17(2):148–153.

149. Sarter M, Hasselmo ME, Bruno JP, Givens B. Unraveling the attentional functions of cortical cholinergic inputs: Interactions between signal-driven and cognitive modulation of signal detection. *Brain Research Reviews* 2005;48(1):98–111.

150. Rokem A, Silver MA. Cholinergic enhancement augments magnitude and specificity of visual perceptual learning in healthy humans. *Current Biology* 2010;20(19):1723–1728.

151. Arnold H, Burk J, Hodgson E, Sarter M, Bruno J. Differential cortical acetylcholine release in rats performing a sustained attention task versus behavioral control tasks that do not explicitly tax attention. *Neuroscience* 2002;114(2):451–460.

152. Silver MA, Shenhav A, D'Esposito M. Cholinergic enhancement reduces spatial spread of visual responses in human early visual cortex. *Neuron* 2008;60(5):904–914.

153. Greuel JM, Luhmann HJ, Singer W. Pharmacological induction of use-dependent receptive field modifications in the visual cortex. *Science* 1988;242(4875):74–77.

154. Kilgard MP, Merzenich MM. Cortical map reorganization enabled by nucleus basalis activity. *Science* 1998;279(5357):1714–1718.

155. Rokem A, Silver MA. The benefits of cholinergic enhancement during perceptual learning are long-lasting. *Frontiers in Computational Neuroscience*, 2013;7:66,1–7.

156. Cahill L, McGaugh JL. Mechanisms of emotional arousal and lasting declarative memory. *Trends in Neurosciences* 1998;21(7):294–299.

157. Beer AL, Vartak D, Greenlee MW. Nicotine facilitates memory consolidation in perceptual learning. *Neuropharmacology* 2013;64:443–451.

158. Holroyd CB, Coles MG. The neural basis of human error processing: Reinforcement learning, dopamine, and the error-related negativity. *Psychological Review* 2002;109(4):679–709.

159. Noudoost B, Moore T. Control of visual cortical signals by prefrontal dopamine. *Nature* 2011;474(7351):372–375.

160. Leguire LE, Walson PD, Rogers GL, Bremer DL, McGregor ML. Levodopa/carbidopa treatment for amblyopia in older children. *Journal of Pediatric Ophthalmology and Strabismus* 1995;32(3):143–151.

161. Leguire LE, Walson PD, Rogers GL, Bremer DL, McGregor ML. Longitudinal study of levodopa/carbidopa for childhood amblyopia. *Journal of Pediatric Ophthalmology and Strabismus* 1993;30(6):354–360.

162. Price A, Shin JC. The impact of Parkinson's disease on sequence learning: Perceptual pattern learning and executive function. *Brain and Cognition* 2009;69(2):252–261.

163. Rösser N, Heuschmann P, Wersching H, Breitenstein C, Knecht S, Flöel A. Levodopa improves procedural motor learning in chronic stroke patients. *Archives of Physical Medicine and Rehabilitation* 2008;89(9):1633–1641.

164. Bao S, Chang EF, Woods J, Merzenich MM. Temporal plasticity in the primary auditory cortex induced by operant perceptual learning. *Nature Neuroscience* 2004;7(9):974–981.

165. Dinse HR, Ragert P, Pleger B, Schwenkreis P, Tegenthoff M. Pharmacological modulation of perceptual learning and associated cortical reorganization. *Science* 2003;301(5629):91–94.

166. Roelfsema PR, van Ooyen A, Watanabe T. Perceptual learning rules based on reinforcers and attention. *Trends in Cognitive Sciences* 2010;14(2):64–71.

167. Ashby FG, Alfonso-Reese LA. A neuropsychological theory of multiple systems in category learning. *Psychological Review* 1998;105(3):442–481.

168. Ashby FG, Ennis JM, Spiering BJ. A neurobiological theory of automaticity in perceptual categorization. *Psychological Review* 2007;114(3):632–656.

169. Maddox WT, Ashby FG. Dissociating explicit and procedural-learning based systems of perceptual category learning. *Behavioural Processes* 2004;66(3):309–332.

170. Poldrack RA, Foerde K. Category learning and the memory systems debate. *Neuroscience & Biobehavioral Reviews* 2008;32(2):197–205.

171. Dosher BA, Jeter P, Liu J, Lu Z-L. An integrated reweighting theory of perceptual learning. *Proceedings of the National Academy of Sciences* 2013;110(33):13678–13683.

172. Petrov AA, Dosher BA, Lu Z-L. The dynamics of perceptual learning: An incremental reweighting model. *Psychological Review* 2005;112(4):715–743..

173. Petrov AA, Dosher BA, Lu Z-L. Perceptual learning without feedback in non-stationary contexts: Data and model. *Vision Research* 2006;46(19):3177–3197.

174. Herzog MH, Aberg KC, Frémaux N, Gerstner W, Sprekeler H. Perceptual learning, roving and the unsupervised bias. *Vision Research* 2012;61:95–99.

175. Roelfsema PR, van Ooyen A. Attention-gated reinforcement learning of internal representations for classification. *Neural Computation* 2005;17(10):2176–2214.

176. Neftci EO, Augustine C, Paul S, Detorakis G. Event-driven random back-propagation: Enabling neuromorphic deep learning machines. *Frontiers in Neuroscience* 2017;11:324,1–18.

V

Comparisons, Applications, and Optimization

10

Forms of Plasticity and Other Modalities

The plasticity of sensory processes occurs not just in visual perceptual learning but also at very different timescales, from evolution and development to immediate adaptation. Learning also occurs in modalities other than vision, including hearing, touch, taste, and smell, as well as multimodal interactions. Perhaps unsurprisingly, plasticity in these domains has many parallels with those in visual perceptual learning, but also some key differences. Though visual perceptual learning also differs from other forms of learning, such as category learning, even for seemingly similar stimuli, the key theoretical concept of reweighting is important in all these domains and may play a similar function in each in adjudicating between plasticity and stability of system behavior.

10.1 Learning and Plasticity

Visual perceptual learning is only one of many examples of our remarkable ability to adapt to our surroundings. Whether over thousands of generations, the life span of an individual, or the first few hours in a new perceptual environment, this ability to adapt has been key to the success not only of humans but of all biological systems.

This chapter examines the concept of plasticity—the mechanisms for change and adaptation—on multiple timescales. In this sense, perceptual learning can be seen as fitting within an evolutionary framework of general fitness just as that broader framework of interlocked changes can be brought to bear on our understanding of it. Next, we turn to a consideration of perceptual learning in modalities other than vision: hearing, touch, taste, smell, and multimodal. Finally, we compare learning in these modalities with similar forms of category learning often thought of as cognitive or conceptual.

Of course, entire books, or at least long papers, could be written about any of these topics. Our goal in this chapter is to identify common principles in different forms of plasticity, different empirical approaches that could profitably be extended to other domains, and where theoretical ideas and models that have been developed in one domain may be (or have been) applied in others. Broad similarities, even across significant differences, can help point to more fundamental principles.

10.2 Different Timescales of Plasticity

The visual system is a powerful processing engine. Its many modules and regions orchestrate the complex flow of perceptual information allowing us to interface so effectively with the world. Like other sensory systems, the human visual system has evolved to support the processing of cues in the stimulus environment. Still, even after millions of years of evolution, human visual functions continue to develop and improve over a significant period in early childhood, given exposure to normal visual experiences. The visual processing of an adult may be further fine-tuned with lengthy experience or exposure to stimuli in particular tasks. One form of fine-tuning occurs through very rapid sensory adaptations to immediately preceding stimuli. Another is the improvement through training or practice, or perceptual learning.

These different forms of plasticity operate on vastly different timescales, from multigenerational change to modulation of responses over a second or less. Understanding the range of plasticity at all these levels may provide insights into the special role—or niche—of visual perceptual learning and lead us to consider interactions between perceptual learning and development as well as adaptation (see table 10.1).

Plasticity through evolution involves changes in response to environmental demands at a generational level. In this context, the unique success of humans is truly remarkable. The resulting genetic codes passed to the individual are plastic insofar as environmental factors or experience change epigenetic expression during development or under environmental stress.[1–3]

Within the individual life span, humans express a relatively long period of postnatal development, sometimes explained as a trade-off between enhanced brain capability and initial vulnerability (relative to earlier developmental maturity of newborns in other species).[4, 5] Visual development continues from birth through adolescence. The period of plasticity varies considerably from one visual function to another, with some abilities achieving essentially adult levels within a year or two, while others continue to develop through early adulthood. Of course, at the other end of the life span, perceptual capabilities can diminish as a result of aging.

At a very rapid timescale, adaptation to environmental stimuli can make the observer exquisitely sensitive to recent inputs. Some forms of sensory adaptation either decay or

Table 10.1
Forms of visual plasticity

	Timescale	Duration	Primary basis
Evolution	Millions of years	Generations	Genetic
Development	Years	Life span	Neural anatomical
Perceptual learning	Minutes to hours	Years	Neural plasticity
Adaptation	Seconds to years	Seconds to years	Neural sensitivity

are reversed within a few seconds, while those following longer-term sensory induction may last for longer periods—seasons or even years.

As seen in the previous chapters, visual perceptual learning may improve perceptual judgments with training on the scale of hundreds or thousands of trials, often over several hours or days. Research in this area has overwhelmingly emphasized learning in adults (usually young adults), yet perceptual learning may have important applications during development and aging. Periods of early development are seen as unusually susceptible to experience, and experience can influence when and how well certain visual functions emerge.[6] Similarly, visual perceptual training may interact with, or be diminished or otherwise altered by, adaptation to recent stimuli.[7–9]

To recapitulate, the visual system is not stationary but rather a dynamic system that changes as a result of development, learning, and adaptation. At the limit, especially in response to significant challenges, it may also reflect its evolutionary legacy. What follows is a brief sketch of visual evolution, development, and adaptation for nonspecialists.

10.2.1 Visual Evolution

The visual system in humans evolved to support successful interaction with the world. We can recognize objects in scenes, estimate their speed, or interpret their motion on the retina if we are moving through a stationary environment. We use these visual inputs to guide our motor actions. Notwithstanding Darwin's famous caveat—"To suppose that the eye, with all its inimitable contrivances for adjusting the focus to different distances, for admitting different amounts of light, and for the correction of spherical and chromatic aberration could have been formed by natural selection, seems, I freely confess, absurd in the highest possible degree. Yet reason tells me…,"[10] p. 168)—the dominant hypothesis is that the human visual system evolved to extract meaningful information most likely to occur in natural scenes.

Early ancestral primates emerged as an evolutionary branch about 80 million years ago, differentiating from the rodents, flying lemurs, tree shrews, lagomorphs, and other members of the Euarchontoglire superclade (superclass), which had diverged from other placental mammals over 90 million years ago.[11] Primates diversified into euprimates (lemurs, galagos, etc.) and anthropoid primates (monkeys, apes, and humans), with the precursors to modern humans estimated to have diverged perhaps 6 million to 8 million years ago, possibly as part of an adaptation to grasslands in drier climates.[12]

One interpretation of the available data is that evolutionary adaptation to the behavioral demands of this new environment supported the development of a primate brain densely packed with a variety of neurons. In one well-known comparison, the brain of the owl monkey was contrasted with that of the agouti, a large South and Central American rodent. The owl monkey has a brain weighing 16 grams, with approximately 1.5 billion neurons, while the slightly larger 18 gram brain of the agouti has only 0.9 billion neurons. The contrast with larger animals is even greater.[12] The adult human brain, meanwhile, has been estimated to contain about 86 billion neurons.[13] (The higher neural density of the primate brain derives from neurons that are on average more compact, though they also have a greater range of sizes, shapes, and functions that support a varied set of computations.)[13]

The visual systems of primates have evolved key functional characteristics that adapt them for visual performance.[14] Even early primates, thought to be nocturnal, showed increased emphasis on central vision, with front-facing eyes to better support depth perception (though a reflective surface in the eye, which added reflected light stimulation to direct light stimulation, has since disappeared). As anthropoid primates became diurnal, the ratio of cones to rods shifted to support color vision in higher light levels as well as luminance at night. The existence of multiple differentiated cone receptors is thought to have produced an advantage in food selection.[15, 16] Other adaptations altered the function and structure of the lateral geniculate nucleus (LGN) to emphasize central vision and binocular depth.[17, 18]

Primates are also distinguished by their larger visual cortex.[19] The primary visual cortex, the first way station of visual representation and processing in the cortex, is two or three times larger than that of other mammals of similar size, although somewhat smaller in humans than in other primates. The visual cortex of primates also evolved a more complex architecture of visual areas that, together with inputs from the auditory and somatosensory systems, were integrated to help localize the body surrounding objects in the environment. Meanwhile, the posterior parietal cortex evolved to support planning and articulation of movements.[20] The human eye is not a perfect optical instrument, yet the optics of the eye transmit images that are as good as they need to be to match the neural coding in the early visual system.[21, 22]

It has been argued that visual system evolution developed in special relation to cues in the natural environment—"natural scene statistics."[23] Recent work suggests that the neural responses to features such as luminance or contrast do a good job of spanning the range of those features, including the statistics of light and dark patterns in the visual world.[23–26] The three human cone receptor types are furthermore thought to provide a very good, though not perfect, representation of natural variations in color.[27–30] Representations of visual inputs from natural scenes may enable a sparse cortical representation that concentrates responses in fewer neurons in order to conserve energy in the neural system.[31–33] While carrying out higher-level perceptual tasks, humans can still make perceptual inferences that are quite sensitive to the natural scene statistics and to the response properties of the visual neurons.[23]

Over all these examples, it is clear that, to a remarkable degree, the human visual system, along with the visual systems of primates more generally, has evolved over millennia to optimize visual function and the ability to act in the environment.

10.2.2 Visual Development

The native visual processes of humans and other primates undergo further significant morphological and functional changes during the first months and years of life, which are associated with improved ability to detect and categorize visual cues. Although newborns respond to changes in light and color and to moving objects almost immediately, these abilities are considerably less sensitive than in adults. The eyes, and the muscles that control them, are immature and improve with growth and development. The neural systems of the retina in the eye and the neural circuitry of the visual system also experience rapid change.

The majority of postnatal visual development, however, is thought to reflect development of functions carried out in the visual cortex. Certain cortical functions show significant development over the first few months of life, while others improve through early childhood, and some develop well into adolescence. What follows is meant to be only an approximate snapshot of the fantastically articulated postnatal visual development measured behaviorally in humans and other primates (for more details, see Boothe, Dobson, and Teller[34]).

One limiting factor in newborn vision is directly related to the reduced information received at the eye relative to that in adults—reduced light because of the small pupil size and reduced sensing because of the smaller cone density. Pupil size and regulation, as well as cone density, improve during the early stages of postnatal development. Another important factor is muscle control of the two eyes, helping to focus an image on the retinas and coordinate their position together to look at target locations. These improve in the first two or three months after birth. In particular, the coordination between the two eyes is necessary for fixation on relevant stimuli, for the encoding of stereoscopic information, and thus for perception of depth in the near field.[34–36]

Despite these physical changes, some researchers have estimated that postnatal changes within the eye account for no more than 25% of the improvement in visual function during early development.[37–39] The remaining 75% has been attributed to changes in the visual cortex and its afferent connections. Early neonatal cortical development exhibits a spurt of rapid and consequential growth and reorganization of the visual cortex during the last third of the first year. Although connections from the retina to the LGN, along with the volume of the primary visual cortex, are established shortly after birth, the synaptic density changes significantly throughout the first year, with development in connectivity along this pathway likely the most important factor in spatial and temporal visual acuity.[40–42] The development of long-range cortical interactions continues well into adolescence. These longer-range interactions help to support the perception of patterns or textures that integrate information over larger regions of the visual field.

The research on visual performance in infants and very young children has been transformed since the 1970s, with the measurements of preferential looking and visual evoked potentials.[36, 38] Functions such as orientation selectivity, spatial-frequency selectivity, motion-direction sensitivity, visual acuity, stereo acuity, segmentation of different visual textures, and a set of other tasks requiring integration over larger visual regions have all been evaluated in children at different stages of development. What has emerged from this research enterprise is a fascinating and complex set of stages in developmental visual function.[38] A few of these are illustrated in figure 10.1, based on a selective sampling of estimates from the literature.

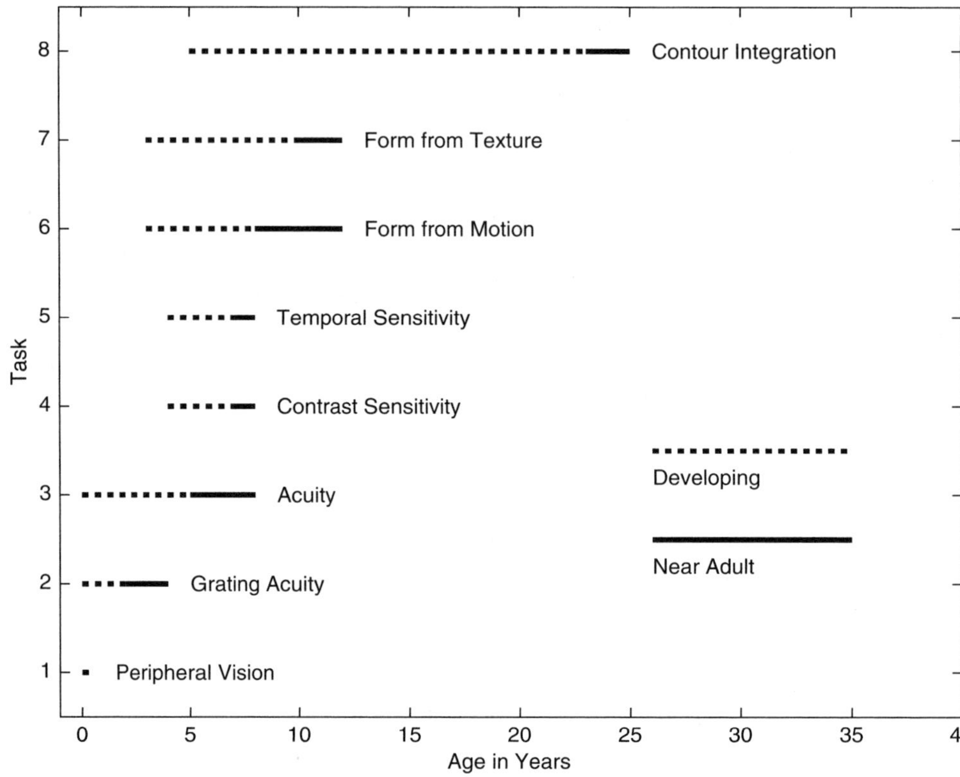

Figure 10.1
Developmental age ranges of some visual functions estimated from the literature. Dotted lines are periods of change; solid lines indicate near-adult performance. Approximate learning periods are estimated from the literature; see the text.

The approximate sequence of visual function development, estimated from behavioral or electrophysiological measures, shows maturity in basic features early in development (figure 10.1). Signals related to orientation (i.e., left or right diagonal stripes) and spatial frequency (i.e., fine versus coarse patterns) have been seen in the visual evoked potentials of infants by three to six weeks.[43, 44] Signals related to motion direction have been estimated to develop after 10 to 12 weeks,[44–46] binocular interactions at around three to four months,[47–49] and depth disparity from the stereo cues between the two eyes at between 11 and 18 months.[47, 49–53] Visual acuity for fine detail in patterns took until five to seven years of age to fully develop,[54, 55] while contrast sensitivity seemed to approach adult levels by about 7 years of age.[42]

The ability to process rapidly changing visual inputs was shown to increase to nearly adult levels by four years of age, with improvements in processing of slower-changing inputs continuing until about age seven. Sensitivity to motion and texture also appears in early childhood.[56] Motion-direction sensitivity and orientation sensitivity, the building blocks of shape perception, have been seen fairly early, together with figure-ground segregation and shape identification. More complex motion stimuli that require integration of global motion patterns from motion of multiple elements[57, 58] seemed to approach maturity slightly earlier (at seven to eight years) than forms defined by orientation textures (11–12 years).

The development of higher-level visual functions, especially those involving longer-range interactions, has been measured into early adulthood. One example task measuring this, the perception of patterned contours integrated from local orientation elements (e.g., long-range contour integration), has shown that development continues through late adolescence.[59]

These remarkable examples of visual development suggest a cascade of functions. Some may reach adult levels very early on, while others take a few years, and still others continue considerably longer. During certain critical periods, it is important that these developments occur in the presence of natural visual experience; they can be abnormal if visual inputs are of poor quality or experience is limited.[34, 38] The brain seems to be uniquely plastic during these early periods, with a range of medical implications. Amblyopia, for example, can develop in children suffering from cataracts during early critical periods.[60] Work using animal models is beginning to illuminate the complex interactions of neural circuits and molecular mechanisms that regulate the impact of experience on brain systems during the critical periods of early development[61] and that may interact with other developmental processes for different visual functions.

10.2.3 Adaptation

Evolution and early individual development encompass long-lasting changes that presumably determine the stable adult state of the observer. Adaptation, by contrast, is a form of plasticity that can be quite transient, though in some cases it may last for a longer interval. It includes responses to light level[62] but also visual features such as orientation, color, or pattern.[63–65] Adaptation alters the system such that the visual response depends not only on the current stimulus but also on recent experiences with similar stimuli.[65]

Adaptation is thought to have a functional goal of enhancing responses to novel features while diminishing responses to repeated stimuli. Figure 10.2 shows some estimated effects of adaptation on hypothetical responses to a stimulus before and after adaptation—in this instance, for adaptation caused by long exposure sequences. Adaptation may also be useful in maintaining constancy of what is perceived when there are significant fluctuations in illumination, thereby maintaining equilibrium in the system. Longer-term adaptation may also be important for maintaining perceptual constancy as the eye changes because of aging.

The phenomenon of adaptation, along with its associated illusory aftereffects, has been studied ever since initial reports of the waterfall illusion in 1834.[66] In this illusion, stationary rocks on the side of a waterfall appear, after about a minute, to be moving upward in a so-called motion aftereffect. Similar effects have been reported at essentially every level of the visual system, from the retina to the highest level of the visual cortex.

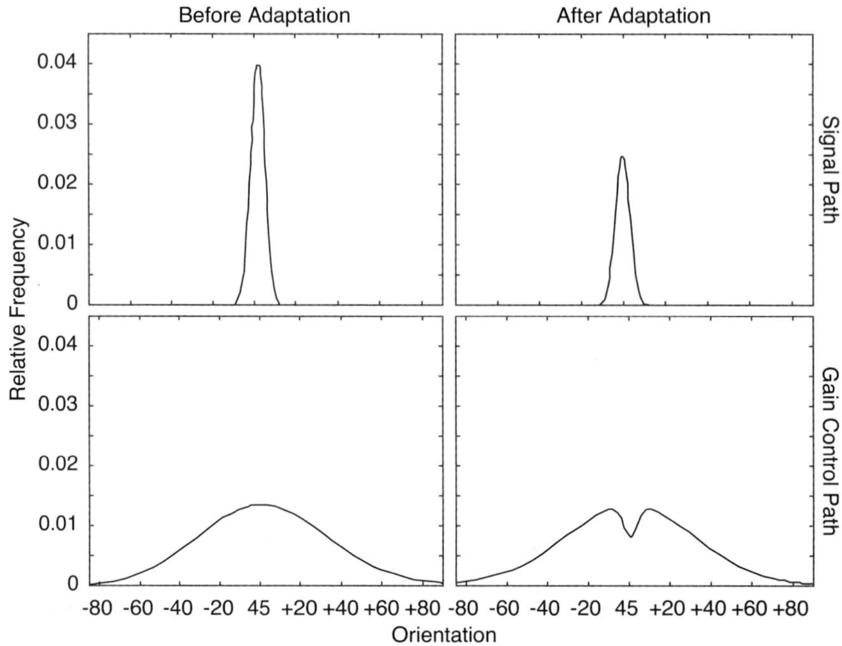

Figure 10.2
Responses to signal and noises after adaptation to a stimulus with orientation of 45°. Responses near the adapter are reduced in both the signal path and the broader gain-control normalization pool. After Dao, Lu, and Dosher,[68] figure 9.

One of the most important functions of the retina is to adjust to the ambient light level. This adaptation seems to adjust the sensitivity of the visual system to allow the perception in a dark environment of weak light that otherwise would be very difficult to see. Prolonged exposure to a colored stimulus affects the subsequent perception of color, and prolonged exposure to an orientation stimulus affects the subsequent perception of orientation.[67, 68] In most of these classic adaptation demonstrations, perception is shifted away from the adapting stimulus in "repulsion" aftereffects. For example, the McCollough effect shifts perception away from a red adapter toward a green percept. Repulsion aftereffects have been observed following adaptation to a single color,[69] to a single spatial frequency,[63] and for many other features.[70–73] In addition to biasing the percept, adaptation can impact the ability to discriminate or classify the stimuli near the adapter.[64, 74]

Although the primary focus of research in this area has been on adaptation that occurs over very short intervals (seconds to minutes), some adaptation persists for weeks or even longer.[75, 76] Very long-term exposure to environments, such as seasonal shifts in natural environments or the use of colored contact lenses, may result in this longer-term adaptation.[77] There are other examples, too. Curved or tilted lines, for example, have been shown to appear straightened and more vertical or horizontal depending on the long-term context.[78, 79] Such shifts in perception following adaptation, often to a broad range of stimuli, have been called "renormalization"—the creation of a new normal. This kind of process may serve as a stability mechanism, calibrating current responses in relation to current environmental statistics. (An example of hypothetical changes in color sensitivity resulting from long-term exposure to distinct distributions of color in lush and arid environments is illustrated in figure 10.3.)

Aging of the visual system also produces systematic long-term changes that may require long-term adaptation to support perception. For example, there is a shift of pigment density in the lens with aging. Renormalization through adaptation may compensate for the "yellowing" of color information, and in clinical applications, such as following cataract surgery, perception may take somewhat longer to settle into the new normal.[80] Although some age-related changes in the visual system cannot be stabilized and may result in functional declines, mechanisms such as adaptation play an important compensatory role.

Figure 10.3
Simulations illustrate perceived shifts in color appearance following adaptation to the color distributions in lush or arid environments. After Webster,[73] figure 2, with permission. (See plate 9.)

It is truly remarkable how adaptation and development seem to work together to stabilize the response of the visual system through all these changes. Adaptation may have a role at every stage of visual processing: it may cause both short- and long-term effects on perception;[65] serve as a mechanism to normalize the system to changing environments; and make the system more sensitive to novel information.[77]

10.2.4 Discussion

Plasticity occurs at multiple timescales. In this abbreviated synopsis, we have sketched several features of plasticity occurring during evolution, development, and adaptation. Taken together, these different scales of plasticity relate to at least three principles of visual perceptual learning: the plasticity of perceptual systems; the occurrence of plasticity within a whole-brain context; and the importance of balancing plasticity with stability, or the virtue of homeostasis.

Examining visual perceptual learning in relation to other forms of plasticity raises a number of compelling questions about how these different forms of plasticity might interact: In what ways has evolution constrained or guided the properties of visual perceptual learning?[81] Does effective perceptual learning contribute to evolutionary selection?[82] To the degree that there is more plasticity during early development, can perceptual learning advance perceptual competencies in children or be used to treat or ameliorate developmental delays or diseases?[83] Are there interventions that reopen plasticity or reopen the critical period?[84–86] Can such interventions be combined with perceptual learning to enhance treatment outcomes? Are there interactions between adaptation and visual perceptual learning? Does adaptation enhance or reduce perceptual learning and generalization, as proposed in learning texture tasks?[7, 87] If so, should adaptation be controlled in learning protocols in order to optimize visual perceptual learning?

Such questions move across distinct timescales within the human life span. At present, there have been relatively few studies, and many of these questions remain open. Further research in this area may suggest important extensions or elaborations to existing models of learning.

10.3 Learning in other Sensory Modalities

Although our focus in this book is on learning in vision, perceptual learning occurs widely across all sensory modalities, including hearing, touch, taste, and smell. It may also be multisensory. In what follows, we highlight some examples of perceptual learning in these different modalities with an eye toward identifying correspondence between principles and findings. Our purpose is to cross-fertilize methods and models and to extract common principles and conclusions over the different domains. Auditory learning in humans has perhaps the most extensive literature of the nonvisual modalities. It also has the most parallels to visual perceptual learning, so our analysis begins there.

10.3.1 Auditory Perceptual Learning

Just as in the visual domain, perceptual learning occurs for many auditory stimuli and tasks. It occurs for features of simple stimuli; for intermediate-level stimuli, such as synthesized sine-wave complexes; and for naturalistic stimuli, such as speech. Given the complexity of the auditory domain, learning in audition, especially in relation to speech, is a massive area of research, with a correspondingly extensive number of books and papers.

In what follows, we sketch a sample of these findings. To parallel the organization of visual perceptual learning in chapters 2 and 3, we consider the evidence for auditory training effects in basic, mid-level, and high-level auditory stimuli and judgments. Next, we examine the specificity and transfer of these learning effects, including the different emphases on generalization that have been common in auditory learning. We then examine the mechanisms of learning, including experiments focused on measuring internal-noise reduction and the use of reweighting models analogous to models that we and others have developed in the visual domain. Finally, we consider reports of physiological changes as early as the primary auditory cortex (A1) as possible substrates for plasticity in auditory perceptual learning, where the evidence has suggested significant though task-dependent retuning.

As in vision, perceptual learning occurs for many *basic auditory judgments*. It has been reported in judgments related to spectral properties such as frequency or harmonics; temporal properties such as duration of a sound or asynchrony between two sounds; and localization based on relative intensity or relative phase between signals to the two ears. In the relevant experiments, these judgments have generally required that the observer discriminate between stimuli presented in different temporal intervals. In a typical example, an observer might be asked to judge which interval contained the higher-frequency tone or which of three tones in sequential intervals was different from the two others. Indeed, two-interval and three-interval discrimination tasks using very brief auditory stimuli have been among the most commonly used in audition and therefore, naturally, also in auditory perceptual learning.

One significant early study used a frequency-discrimination task. Researchers presented tone pairs in each of two intervals and asked observers to choose the interval that contained the tone pair in which the higher-frequency tone was in the second position.[88] The relative discrimination-threshold difference was reduced substantially with practice (see figure 10.4), with substantial differences between a pretest and a posttest (the $\Delta f_n/\Delta f_2$ was about 0.4 for a 200 Hz standard tone, or a pretest threshold about 2.6 times that at posttest). Training with several different tone standards below 6,000 Hz also produced learning at 200 Hz, indicating some generalization (figure 10.5). Other studies have shown similar learning effects. Difference thresholds improved by more than 50% after training with standard tones of 5 Hz and 8 Hz in a three-interval task[89] and 1,000 Hz in a two-interval task.[90] The improvements in these basic tasks were sufficiently large that behaviorally they could be quite significant. As in visual perceptual learning, auditory learning occurred over many hundreds of trials, though early rapid learning has also sometimes been observed.[91] Indeed, one report even showed robust learned improvements from a pretest to a posttest in a three-interval paradigm where observers picked the odd frequency even when all tones were essentially identical during practice—perhaps because something was being learned about the standard tone.[92]

Learning can also focus on temporal features, including training of temporal-interval discrimination, sometimes called "auditory interval discrimination." In temporal-interval tasks, one interval contains a tone of a standard duration t; the other contains a tone that is slightly longer (or shorter), or $t+\Delta t$. and the observer chooses the interval with the longer (or shorter) sound. Training auditory interval discrimination similarly reduced interval thresholds by 50% or more, as shown for a 100 ms standard interval and a 1 kHz (1,000 Hz) tone (figure 10.4).[93]

Learning has also been reported in tasks requiring the localization of sounds in space. This can be based on intensity differences between the two ears ("interaural level differences") or time differences between when the sounds are heard in the two ears ("interaural time differences"),[94, 95] although learning has been reported to be more robust for level differences.[96] Learning of other complex discriminations based on amplitude modulation of (usually sinusoidal) carrier tones has also been reported, though learning appeared to occur more slowly.[95]

Auditory perceptual learning has also been reported for *mid-level tasks* using stimuli that are more complex (but still less complex than speech). For example, identification of some emergent properties of tone complexes, such as the fundamental frequency, has in some cases been reported as susceptible to training. Here, researchers have distinguished

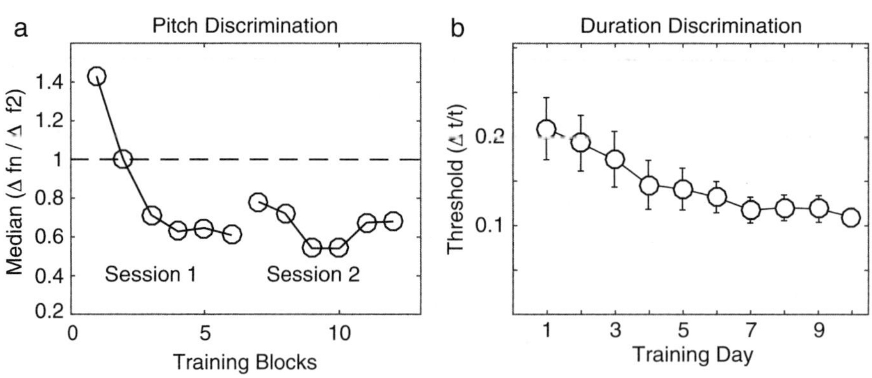

Figure 10.4
Examples of auditory perceptual learning for (a) auditory pitch frequency discrimination thresholds (normalized to 1 at block 2) and (b) for temporal-interval or duration discrimination. Redrawn from data in Demany,[88] figure 1, and Wright et al.,[93] figure 2b, with permission.

between learning in different perceptual regimes: those in which the fundamental and the harmonics are in the same frequency channels in the cochlea or those in which they are not. If the fundamental and the harmonics were widely separated in early channels, the tasks were labeled "resolvable," while if they shared early channels they were labeled "unresolvable." Training was shown to improve difference thresholds for resolvable stimuli but to improve only weakly or not at all for stimuli requiring the discrimination of unresolvable harmonics.[97]

The most widely studied cases of auditory perceptual learning for *high-level tasks* and *stimuli* have involved speech. Because of its obvious practical importance and its relevance to philosophical questions about the etiology of language in humans, speech recognition has long been among the most studied of auditory tasks. One set of learning research focused on improvements in identification of new or nonstandard forms of speech, such as phonetic contrasts in a foreign language, unknown accents or dialects, or degraded speech. Another set examined shifts in classification that are sensitive to local experience, as reviewed elsewhere.[98]

A few sample studies can stand in for the larger field, giving a flavor of cross-language speech training, with clear practical implications. Several experiments have trained phonetic contrasts that were typical of a second language but not used in the native language of the speaker. In one, the difficulty in discriminating the sounds /r/ and /l/ for native speakers of Japanese was measured along with different forms of training. Japanese speakers studying at an English-speaking university were trained on minimal word pairs with an /r/ and /l/ contrast in the initial position, the final position, and other varied contexts (e.g., *light-right*, *collect-correct*, *real-rear*). Training with tokens (samples) from several talkers led to reasonable generalization to new examples on a minimal-pairs test, although the learning effect itself was modest (only increasing performance from 80% correct to 86% correct),[99] while training with tokens from a single talker generalized primarily to other words spoken by the same talker.[100] These training effects may have been smaller than expected, reflecting the fact that the listeners already had significant training in English. Similar training of Japanese listeners untrained in English led to larger gains.[101] As in the visual domain, trained benefits were still available months later.[101] Training that improved speech production of /r/ and /l/ also seemed to have practical benefits.[102] Similar results were extended to other phonetic distinctions and to tonality in other language contrasts.[103] Training on accented speech, measured by the transcription accuracy for sentences spoken by native talkers of another language (e.g., native Chinese hearing English) led to modest, perhaps 10%, improvements, which generalized to new speakers only if the training included several talkers.[104] Although some studies used significant amounts of practice, others have reported significant improvements even with exposures of only a few minutes.[105]

In a separate subdomain, applicable in distinct operational contexts, training has been shown to improve the identification of speech in noise, compressed speech, and spectrally reduced speech. Spectrally reduced speech is of special interest because of its relationship to the signals used in cochlear implants, where performance with the implants often improves with experience.[106] In other examples, training in transcription of the last word in low- and high-predictability sentences or identification of the gender of the talker in auditory noise were reported as modestly improved in posttraining performance.[106] Experience was also shown to improve the discrimination of simplified (spectrally reduced) natural environmental sounds (e.g., mechanical, aerodynamic, or bodily sounds) and spectrally reduced speech, which are challenging for individuals with cochlear implants.[107] Furthermore, special training improved the perception of "noise-vocoded" speech, a form of sound transformation used in some cochlear implants that translates the energy in a spoken sentence into a corresponding amplitude profile pattern on an auditory noise carrier. In this case, learning seemed to require top-down knowledge of the content of the sentences during training.[108]

As in visual perceptual learning, the *specificity* and *generalizability* of auditory perceptual learning may diagnose the mechanisms involved. Training on auditory tasks often yields some mixture of specificity and transfer (generalization), as summarized in a recent review.[109] Without diving into details, this process seems to parallel the mixture of specificity and generalizability found in the training of visual tasks, though the relative prominence of specificity and transfer does seem to differ in the two modalities.

A few examples will make the comparison with vision more concrete. Frequency-discrimination training, for example, often showed relatively high generalizability over training frequencies, although there was also some residual specificity even for similar training frequencies. One study (figure 10.5a) measured improvements in frequency discrimination in a two-interval four-tone task for a 200 Hz standard following training at different standard frequencies and found substantial and similar benefits from training at 200 Hz, 360 Hz, and

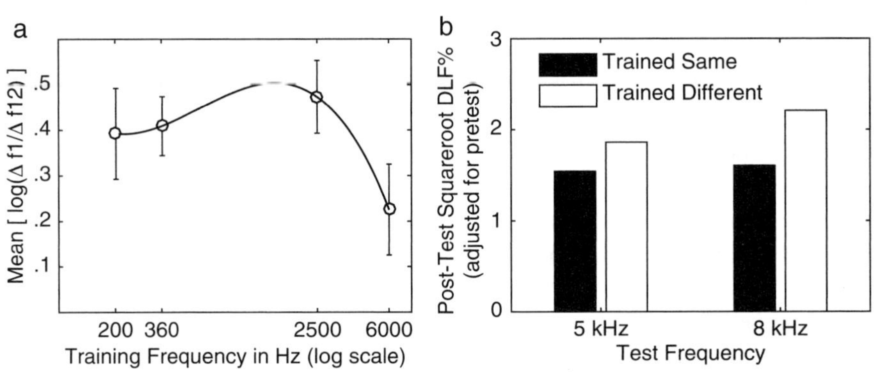

Figure 10.5
Generalizability (a) and specificity (b) of learning in auditory frequency discrimination. After Demany,[88] figure 3, and Irvine et al.,[89] figure 2c.

even 2,600 Hz but significantly less from training at 6,000 Hz.[88] In other examples, training on either 5 Hz or 8 Hz in an odd-tone three-interval paradigm showed larger performance improvements when the training stimuli matched the posttest, although cross training was also substantial (figure 10.5b);[89] findings in other studies were similar.[110, 111]

Frequency-discrimination training has often been reported to generalize substantially across ears[109, 111] but also to be at least partially specific to the training duration.[110] In the case of temporal-interval discrimination, training benefits have also tended to generalize over some orthogonal dimensions such as tone frequency but not to other interval durations.[93, 109, 112] Training at a 100 ms base duration, for example, generalized from a 1 kHz to a 4 kHz carrier but not to durations of 50 ms or 200 ms.[93] In a curious phenomenon, temporal-interval discrimination has sometimes even been shown to transfer from the auditory domain to other sensory domains.[109] Intermixing training on different temporal standards can be used to promote more generalization if learning occurs despite the uncertainty or roving inherent in using different standards.[112]

Several tentative conclusions can be drawn from these observations. Both visual and auditory perceptual learning have resulted in a combination of specificity and generalizability of that learning. If one were to focus on differences, it might be speculated that generalizability is more characteristic in the auditory domain, while visual learning tends to be more specific, a difference for which there may be procedural as well as structural reasons. Many of the auditory assessments, for example, used designs that interposed training between a pretest and a posttest, including cases in which the tasks showed relatively rapid initial learning. (Controlled analyses have argued that this rapid auditory learning represents perceptual rather than procedural learning.[91]) If a task does have a rapid-learning component, then improvements from a pretest to a posttest might have occurred even without intervening training, thus contributing to potential overestimation of generalization in some studies (alternative design approaches to measuring specificity appear in section 3.8 of chapter 3).

Despite these possible differences between modalities, and the corollary distinctions regarding task procedures, it should be emphasized that the primary theoretical and methodological issues of learning show broad parallels across the visual and auditory domains. In addition to such homologies, auditory learning exhibits further phenomena that have been observed in the visual domain, including *stimulus uncertainty* and *roving effects*.

In one study, frequency discrimination was easily learned when training with a fixed standard tone, but when the standard tone was varied modestly or roved, the listener learned much more slowly. As with the visual analog, learning was released for widely separated standards in the case of better listeners.[90] Training with roved standards also led to somewhat more transfer, while training with a single standard tone led to more specificity, and roving of the standard damaged both learning and transfer in poor listeners.

Another recent paper reviewed the now considerable literature on target uncertainty and roving on perceptual learning of speech and of nonspeech sounds.[113] These findings in auditory tasks are very similar to those in visual tasks. In the case of roving and visual training, reweighting models have been developed to account for quite similar results of roving experiments in vision (see chapter 8). Indeed, this might inspire homologous models of transfer in auditory perceptual learning (though this would of course require an appropriate representation module and decision structure for the relevant auditory tasks).

Alongside these functional similarities, vision and audition seem to share similar *mechanisms* for learning, with recent analyses of auditory learning using external-noise manipulations and models that closely parallel the external-noise manipulations and the perceptual template model developed for vision (see chapter 4). In fact, the use of external noise and noise carriers has a much longer history in the study of audition than in vision,[114] with the use of external noise to specify observer models first originating in the auditory domain.[115] All these methods were designed to model the observer, especially with respect to limitations resulting from internal and external noises, though using external-noise methods to understand the *mechanisms* of observer *change* (e.g., resulting from attention or learning) were first developed in visual applications.

As in the visual domain, the analysis of noise can also be related to physiological responses. Physiological studies have analyzed sources of internal noise (sometimes called intrinsic noise) in the ascending processing pathways of the auditory system, including stochastic processes in the transduction at the hair cells, neural encoding and transmission in the periphery, and more central noise; other studies have also focused on top-down processes that may alter neural responses or even the muscle engagement in the ear. A recent paper that reviewed some of these data[116] focused on the potential role of internal noise in particular as a limiting factor in auditory performance. This was concretely illustrated in an experiment showing that the listener's selection of one of three intervals as the loudest when all three intervals played identical stimuli (without external noise) was directly related to fluctuations in the EEG responses in the three intervals.[117]

In the external-noise studies, like those in vision, perceptual learning has often been shown to improve performance both for stimuli with external noise and for those without it (in external and internal noise regimes, respectively). In one study, for example, listeners' ability to discriminate a tone followed by a backward noise mask from a noise mask alone improved with training in a variety of tasks.[118] Performance also improved in the absence of external noise. Another study, using a different form of external noise, showed that training improved frequency discrimination when listeners chose which of two intervals had the highest tone.[119] (In this experiment, external noise was manipulated by selecting frequencies in different trials from two frequency distributions with different levels of overlap.) Using a model of the observer, the study's authors concluded that learning in these tasks corresponded with decreases in internal noise. In our analysis of the mechanisms of (visual) learning, this would be labeled stimulus enhancement (internal-additive-noise reduction) in the perceptual template model (see chapter 4). Different kinds of studies would be required to reveal the precise mechanism of external-noise exclusion in auditory learning. Because of this, the association of learning with internal-noise reduction should be seen only as documenting one mechanism, not ruling out the other.

Several other studies in auditory learning were directly inspired by the transfer asymmetries that were observed in visual learning.[120] One analogous auditory study found asymmetric transfer of training between auditory frequency discrimination of short and long tones:[121] short-tone training transferred to long-tone discrimination but not the reverse. The theoretical interpretation of these results was that long-tone training promoted averaging over internal noise during the long duration of the tone, a strategy unavailable for short tones, while short-tone training reduced phase-locked internal noise that could be used to improve both conditions. In short, the results in auditory learning were analogous to those of visual perceptual learning in zero or high noise, with explanations that are adapted to the nature of the stimuli and the task.

All this collected evidence has led several researchers to propose a *reweighting model* for perceptual learning in audition (figure 10.6). This model is analogous to the reweighting models of perceptual learning in vision and has been tested with analogous experiments.[116, 122, 123] In this model, the stimulus combines an auditory signal and external auditory noise; the input is analyzed by a set of auditory channels that code for task-relevant features of the stimuli, to which internal noise is added; and then the resulting activities in these channels are weighted and combined with decision noise, and the decision drives the behavioral response. Finally, learning may alter the weights of the model (or the priors), proceeding from a general framework that is essentially isomorphic to visual observer models (see figure 6.5).

Of course, the representation module—the auditory channel analysis—will be specialized for any given auditory task. In the detection of tones, for instance, the channels would likely be a bank of auditory band-pass filters tuned to different frequency bands from high to low (base to apex of the cochlea), as in classic models of cochlear transduction.[124] For interval detection, the channels would likely be tuned to different temporal periods or counts. The associated experimental studies of these models examined learning of frequency discrimination only in the absence of external noise, following from the inference that "neural-network simulations ... suggested that noise reduction was achieved through reweighting the frequency specific channels affecting early sensory representations ... consistent with conclusions from learning of visual tasks" (p. 71).[116] In addition, attention to an ear[125] or a frequency range[126] may also influence performance, possibly by enhancing the responses to the attended auditory inputs (see chapter 9). Here, too, had this model been tested with experiments that manipulated external noise, we expect that external noise exclusion would have also been improved by training.

Investigations of auditory learning using *physiological measures of plasticity* in the auditory cortex also have a long history in studies in rats, cats, and other animals. A few studies have used EEGs or fMRIs to examine auditory perceptual learning in humans. We

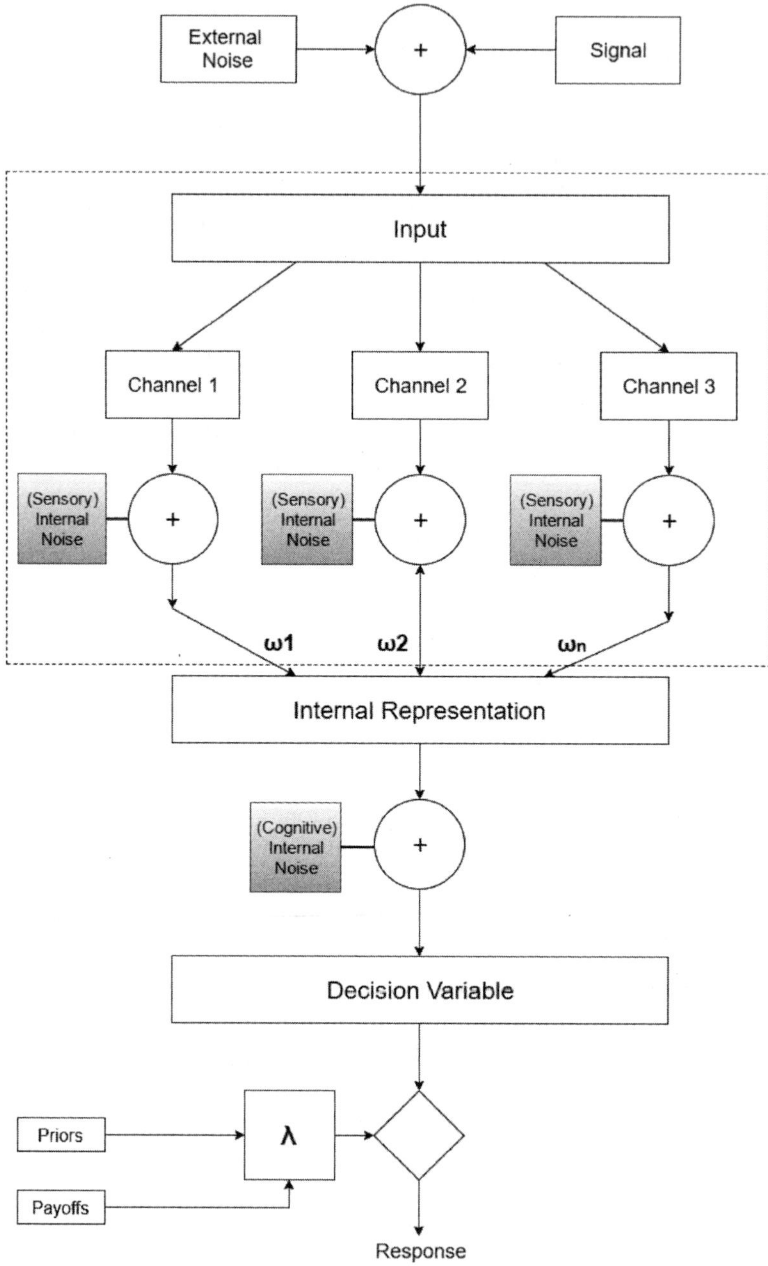

Figure 10.6
An outline model of auditory decision making and perceptual learning analogous to reweighting models of visual perceptual learning.[122, 123] After Amitay et al.,[116] figure 1, with permission.

consider only a few representative findings from animal studies here and provide a more complete review of the work in humans. To give an overview in advance, during active task performance, researchers have observed posttraining changes in auditory sensory responses, especially when using brain imaging in humans. This is similar to the visual domain, where the strongest effects of learning in the visual cortex were also found under active task conditions (see chapter 5); however, changes in the responses of A1 neurons (primary auditory cortex) seem to occur more widely than plastic changes in V1.

Let's begin with a sampling of some classic reports. Many of the first demonstrations of plasticity in the early sensory cortices were from auditory studies in rodents showing that training either increased or decreased the firing probability of neurons or produced a shift in the best frequency of neurons toward a reinforced frequency.[127–129] One influential study documented changes in the cortical area representing trained frequencies in the primary auditory cortex of adult owl monkeys and showed that these changes correlated with behavioral performance in a frequency discrimination task.[130] In this experiment's training task, a series of up to 12 tone pairs were presented on each trial, and the monkey detected the pair in which there was a small frequency difference from the standard tone. Training reduced the threshold frequency increment ($\Delta f/f$) from about 8% to less than 2% for some monkeys, and the false alarm rate changed somewhat (< 15%), while the slope of the psychometric function increased, corresponding with modest increases in d'.

The primary auditory cortex (A1) is organized in a tonotopic map in which different "bands" are sensitive to different auditory frequencies, and such maps have been an especially fertile index for assessing representation change resulting from learning in animals. In most studies, such as the one just described, training produced alterations in the tonotopic regions sensitive to the trained frequency measured during passive listening under anesthesia, and the relationship to frequency-discrimination thresholds was evaluated, as summarized in a recent review.[131] In a study that varied both the intensity and the frequency of stimuli, training frequency judgments led to expanded representations of trained frequencies without altering the coding for sound intensity, while training intensity discrimination altered the responses tuned to the trained intensity range without changing frequency coding, compared to untrained controls (see figure 10.7).[132] The changes in the cortical representation were then correlated with the amount of perceptual learning. This study concluded that plastic changes in the auditory cortex were driven top-down by the task-relevant dimension of judgment, resulting in a change to only those codes most relevant to a task (see chapter 9 for a discussion of related top-down effects in vision).

On the other hand, a separate study found that perceptual learning of frequency discrimination occurred *without* changes in the maps in A1 in anesthetized cats,[133] suggesting that behavioral improvements may also occur in the absence of persistent changes that early in the auditory cortex. These authors speculated that the correlation of changed maps in A1 with improved behavioral performance will likely depend on the task,[130] (though there may also be other explanations).[134] These mixed results concerning the engagement

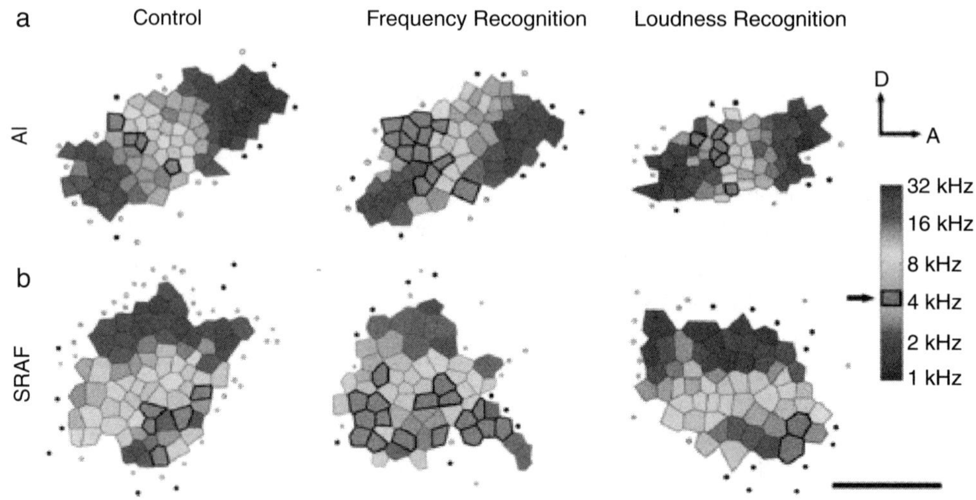

Figure 10.7
Plastic changes in tonotopic frequency maps in the primary auditory cortex A1 and secondary auditory cortex (SRAF) in rats. Increased representation of trained frequencies near 4 kHz is seen only in animals trained in a frequency task, even though the stimuli were heard in an intensity task. After Polley, Steinberg, and Merzenich,[132] figures 3a, b. Copyright (2006) Polley, Steinberg, and Merzenich.

of A1 plasticity in learning seem also to parallel the mixed results of the impact of learning on V1 (see chapter 5).

One provocative claim is that cortical plasticity observed during early stages of auditory learning is subject to "renormalization" later in learning.[135] This gives a name to the findings that the physiological changes in responses that emerged early in training disappeared later—returning to their pretraining characteristics—while the learning remained. There are several relevant studies that support this idea. One found induced plasticity in the tonotopic map using neural stimulation (of the cholinergic nucleus basalis) without a training task, with changes that were correlated with improvements in performance; however, the map returned to earlier patterns over several weeks, while the behavioral benefits of the stimulation remained. Another contradictory study found that tonotopic map changes induced by microstimulation without a training task failed to correlate with behavioral performance.[136]

The conclusion to draw might be that while representation modification in the early auditory cortex can occur, it is neither necessary nor sufficient to explain improved perceptual performance. This view was detailed in a review paper citing mixed effects of training on plasticity in A1 in animals, which concluded that "the general relevance of increased map representations for improved [performance] remains unclear"[137] (p. 471). Similar findings and statements appear in early influential studies of the relationship between motor maps altered by vibrotactile training.[138] Such a view seems also to be consistent with the conclusions drawn from reviews of brain imaging studies of auditory learning,[139] where responses to trained stimuli in the auditory cortex have shown increases in some studies and decreases in others. For example, one study of standard and oddball discrimination with positron-emission tomography (PET) showed higher activity in the auditory cortex following training,[140] while another fMRI study showed reduced responses over the auditory cortex after training.[141] Again, such mixed brain imaging results parallel similarly mixed results found in visual perceptual learning (see chapter 5). Apparently, the engagement of plasticity in early auditory cortical responses in learning may depend on the nature of the task, the phase of training in which the responses are measured, or even the species being trained.[9, 142]

Our purpose here is not to claim only uniformity across domains but rather to suggest that analyses linking physiology to behavior in any given study might be given context and perspective by related analyses in other modalities. Such analyses of the relationships between cortical changes and the corresponding behavioral changes resulting from perceptual learning in audition have often relied on correlation, while recent physiological studies of visual learning have increasingly used population response models or machine-learning decoders to relate physiology to behavioral choice. These model-based approaches may provide better estimates of this relationship in single-cell recording, multicellular recording, EEG, or fMRI to provide more substantial accounts of the behavioral improvements in performance resulting from learning. Alternatively, regardless of the sophistication of the methods used, it may also be that changes in the first cortical representations are only weakly or modestly predictive of the improvement, suggesting changes elsewhere in the brain network. (For example, changes in cortical responses early in learning might reflect top-down effects of attention, which subsequently disappear as attention is less necessary to performance after training.)

As we have seen, the parallels between auditory and visual perceptual learning are remarkable. Although the details of learning in each modality are unique, the overall similarities are substantial (see table 10.2). Both domains express learning of low-level features, mid-level patterns, and higher-level natural stimuli. Both modalities have likewise shown partial specificity and partial transfer of training (although auditory perceptual learning may show more generalization than in visual tasks). Both are susceptible to

Table 10.2
Empirical phenomena in both auditory and visual perceptual learning

Perceptual learning at low, intermediate, and high levels
Partial specificity and partial transfer
Reductions in learning resulting from stimulus roving or uncertainty
Improved limiting internal and external noise and observer model mechanisms
Plasticity in the sensory cortices and top-down modulation

disruption of learning by stimulus roving or uncertainty. Both associate learning with reductions of the effective impacts of internal and/or external noise. Both showed some changes in sensory representations (although changes in A1 may be stronger than in V1).

It seems likely that the differences between the two modalities, when they do occur, reflect intrinsic differences, either quantitative or qualitative, in the respective representations and processes of each. Yet it is also the case that the commonly used experimental tasks and techniques often differ and that such methodological differences may in fact produce the observed differences in results. What is clear regardless is that the two domains overwhelmingly share a number of principles and phenomena of perceptual learning. For this reason, increased interchange of techniques and models has the potential to enrich both domains of study.

10.3.2 Tactile Perceptual Learning

Another modality demonstrating learning and plasticity is touch. Sometimes called tactile or somatosensory learning, this domain was famously studied early in the history of psychology by using two-point discrimination tasks.[143] Somatosensory plasticity in animals was also among the earliest reported, and there is a correspondingly large amount of literature on somatosensory organization and its sensitivity to experience in animals. We touch on a few classic studies here, while focusing our analysis on the smaller number of studies focused on tactile discrimination in humans. In many ways, the results in human tactile learning parallel results found in vision and audition, and can be categorized correspondingly. Unlike the research in other modalities, however, the research on learning in humans has focused largely on the topography of generalization.

Let us first broach the extensive literature on the topographic representation of tactile stimuli in the cortex in animals, as well as the generalization of learning. Some of the earliest evidence of learned plasticity involved training of whisker sensations in rats. Pressure on the whiskers in this species is reflected in responses in a topographically organized representation in the primary somatosensory cortex, known as the whisker cortex (or the "barrel cortex"), with regions organized similarly to cortical columns in other modalities. These representations seemed to underlie generalization gradients in learning as a function of whisker position.[144, 145] The topographic organization and brain regions involved in any given task have also been found to depend on the particular form of tactile discrimination being tested.[146]

One of the first highly cited studies of the impact of experience in primates focused on plasticity of the hand representation in owl monkeys.[138] Topographical organization in an early somatosensory cortical area (area 3b) changed after the animal was trained to discriminate differences in tactile vibration frequency delivered to one finger (e.g., differences in a rapid pattern of pressure relative to a 20 Hz standard). Behavioral discrimination improved over many sessions, ending with thresholds of about 2.3 Hz in the trained finger compared with about 4.35 Hz in the adjacent finger and 6 Hz or more elsewhere. The neural response maps in the somatosensory cortex were also changed in several ways. Trained maps were more complex; representations for the trained skin location on the trained finger were 1.5–3 times larger in the cortical area compared to controls (although the total cortical region representing that finger was not larger); and receptive fields for neurons representing the trained region of the trained finger sometimes extended to the representations for the adjacent finger. Curiously, though, while training improved performance and changed cortical maps, the changes in the two were not highly correlated. Furthermore, the results suggested that if representation for a small part of the trained finger were traded off with representation for other parts of that finger (if this affects behavior), then these other regions should be disadvantaged, and any retuning may be transient. (Frequency discrimination is coded in the early somatosensory cortex, unlike some other vibrotactile tasks.)

Investigations of tactile learning in humans have emphasized patterns of generalization following training, with implications for the localization of plasticity. One classic study examined generalization across fingers after learning in three forms of tactile discrimination: vibration frequency, punctate pressure, and roughness.[147] In each regime, observers judged a stimulus as higher or lower along the task-relevant dimension (e.g., higher or lower frequency, pressure, or roughness). Starting with performance just above chance (62%–65% correct), observers were trained until they met a performance criterion (80%–85% correct). Learning was generally rapid, with training for hundreds, not thousands, of trials. The patterns of generalization differed, however: frequency discrimination was specific to the trained finger; punctate pressure discrimination generalized to the adjacent finger and partially to the corresponding fingers of the untrained hand; and roughness discrimination generalized to the adjacent finger, partially to the other fingers of the same hand, and almost completely to the homologous fingers of the untrained hand. Relating these results to known physiology led to two inferences. First, frequency discrimination seemed to use representations in the early somatosensory cortex, where almost all cells are sensitive to a single finger. Second, punctate pressure or roughness discrimination instead seemed to rely on information coded in the secondary somatosensory cortex, where receptive fields are often sensitive to more than one adjacent finger and have projections to the corresponding cortical region of the opposite hand. (These correspond to Brodman area 3b and area II; for details, see the review in Harris, Harris, and Diamond.[147])

As a result of this study and others, the dominant view has been that training changes the topological representations in the somatosensory cortex. An alternative view, however, and one that we favor, is that learning improved readout (reweighting) from these representations, or the top-down task reweighting temporarily altered these topological representations. This seems especially plausible given the relatively short duration of the training and the possibly transitory nature of the topological changes.

Related patterns of generalization were found after learning coarse tactile orientation discrimination at different spatial scales.[148] In this study, observers were trained with a set of brass domes embossed with horizontal or vertical line gratings of different scales (line widths and spacing), often used in testing the blind. Blindfolded seeing subjects were trained to discriminate using one index finger, and pre- and posttraining thresholds were measured for the trained (T), adjacent (A), and corresponding homologous (H) and other (O) fingers of the other hand. As with pressure and roughness in the previous experiment, learning to discriminate spatial orientation at different scales generalized to the adjacent and homologous fingers.[147] The common interpretation of these results, again, is that learning reflects changed cortical representations; contrary to this, we suggest that the data are also compatible with learned readout (reweighting) of information from those regions.

Consistent with the reweighting explanations, brain imaging of humans using fMRI has primarily found response changes in higher brain regions following training; as indicated, for example, in a tactile acuity task analogous to a three-point Vernier task in which behavioral thresholds changed from about 1.2 mm to less than 0.2 mm of offset.[149] Increased brain activations after training were found in the pre-supplementary motor area (pre-SMA), which is associated with the decision network, but not in the somatosensory cortices. The weights in a connectivity analysis from the somatosensory cortex and the frontal eye fields into decision regions also changed. The authors concluded that learning occurred through reweighting of perceptual readout from the (unchanged) sensory response cortices. (Note, however, that stimuli in the pre- and posttraining imaging sessions were chosen to approximately equate behavioral accuracy, a practice also used in some visual learning fMRI studies, with arguable advantages; see chapter 5.)

From these studies, several key points emerge. The early studies of tactile learning in animals reported changes in the somatosensory cortices that altered the topological representations, although the relationship between these changes and the behavior varied. Studies in humans reported different patterns of behavioral generalization within and between digits, which makes sense when mapped onto known properties of representations in the secondary somatosensory cortices. Only in vibrotactile frequency discrimination were learning effects limited to the trained finger, consistent with properties of the primary somatosensory cortex. The human fMRI study also found that learning was largely associated with changes in activation in decision areas that seemed related to changes in connectivity from the somatosensory cortex to decision. While the human behavioral studies might be consistent either with learned changes in the somatosensory cortices or with reweighting of evidence from these representations to decision, this fMRI study favored the latter. Obviously, more work will be required to draw definitive conclusions.

What is again clear is that there are many significant parallels between learning in the tactile domain and visual perceptual learning. The shared phenomena are striking, and methodological interchange promises to cross-fertilize the field. To our knowledge, for example, neither uncertainty manipulations nor external-noise manipulations have been carried out for the tactile domain. Likewise, models and methods that are more sophisticated, including connectivity analyses in fMRI, could be used to quantify how brain activation (or cellular responses in animals) accounts for behavior in tactile learning. Interdisciplinary dialogue could be beneficial. It is significant that the early physiological

reports about the impact of learning had been carried out in the tactile domain, and only later were pursued in the visual domain. Just as the subsequent study of visual and auditory learning had much to glean from the early study of touch, that might also work in reverse.

10.3.3 Perceptual Learning in Taste and Odor

Taste and odor (olfaction) also express strong forms of plasticity. The receptors in both systems are sensitive to the molecular structure of chemicals, with both highly responsive to food. Each modality operates both separately and together (although many argue that a huge part of what we think of as taste is actually mediated by smell).

The sensory systems corresponding to taste and odor have been studied extensively, especially in rodents.[150–152] Indeed, the physiology of both taste and odor in animals is associated with a vast body of literature that can be pointed to only briefly here. In humans, by contrast, there are only a few studies that consider learned improvements in the discrimination or detection of tastes or odors. As in previous modalities, these human studies will be the focus of our review.

Taste perception is key to food selection and the identification of toxic substances. Current doctrine says that mammals sense four basic kinds of tastes: salty, sour, sweet, and bitter (along with umami, a savory sensation related to glutamate). This domain reflects the sensing of differences in specific protein binding and ion specification in different taste receptors.[153] These sensory encodings are then represented in the primary olfactory and gustatory cortex and further processed in higher cortical regions. Taste and smell also converge and interact in privileged ways, each altering the perception of the other,[154–157] possibly through interactions coded in the orbital frontal cortex (OFC).[156]

Experience substantially affects taste judgments. There are many examples to choose from. A series of taste tests, for example, has been shown to improve the discrimination of amounts of glucose in subsets of glucose tasters.[158] Concentration thresholds for detection of monosodium glutamate (MSG) are known to be affected by both recent and long-term experience. (Interestingly, thresholds in American and European tasters were lower after 10 days of exposure to dietary MSG while threshold detection in Japanese tasters with long-term cultural exposure were even lower, and these short-term exposure effects disappeared when an MSG diet was discontinued.[159, 160]) Experience has also been shown to alter the detection of many other chemical substances, even very common ones, such as sugar.[161, 162] In one study, different chemicals that were tested repeatedly over many sessions required systematically lower concentrations in order to be judged as isointense with a standard, with the corresponding sensitivity functions appearing similar to learning curves in other sensory modalities. Correspondingly, increases in rated intensity were correlated with increased activation over sensory cortices as measured in fMRI.[163]

In each of these examples, taste perception was affected by experience. While some cases have been interpreted as a form of sensitization (because lower quantities were required for perception), others seem to reflect more traditional forms of learning. Unlike other forms of perceptual learning, however, these changes often show a rapid return to baseline sensitivity, standing in clear contrast to the long-lived persistence of perceptual learning in vision, audition, and touch.

Training and experience have also been shown to be important to the discrimination of more complex and everyday substances such as beers or wines, especially in so-called multisip protocols. (Although presented as studies in taste, odor often contributes to identification of food and drink, and the same studies are also sometimes cited in reviews of olfactory perceptual learning.) One early experiment contrasted the impact of different training protocols on the discrimination of two wine samples.[164] In this study, experience was reported to sometimes improve judgments of whether two wine samples were the same or different for certain wines, though less so for others; meanwhile, other forms of training seemed to have only small effects. In another study, beer novices received either tasting experience, instruction in labeling (beer tasters, like wine tasters, have developed a system of description and classification), or both.[165, 166] Training in any condition that exposed observers to the taste of the beers increased the similarity ratings and matching identification for identical beer samples; a related study showed increases in same/different discrimination after training with four white wines.[167] Although training effects in these experiments were mixed, especially for complex tastes, there were also many examples in which experience did improve discrimination.

Following principles similar to those in taste, olfactory sensations, or smells, occur when the *odorant molecules* of a volatile substance bind to receptors in the olfactory epithelium of the nasal cavity. Activity in these receptors is passed to the glomeruli and mitral cells of the olfactory bulb and then onward for further processing in the olfactory cortex and other areas.[168] Outputs from the olfactory bulb synapse on several regions, including the piriform cortex, which codes odors, and the amygdala and entorhinal cortex, which are related to affect and memory. The chemical structure of the odorant is encoded in the anterior piriform cortex, while the posterior piriform cortex is thought to be more involved in categorizing and discriminating odors. Information from these areas is then projected to the orbitofrontal cortex, which acts as the representational basis for the more complex perceptions of odor and in multisensory integration. (A second auxiliary olfactory reception system specialized to detect pheromones is thought to possibly still function in humans.)[169]

In one classic view, olfaction directly codes the chemical structure of odorants.[170] An alternative view is that the initial chemical coding of odorants is "not behaviorally/consciously accessible, but rather is a first necessary stage for subsequent cortical synthetic processing which in turn drives olfactory behavior" in which complex ensembles of chemical features are synthesized into odor "objects" and encoded in the piriform cortex based on experience.[168] This view is supported by the relative inability to recognize or discriminate individual components in a mixture of more than two odorants—even though almost all real odors are complex chemical combinations.[168, 171] (This seems to correspond exactly with the case of the experience-based creation of new neural representations for compound objects discussed in chapter 2.)

Familiarity with component odors has been shown to increase the ability to discriminate odor mixtures.[172] In one study, observers were trained to label seven unfamiliar intensity-matched odorants and then tested for same/different discrimination between two sniffs separated by a delay. Training on labels and to a lesser degree "profiling" (operationalized as rating odorants on a series of adjectives) improved same/different discrimination compared with learning to label another set of odorants or no training. Even mere exposure can change detection based on the complex percepts. In one study, just three familiarization exposures on different days more than doubled the discrimination (d' of about 4 compared to 1.6) and also altered the perceived similarity of odorants.[173] Another study, using pheromones, showed that exposure to pemenone reduced the threshold for androstenone in many observers, including some with very poor initial detection of this chemical.[174, 175]

A few imaging investigations have demonstrated effects of experience with odor on the cortical responses in humans. One fMRI study investigated behavioral intensity thresholds following exposure to an odorant and associated fMRI activity in the piriform cortex (thought to reflect odor quality and structure) and orbitofrontal cortex (thought to express perception of learned odors).[171] See figure 10.8 for some of the data. During a several-minute exposure to a target odorant, intensity ratings declined approximately exponentially. This exposure enhanced the fMRI response to a quality-related odorant but not to a functionally similar odorant from a different functional group or an unrelated odorant. In contrast, signals were enhanced in the left olfactory orbitofrontal cortex for both related odorants. Only the changes in fMRI signals in the orbitofrontal cortex correlated with discrimination or with the change between pre-exposure and postexposure ratings of similarity. These effects of exposure were classified as perceptual learning rather than habituation because they were still visible 24 hours after exposure. The authors concluded that "the magnitude

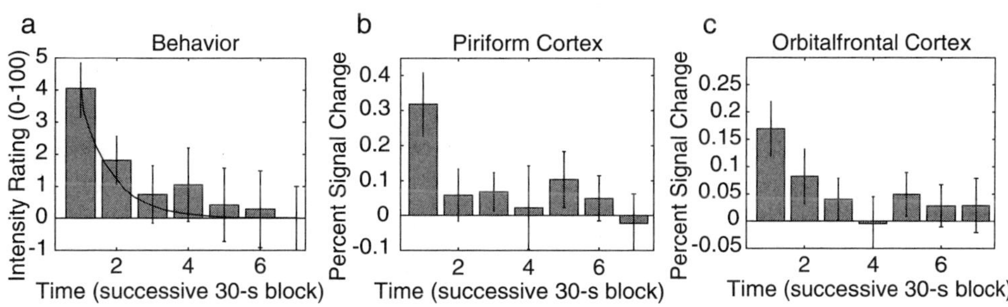

Figure 10.8
Exposure to an odor affects behavior and brain activity. (a) Perceived intensity ratings decreased during a several-minute exposure to an odor, while activity in the (b) piriform cortex and (c) orbitofrontal cortex also declined. Response changes in the orbitofrontal cortex were correlated with changes in behavior and rated discriminability of stimuli. From Li et al.,[171] parts of figures 3 and 5, with permission.

of learning-induced activation in OFC directly predicted the degree of perceptual enhancement on the similarity judgment task, ... suggesting a critical role of the olfactory OFC in perceptual learning"[171] (p. 1103).

This fMRI study supports the idea that odor perception reflects learned experience-dependent codes for odor *objects* in higher levels of cortex.[168] While not ruling out some experience-dependent changes in the lower-level representations coding the chemical properties of the stimuli, the results highlight the role of learning in creating responses in the higher cortex that are the basis of complex odor percepts and promote the ability to differentiate between similar odors. It seems that neural representations of odors are a dynamic product of lower-level coding in olfactory bulb and higher-level cortical inputs regulated by learning and experience.[171]

After the early cortical areas encode taste and odor approximately independently, the information streams from the two modalities *converge* in the orbitofrontal cortex to act jointly on neurons.[156, 176] These convergent inputs influence the representations of odors[154] and flavors.[156] In olfactory learning, only a few trials that pair a taste with an odor have been shown to change perception. For example, pairing sucrose with a tasteless odorant such as lychee increased ratings of sweetness.[155] In another example, pairing an odor and flavor caused the odor to be rated as sweeter if it was paired with sucrose and sourer if it was paired with citric acid.[154, 155] Although the influence of pairing a taste with an odor does not seem to depend on conscious awareness, the cross-influences were reduced or inhibited when attention to separate elements was required, emphasizing the potential role of top-down influences in perception.[157] Using fMRI and other imaging methods, a number of brain regions were discovered to jointly respond to both odor and taste, including the caudal orbitofrontal cortex, amygdala, and several other cortical areas.[156] Furthermore, the activation patterns in one part of the orbitofrontal cortex were correlated with consonance (agreement) ratings for smell/taste combinations and for rated pleasantness. Perhaps most striking, even pairing the stimulus with a visual image was shown to influence the responses to olfactory inputs.[177]

As these studies demonstrate, the perceptions of taste and odor, as well as their interactions, reflect prior experience and are coded in a variety of brain regions. The same is true for the visual, auditory, and tactile modalities. In taste and smell, however, mere exposure, as well as training, contingency, and semantic labeling, have been shown to induce perceptual alterations. Although some changes in the low-level representations of the chemical components of these stimuli cannot be ruled out, the most important influences seem to be hedonic qualities, with the discriminability of tastes and odors driven by so-called synthetic representations upstream of stimulus registration. Exposure, behavioral training, or attention can in turn influence these higher-order codes.

Perceptual learning of taste and smell in humans has properties consistent with several general learning principles. There are also unique differences from other modalities, however. Plasticity related to the lower-level representations seems to be quite rapid in some cases, occurring after relatively brief exposure or training periods, and in many cases may return to steady state relatively quickly—over the course of a few hours or a day or two. On the other hand, the experience-dependent perceptions of taste or odor *objects* at higher, synthetic levels, once learned, can be quite stable. Here, too, there are parallels. The development of codes for such objects or categories seems similar to the development of new representations for objects such as faces or objects in visual high-level perceptual learning, and while lower-level representations of the chemical properties of the stimuli may or may not change with experience, these more synthetic higher-order representations seem to dominate the conscious perceptions of humans. The higher-level representations can integrate information from odor and taste, and even visual and semantic cues in some cases, reflecting a higher-level convergence of multiple sources of information. The convergence or influence of multiple cues to these representations from more basic sensory evidence, itself influenced by top-down classifications, is yet another process that naturally emerges from reweighting. In these senses, the emergence of experience-dependent synthetic representations of complex sensory objects seems to dominate higher-level forms of sensitivity. The process by which these complex synthetic representations emerge seems more likely to reflect the creation of new representations to represent special combinations of features rather than simple selection of preexisting low-level representations (see chapter 2).

10.3.4 Multisensory Perceptual Learning

Perceptual learning always involves the interaction of brain networks, and some of these networks seem to be wired to incorporate information from multiple sensory modalities. A predator in the bush may be seen as well as heard. We taste and smell an orange simultaneously. Objects may be touched and seen at the same time.

Multisensory processing, or the processing of multiple senses together, has been an active area of research in recent decades. This research has generally focused on the degree to which inputs in two (or more) sensory modalities interact to generate a behavioral response.[178, 179] In what follows, we focus on questions concerning the impact of training on these multisensory effects and, conversely, the effects of multisensory experience on perceptual learning. Does perceptual training change the interactions of cues in multiple modalities? Can we improve the learning in one modality by adding other sensory cues during training? Does training in one modality transfer to analogous judgments in another? Can we use trained judgments in one sensory modality to substitute for information that normally is processed in another? Despite the long-standing research on multisensory effects, certain of these questions are only now beginning to be explored.

One idea in this research area has been that the interaction or integration of inputs from two sensory modalities depends on events occurring within a temporal window of integration. Auditory and visual stimuli, for example, would be bound together and perceived as part of the same audiovisual event if they occurred within close temporal proximity. In one study, perceptual learning significantly reduced the width of the integration window by about half (as measured by changes in the probability of judging an auditory tone and visual flash as simultaneous when the tone either led or followed the flash by some temporal lag).[180] In this case, perceptual learning led to more accurate labeling of lagged tones, while a control that merely exposed the stimuli did not affect the integration window.

Perceptual learning can also alter how cross-sensory attention cues operate. In one experiment, irrelevant training with visually misaligned but simultaneous auditory and visual stimuli was shown to modify the spatial calibration of auditory precuing in visual discrimination.[181] Performance was measured in five spatial locations to the left or right of fixation. Before training, an auditory precue whose source was spatially colocated with the central position resulted in better performance in that location, but extensive training with misaligned visual and sound cues changed the natural associations and shifted the benefits of valid precuing to a proximal location rather than the location coincident with the sound (while leaving late inhibitory effects of the auditory cues unchanged). A related study found that task-irrelevant learning from cross-modal cuing of an auditory task also led to task-irrelevant learning of visual motion direction.[182]

This relationship also worked in reverse: cross-sensory cues may speed perceptual learning.[183] For example, adding a congruent auditory motion cue during training improved learning in a low-coherence dot-motion detection task, even for tests of visual stimuli alone; meanwhile, incongruent auditory cues (created with different intensities in speakers to the right and left of the screen) did not affect learning.[184, 185] These findings were interpreted as a form of multisensory learning, but the benefit accrued to visual motion discrimination even in the absence of an auditory stimulus, so it seems unlikely that a specifically multisensory representation was trained. Alternatively, the congruent auditory cue may provide additional information, such as feedback during learning that was only effective when it was congruent with the visual motion direction.

Another line of research has focused on cross-modal *transfer* of training, which some have suggested may also reflect multimodal learning. In one study, separate groups of observers were trained in visual, auditory, or auditory and visual temporal-order judgments (TOJs), in which observers indicated which of two events (two auditory, two visual, or one of each) occurred first.[186] The threshold time separation between visual events was longer, but learning was faster, while the threshold separation between auditory events was much shorter, yet learning was slower. The only modest transfer was from visual training to an auditory and visual test condition (see figure 10.9). Similarly, training in auditory duration discrimination did not transfer to improved performance in visual duration discrimination.[187] By contrast, temporal-interval training or motor-interval training has been reported to transfer more generally,[188, 189] leading several researchers to conclude that more salient characteristics will be more likely to be generalized, and that timing of events may be one such salient aspect of real-world multisensory events.[190] An alternative interpretation, however, may simply be that training improves whatever is limiting performance, which for temporal properties are more limited for visual cues than for auditory ones.

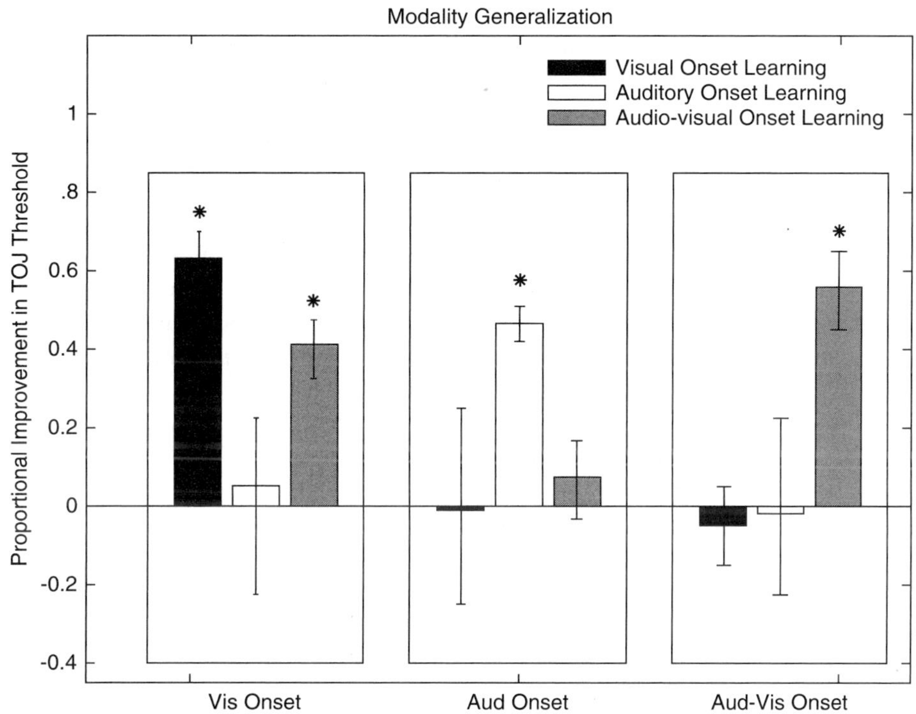

Figure 10.9
Trained improvements in temporal-order judgments in visual, auditory, and auditory and visual training conditions, and the transfer to other modalities. The only transfer of learning is from visual training to auditory and visual temporal-order judgments. After Alais and Cass,[186] figure 2. Creative Commons, copyright (2010) Alais and Cass.

The insights here may help to guide possible practical applications of learning, such as sensory substitution devices, which create representational proxies for sound or touch, allowing the blind to "see" through audition, the deaf to "hear" through vision, and so on.[190, 191] Training with visual-to-auditory substitution devices that code visual pixels into auditory signals of frequency or frequency and time have found some success, as outlined in a recent review.[192] Sighted individuals were able to associate auditory-coded patterns with related visual patterns varying in position, orientation, or size with minimal training, while even more training produced benefits in interpreting familiar stimuli.[192] One idea this research suggests is that many important functions, such as object recognition or classification, occur in brain regions that predominantly receive inputs from one modality but either already receive or could be trained to receive inputs from others.[190, 193] Indeed, one imaging study concluded that connections between inputs in the secondary modality and higher-level brain regions were strengthened during a training task.[193]

In summary, there seem to be a number of situations in which the brain routinely integrates inputs from multiple modalities. Inputs are combined to understand speech (the movement of the speaker's lips, facial and other gestures, and other cues).[194] Objects may be interpreted not only through sight but also through touch or hearing. The studies reviewed here used either multisensory stimuli or multisensory training protocols, leading to a common conclusion that perceptual learning often affected multisensory representations and/or their accessibility. It was difficult, however, to conclude definitively that this effect relied on truly multisensory processing or multisensory representations. There may be alternative explanations related to dual inputs to decision, associative learning, or secondary information such as feedback during learning. Notwithstanding these caveats, multisensory perceptual learning presents a fascinating research topic that deserves further investigation.

10.3.5 Summary

In this section, we analyzed the phenomena of learning in sensory modalities other than vision—audition, touch, smell, taste, and multimodal combinations—with an emphasis on human data. Many empirical phenomena of learning arose in all of them. Learning occurred in each modality, albeit at rather different rates. A mixture of specificity and generality occurred, depending on the stimuli and the task. The mechanisms of learning, measured with external-noise methods, have so far only been examined for visual and auditory domains. In the latter, the indicated changes in performance-limiting internal noise (and likely external noise) parallel the findings in vision. Similar analyses of auditory and visual forms of multimodal learning, and possibly even of tactile learning, may be possible, but these would seem far more complicated for the chemical senses, for a variety of reasons.

The physiological basis of plasticity and its relation to behavioral measures of learning have been examined in most modalities. In almost all of them, the patterns of learning implicated both low- and high-level sensory cortices in ways that depended systematically on the task. A number of studies drew attention to a significant and sometimes dominant role for attention, task context, and other top-down processes. The similarities go beyond these phenomenological parallels, however, to include potential crossover of methods and models.

Whenever judgment accuracy improves with training, regardless of modality, this—essentially by definition—must reflect experience-dependent improvements in the signal-to-noise ratio limiting those perceptual judgments. Regardless of the domain, formal models can play a critical role in testing ideas about the mechanisms of this learning. In addition, the models may provide a context for understanding physiology. We anticipate that a common theme emerging across modalities will be the role of reweighting in learning at multiple levels of representation. The specifics may differ, but the general idea of an improved readout of evidence from one representation level to the next will likely play a fundamental role in future models describing a number of these modalities.

In the history of the science of perceptual learning, some of the earliest studies arose in the context of somatosensory and auditory improvements. These seminal studies inspired and informed later studies in vision. They certainly led the way in the physiological analysis of underlying brain substrates. Nevertheless, there are notable differences between modalities. The receptors used to register the sensory inputs are obviously unique to each sense, while the organization of the representations and processing in the brain will have correspondingly unique systems for representing the sensory information. Beyond these straightforward differences, however, the preceding analysis of the literature has highlighted several other apparent differences in the hierarchical balance of plasticity (changes in low-level representations) versus stability (changes in connection to decision). The rate of learning seems to differ substantially between the different modalities as well, as does the specificity of what is learned. At present, there are still relatively few experimental studies whose data serve as the basis for these conclusions, however, and more for some modalities than for others. Even certain classic studies can give rise to debates on the level of explanation and interpretation.

In the near future, there should be opportunities to use paradigms, methods, and models developed in one modality to cross-fertilize research in others. In so doing, researchers may be able to identify additional general principles of learning or further demarcate their respective scopes of application. For instance, there are a number of phenomena that have been found in vision and audition that have yet to be studied in other modalities. These include (but are not limited to) the effects of task roving or task mixture on learning, the signal-and-noise mechanisms of learning, and the development of computational models of learning in certain modalities.

This research could conceivably also begin to shed light on the evolutionary developments of sensory learning in the brain. Did a certain functional circuit evolve first in one modality only to be transferred to another? Or did similar functional circuits emerge independently in each modality? Or, as a third option, did the need to coordinate and integrate modalities in perception constrain evolution of learning systems in each modality such that they came to resemble one another? Comparing learning in the different modalities with behavioral and physiological methods may contribute to fundamental inferences about the evolution of human sensory abilities.

10.4 Category Learning

How does perceptual learning differ from what appears on the surface to be the very similar phenomenon of category learning, or the ability to classify potentially varied sensory objects into categories? The potential relationship between the two is nuanced. Though perceptual and category learning are generally believed to be distinct forms of learning, they nevertheless share many features, especially in the visual domain. The stimulus representations and decision rules seem similar, for example, though the experiments typically used to study these two forms have tended to use different kinds of stimulus distributions and different paradigms. Some researchers have even proposed that the two domains of learning rely on entirely separate physiological substrates.[195]

Category learning with visual stimuli has been the subject of extensive experimentation, theory development, and physiological investigation. There are a number of important paradigmatic differences that distinguish this research project from that of perceptual learning, most significantly in experimental method. In visual perceptual learning experiments, stimuli often vary only in the dimension of the decision (e.g., vary only in orientation for orientation judgments), and in many cases there is little stimulus variety from trial to trial. In category learning experiments, stimuli have tended to vary in two or more dimensions simultaneously (e.g., orientation and spatial frequency; color, shape, and number), and varied stimuli are tested in different trials. In perceptual learning studies, observers are informed directly about the desired classification (e.g., they are told that the judgment is whether the stimulus was oriented clockwise or counterclockwise of vertical, and some stimulus examples may even be shown), while in the standard category-learning task, observers are not informed about the nature of the judgments and must infer the intended basis of categorization from feedback (e.g., stimuli varying in orientation and spatial frequency are classified into arbitrary categories A and B that must be inferred from feedback). The categories might be based on orientation, spatial frequency, or some combination of the two. In perceptual learning, the stimuli have often been hard to see because of low contrast or have involved fine discriminations, while in categorization the differences in stimuli have typically been easy to see.

Several theoretically distinct forms of category learning have been identified. These forms include so-called *rule-based* categories and *information-integration* categories, and a third kind of category based on *prototype learning.* Figure 10.10 illustrates exemplary stimuli for versions of rule and information-integration categorization tasks, along with decision boundaries. In the rule-based case, categorization depends only on spatial frequency (bar size, high versus low) and not at all on orientation. The two categories are separated by a linear boundary that is perpendicular to the spatial-frequency dimension. In the information-integration case, categorization reflects a combination of spatial frequency and orientation. The two categories are still separated by a linear boundary; however, that (diagonal) boundary is now perpendicular to neither axis. These stimuli come from an experiment that, more than in many others, controlled the stimulus sets; in this case, they were identical except for rotation, so in principle the categorization should have presented the same difficulty to an ideal observer in a signal detection regime. Examples from prior experiments also included other kinds of dimensions, such as number, color, size, and/or background color; in many of these, the sensory properties of the stimuli were not as well controlled, yet the results were quite similar. The third form of prototype category learning has generally been tested using random dot patterns (figure 10.10c), in which

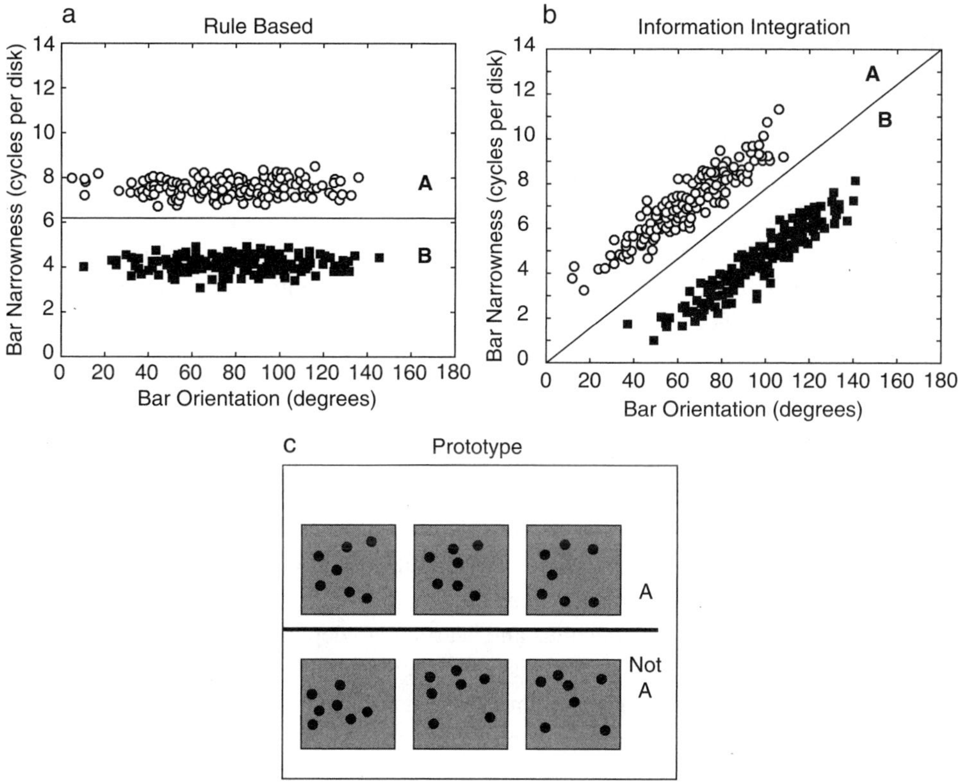

Figure 10.10
Two forms of category learning based on dimensional variation—(a) rule based and (b) information integration—and a third form (c) based on prototype plus variation. After Ashby and Valentin,[195] figure 1, and Ashby and Ell,[196] box 3, with permission.

category A includes a prototype pattern with minor variations, while examples of category B are random dot patterns that are more different from the prototype.

Rule-based categorization tasks have been shown to be the easiest of the three classes to learn. This was partly definitional, as the rule-based designation generally referred to those tasks that could be solved by explicit reasoning about an easily describable rule. In some cases, however, rule-based tasks involved several dimensions, if the decision boundaries were easy to construct from simple rules, such as "high spatial frequencies and left orientations." (Note that several standard neuropsychological tests, such as the Wisconsin Card Sorting task [PAR Corporation, D. A. Grant and E. A Berg, developers], itself a common measure of frontal-lobe function, examine how easy it is to switch from one such basis of categorization to another in a sequence of stimuli; for example, switching from categorization based on color to one based on shape.) Information-integration tasks require combining information from two or more dimensions in less obvious ways and are typically learned more slowly and are also more difficult to verbalize.[196] If there are just a few examples, memorization may suffice, but if there are many stimuli, then a true integrated decision boundary must be developed. Prototype learning may be more similar to information-integration category learning, as it typically does not involve a dimensional decision rule, while some theories attribute prototype category learning to the learning of exemplars.[197, 198]

Rule-based learning has often been contrasted with information-integration learning, on both empirical and neuropsychological grounds. One strong claim is that the two are supported by distinct neural systems, a declarative learning system and a procedural learning system, respectively.[199–202] The COVIS (competition between verbal and implicit systems) model proposes that a frontal-based declarative system learns explicit rules, while a procedural system based on basal ganglia learns more complex category structures.[199]

Such systems differ in a number of observable ways. The declarative system is speedy, while the procedural system is slow, incremental, and requires consistent and immediate feedback during learning.[195, 196] The declarative system has been associated with the prefrontal cortex, decision centers in the anterior cingulate cortex (ACC), and the hippocampus, while the procedural system has been associated with the supplementary motor area, the striatum, and nuclei of the thalamus and is argued to depend critically on dopaminergic fiber projections related to reward. This two-system model has been challenged by other researchers, and alternative single-system models have also been proposed.[198, 203, 204]

One key prediction of the two-system account is that the information-integration (procedural) system should rely on timely reward or feedback for learning, while the rule-based (declarative) system need not. If learning information-integration categorization is mediated by the basal ganglion reward system, then, the reasoning goes, it should depend on timely feedback, usually operationalized as within 2 s after the response. By contrast, if rule-based learning is mediated by the frontal declarative system, it should be far less sensitive to feedback delays. As expected, information-integration learning has been found to be more successful with feedback delays of 0.5 s, compared to 0 s or 1 s; indeed, any feedback delays of 2.5 s or longer lead to impaired information-integration category learning.[205] On the other hand, rule-based category learning survived feedback delays as long as 10 s.[206–208] These results have been cited in support of the declarative and procedural neural systems involved in category learning.[195] Distinctions between the two classes of category learning have been further supported by a dissociation of effects under different empirical manipulations. For example, rule-based category learning, which may rely on logical reasoning about possible rules, was more affected by concurrent working-memory demands,[209] the number of possible categories,[210] or sleep deprivation.[211] On the other hand, information-integration category learning, which may be more dependent on basal ganglia reward structures, was more susceptible to disruption through delay of feedback,[207] changes in response mapping,[212] and separation of the categories in stimulus space,[213] as summarized in a review by Ashby and Valentin.[195]

Though usually treated separately, category learning and perceptual learning share a number of features. The stimuli can be placed in a multidimensional space, the classifications can often be represented in this space as a decision boundary, and learning in both involves the evolution of improved decision boundaries (or, alternatively, reductions in internal variability). Beyond these formal similarities, however, the two domains are quite different. In perceptual learning, the ability to verbalize a rule such as choosing the more clockwise orientation or the highest frequency—which are typically provided in the instructions—does not guarantee rapid learning or a simple dimensional interpretation of what is learned. Perceptual learning often takes place even in the absence of feedback, and there are conditions in which the addition of feedback is unimportant (see chapter 7). Indeed, certain researchers have recently argued that perceptual learning might be distinct from both forms of category learning and supported in early sensory cortices.[199]

One paper explicitly studied the relationship between perceptual and category learning in patients treated for Wilson's disease and controls, concluding that the relationship was especially complex.[214] Wilson's disease involves damage to the basal ganglia, which is proposed to be important in information-integration category learning. Such a patient population allowed us to carry out a correlational study measuring rule-based category learning, information-based category learning, and visual perceptual learning in different external-noise levels in the same subjects. The measures included the learning rate and ultimate accuracy in rule-based category tasks and in information-integration category tasks, and the magnitude of perceptual learning in different levels of external noise. Patients with Wilson's disease showed deficits in both forms of category learning and in perceptual learning in high external noise but not in low external noise. However, only perceptual learning in high external noise was correlated with information-integration category learning; no other correlations were significant.

This correlational analysis suggests a relationship between visual perceptual learning in high external noise and information-integration, but not rule-based, category learning. Such results do not, however, explain the nearly intact form of perceptual learning in conditions of zero or low external noise in patients with Wilson's disease.[214] This is only one study, of course, and more research will be required to fully understand any relationship between the neural substrates of categorization and perceptual learning.

More generally, category learning and perceptual learning might, in principle, share both a conceptual stimulus space and response categories. They may share the representation of the stimulus, including the dimensions of variation, and the final category boundaries that determine stimulus classification and responses. In our view, even if these two domains can be conceptualized within the same dimensional structure, they differ fundamentally in the nature of what is learned. The limiting factors in the two cases seem to be different.

In category learning, it is the very general position of the category boundary that must be learned—because the observer begins with undefined categories and must infer them from feedback over a series of trials. Stimuli are generally relatively easy to see and unambiguous in their sensory representation. Often, there is significant stimulus variation on one or more dimensions. The limiting factor in the primary measure, the number of trials to correctly classify a criterion number of stimuli or infer the category rule, may not be limited by the variability or relative signal-to-noise in the sensory representation. By contrast, in perceptual learning, the observer generally begins with a clear conceptual understanding of the task from the experimenter's instructions or the situation, and the discriminations are then limited by noise (internal or external) and/or by the close similarity of the stimuli to be discriminated. More often than not, the same or a relatively small set of training stimuli will have been used, such that the uncertainty resides not in determining what task is to be performed but rather in correctly weighting sensory representations in order to carry out a judgment. In other words, the limiting factors are the signal-to-noise ratio in the sensory information as well as the optimization of the decision boundary.

To recapitulate, in category learning experiments, observers must discover the rule that defines the categorization, and typically the stimuli themselves are easily visible but variable. In perceptual learning experiments, observers are typically preinstructed about the response rule but must discover how to interpret noisy, weak, or similar sensory information. These complementary approaches to understanding learning have the potential to be combined and experimentally tested in a variety of ways that would enrich our understanding of the boundary conditions of learning in multiple situations. Model elaborations that include uncertainty about the nature of the correct classification and internal and external variation in the stimuli themselves may best reflect perceptual learning in a naturalistic context.

10.5 Conclusions

Plasticity occurs at several temporal scales and in many modalities to better adapt behavior to real-world challenges. This chapter began by considering the multiple forms of plasticity of the visual system at different timescales. It then progressed to an analysis of perceptual learning in different sensory modalities and ended with a comparison between related forms of learned categorization. For a number of modalities, the related literature in animal models is enormous, so our survey was necessarily partial, with an emphasis on findings in human observers. In other cases, the research literature is more sparse and preliminary. In some modalities, certain tests of learning, retention, transfer, mechanisms, and models of perceptual learning and tasks remain to be explored.

Despite the varying stages of research, it is particularly striking to note the commonalities in phenomenology of perceptual learning across modalities. Three principles found in visual perceptual learning seem to occur elsewhere: first, that learning occurs within the context of a complex brain network at different levels of representation and processing; second, that these learning phenomena critically balance the advantages of plasticity and the need for stability in neural and information systems; and third, that reweighting sensory inputs to drive decision and behavior is almost surely one ingredient in learning by finding the best signal among noise. These principles can be summarized as multilevel complexity, the stability/plasticity balance, and the principle of reweighting. All of them seem to transcend domain-specific differences.

Such differences also need to be taken into consideration, of course, because they reveal significant variations across perceptual modalities. These involve the level, rate, persistence, and specificity of learning, where each may have been optimized to best suit the different uses of sensory information for evolutionary fitness. They may also correspond with the somewhat different emphases that learning in each modality places on early versus late cortical areas, although all surely involve some common circuits related to decision and motor control.

There are a number of other forms of learning that may have fundamentally analogous properties. Though outside the scope of this chapter, motor learning, like perceptual learning, is limited by internal noise in the signals and often integrates multiple cues to guide behavior. Likewise, classical conditioning engages in the learning process reward mechanisms not unlike those found in perceptual learning. It seems possible that many of the same core mechanisms are embodied in all forms of learning that involve sensory inputs and effector outputs. Future research promises to further elaborate these similarities and differences.

We began the chapter by situating perceptual learning in relation to other biological processes with different timescales: species evolution and early cognitive development at one extreme and moment-to-moment situational adaptation at the other. The possible relationship between these scales of plasticity remains an open area for both research and possible application. How can the power of perceptual training in development be harnessed? How might the interactions between short-term adaptation and perceptual learning be synthesized?

Another important focus of investigation concerns the function of multiple sensory modalities co-occurring closely in time. The role of learning in fine-tuning the interpretations of multisensory events and, conversely, the possible role of multisensory cues in perceptual learning both deserve further analysis. The relative importance of synthetic multifeature or multicue object representations should also be examined further. In order to integrate perceptual and other related forms of sensory and motor learning into our understanding of the larger human cognitive system, future science will need to examine the degree to which each form of learning proceeds independently of the others or else works to develop multimodal synergies.

References

1. He S, Li X, Chan N, Hinton DR. Review: Epigenetic mechanisms in ocular disease. *Molecular Vision* 2013;19:665–674.

2. Keverne EB, Curley JP. Epigenetics, brain evolution and behaviour. *Frontiers in Neuroendocrinology* 2008;29(3):398–412.

3. Keverne EB, Pfaff DW, Tabansky I. Epigenetic changes in the developing brain: Effects on behavior. *Proceedings of the National Academy of Sciences* 2015;112(22):6789–6795.

4. Dye M. The advantages of being helpless: Human brains are slow to develop—a secret, perhaps, of our success. *Scientific American* February 2010.

5. Isler K. Brain size evolution: How fish pay for being smart. *Current Biology* 2013;23(2):R63–R65.

6. Gottlieb G. The roles of experience in the development of behavior and the nervous system. *Studies on the development of behavior and the nervous system*. Vol. 3. Elsevier;1976:25–54.

7. Censor N, Karni A, Sagi D. A link between perceptual learning, adaptation and sleep. *Vision Research* 2006;46(23):4071–4074.

8. Harris H, Gliksberg M, Sagi D. Generalized perceptual learning in the absence of sensory adaptation. *Current Biology* 2012;22(19):1813–1817.

9. Yotsumoto Y, Watanabe T, Sasaki Y. Different dynamics of performance and brain activation in the time course of perceptual learning. *Neuron* 2008;57(6):827–833.

10. Darwin C. *On the origins of species by means of natural selection*. Murray;1859.

11. Meredith RW, Janečka JE, Gatesy J, Ryder OA, Fisher CA. Impacts of the Cretaceous Terrestrial Revolution and KPg extinction on mammal diversification. *Science* 2011;334(6055):521–524.

12. Kaas JH, ed. *The evolution of the visual system in primates*. The New Visual Neurosciences. MIT Press;2013.

13. Azevedo FA, Carvalho LR, Grinberg LT, Farfel JM, Ferretti REL, Leite REP, Filho WJ, Lent R, Herculano-Houzel S.. Equal numbers of neuronal and nonneuronal cells make the human brain an isometrically scaled-up primate brain. *Journal of Comparative Neurology* 2009;513(5):532–541.

14. Ross C, Martin R. The role of vision in the origin and evolution of primates. *Evolution of Nervous Systems* 2007;4:59–78.

15. Wikler KC, Rakic P. Distribution of photoreceptor subtypes in the retina of diurnal and nocturnal primates. *Journal of Neuroscience* 1990;10(10):3390–3401.

16. Regan BC, Julliot C, Simmen B, Vienot Fo, C-D P, Mollon JD. Fruits, foliage and the evolution of primate colour vision. *Philosophical Transactions of the Royal Society B: Biological Sciences* 2001;356(1407):229–283.

17. Weller RE, Kaas JH. Parameters affecting the loss of ganglion cells of the retina following ablations of striate cortex in primates. *Visual Neuroscience* 1989;3(4):327–349.

18. Lane R, Allman J, Kaas J, Miezin F. The visuotopic organization of the superior colliculus of the owl monkey (*Aotus trivirgatus*) and the bush baby (*Galago senegalensis*). *Brain Research* 1973;60(2):335–349.

19. Felleman DJ, Van Essen DC. Distributed hierarchical processing in the primate cerebral cortex. *Cerebral Cortex* 1991;1(1):1–47.

20. Kaas JH, Gharbawie O, Stepniewska I. The organization and evolution of dorsal stream multisensory motor pathways in primates. *Frontiers in Neuroanatomy* 2011;5:34:1–7.

21. Liang J, Williams DR. Aberrations and retinal image quality of the normal human eye. *Journal of the Optical Society of America A* 1997;14(11):2873–2883.

22. Wandell BA. *Foundations of vision*. Sinauer Associates;1995.

23. Geisler WS. Visual perception and the statistical properties of natural scenes. *Annual Review of Psychology* 2008;59:167–192.

24. Laughlin SB. A simple coding procedure enhances a neuron's information capacity. *Zeitschrift fuer Naturforschung* 1981;36(9–10):910–912.

25. Frazor RA, Geisler WS. Local luminance and contrast in natural images. *Vision Research* 2006;46(10):1585–1598.

26. Cormack LK, Liu Y, Bovik AC. Disparity statistics in the natural environment. *Journal of Vision* 2005;5(8):604.

27. Maloney LT. Evaluation of linear models of surface spectral reflectance with small numbers of parameters. *Journal of the Optical Society of America A* 1986;3(10):1673–1683.

28. Maloney LT, Wandell BA. Color constancy: A method for recovering surface spectral reflectance. *Journal of the Optical Society of America A* 1986;3(1):29–33.

29. Brainard DH, Freeman WT. Bayesian color constancy. *Journal of the Optical Society of America A* 1997;14(7):1393–1411.

30. D'Zmura M, Iverson G. Color constancy. I. Basic theory of two-stage linear recovery of spectral descriptions for lights and surfaces. *Journal of the Optical Society of America A* 1993;10(10):2148–2165.

31. Barlow H. Single units and sensation: A neuron doctrine for perceptual psychology? *Perception* 1972;1:371–394.

32. Field D. What is the goal of sensory coding? *Neural Computation* 1994;6(4):559–601.

33. Olshausen BA, Field DJ. Sparse coding with an overcomplete basis set: A strategy employed by V1? *Vision Research* 1997;37(23):3311–3325.

34. Boothe RG, Dobson V, Teller DY. Postnatal development of vision in human and nonhuman primates. *Annual Review of Neuroscience* 1985;8(1):495–545.

35. Teller DY, Movshon JA. Visual development. *Vision Research* 1986;26(9):1483–1506.

36. Dobson V, Teller DY. Visual acuity in human infants: A review and comparison of behavioral and electrophysiological studies. *Vision Research* 1978;18(11):1469–1483.

37. Banks MS, Bennett PJ. Optical and photoreceptor immaturities limit the spatial and chromatic vision of human neonates. *Journal of the Optical Society of America A* 1988;5(12):2059–2079.

38. Braddick O, Atkinson J. Development of human visual function. *Vision Research* 2011;51(13):1588–1609.

39. Kiorpes L, Movshon JA. Neural limitations on visual development in primates. *Visual Neurosciences* 2003;1:159–173.

40. Blakemore C, Vital-Durand F. Organization and post-natal development of the monkey's lateral geniculate nucleus. *Journal of Physiology* 1986;380(1):453–491.

41. Blakemore C. Maturation and modification in the developing visual system. *Perception.* Springer;1978:377–436.

42. Ellemberg D, Lewis TL, Liu CH, Maurer D. Development of spatial and temporal vision during childhood. *Vision Research* 1999;39(14):2325–2333.

43. Braddick O, Wattam-Bell J, Atkinson J. Orientation-specific cortical responses develop in early infancy. *Nature*, 1986:320(6063);617–619.

44. Braddick O. Orientation- and motion-selective mechanisms in infants. In Simons K, ed., *Early visual development: Normal and abnormal.* Oxford University Press;1993:163–177.

45. Wattam-Bell J. Development of motion-specific cortical responses in infancy. *Vision Research* 1991;31(2):287–297.

46. Braddick O, Birtles D, Wattam-Bell J, Atkinson J. Motion- and orientation-specific cortical responses in infancy. *Vision Research* 2005;45(25):3169–3179.

47. Braddick O, Atkinson J, Julesz B, Kropfl W, Bodis-Wollner I, Raab E. Cortical binocularity in infants. *Nature* 198;288(5789):363–365.

48. Julesz B, Kropfl W, Petrig B. Large evoked potentials to dynamic random-dot correlograms and stereograms permit quick determination of stereopsis. *Proceedings of the National Academy of Sciences* 1980;77(4):2348–2351.

49. Petrig B, Julesz B, Kropfl W, Baumgartner G, Anliker M. Development of stereopsis and cortical binocularity in human infants: Electrophysiological evidence. *Science* 1981;213(4514):1402–1405.

50. Birch EE, Gwiazda J, Held R. Stereoacuity development for crossed and uncrossed disparities in human infants. *Vision Research* 1982;22(5):507–513.

51. Braddick O, Wattam-Bell J, Day J, Atkinson J. The onset of binocular function in human infants. *Human Neurobiology* 1982;2(2):65–69.

52. Held R, Birch E, Gwiazda J. Stereoacuity of human infants. *Proceedings of the National Academy of Sciences* 1980;77(9):5572–5574.

53. Fox R, Aslin RN, Shea SL, Dumais ST. Stereopsis in human infants. *Science* 1980;207(4428):323–324.

54. Mayer D, Beiser A, Warner A, Pratt E, Raye K, Lang J. Monocular acuity norms for the Teller Acuity Cards between ages one month and four years. *Investigative Ophthalmology & Visual Science* 1995;36(3):671–685.

55. Maurer D, Lewis TL. Visual acuity: The role of visual input in inducing postnatal change. *Clinical Neuroscience Research* 2001;1(4):239–247.

56. Parrish E, Giaschi D, Boden C, Dougherty R. The maturation of form and motion perception in school age children. *Vision Research* 2005;45(7):827–837.

57. Morrone M, Burr D, Vaina LM. Two stages of visual processing for radial and circular motion. *Nature* 1995;376(6540):507–509.

58. Newsome WT, Pare EB. A selective impairment of motion perception following lesions of the middle temporal visual area (MT). *Journal of Neuroscience* 1988;8(6):2201–2211.

59. Kovács I. Human development of perceptual organization. *Vision Research* 2000;40(10):1301–1310.

60. Kalia A, Lesmes LA, Dorr M, Gandhi T, Chatterjee G, Ganesh S, Bex PJ, Sinha P. Development of pattern vision following early and extended blindness. *Proceedings of the National Academy of Sciences* 2014;111(5):2035–2039.

61. Takesian AE, Hensch TK. Balancing plasticity/stability across brain development. *Progress in Brain Research* 2013;207:3–34.

62. Shapley R, Enroth-Cugell C. Visual adaptation and retinal gain controls. *Progress in Retinal Research* 1984;3:263–346.

63. Blakemore C, Sutton P. Size adaptation: A new after-effect. *Science* 1969;166(3902):245–247.

64. Clifford CW. Perceptual adaptation: Motion parallels orientation. *Trends in Cognitive Sciences* 2002;6(3):136–143.

65. Bao M, Engel SA. Distinct mechanism for long-term contrast adaptation. *Proceedings of the National Academy of Sciences* 2012;109(15);5898–5903.

66. Adams R. An account of a peculiar optical phenomenon seen after having looked at a moving body, etc. *The London, Edinburgh, and Dublin Philosophical Magazine and Journal of Science* 1834;5(29):373–374.

67. Mitchell DE, Muir DW. Does the tilt after-effect occur in the oblique meridian? *Vision Research* 1976;16(6):609–613.

68. Dao DY, Lu Z-L, Dosher BA. Adaptation to sine-wave gratings selectively reduces the contrast gain of the adapted stimuli. *Journal of Vision* 2006;6(7):739–759.

69. Webster MA, Mollon J. The influence of contrast adaptation on color appearance. *Vision Research* 1994;34(15):1993–2020.

70. Calder AJ, Jenkins R, Cassel A, Clifford CW. Visual representation of eye gaze is coded by a nonopponent multichannel system. *Journal of Experimental Psychology: General* 2008;137(2):244.

71. Lawson RP, Clifford CW, Calder AJ. About turn the visual representation of human body orientation revealed by adaptation. *Psychological Science* 2009;20(3):363–371.

72. Fang F, He S. Viewer-centered object representation in the human visual system revealed by viewpoint aftereffects. *Neuron* 2005;45(5):793–800.

73. Webster MA. Adaptation and visual coding. *Journal of Vision* 2011;11(5):3;1–23.

74. Regan D, Beverley K. Postadaptation orientation discrimination. *Journal of the Optical Society of America A* 1985;2(2):147–155.

75. Vul E, Krizay E, MacLeod DI. The McCollough effect reflects permanent and transient adaptation in early visual cortex. *Journal of Vision* 2008;8(12):4:1–12.

76. Zhang P, Bao M, Kwon M, He S, Engel SA. Effects of orientation-specific visual deprivation induced with altered reality. *Current Biology* 2009;19(22):1956–1960.

77. Juricevic I, Webster MA. Variations in normal color vision. V. Simulations of adaptation to natural color environments. *Visual Neuroscience* 2009;26(1):133–145.

78. Gibson JJ. Adaptation, after-effect and contrast in the perception of curved lines. *Journal of Experimental Psychology* 1933;16(1):1–31.

79. Gibson JJ, Radner M. Adaptation, after-effect and contrast in the perception of tilted lines. I. Quantitative studies. *Journal of Experimental Psychology* 1937;20(5):453–467.

80. Delahunt PB, Webster MA, Ma L, Werner JS. Long-term renormalization of chromatic mechanisms following cataract surgery. *Visual Neuroscience* 2004;21(3):301–307.

81. Bedford FL. Constraints on perceptual learning: Objects and dimensions. *Cognition* 1995;54(3):253–297.

82. Rozin P, Schull J. The adaptive-evolutionary point of view in experimental psychology. In Atkinson RC, Herrnstein RJ, Lindzey G, Luce RD, eds., *Handbook of experimental psychology*. Wiley-Interscience;1988:503–546.

83. Gibson EJ. Perceptual learning in development: Some basic concepts. *Ecological Psychology* 2000;12(4):295–302.

84. Fagiolini M, Fritschy J-M, Löw K, Möhler H, Rudolph U, Hensch TK. Specific GABAA circuits for visual cortical plasticity. *Science* 2004;303(5664):1681–1683.

85. Hensch TK. Critical period plasticity in local cortical circuits. *Nature Reviews Neuroscience* 2005;6(11):877–888.

86. Hensch TK, Fagiolini M, Mataga N, Stryker MP, Baekkeskov S, Kash SF. Local GABA circuit control of experience-dependent plasticity in developing visual cortex. *Science* 1998;282(5393):1504–1508.

87. Sagi D. Perceptual learning in vision research. *Vision Research* 2011;51(13):1552–1566.

88. Demany L. Perceptual learning in frequency discrimination. *Journal of the Acoustical Society of America* 1985;78(3):1118–1120.

89. Irvine DR, Martin RL, Klimkeit E, Smith R. Specificity of perceptual learning in a frequency discrimination task. *Journal of the Acoustical Society of America* 2000;108(6):2964–2968.

90. Amitay S, Hawkey DJ, Moore DR. Auditory frequency discrimination learning is affected by stimulus variability. *Perception & Psychophysics* 2005;67(4):691–698.

91. Hawkey DJ, Amitay S, Moore DR. Early and rapid perceptual learning. *Nature Neuroscience* 2004;7(10):1055–1056.

92. Amitay S, Irwin A, Moore DR. Discrimination learning induced by training with identical stimuli. *Nature Neuroscience* 2006;9(11):1446–1448.

93. Wright BA, Buonomano DV, Mahncke HW, Merzenich MM. Learning and generalization of auditory temporal-interval discrimination in humans. *Journal of Neuroscience* 1997;17(10):3956–3963.

94. Wright BA, Zhang Y. A review of learning with normal and altered sound-localization cues in human adults. *International Journal of Audiology* 2006;45(supl):92–98.

95. Zhang Y, Wright BA. An influence of amplitude modulation on interaural level difference processing suggested by learning patterns of human adults. *Journal of the Acoustical Society of America* 2009;126(3):1349–1358.

96. Zhang Y, Wright BA. Similar patterns of learning and performance variability for human discrimination of interaural time differences at high and low frequencies. *Journal of the Acoustical Society of America* 2007;121(4):2207–2216.

97. Grimault N, Micheyl C, Carlyon RP, Collet L. Evidence for two pitch encoding mechanisms using a selective auditory training paradigm. *Perception & Psychophysics* 2002;64(2):189–197.

98. Samuel AG, Kraljic T. Perceptual learning for speech. *Attention, Perception, & Psychophysics* 2009;71(6):1207–1218.

99. Logan JS, Lively SE, Pisoni DB. Training Japanese listeners to identify English /r/ and /l/. A first report. *Journal of the Acoustical Society of America* 1991;89(2):874–886.

100. Lively SE, Logan JS, Pisoni DB. Training Japanese listeners to identify English /r/ and /l/. II. The role of phonetic environment and talker variability in learning new perceptual categories. *Journal of the Acoustical Society of America* 1993;94(3):1242–1255.

101. Lively SE, Pisoni DB, Yamada RA, Tohkura Y, Yamada T. Training Japanese listeners to identify English /r/ and /l/. III. Long-term retention of new phonetic categories. *Journal of the Acoustical Society of America* 1994;96(4):2076–2087.

102. Bradlow AR, Pisoni DB, Akahane-Yamada R, Tohkura Y. Training Japanese listeners to identify English /r/ and /l/. IV. Some effects of perceptual learning on speech production. *Journal of the Acoustical Society of America* 1997;101(4):2299–2310.

103. Wang Y, Spence MM, Jongman A, Sereno JA. Training American listeners to perceive Mandarin tones. *Journal of the Acoustical Society of America* 1999;106(6):3649–3658.

104. Bradlow AR, Bent T. Perceptual adaptation to non-native speech. *Cognition* 2008;106(2):707–729.

105. Clarke CM, Garrett MF. Rapid adaptation to foreign-accented English. *Journal of the Acoustical Society of America* 2004;116(6):3647–3658.

106. Loebach JL, Bent T, Pisoni DB. Multiple routes to the perceptual learning of speech. *Journal of the Acoustical Society of America* 2008;124(1):552–561.

107. Shafiro V, Sheft S, Gygi B, Ho KTN. The influence of environmental sound training on the perception of spectrally degraded speech and environmental sounds. *Trends in Amplification* 2012;16(2):83–101.

108. Davis MH, Johnsrude IS, Hervais-Adelman A, Taylor K, McGettigan C. Lexical information drives perceptual learning of distorted speech: Evidence from the comprehension of noise-vocoded sentences. *Journal of Experimental Psychology: General* 2005;134(2):222–241.

109. Wright BA, Zhang Y. A review of the generalization of auditory learning. *Philosophical Transactions of the Royal Society B: Biological Sciences* 2009;364(1515):301–311.

110. Delhommeau K, Micheyl C, Jouvent R. Generalization of frequency discrimination learning across frequencies and ears: Implications for underlying neural mechanisms in humans. *Journal of the Association for Research in Otolaryngology* 2005;6(2):171–179.

111. Demany L, Semal C. Learning to perceive pitch differences. *Journal of the Acoustical Society of America* 2002;111(3):1377–1388.

112. Karmarkar UR, Buonomano DV. Temporal specificity of perceptual learning in an auditory discrimination task. *Learning & Memory* 2003;10(2):141–147.

113. Banai K, Amitay S. Stimulus uncertainty in auditory perceptual learning. *Vision Research* 2012;61:83–88.

114. Green D, Swets J. *Signal detection theory and psychophysics.* Wiley;1966.

115. Ahumada A, Watson A. Equivalent-noise model for contrast detection and discrimination. *Journal of the Optical Society of America A* 1985;2(7):1133–1139.

116. Amitay S, Zhang Y-X, Jones PR, Moore DR. Perceptual learning: Top to bottom. *Vision Research* 2014;99:69–77.

117. Amitay S, Guiraud J, Sohoglu E, Zobay O, Edmonds BA, Zhang, Y-X, Moore DR. Human decision making based on variations in internal noise: An EEG study. *PloS One* 2013;8(7):e68928.

118. Amitay S, Irwin A, Hawkey DJ, Cowan JA, Moore DR. A comparison of adaptive procedures for rapid and reliable threshold assessment and training in naive listeners. *Journal of the Acoustical Society of America* 2006;119(3):1616–1625.

119. Jones PR, Moore DR, Amitay S, Shub DE. Reduction of internal noise in auditory perceptual learning. *Journal of the Acoustical Society of America* 2013;133(2):970–981.

120. Dosher BA, Lu Z-L. Perceptual learning in clear displays optimizes perceptual expertise: Learning the limiting process. *Proceedings of the National Academy of Sciences* 2005;102(14):5286–5290.

121. Amitay S, Zhang Y-X, Moore DR. Asymmetric transfer of auditory perceptual learning. *Frontiers in Psychology* 2012;3:508;1–8.

122. Dosher BA, Lu Z-L. Perceptual learning reflects external noise filtering and internal noise reduction through channel reweighting. *Proceedings of the National Academy of Sciences* 1998;95(23):13988–13993.

123. Dosher BA, Lu Z-L. Mechanisms of perceptual learning. *Vision Research* 1999;39(19):3197–3221.

124. Patterson R, Robinson K, Holdsworth J, McKeown D, Zhang C, Allerhand M. *Auditory physiology and perception.* Oxford;1992.

125. Woldorff MG, Gallen CC, Hampson SA, Hillyard SA, Pantev C, Sobel D, Bloom FE. Modulation of early sensory processing in human auditory cortex during auditory selective attention. *Proceedings of the National Academy of Sciences* 1993;90(18):8722–8726.

126. Hansen JC, Hillyard SA. Endogenous brain potentials associated with selective auditory attention. *Electroencephalography and Clinical Neurophysiology* 1980;49(3):277–290.

127. Bakin JS, Weinberger NM. Classical conditioning induces CS-specific receptive field plasticity in the auditory cortex of the guinea pig. *Brain Research* 1990;536(1–2):271–286.

128. Weinberger NM. Specific long-term memory traces in primary auditory cortex. *Nature Reviews Neuroscience* 2004;5(4):279–290.

129. Weinberger NM, Hopkins W, Diamond DM. Physiological plasticity of single neurons in auditory cortex of the cat during acquisition of the pupillary conditioned response. I. Primary field (AI). *Behavioral Neuroscience* 1984;98(2):171–188.

130. Recanzone GH, Schreiner C, Merzenich MM. Plasticity in the frequency representation of primary auditory cortex following discrimination training in adult owl monkeys. *Journal of Neuroscience* 1993;13(1):87–103.

131. Weinberger NM, Ashe JH, Metherate R, McKenna TM, Diamond DM, Bakin JS, Lennartz RC, Cassady JM. Neural adaptive information processing: A preliminary model of receptive-field plasticity in auditory cortex during Pavlovian conditioning. In Gabriel M, Moore J, eds. *Learning and computational neuroscience: Foundation of adaptive networks*. MIT Press;1990:91–138.

132. Polley DB, Steinberg EE, Merzenich MM. Perceptual learning directs auditory cortical map reorganization through top-down influences. *Journal of Neuroscience* 2006;26(18):4970–4982.

133. Brown M, Irvine DR, Park VN. Perceptual learning on an auditory frequency discrimination task by cats: Association with changes in primary auditory cortex. *Cerebral Cortex* 2004;14(9):952–965.

134. Recanzone GH. Perception of auditory signals. *Annals of the New York Academy of Sciences* 2011;1224(1):96–108.

135. Reed A, Riley J, Carraway R, Carrasco A, Perez C, Jakkamsetti V, Kilgard MP. Cortical map plasticity improves learning but is not necessary for improved performance. *Neuron* 2011;70(1):121–131.

136. Talwar SK, Gerstein GL. Reorganization in awake rat auditory cortex by local microstimulation and its effect on frequency-discrimination behavior. *Journal of Neurophysiology* 2001;86(4):1555–1572.

137. Ohl FW, Scheich H. Learning-induced plasticity in animal and human auditory cortex. *Current Opinion in Neurobiology* 2005;15(4):470–477.

138. Recanzone GH, Merzenich MM, Jenkins WM, Grajski KA, Dinse HR. Topographic reorganization of the hand representation in cortical area 3b owl monkeys trained in a frequency-discrimination task. *Journal of Neurophysiology* 1992;67(5):1031–1056.

139. Binder JR, Liebenthal E, Possing ET, Medler DA, Ward BD. Neural correlates of sensory and decision processes in auditory object identification. *Nature Neuroscience* 2004;7(3):295–301.

140. Holcomb H, Medoff D, Caudill P, Zhao Z, Lahti AC, Dannals RF, Tamminga CA. Cerebral blood flow relationships associated with a difficult tone recognition task in trained normal volunteers. *Cerebral Cortex* 1998;8(6):534–542.

141. Jäncke L, Gaab N, Wüstenberg T, Scheich H, Heinze H-J. Short-term functional plasticity in the human auditory cortex: An fMRI study. *Cognitive Brain Research* 2001;12(3):479–485.

142. Kang D-W, Kim D, Chang L-H, Kim Y-H, Takahashi Em Cain MS, Watanabe T, Sasaki Y. Structural and functional connectivity changes beyond visual cortex in a later phase of visual perceptual learning. *Scientific Reports* 2018;8(5186):1–9.

143. James, W. *The principles of psychology*. Vol. 1. Holt;1890.

144. Harris JA, Petersen RS, Diamond ME. Distribution of tactile learning and its neural basis. *Proceedings of the National Academy of Sciences* 1999;96(13):7587–7591.

145. Petersen RS, Diamond ME. Spatial-temporal distribution of whisker-evoked activity in rat somatosensory cortex and the coding of stimulus location. *Journal of Neuroscience* 2000;20(16):6135–6143.

146. Kaas JH. The functional organization of somatosensory cortex in primates. *Annals of Anatomy Anatomischer Anzeiger* 1993;175(6):509–518.

147. Harris JA, Harris IM, Diamond ME. The topography of tactile learning in humans. *Journal of Neuroscience* 2001;21(3):1056–1061.

148. Harrar V, Spence C, Makin TR. Topographic generalization of tactile perceptual learning. *Journal of Experimental Psychology: Human Perception and Performance* 2014;40(1):15–23.

149. Sathian K, Deshpande G, Stilla R. Neural changes with tactile learning reflect decision-level reweighting of perceptual readout. *Journal of Neuroscience* 2013;33(12):5387–5398.

150. Alberts JR. Olfactory contributions to behavioral development in rodents. In Doty RL, ed., *Mammalian olfaction, reproductive processes, and behavior.* Academic Press;1976:67–94.

151. Buck LB, Bargmann C. Smell and taste: The chemical senses. *Principles of Neural Science* 2000;4:625–647.

152. Gottfried JA, Zald DH. On the scent of human olfactory orbitofrontal cortex: Meta-analysis and comparison to non-human primates. *Brain Research Reviews* 2005;50(2):287–304.

153. Matsunami H, Montmayeur J-P, Buck LB. A family of candidate taste receptors in human and mouse. *Nature* 2000;404(6778):601–604.

154. Stevenson RJ, Prescott J, Boakes RA. The acquisition of taste properties by odors. *Learning and Motivation* 1995;26(4):433–455.

155. Stevenson RJ, Boakes RA, Prescott J. Changes in odor sweetness resulting from implicit learning of a simultaneous odor-sweetness association: An example of learned synesthesia. *Learning and Motivation* 1998;29(2):113–132.

156. De Araujo IE, Rolls ET, Kringelbach ML, McGlone F, Phillips N. Taste-olfactory convergence, and the representation of the pleasantness of flavour, in the human brain. *European Journal of Neuroscience* 2003;18(7):2059–2068.

157. Prescott J, Murphy S. Inhibition of evaluative and perceptual odour–taste learning by attention to the stimulus elements. *Quarterly Journal of Experimental Psychology* 2009;62(11):2133–2140.

158. Eylam S, Kennedy L. Experience-induced increases in sensitivity for glucose in human glucose-hypogeusics. *Chemical Senses* 1998;23:588–589.

159. Kobayashi C, Kennedy LM. Experience-induced changes in taste identification of monosodium glutamate. *Physiology & Behavior* 2002;75(1):57–63.

160. Kobayashi C, Kennedy LM, Halpern BP. Experience-induced changes in taste identification of monosodium glutamate (MSG) are reversible. *Chemical Senses* 2006;31(4):301–306.

161. Cain WS, Schmidt R. Sensory detection of glutaraldehyde in drinking water—emergence of sensitivity and specific anosmia. *Chemical Senses* 2002;27(5):425–433.

162. Gonzalez KM, Peo C, Livdahl T, Kennedy LM. Experience-induced changes in sugar taste discrimination. *Chemical Senses* 2008;33(2):173–179.

163. Faurion A, Cerf B, Pillias A-M, Boireau N. Increased taste sensitivity by familiarization to novel stimuli: Psychophysics, fMRI and electrophysiological techniques suggest modulations at peripheral and central levels. In Rouby C, Schaal B, Dubois D, Gervais R, Holley A, eds., *Olfaction, Taste, and Cognition,* Cambridge University Press;2002:350–366.

164. Walk RO. Perceptual learning and the discrimination of wines. *Psychonomic Science* 1966;5(2):57–58.

165. Meilgaard M, Dalgliesh C, Clapperton J. Beer flavour terminology 1. *Journal of the Institute of Brewing* 1979;85(1):38–42.

166. Meilgaard M, Reid D, Wyborski K. Reference standards for beer flavor terminology system. *Journal of the American Society of Brewing Chemists* 1982;40(4):119–128.

167. Owen DH, Machamer PK. Bias-free improvement in wine discrimination. *Perception* 1979;8(2):199–209.

168. Wilson DA, Stevenson RJ. The fundamental role of memory in olfactory perception. *Trends in Neurosciences* 2003;26(5):243–247.

169. Trotier D, Eloit C, Wassef M, Talmain G, Bensimon JL, Døving KB, Ferrand J. The vomeronasal cavity in adult humans. *Chemical Senses* 2000;25(4):369–380.

170. Laurent G, Stopfer M, Friedrich RW, Rabinovich MI, Volkovskii A, Abarbanel HD. Odor encoding as an active, dynamical process: Experiments, computation, and theory. *Annual Review of Neuroscience* 2001;24(1):263–297.

171. Li W, Luxenberg E, Parrish T, Gottfried JA. Learning to smell the roses: Experience-dependent neural plasticity in human piriform and orbitofrontal cortices. *Neuron* 2006;52(6):1097–1108.

172. Rabin MD. Experience facilitates olfactory quality discrimination. *Perception & Psychophysics* 1988;44(6):532–540.

173. Jehl C, Royet J, Holley A. Odor discrimination and recognition memory as a function of familiarization. *Perception & Psychophysics* 1995;57(7):1002–1011.

174. Stevens DA, O'Connell RJ. Semantic-free scaling of odor quality. *Physiology & Behavior* 1996;60(1):211–215.

175. Stevens DA, O'Connell RJ. Enhanced sensitivity to androstenone following regular exposure to pemenone. *Chemical Senses* 1995;20(4):413–419.

176. Sewards TV, Sewards MA. Cortical association areas in the gustatory system. *Neuroscience & Biobehavioral Reviews* 2001;25(5):395–407.

177. Gottfried JA, Dolan RJ. The nose smells what the eye sees: Crossmodal visual facilitation of human olfactory perception. *Neuron* 2003;39(2):375–386.

178. Diederich A, Colonius H, Schomburg A. Assessing age-related multisensory enhancement with the time-window-of-integration model. *Neuropsychologia* 2008;46(10):2556–2562.

179. Rach S, Diederich A, Colonius H. On quantifying multisensory interaction effects in reaction time and detection rate. *Psychological Research* 2011;75(2):77–94.

180. Powers AR, Hillock AR, Wallace MT. Perceptual training narrows the temporal window of multisensory binding. *Journal of Neuroscience* 2009;29(39):12265–12274.

181. Beer AL, Batson MA, Watanabe T. Multisensory perceptual learning reshapes both fast and slow mechanisms of crossmodal processing. *Cognitive, Affective, & Behavioral Neuroscience* 2011;11(1):1–12.

182. Beer AL, Watanabe T. Specificity of auditory-guided visual perceptual learning suggests crossmodal plasticity in early visual cortex. *Experimental Brain Research* 2009;198(2–3):353–361.

183. Shams L, Seitz AR. Benefits of multisensory learning. *Trends in Cognitive Sciences* 2008;12(11):411–417.

184. Kim RS, Seitz AR, Shams L. Benefits of stimulus congruency for multisensory facilitation of visual learning. *PLoS One* 2008;3(1):e1532.

185. Seitz AR, Kim R, Shams L. Sound facilitates visual learning. *Current Biology* 2006;16(14):1422–1427.

186. Alais D, Cass J. Multisensory perceptual learning of temporal order: Audiovisual learning transfers to vision but not audition. *PLoS One* 2010;5(6):e11283.

187. Lapid E, Ulrich R, Rammsayer T. Perceptual learning in auditory temporal discrimination: No evidence for a cross-modal transfer to the visual modality. *Psychonomic Bulletin & Review* 2009;16(2):382–389.

188. Bartolo R, Merchant H. Learning and generalization of time production in humans: Rules of transfer across modalities and interval durations. *Experimental Brain Research* 2009;197(1):91–100.

189. Planetta PJ, Servos P. Somatosensory temporal discrimination learning generalizes to motor interval production. *Brain Research* 2008;7(81):51–57.

190. Proulx MJ, Brown DJ, Pasqualotto A, Meijer P. Multisensory perceptual learning and sensory substitution. *Neuroscience & Biobehavioral Reviews* 2014;41:16–25.

191. Bach-Y-Rita P, Collins CC, Saunders FA, White B, Scadden L. Vision substitution by tactile image projection. *Nature*, 1969;221(5184);963–964.

192. Kim J-K, Zatorre RJ. Generalized learning of visual-to-auditory substitution in sighted individuals. *Brain Research* 2008;1242:263–275.

193. Kim J-K, Zatorre RJ. Tactile–auditory shape learning engages the lateral occipital complex. *Journal of Neuroscience* 2011;31(21):7848–7856.

194. McGurk H, MacDonald J. Hearing lips and seeing voices. *Nature* 1976;264(5588):746–748.

195. Ashby FG, Valentin VV. Multiple systems of perceptual category learning: Theory and cognitive tests. In Cohen H, Lefebvre C, eds., *Handbook of categorization in cognitive science*. Elsevier; 2005:547–572.

196. Ashby FG, Ell SW. The neurobiology of human category learning. *Trends in Cognitive Sciences* 2001;5(5):204–210.

197. Posner MI, Keele SW. On the genesis of abstract ideas. *Journal of Experimental Psychology* 1968;77(3,part 1):353–363.

198. Nosofsky RM, Stanton RD, Zaki SR. Procedural interference in perceptual classification: Implicit learning or cognitive complexity? *Memory & Cognition* 2005;33(7):1256–1271.

199. Ashby FG, Alfonso-Reese LA. A neuropsychological theory of multiple systems in category learning. *Psychological Review* 1998;105(3):442–481.

200. Ashby FG, Crossley MJ. A computational model of how cholinergic interneurons protect striatal-dependent learning. *Journal of Cognitive Neuroscience* 2011;23(6):1549–1566.

201. Ashby FG, Ennis JM, Spiering BJ. A neurobiological theory of automaticity in perceptual categorization. *Psychological Review* 2007;114(3):632–656.

202. Ashby FG, Waldron EM. On the nature of implicit categorization. *Psychonomic Bulletin & Review* 1999;6(3):363–378.

203. Nosofsky RM, Kruschke JK. Single-system models and interference in category learning: Commentary on Waldron and Ashby (2001). *Psychonomic Bulletin & Review* 2002;9(1):169–174.

204. Kruschke JK. ALCOVE: An exemplar-based connectionist model of category learning. *Psychological Review* 1992;99(1):22–44.

205. Worthy DA, Markman AB, Maddox WT. Feedback and stimulus-offset timing effects in perceptual category learning. *Brain and Cognition* 2013;81(2):283–293.

206. Dunn JC, Newell BR, Kalish ML. The effect of feedback delay and feedback type on perceptual category learning: The limits of multiple systems. *Journal of Experimental Psychology: Learning, Memory, and Cognition* 2012;38(4):840–859.

207. Maddox WT, Ashby FG, Bohil CJ. Delayed feedback effects on rule-based and information-integration category learning. *Journal of Experimental Psychology: Learning, Memory, and Cognition* 2003;29(4):650–662.

208. Filoteo JV, Maddox WT, Ing AD, Zizak V, Song DD. The impact of irrelevant dimensional variation on rule-based category learning in patients with Parkinson's disease. *Journal of the International Neuropsychological Society* 2005;11(5):503–513.

209. Waldron EM, Ashby FG. The effects of concurrent task interference on category learning: Evidence for multiple category learning systems. *Psychonomic Bulletin & Review* 2001;8(1):168–176.

210. Maddox WT, Ashby FG, Ing AD, Pickering AD. Disrupting feedback processing interferes with rule-based but not information-integration category learning. *Memory & Cognition* 2004;32(4):582–591.

211. Maddox WT, Glass BD, Wolosin SM, Savarie ZR, Bowen C, Matthews MD, Schnyer DM. The effects of sleep deprivation on information-integration categorization performance. *Sleep* 2009;32(11):1439–1448.

212. Ashby FG, Ell SW, Waldron EM. Procedural learning in perceptual categorization. *Memory & Cognition* 2003;31(7):1114–1125.

213. Ell SW, Ashby FG. The effects of category overlap on information-integration and rule-based category learning. *Perception & Psychophysics* 2006;68(6):1013–1026.

214. Xu P, Lu Z-L, Wang X, Dosher B, Zhou J, Zhang D, Zhou Y. Category and perceptual learning in subjects with treated Wilson's disease. *PLoS One* 2010;5(3):e9635.

11

Applications

Visual perceptual learning has the exciting potential to be applied to real-world problems. In this chapter, we consider how the application of visual training methods has been engaged in two broad domains: education and remediation. We first consider a number of protocols used in mathematics and reading pedagogy and then explore the potential for remediation in amblyopia, myopia, low vision, cataracts, and higher function conditions such as dyslexia and ADHD. Training may play defined roles after surgical intervention or when integrating auxiliary devices into everyday tasks. In each domain, there are unique challenges as well as special opportunities for bringing training interventions to the marketplace and/or to clinical applications.

11.1 Perceptual Learning from the Laboratory to the World

From the earliest observations of perceptual learning in humans came the recognition that extensive practice in real-world domains led to very specialized expertise. The expert can perceive and classify aspects of the world not easily accessible to novices and can often translate these perceptions into expert actions as well. Wine tasters, musicians, wool graders, and pilots are examples. All must possess perceptual expertise in order to excel at their professions.

With perceptual learning having begun in real-world examples and only later having migrated into the laboratory, how can we take research back out into the world? Real-world expertise is complex and can involve many levels of learning, from the perceptual and motor to the cognitive. The different aspects of expertise are often difficult to distinguish. Laboratory experiments, by contrast, are well controlled precisely to tease apart these different components. Decades of laboratory research, often using simplified stimuli and task situations, have produced numerous insights into the functions and mechanisms of learning, models of the learning processes, and the possible substrates of the underlying brain plasticity. But how can these insights and developments be translated back into real-world situations?

In this chapter, we focus specifically on applications of perceptual learning. Most of these will relate to the visual domain, although decision, attention, cognitive strategy, and perception may all be involved. The survey that follows considers a range of interventions and training protocols. These include programs to improve reading and mathematics education; the use of video games in enhancing perceptual performance; and programs designed to ameliorate visual function deficits such as amblyopia, myopia, and many others.

To varying degrees, these applications exist at the boundary of theory and practice. Some are similar to laboratory interventions, while others seek to bring visual training procedures to the consumer marketplace in the form of devices or apps. Commercialization in particular introduces challenges that may be new to the academic researcher. No longer in the realm of review boards for research on human subjects, commercialization requires that the practitioner navigate a sometimes difficult regulatory environment, design effective delivery packages, and determine the subpopulations most likely to benefit from their training interventions. The leading edge of some of this work is likely to involve devices aimed at augmenting or providing "bionic" replacements for damaged sensory inputs. With the advent of devices for augmented sensing, along with the emergence of digital monitoring, digital delivery, and artificial intelligence and machine-learning methods, these and other training technologies will almost certainly be incorporated into education, expertise, and remediation. The movement from the laboratory to the world will only grow.

11.2 Perceptual Training and Expertise

Expertise is critical to success in most real-world situations. There are examples in almost any context in which a complex task must be performed to achieve a desired outcome. Specialists in reading images operate as screeners in airports or as x-ray pathologists in hospitals. Agricultural work schedules follow the forecasts of meteorologists expert at reading weather satellite images. Some sports professionals are selected in part for their perceptual expertise. Surgeons rely on perceptual and motor skills in their technical practices. As these examples illustrate, expertise is at the foundation of an advanced, civil society.

How does expert performance differ from that of novices?[1] One core principle is that experts perceive things that cannot be seen or understood by others. Perceiving specialized patterns improves domain-specific memory, thus reducing the cognitive workload. The idea that experts recognize complex patterns that novices can only work out with effort was first highlighted in early analyses of chess masters.[2–5] A chess master looking at a board from a real game will perceive a meaningful arrangement of pieces and their relations to one another far more than an amateur player would. Once this is recognized, the pattern naturally leads to ideas about what moves might be useful, while also supporting memory and reconstruction of the board and play. In this sense, expert perception is often an integral part of expert strategy.

The detection of an integrated set or pattern of features leads almost automatically to inferences or next steps. This is true whether reading weather maps or x-rays,[6, 7] in much the same way that a certain significant number might jump out to a mathematician, such as 1728 being recognized as a perfect cube or 65,537 as the largest known Fermat prime. (Beyond the simple ability to recognize a pattern, there is also a role for fluency, roughly understood as the speed at which something can be recognized or processed.)[8, 9]

At least in real-world activities, expertise seems to require an enormous amount of practice. This truism has led researchers and pundits to make several rule-of-thumb claims about just how much practice is needed. One such claim is that it takes at least a decade to become an expert in many complex domains.[10] Another is that it requires 10,000 hours of practice.[11, 12] (The two values seem roughly consistent: putting in 10,000 hours over 10 years calculates to almost three hours per day.) While perceptual learning, at least as studied by the majority of working scientists, tends to be measured over thousands of *trials*, not hours, what is learned by the subject may nevertheless provide notable improvements in performance.

While recognizing the value of raw practice, other researchers have suggested that time alone accounts for only about half the variance in achievement. Other critical factors may include starting age, innate talent, or some other individual characteristic.[13] In music, for example, it has been claimed that the ability to benefit from practice is itself a heritable trait.[14] Individual talent is, of course, undeniably important, but even when taking the value of talent into account, a large amount of deliberate practice is almost always necessary in order to attain significant expertise. Alongside this, the natural progression of performance may also be important, whether because of the accumulation of varied experience (as with naturally occurring events in weather forecasting) or years of instruction and practice with increasingly difficult task variants.

This raises the question of whether organized perceptual training can shorten the road to expertise. There are several examples from naturalistic domains to suggest that the answer may be yes. One such domain involves natural stimuli or synthetic approximations of them. In this vein, a commercially inspired study trained novices to determine the sex of baby chickens (a specialized agricultural expertise) to near-expert performance by focusing training on diagnostic features.[15] In another study, training classification at the perceptually detailed species level ("great blue crown heron") rather than a more general class level ("wading birds" or "owls") improved bird identification more rapidly and led to better transfer to novel exemplars and the untrained category.[16] Other examples of perceptual learning in the recognition and classification of faces[17] and strange artificial entities called "greebles"[18] were discussed in chapter 2. (Greebles are synthetic entities or avatars meant to replicate visually some of the complexity of natural objects.)

The reading of medical images is another significant domain where visual training contributes to expertise. Medical x-ray experts have been shown to possess better sensitivity to low-contrast dots in medical x-rays than novices, while training novices to detect the dots in artificial x-ray images improved their ability to later detect abnormalities in real medical x-rays.[19] In another example, a perceptual learning module based on adaptive methods worked to improve the classification of skin histopathology through exposure to sample images related to injury, inflammation, or other disease processes.[20] Repeated viewing of video clips of surgery likewise improved relevant pattern recognition,[21] while applications of perceptual training have been used in nursing education as well.[22]

These examples illustrate the potential for perceptual training to enhance or speed training toward domain-specific expertise. Alongside its medical applications, another set of specialized task domains in which perceptual learning may prove useful is in complex operator environments. Although such domains remain only selectively explored, there are several interesting studies. One used perceptual training modules for pilots by exposing trainees to variants of ground terrain, aeronautical chart patterns, and flight instruments. The speed and accuracy of the corresponding judgments of nonpilots after training approached those of experienced pilots before additional training, leading the authors to propose that perceptual learning modules could automatize and improve the fluency of certain components of the pilot's operational activities.[8] A similar study showed that training on motion displays improved the estimation of car collision trajectories in college students.[23]

Visual training interventions have also been implemented in sports. Giving college baseball players practice in detecting targets in a range of visual search tasks improved the contrast-sensitivity functions of these players. The authors of this study even suggested that this training might be credited for the improved overall record of wins that the players' team recorded that season.[24] Other research groups have investigated the potential benefits of visual training with Nike glasses, devices that interrupt the view of the environment to create a stroboscopic experience. In one study, players practiced visual-motor training tasks with the glasses, which led to improvements in motion sensitivity and attention to central vision but improved neither in the periphery.[25, 26] In another study, ice hockey players showed a significant improvement in on-ice performance after training, while a control group showed no improvement. Such examples suggest that training under difficult viewing conditions might provide incremental but perhaps useful improvements in visual performance that in turn contribute to overall athletic performance.

The logical conclusion to draw across all these spheres is that the potential applications for perceptual training are consequential. Training modules might be inspired by, or even simply import, protocols developed in the laboratory. Though there are considerable reasons to be excited by this prospect, it must also be remembered that real-world tasks often involve the selection or categorization of complex cues in dynamic, complex environments. This is especially true when they are compared to typical laboratory studies. Furthermore, an additional characteristic of real-world expertise is its robustness in the face of environmental variation. This almost surely requires building up representations for complex cues within and across contexts. With this in mind, perceptual training could start by aiming for something more modest: the improvement of one component out of a complex set of interacting skills. If perceptual ability is an important factor limiting performance, then training may be a key ingredient for expertise. As proposed in our analysis of learning and transfer, this means training the limiting factor.[27]

A further insight suggests that perceptual training modules might *accelerate* learning simply by arranging exposure to training examples that in real life occur infrequently, artificially accelerating the experience of rare events. For example, a study on visual training in skin histology presented a long series of images associated with injury or disease over a short time—an exposure history that standard clinical training experience could not hope to match in such a short time. Similarly, replaying video clips of relevant surgical images provides many repetitions that would take countless real surgeries to encounter. The mere repetition outside standard real-world practice has potential benefits for fluency of the relevant perceptual classifications though, of course, protocols must take care not to overtrain exceptionally rare instances or give a false picture of their frequency.[8]

This caveat reflects an important downside to consider when devising training paradigms. Providing extensive exposure to low-probability examples that ordinarily would take thousands of hours to experience in a natural setting could lead to unrealistic base-rate estimates for these instances and thereby damage ultimate decision performance.[6] For this and other reasons, finding an optimal perceptual learning program presents a challenging design problem. Development of a successful intervention requires understanding the problem domain, including the variation in perceptual stimuli and the complex integrated sets of cues that normally drive behavioral discrimination and action. In some cases, a successful design might require assessing the relative frequencies of different perceptual configurations so that base rates can be built into training. The use of computer algorithms to simulate key aspects of natural situations may prove especially important as learning protocols hope to speed the achievement of expertise.

11.3 Perceptual Learning in Education

Perceptual learning is increasingly thought to have a potentially transformative role to play in the domain of education. Such a belief has given rise to the field of educational neuroscience, or *neuroeducation*, which has recently become a growing focus of a number of educational research programs. The confluence of education with perceptual training and plasticity—traditionally the domain of cognitive neuroscience—has provoked new research in this burgeoning field.[28–33] (Ironically, in some ways this could be seen as a return to the days of repetition and drills, though one would hope with more effective protocol design.)

Achieving competence in educational domains has historically been defined in relation to learning facts or solving problems. In this view, education primarily relies on the acquisition of new information or new concepts.[34] However, the *fluency* of using that information is often another goal of education, and perceptual expertise acquired through training could have a legitimate role in supporting fluent performance.

Perceptual training can be used to improve the extraction of relevant information from sensory inputs, to develop efficient encoding of input patterns that are more complex, and to improve pattern recognition, among other skills. The development of perceptual recognition routines has thus been proposed as one important factor in improving core skills of early education, such as reading and mathematics.[34,35] It has also been suggested that perceptual training could, in principle, affect more general abilities, such as working memory, that are in turn employed by more complex skills.[36, 37] In this section, we consider some of the initial research on the role of perceptual learning in several classic educational applications.

11.3.1 Training Auditory Perception to Improve Language and Reading

Can perceptual training improve reading and language processing? Two high-profile studies documenting apparent improvements in these domains went on to inspire a line of research on the effects of perceptual training interventions in education.[35, 38] Some of the potential benefits seemed to derive from surprising methods, such as training children to discriminate rapid auditory sequences in order to improve auditory language perception, sparking further interest. The proposed relationship between these listening skills and reading followed from the view that "studies of non-verbal auditory, visual and cross-modal processing have suggested that … reading impaired children may have some very basic non-verbal perceptual difficulties" and in particular that "reading impaired children have difficulty in processing temporal patterns sequentially"[39] (p. 171) (see also International Dyslexia Association, https://dyslexiaida.org/definition-of-dyslexia/).

The original studies in this line of research used computer games to train children with language-learning impairments.[35, 38] The training protocols included nonspeech auditory stimuli that nonetheless had characteristics related to speech and used frequencies in the range of formants (concentrations of energy at a frequency) of English consonants. Another task involved temporal-order judgments in rapidly presented consonant-vowel pairs excised from speech samples. For those children (ages seven to ten) identified as having a deficit, training about 20 hours over the course of a month reportedly improved several measures of receptive language. This in turn led the researchers to claim that a few hours of training with their method caused children to progress by about a year of otherwise normal advancement as measured by language assessments.[35, 38] The idea was that, for these individuals, processing rapid speech transitions was a *bottleneck* that, once remedied, permitted them to express age-appropriate levels of competency in vocabulary and grammar.[38]

Inspired by these findings, a training program was developed and commercialized as Fast ForWord (Scientific Learning Corporation). However, subsequent studies and a meta-analysis led to more cautionary conclusions.[40] The claim that the critical deficit to be trained was in temporal sequencing (as distinct from discrimination per se) has also been challenged.[41] The controversy is of special interest, as it hinged on many technical questions at the center of this book. A randomized field trial in the second and seventh grades in urban schools for students at risk of poor reading and language skills concluded that the program "did not, in general, help students … improve their language and reading comprehension test scores" (although the authors reported some implementation problems in the field setting that could in turn be used to question their results).[42] The authors of this study go on to state that, "Our supplementary analyses, which examined the causal effects of participation, revealed that when the middle school teachers and students remained committed and more faithfully achieved the completion standards set by Scientific Learning

Corporation, the students exhibited statistically significant improvements in reading comprehension"[42] (p. 99). A meta-analysis that examined six studies that used standardized tests of reading or oral language concluded: "There is no evidence ... that Fast ForWord is effective as a treatment for children's oral language or reading difficulties"[40] (p. 224). It should be noted that many of the original studies of these training protocols compared pre- and posttraining scores on special research tests of receptive language rather than broad standardized tests. While these methods are still seen as potentially promising by many, the history of testing suggests the importance of incorporating multiple assessments even early in development of a training protocol (when practically possible).

There have been several tests of related training programs aimed at other demographics with reading challenges. One study showed that individuals with dyslexia or other learning difficulties (labeled DLDs) who also showed worse performance on simple auditory discrimination could be trained with standard auditory tasks. A battery of tests likewise showed improvement in verbal working memory, though not in reading or nonverbal cognitive tasks.[43] These researchers concluded that auditory training might be one tool for "improving general working memory skills, whose underlying mechanisms seem to be shared by simple tones and complex speech sounds"[43] (p. 115).

To date, however, this research history presents a cautionary tale about the challenges faced when translating initial laboratory tests into real-world educational, commercial, or clinical applications. Many early laboratory studies reported findings in which targeted training in difficult temporal auditory phoneme tasks seemed to produce remarkable improvements. This led researchers to expect the same training scheme to have far more general consequences for reading than perhaps was the case. The important point here is not to challenge the legitimacy of the laboratory observations, or the more targeted tests of training protocols, but rather to underscore the thorny question of generalizability and the overall difficulty of extending these protocols into the actual world of applied education (a topic further discussed in section 11.6).

11.3.2 Training Visual Perception in Math Education

When we think of learning math, we generally think of learning concepts, procedures, algorithms, or analogies to solve formal problems. But on the way to gaining expertise in these domains, students also must gain fluency in recognizing when certain concepts apply and in applying the algorithms. For this reason, training in pattern identification and selection has been investigated as a possible way to enhance performance as well as provide skills-based practice on operations, with an eye to a possible role in standard math education.

This idea was pursued in some recent proof-of-concept research in which so-called *perceptual learning modules* (PLMs) were proposed to have "the potential to address crucial, neglected dimensions of learning, including discovery and fluent processing of relations"[8] (pp. 301). These principles were even seen as possibly applying to complex symbolic tasks.[8]

The situation in mathematics education might in some ways be analogous to the role of perceptual expertise in other conceptual skills such as chess or Go, where the expert "sees" patterns in the input and knows the right move or moves.[5, 10, 44–47]

The central claim in this literature is that conceptual knowledge relies on procedural knowledge, usually involving pattern recognition of input problems. From this it follows that perceptual learning could improve the likelihood and fluency of finding a problem-relevant pattern.[48, 49] One study predicated on this reasoning examined the effectiveness of a PLM for training the mapping between linear relationships, a topic in middle school mathematics.[8] These linear relations could be expressed in word problems, equations, and/or graphs. For example, for a linear equation such as $y = (50/2)x + 10$, students could be asked to choose the corresponding graph from three graphs or conversely might be given a graph and asked to choose the corresponding equation or word problem (see figure 11.1). Pre- and posttraining tests presented a word problem, graph, or equation, and the student then generated either a graph or an equation in response. Training with the PLM led to larger improvements in subsequent problem accuracy than a control practice condition, although both interventions yielded improvements. PLM training on these problems even improved performance in twelfth graders (high school), who should already have mastered them (although this might reflect the need for a refresher course). Other studies, for example, showed large improvements in response times associated with fluency, rather than accuracy, in algebraic transformations after training with a perceptual module.

In summary, perceptual training to promote the recognition of patterns or mapping structures has been shown to augment standard methods of middle school mathematics education. A question that emerges from the literature, however, is to determine the proper control interventions and the best design to use, and the degree to which other forms of general problem practice may already involve some perceptual training. A related question concerns the degree to which perceptual module training produced improvements specific to the trained formats and/or potentiated deeper conceptual understanding.

11.4 Using Video Games to Train Visual Perception

One particularly compelling approach to learning that has garnered much recent mainstream attention sees video games as a form of training intervention.[50] Video games have been said to improve everything from attention and decision to perceptual function. It is the use of games as a form of visual training that is of special interest here.

Video game training has been contrasted with other forms of visual training primarily in the claims of broad generalizability of the training effects, such as "playing certain types of video games, so called 'action video games,' leads to improvements in a broad set of behavioral abilities that extend well beyond the confines of the games themselves"[51] (p. 103). Other researchers have raised questions about these conclusions, however. Skeptics' doubts have often focused on comparisons between "gamers," who may be self-selected for high

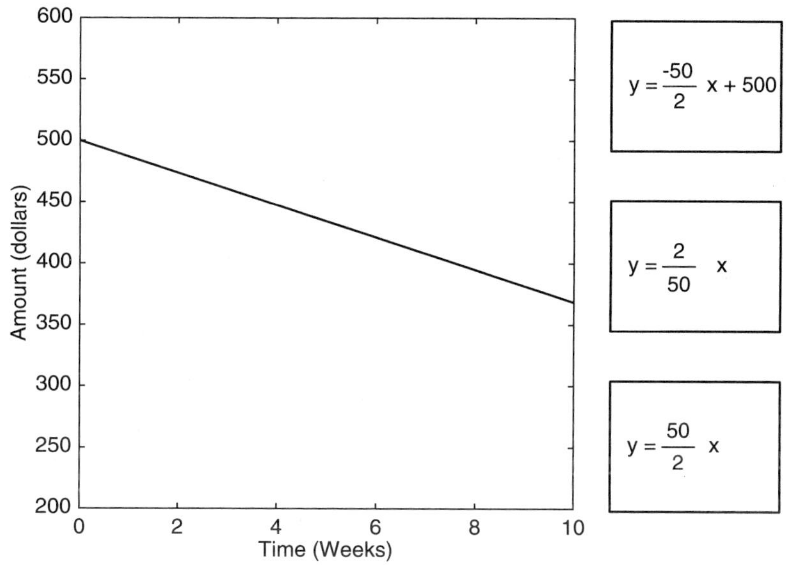

Figure 11.1
A perceptual learning module (PLM) is used to train understanding of different representations of linear relations in mathematics. The example here is similar to examples in Kellman, Massey, and Son.[8]

visual function, and "non-gamers."[50, 52] This caveat notwithstanding, the possibility of a unique role for video game training, in part because of its unique ability to motivate and reward users, is provocative.

It has been claimed that playing action video games benefits everything from low-level perceptual tasks to higher-level activities involving cognitive control. Video game expertise has been associated with improved visual sensitivity and standard visual perimetry testing,[53–56] speed of processing,[57, 58] and temporal-order judgments.[59] Researchers have suggested that gaming may also improve the ability to attend to relevant details in rapidly changing displays.[60–65] It has also been associated with improved performance in higher-level functions such as tracking moving objects,[60, 66, 67] decision-making,[68] remembering,[69, 70] cognitive control,[71] and task switching or dual task performance.[72–78]

The range of tasks reportedly affected by video game experience is vast, but it is not unlimited or indiscriminate.[79–82] Nevertheless, understanding the many influences through which gaming might affect performance remains elusive. Are improvements the result of training per se, or do they instead reflect increases in expectancy, motivation, or arousal? To what extent are the observed improvements correlative rather than causal? Responding to this last question, many studies have been cross-sectional, comparing long-standing video game players with non-video-gamers, such that the observed differences may at least partially reflect the demographic self-selection of gamers.

One recent review of the literature listed 22 cross-sectional studies of video gaming (with 18 reporting significant effects) and only 9 explicit training studies (with 8 reporting significant improvements).[50] (It should be noted, however, that several of the studies in both categories in fact report data from the same individuals in separate papers, so these observations do no reflect independent samples of subjects.) Furthermore, the authors of the review concluded that "even with optimal recruiting strategies, correlational and cross-sectional evidence for expert/novice differences is only suggestive of gaming benefits.... Claims that gaming causes cognitive improvements require an experimental design akin to a clinical trial; in this case a training experiment" (p. 3)[50].

To demonstrate conclusively that video game training improves other performance measures requires an explicit training study.[51] Such studies have recruited subjects who do not play and then trained one group in an action video game and another group in another game as a control for motivation, engagement, scheduling, interaction with experimenters, and other measures (see subsection 11.6.2). In such studies, care has also been taken to ensure that the testing order in pre- and post-test secondary measures is balanced for groups and games. The overall purpose of such protocols is to guard against epiphenomenal mediators or confounds. If the assessment tests benefit by the strategic use of eye movements, for example, and the experimental but not the control game encourages eye movements, this could favor apparent generalization of the experimental training manipulation.[50]

Another open question concerns the *kinds* of video games that produce broad training effects (including effects on visual perception, attention, and higher cognitive functions). Many key studies used heuristic genre classifications borrowed from fan communities, such as "action games" or "first-person shooter" games. These are "characterized by complex 3D settings, quickly moving and/or highly transient targets, strong peripheral processing demands, substantial amounts of clutter, and the need to consistently switch between highly focused and highly distributed attention"[51] (p. 103). Control games (Tetris or Sim City in many of these studies) may be engaging but may not require the rapid responses or attention switching of the action games. However, to avoid the tautology of assigning different games as training games and control games based on the nature of their training effects, in future studies it will be necessary to develop and test ideas about what makes one game effective and another less so. Not doing this will limit future understanding of the important features of effective training protocols.

The popularity of using and studying video games has spawned many kinds of applications, including harnessing the idea of perceptual training within a game context to inspire the design of experimentally manipulated laboratory-created games for training. The theoretical premise is that video game training is more powerful at releasing plasticity and therefore allows researchers to use it to train special populations. Either off-the-shelf action games or "designer games" created to provide training relevant to some specific population with particular deficits aim to capitalize on the motivation and reward structure of the game environment. One such study trained amblyopic adults with the unaffected fellow eye patched using a standard action video game (Medal of Honor, Pacific Assault) and compared this group to a group that practiced on a nonaction game (chess) and to another that simply patched the fellow eye.[83] Results indicated that 40–80 hours of video game training *of either sort* improved performance relative to passive patching as measured in visual acuity (by 33%), position acuity (16%), spatial attention (37%), and stereoacuity (54%) assessments. Another study presented similar findings.[54] A third study, meanwhile, reported that when dyslexic children played 12 hours of an action video game, the gameplay improved their reading speed by more than one year of spontaneous reading development or traditional reading therapy, an improvement attributed to enhanced visual attention.[84] Other initiatives have developed their own designer games to train visual functions as either training modules or apps, including a number of the educational programs and other commercial applications aiming to improve vision in aging or in populations with specific visual dysfunctions.

Given the size of the video game industry and the amount of time we now spend in front of screens, the use of video game platforms for perceptual training is only bound to increase. Amid this inevitable increase, the field may discover and codify the critical ingredients of success for video game training. Is it the demand for rapid visual analysis, decision, and action that is critical? Does the training improve general functions of selective attention?[85] Is it the fact that these games often vary stimulus sets that influences the generality of training effects? Do action video games tap into reward and punishment circuits in a special way, thus generating high levels of motivation and engagement? Or does video game expertise influence the ability to learn?[86]

Speculations, both affirmative and skeptical, have been voiced about all the potential mediators. Definitive conclusions will require further research and data. For this project to be successful, the experiments will need to be targeted, including a developed and principled classification of relevant video game characteristics and an empirical program of testing to identify the advantages and disadvantages of each of these features.

11.5 Training Limits in Visual Conditions

So far, we have looked at research detailing the development of special expertise, the use of visual training in mathematics and language education, and the use of video gaming to enhance visual learning and attention. Now we turn to the potential role visual training might play in ameliorating visual deficits.

There is a growing body of literature that uses training to remediate visual deficits in clinical populations. These include people with amblyopia, myopia, aging or presbyopia, low vision, readjustment after cataract surgery, and cortical blindness. Some of these conditions are acquired during early visual development, others can result from aging and/or long-term experience with visual correction, and still others reflect an active disease process or injury. In what follows, we examine the effects of visual training in each of these cases, beginning with amblyopia, which has been studied the most extensively.

11.5.1 Amblyopia

Amblyopia, a visual condition thought to result from cortical deficits in processing, usually in one eye, has been a significant epicenter of remedial visual training research. Sometimes called "lazy eye," amblyopia is characterized by loss of spatial vision, often because of abnormal visual development, and is estimated to affect about 2%–4% of the North American population. There are three typical varieties: anisotropic, strabismic, and so-called deprivation amblyopia. In anisotropic amblyopia, there is a large difference in refractive error between the two eyes because of nearsightedness, farsightedness, or major astigmatism. In strabismic amblyopia, the eyes cannot be properly aligned (either cross-eyed or diverged), and the images in the two eyes thus cannot be combined binocularly. In deprivation amblyopia, some process, such as childhood cataracts, limits vision in one or sometimes both eyes, almost always leading to the selection of a dominant eye and thus deficits in binocular function. Because of developmental deficits during critical periods in the neural or cortical processing, the amblyopic eye cannot simply be corrected using refractive lenses or by corrective surgery.[87, 88]

A clinical diagnosis of amblyopia is often triggered by measured differences in visual acuity between the two eyes, such as 20/20 or 20/25 in the good (fellow) eye and 20/40 to 20/200 in the amblyopic eye. If this is detected in childhood (up to seven to eight years of age), the standard treatment is to patch the good eye and thus force use of the weak eye. Another alternative treatment blurs the image in the good eye with atropine drops. In either case, the treatment forces increased reliance on the amblyopic eye, leading to improvements in the visual acuity of the amblyopic eye in about 75% of patients.[89–92]

Although the clinical definition of amblyopia is based on visual acuity, other visual functions are also affected, including contrast sensitivity,[93–95] hyperacuity,[96, 97] motion perception,[98] contour integration,[99] spatial lateral interaction,[100, 101] visual crowding,[102] and stereovision and binocular interactions relying on similar inputs from both eyes.[103] Even after treatment, however, the "good" eye often remains dominant, with binocular function reduced or deficient. Indeed, eye patching itself may negatively impact binocular vision,[104] which some researchers believe is the core deficit in amblyopia.[105] At present, the fundamental limiting factors that cause the broad deficits in performance with the amblyopic eye remain to be completely determined.

Active visual training has mostly been tested as an alternative to patching in individuals over the age of eight to ten, for whom patching is no longer considered clinically applicable. Going back to the first remedial training research, many early monocular training studies showed improvement of the amblyopic eye in the trained task, partial generalizability to other tasks, and some improvements in visual acuity.[106] One review listed 11 studies that reported training effects with improvements of a factor of two or more (i.e., a posttraining threshold half the pretraining threshold).[107] Another meta-analysis came to less sanguine conclusions and estimated that the average improvement in acuity is 0.17 log MAR units (log of the minimum angle of resolution)—equivalent to between one and two lines on an eye chart, although about one-third of individuals showed improvements of 0.2 log MAR units or more. When stereoacuity was measured, almost half the individuals showed improvements of two octaves or more.[108] Exact measurements aside, it is reasonable to conclude that training the amblyopic eye can yield significant improvements in visual performance even in adults long past the critical period, at least for some individuals.

The first influential study of perceptual learning in amblyopia introduced the so-called CAM (Cambridge) training protocol,[106] in which amblyopic children viewed rotating images of different spatial frequencies with the preferred eye patched while they played a tic-tac-toe game on a transparent plate over the display. Significant improvements in acuity were reported following total exposure time of less than an hour, consisting of a few minutes per day over several days. Subsequent studies found that playing the game with the preferred eye patched but without the rotating spatial-frequency images produced comparable improvements in acuity.[109] Together with subsequent studies using a variety of controls, this suggested to some researchers that short-term patching or occlusion plus near visual activity were important to the training recipe, and not the more complicated CAM protocol.[110–112]

Another approach, independently developed, trained tasks that required fine-grained discriminations. Intensively training a Vernier acuity task in the anisotopic-amblyopic eye significantly improved both Vernier performance and Snellen acuity.[113–116] In another example, training detection of Gabor gratings with high-contrast collinear flankers, progressing from lower to higher spatial frequencies, with varied orientations and flanker distances, improved detection of low-contrast Gabor and improved Snellen acuity.[117]

One testable hypothesis about amblyopic deficits is that internal noise and poor perceptual templates limit the amblyopic eye. This suggests that training reduces the impact of limiting noises and thus improves the template. To test this, one study monocularly trained adult anisotropic amblyopes to detect a Gabor embedded in varying amounts of external noise.[118, 119, 126] Training improved contrast thresholds in both low- and high-external-noise regions of TvC functions (signal contrast threshold versus external-noise contrast) (measured at 79.3% and 70.7%, from 3:1 and 2:1 staircases; see figure 11.2).[118] These findings align with the mechanisms of improved external-noise exclusion and stimulus enhancement by reducing internal noise (as discussed in chapter 4).

One of the more efficient training protocols sought to train the limiting factors in performance. One such limit in amblyopia is acuity, corresponding to the perception of high spatial frequencies, which suggested concentrating training at a relatively high spatial frequency near the cutoff limit of the contrast-sensitivity function. One study compared this focused training in one group to that of a group trained at a mixture of spatial frequencies and another with no training by measuring the contrast-sensitivity function before and after training in each eye separately. Concentrated training was more effective than training a mixture of spatial frequencies, compared with no changes in the control group. Improvements to contrast sensitivity and visual acuity in the amblyopic eye generalized somewhat

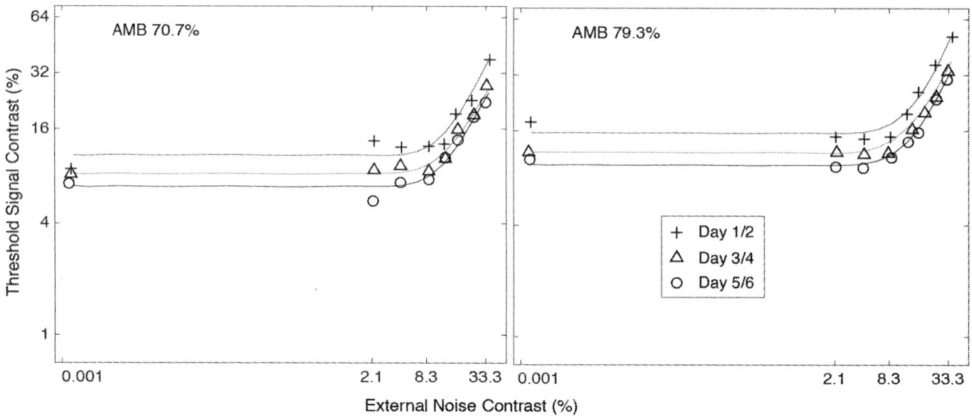

Figure 11.2
An external-noise analysis of the mechanisms of perceptual learning in amblyopes trained in contrast detection. Training improves performance at all levels of external noise, which is a mixture of stimulus enhancement and external-noise exclusion in the perceptual template model (chapter 4). After Huang, Lu, and Zhou,[118] figure 3.

to the fellow eye, and the effects persisted for a year or more.[120] Another study compared concentrated training in amblyopes to concentrated training in individuals with normal vision, with each group trained near their respective high-spatial-frequency cutoffs (10 and 25 cycles per degree, respectively).[121] The magnitude and bandwidth of the training effect showed a larger range of improved spatial frequencies in amblyopes compared to normally sighted observers, whose training effects were more narrowly focused at the single trained spatial frequency (see figure 11.3). Training in the amblyopic population generated relatively broad benefits. This concentrated spatial-frequency training also improved visual motion detection over a large range of spatial and temporal frequencies.[122]

Such improvements in acuity are promising, but leave open the question: can deficits in binocular vision be similarly improved? Binocular vision, considered by some to be the core deficit in amblyopia, is the basis for depth perception, hand-to-eye coordination, and camouflaged object recognition. In practice, other tasks do not require inputs from both eyes, so amblyopes can simply rely on the good eye.[123, 124] Standard patching or other treatments that depress the dominant eye may themselves have consequences for binocular vision or stereopsis by reducing effective binocular function.[125] On the other hand, monocular training may ultimately improve binocular function if the inputs from the amblyopic eye become more comparable to those of the fellow eye. (Strictly speaking, monocular training is not aimed at binocular vision as such, and studies have been more likely to employ acuity over binocular performance as the outcome measure because acuity is the basis for the amblyopic diagnosis.) In one study that did target binocular training, a dichoptic protocol was used to reduce relative eye dominance by placing higher-contrast images in

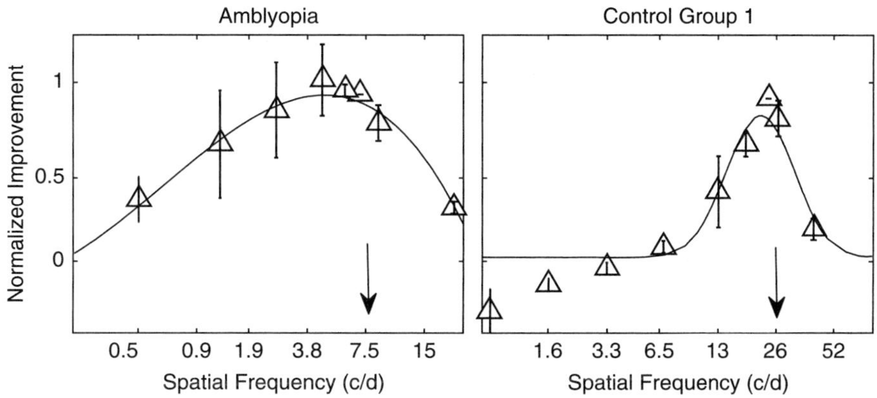

Figure 11.3
Average improvements in contrast sensitivity at different spatial frequencies following single-frequency training (indicated by the arrows) in amblyopes and in a normal control group. Training the amblyopes led to a broader bandwidth of generalization. After Huang, Zhou, and Lu,[121] figure 5.

the amblyopic eye in order to reduce suppression of that eye.[126] Training consisted of judging random-dot-motion direction near coherence thresholds, randomizing which eye contained the signal dots or the noise dots. This sharply improved coherence thresholds until those in the amblyopic eye were only slightly worse than in the fellow eye, and this generalized to improved visual acuity of the amblyopic eye and to stereoacuity. In sum, training methods that focus explicitly on binocular performance seem to have generalizable effects.

Video games were used, with the aim of releasing suppression of the amblyopic eye. One study used Tetris to examine the effect of dichoptic training using contrast-reduced shapes in the dominant eye, which resulted in greater benefits in visual acuity than in a control group trained monocularly with the dominant eye patched.[127] Subsequently shifting the monocular group to dichoptic training provided further improvements as well as a substantial improvement in stereoacuity (see figure 11.4). (The correlation of these improvements with binocular suppression was not measured in this study of anisometropic and strabismic amblyopes.)

A different training study that also used video games and dichoptic displays similarly showed improved visual acuity and stereopsis.[128] Observers played 40 hours of a video game with reduced contrasts in the fellow eye set at the beginning of each training session to equate appearance in the two eyes. This adjustment itself yielded a measure of relative binocular suppression, defined in terms of the interocular ratio (the ratio of contrasts in the fellow and amblyopic eyes). Before training, the interocular ratio was correlated with the poor acuity in the amblyopic eye and with poor stereo sensitivity (more so in anisometropic amblyopes than in strabismic ones). Training reduced suppression of the amblyopic eye, improved Gabor detection, and improved visual acuity. However, the magnitude of the improvement in visual acuity was not predicted by changes in binocular suppression,

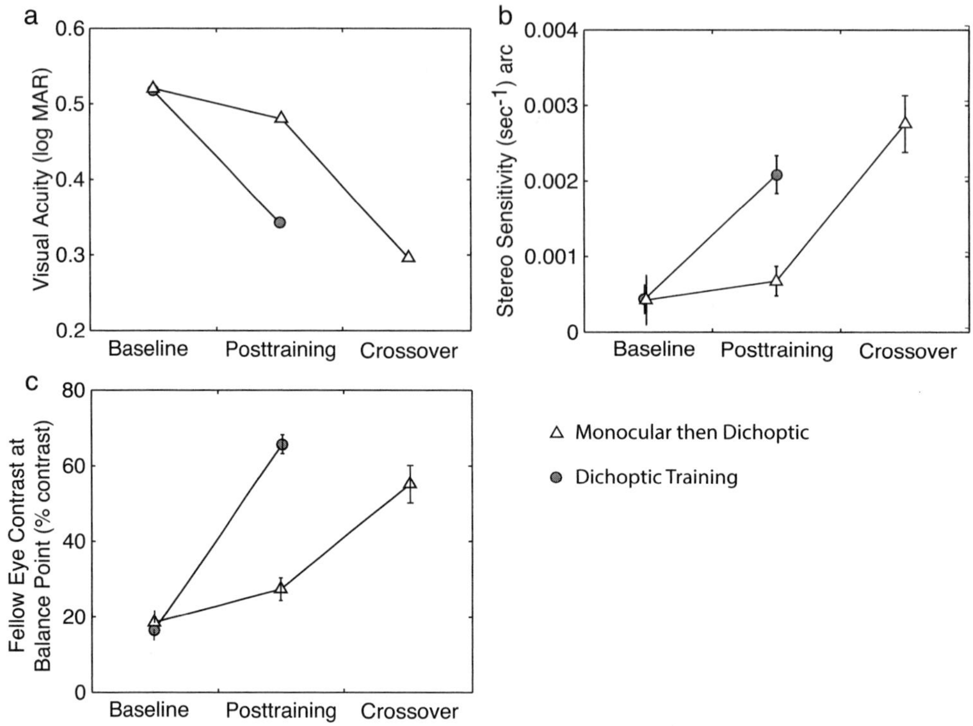

Figure 11.4
Dichoptic training with different images in the two eyes, designed to alter eye dominance, is more effective than monocular training of the amblyopic eye in improving visual acuity (a), stereo sensitivity (b), and balance point measures (c), for groups that received monocular training followed by dichoptic training or those that received dichoptic training first. After Li et al.,[127] figure 1, with permission.

leading the authors to challenge the hypothesis that release from suppression is the key training factor. Another intricate study trained the binocular balance between eyes using a "push-pull" training protocol that excited the weak eye while inhibiting the strong eye, in order to "recalibrate the interocular balance of the excitatory and inhibitory interactions"[129] (p. R309). This improved the contrast threshold in the amblyopic eye and led to improved stereo thresholds without changing the contrast threshold in the fellow eye. Yet another study directly trained depth judgments, which improved disparity thresholds by about 37%, stereoacuity by almost 60%, and visual acuity by about one line, or 19%; these improvements were largely retained when tested five months later.[130]

All these studies demonstrate that, whether observers were trained in visual acuity, contrast sensitivity, or binocular function, there is evident promise for such interventions. Perceptual learning improved visual acuity (the core measurement defining amblyopia) and generalized to other functions of spatial vision in the amblyopic eye, while improving binocular vision and stereovision. Although the resultant training effects may seem modest in terms of acuity (perhaps only about one to two lines on an eye chart) and only slightly more important in terms of binocular function, such improvements may nonetheless be of genuine significance in everyday contexts.

What remains to be fully determined, however, is how to design training so it yields the best outcome for patients. Perhaps a combination of concentrated training in high-frequency detection along with training aimed specifically at binocular rebalancing will present an optimal path. Finding these optimal protocols for rehabilitation of adult amblyopia remains an active area of research. Although most work to date has focused on incremental modifications to existing paradigms, the field has also branched out to explore more exotic and invasive interventions such as transcranial magnetic stimulation,[131, 132] drug therapy that seeks to "reopen the window of plasticity,"[133–135] and multiday light deprivation.

There are many potential avenues that research into amblyopia rehabilitation can take. Identifying core deficits and the way they vary across the heterogeneous patient population remains an ongoing project for the field. Such questions will certainly be important in translating theory into interventions aimed at improving daily visual function. Future research might also assess the *constellation* of deficits in a range of functions, including visual acuity, contrast sensitivity, binocular and stereo functions, and fixation stability or eye coordination. This could help further specify the potentially heterogeneous deficits at work in the different forms of amblyopia, thereby suggesting distinct training interventions that might focus on the particular limiting factors in possible subgroups.

So far, the literature has tended to examine training in adults who are no longer considered candidates for standard patching treatments. The potential for perceptual learning interventions in children deserves more study. Patching typically involves extended periods of deprivation that could potentially be shortened with the use of targeted protocols. However, the standard patching and diminishment protocols are intrinsically monocular and therefore may actually inhibit rather than promote strong binocular visual function that could be countered by training. At a minimum, it would be reasonable to pursue perceptual learning regimens in individuals for whom patching proves unsuccessful. Some combination of patching, visual training, and other methods, such as brain stimulation, may prove to be the optimal therapy. Of course, precisely because children are more susceptible to unforeseen side effects during the critical periods of development, caution is called for when engaging in such interventions.

Broadly speaking, there are a number of ways in which both measurements and training protocols could be improved. These include classifying the type of amblyopia more precisely, recording details of an individual's history and treatment more accurately, and informed use of the best optical correction for the amblyopic eye (which might require frequent corrective adjustments over several months). Adding control groups with alternative treatments would be especially useful, as this sampling more closely approximates the standards of randomized clinical trials. Using larger randomly selected treatment and control groups would further protect against selection biases in training and contribute to the understanding of patient subtypes. Beyond these measures, future work would also benefit from the development of computational models of core deficits and the different forms of suppression and/or interaction between the eyes as well as, perhaps, increased testing with animal models.[87, 136] All these fronts deserve to be pursued. Taken together, it is hoped that they will lead to a more sophisticated understanding of the linkage between core deficits and observed deficits in the wide range of tasks that cannot be generated by phenomenology alone, ultimately improving the chances for successful practical interventions.

11.5.2 Myopia

Some approaches similar to those used in amblyopia have been successful in improving vision in mild to moderate myopia (nearsightedness). A common visual condition usually treated with corrective lenses, myopia involves poor focus and the consequent inability to resolve details at a distance. (This is because of an excessive curvature in the cornea relative to the length of the eyeball, causing the image to be focused in front of rather than on the retina.) The National Eye Institute estimates that myopia now affects almost 43% of the US population below age 50.[137] (The exact incidence depends on a number of factors, including age, race, and region, with even higher rates in China and the Pacific Rim).[137] Although the cause of the rise in myopia is not certain, with hereditary as well as behavioral causes being suspected, it has been widely attributed to increased time in near vision tasks such as reading and computer use. Recent epidemiological studies suggest that increasing the time that children spend outdoors may be a mitigating factor.[137, 138]

Several research groups have tried to devise training to counteract myopia. Since the primary treatment is refractive correction (glasses or contact lenses), the goal has been to eliminate the need for glasses in mild cases by improving the neural processes that interpret blurry images on the retina. Two main approaches have been implemented: first, training detection in the presence of collinear distractors; and second, training contrast sensitivity across a range of spatial frequencies or near the high-frequency cutoff (see the related descriptions in the earlier discussion of amblyopia).

Within these two main training approaches, several protocols have been developed. The NeuroVision method, a packaged protocol, trains participants to detect Gabors in the presence of flankers by using stimuli with different spatial frequencies, orientations, and spatial arrangements. Thirty 30-minute sessions over several months tested and trained participants near the individual contrast threshold. One particular study found moderate effects of training in mildly myopic individuals (–1.75 diopters or less of correction), reporting significant improvements in the contrast-sensitivity function, and an average of 0.22 logMAR units in acuity, with no change in refractive error (and no improvements in the control group).[139] The claim was that training "facilitates neural connections at the cortical level" and "improves neuronal efficiency"[139] (p. 132). Another similar study also reported improvements in contrast sensitivity and mean acuity (2.1 lines logMAR, with −0.5 to −1.5 diopter myopia, retained at 12 months).[140] Subsequently, questions were raised about whether the presence of flankers was an important component of training, leading to a series of studies testing the consequences of spatial-frequency training without flankers. One study trained mild myopes (−0.75 to −2 diopters) to detect low-, medium-, or high-spatial-frequency Gabors without flankers at or near threshold and found only modestly, although significantly, improved visual acuity (0.16 logMAR units) after extensive training, but no changes in Vernier hyperacuity, contrast sensitivity, or lateral interaction tests or in refraction, accommodation, or pupil size.[141] On the other hand, concentrated monocular training at the cutoff frequency in myopes (up to −6 diopters or more of correction) over 10 sessions improved the entire contrast-sensitivity curve and visual acuity in both the trained and untrained eye by the same amount as NeuroVision training (logMAR improvement equivalent to 2.5 lines).[142] These improvements reflected both improved exclusion of external noise and reduction in internal noise, as assessed by comparing TvC curves before and after training. What this study suggested, in effect, was that the same or larger improvement in myopic visual acuity and contrast sensitivity could actually be accomplished in a shorter training period using cutoff-frequency training.

These studies showed that several training methods similar to those used in amblyopia also improved performance in the uncorrected eye in mild to moderate myopia. The results indicate that training does not alter the eye's optical or functional properties but instead improves information use at the cortical level. The best training interventions yielded about two lines of improvement (on a logMAR chart). It should be noted that refractive lenses, contact lenses, and refractive surgery have financial costs and can involve complications (infections with contact lenses, unintended side effects with surgery, and so on). While the clinical relevance of a two-line improvement has been debated, it may make it possible for individuals with mild myopia to forgo lenses more often than not or to mitigate the reasons for surgery in others (for example, military personnel operating in desert conditions often use refractive surgery). Even if perceptual learning cannot entirely eliminate the need for corrective lenses, it might conceivably help to slow the progression of the condition (although this possibility is admittedly speculative).

11.5.3 Aging and Presbyopia

As with most everyday processes, visual function is not immune to aging. Normal age-related visual losses include reduced contrast sensitivity,[143] visual acuity,[144, 145] spatial vision,[146] and motion perception,[147] as well as other deficits.[148] Reductions in contrast sensitivity and acuity are especially important from a public health standpoint, as they correlate with the rates of falls and driving accidents in older individuals.[149, 150] Age-related declines occur throughout the visual pathway, from the cornea to the cortex.[151] They can be found in diminished optical quality on the retina and move upward to changes in inhibitory processes in the early visual cortex or in temporal processing in the primary or secondary visual cortex.[148] Another age-related change occurs in presbyopia, where the eye becomes less able to focus on near objects, reflecting either loss of lens elasticity, changes in lens curvature, or reduced muscular control of curvature.

A number of studies have examined a range of interventions to see whether training might mitigate these age-related declines in vision. Training has been shown to help performance in many ways. It can speed up responses in brightness or letter discrimination,[152] can improve classification of glass patterns[153] or performance in tests of useful field of view,[154] and improve visual discrimination in texture discrimination.[148]

This last study was unusually well controlled and thus deserves further attention. The researchers initially screened to guarantee normal or corrected-to-normal vision, ruling out eye or cognitive diseases (glaucoma, macular degeneration, Parkinson's, etc.). They also arranged the view of the display through a collimation lens to eliminate any differences in accommodation between older and younger adult observers (average ages 72 and 21 years, respectively). Training near threshold SOAs (delays to the mask) in a texture task was shown to improve thresholds in older observers from around 1.5 s to about 0.25 s (close to untrained thresholds in college-age observers), while training on longer times had little effect (see figure 11.5). Improvements, still retained after three months, were specific to

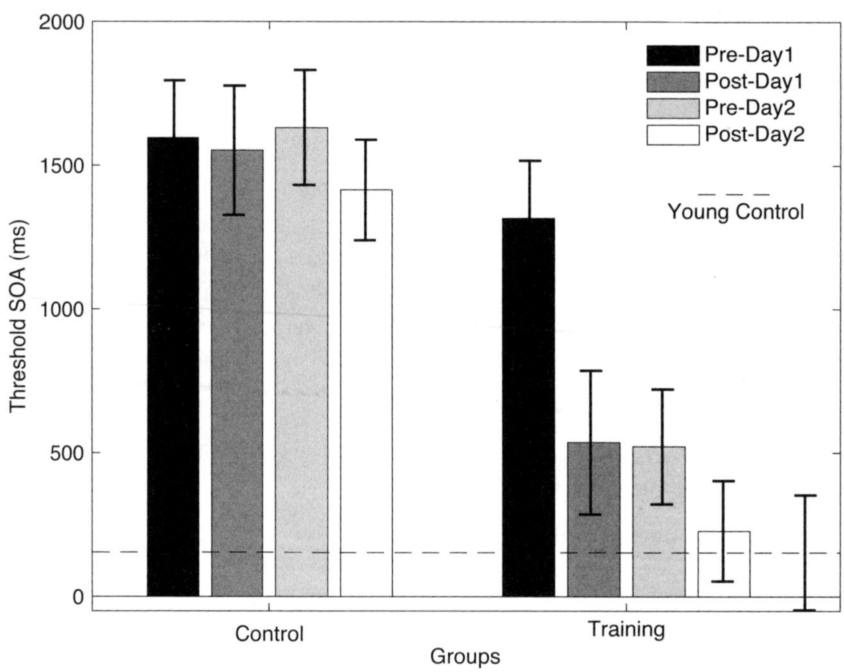

Figure 11.5
Perceptual learning reduces the threshold time between the stimulus display and the mask (SOA) in older individuals trained in the texture-discrimination task near threshold. Redrawn from data in Anderson et al.,[148] figure 2.

the trained visual quadrant and did not improve performance in a test of useful field of view, suggesting selective effects of remediation. In another kind of task, older observers, who were found to be more susceptible to high external noise, showed larger improvements in contrast thresholds in high external noise with training.[155] Here, too, performance of older observers following training approached the level of younger observers before training, while the training slightly improved acuity (equivalent to 0.5 logMAR lines).

Certain training successes have also been reported in presbyopic observers. Training with the NeuroVision method, for example, improved uncorrected near visual acuity (0.22 logMAR units), and contrast sensitivity across multiple spatial frequencies.[139] The protocol of training contrast detection with flankers likewise improved processing speed as well as contrast sensitivity and near visual acuity (about 0.2 logMAR units).[156] In the latter case, training improved contrast detection by 20%–30% and reading speed of the smallest readable print by about 17 words per minute, which is enough to bring many individuals into a more comfortable reading range (see figure 11.6). These interventions in older presbyopes produced functional improvements without changing physical accommodation, pupil size, or other physical properties of the eye. While leaving physical stimulus processing unchanged, training enhances the *use* of visual information. In many cases, the training effects directly translated to improvements in visual acuity, and sometimes even to improvements in contrast sensitivity. Though much remains to be investigated, what is already clear from this research is that some improvements in functional performance are possible based on training interventions. While the magnitude of the improvements has been modest in many cases, they may still have clinical relevance for the affected individuals.

11.5.4 Low Vision

Training-related improvements have extended to low-vision conditions. One of the most common of these is age-related macular degeneration (AMD), a condition that compromises central vision, producing a clinically significant functional loss in reading and other visual activities.[157, 158] Most of the research on central vision loss has focused on reading, with behavioral corrections that include increased lighting, magnifying glasses, or the use of larger fonts displayed in the periphery in single-word formats.[159] Alongside AMD, other diseases, such as diabetic retinopathy, can lead to patches of loss in the visual field, while glaucoma[160] and retinitis pigmentosa[161] tend to show neural-cell loss starting in the periphery and progressing toward tunnel vision near the fovea. For the affected individuals, each of these forms of low vision poses significant challenges in everyday life, most of which are not easily correctable by behavioral modifications or health aids.

At present, there are relatively few studies investigating perceptual learning in low-vision individuals. In one, subjects practiced variants of visual search, a task roughly equivalent to finding a visual object in a complex scene.[162] This task is also less efficient in older populations with normal vision who are slower and more error prone.[163–166] The patients in this study, mostly between the ages of 68 and 81, had profoundly low vision (worse than 20/200

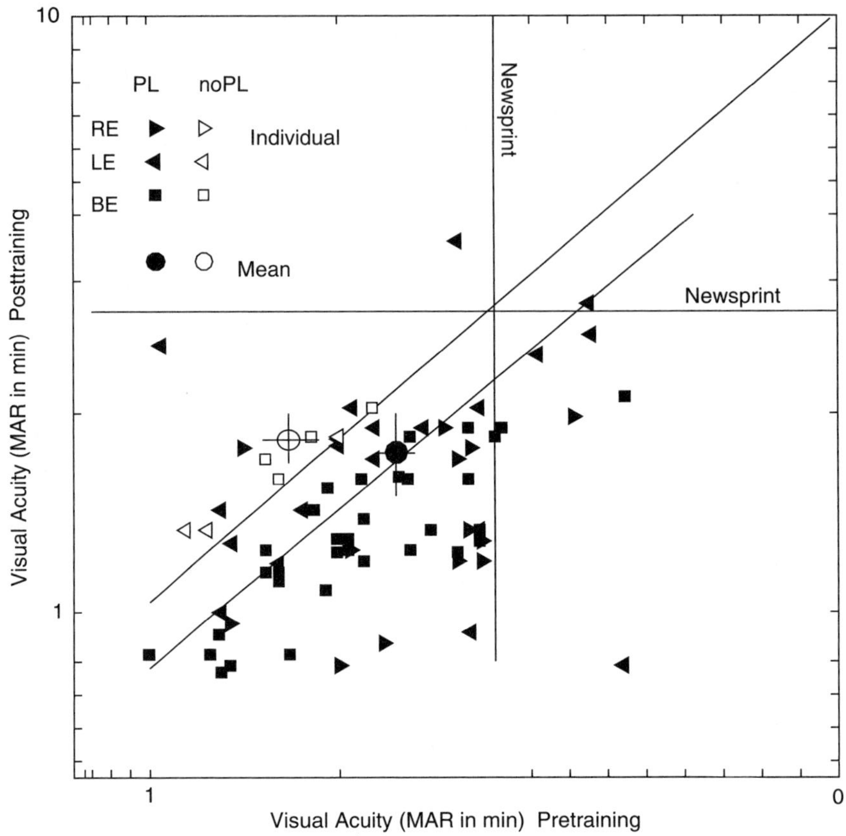

Figure 11.6
Perceptual training in contrast detection with collinear flankers improves visual acuity by about two lines (logMAR, as seen in individual and mean data for observers with perceptual learning (PL) and without (noPL). After Polat et al.,[156] figure 1a. Creative Commons, copyright 2012 Polat et al.

corrected visual acuity or less than 20% of the visual field). About 70% of them suffered from macular degeneration, and there were a few additional patients each with glaucoma, diabetic retinopathy, retinitis pigmentosa, detached retina, and other conditions. Training in visual search improved performance in both the low-vision group and the control group, with the largest improvements observed in the first few sessions in difficult training conditions.

Other projects have aimed to use training to improve reading speed in the periphery. In one study, practice was shown to improve recognition of peripheral letter trigrams in normally sighted older adults (as well as normally sighted young adults), yielding 60% increases in reading performance in the trained font size and trained visual field.[167, 168] Yet other studies trained individuals with central vision loss in RSVP reading in the periphery at the smallest viable print size in the periphery, though these did not succeed in improving reading speed, while leaving the critical print size, preferred retinal locus, and visual acuity essentially unchanged.[159]

Most of these studies have focused on patients with some form of central vision loss. For these patients, blind spots or patches (scotomas) tend to force a reliance on peripheral vision, leading them to develop one or more preferred visual locations in the periphery, or preferred retinal loci (PRLs). In a number of studies, perceptual learning was seen as potentially useful in promoting selection, stabilization, and eye movement control for good PRLs. This PRL development has been studied in normally sighted observers with a simulated scotoma (in which a computer blanks out the stimulus display near fixation), who relatively quickly developed a single PRL with practice, even though many other possible locations were visually equivalent.[169] One of the limitations of perception in the periphery is crowding, in which flankers or clutter around a letter limit its identification; it increases with distance from the fovea and depends on the distance of the flanking letters from the target letter.[102, 170–172] Perceptual training in normally sighted individuals has been shown to mitigate the negative effects of crowding, so this training may also help overcome limitations in peripheral processing, which is critically important in those with central vision loss.[173] Although the investigations to remediate low-vision conditions are just beginning, even modest training benefits may help to improve daily visual functions in these populations.

11.5.5 Adjustment to Surgery, Lenses, and Sensory Implants

New medical technologies and interventions involving vision are becoming increasingly common, and some may benefit from experience or training during adjustment periods. One common example is the insertion of new lenses during cataract surgery. There is also an active research agenda to develop prosthetics such as retinal or cortical implants and to functionalize the inputs from other senses as proxies for vision. In most of these cases, there is a period of adjustment analogous to the adjustment to new prescription glasses. For exotic prosthetics or major transformations of visual input, the adjustment to the more radically transformed input will be more akin to prism or inversion glasses.

In most cases, the protocol during the adjustment period simply involves everyday experience of the world. In a few cases, specific exposure protocols are used to improve or accelerate the transition of the patient. What is undeniable in all these cases is the critical role experience plays in the recovery of visual function. The open question, however, is to what degree specific perceptual learning or training applications might further optimize this adjustment period by capitalizing on visual plasticity.

In the last several years, a number of interesting reports have focused on improvements in vision following surgical interventions, and their dependence on experience. Examples include case studies of visual function recovery following surgical correction of childhood cataracts, case studies of more extensive surgical interventions, measurements of performance with newly inserted multifocal lenses, and studies of simulated prosthetics. In addition to other medical applications, the differential recovery of function following surgical interventions also offers a window into more general questions about visual plasticity. What can cognitive neuroscientists learn from the existing medical literature on visual recovery? In the studies described next, research activities were coupled with humanitarian efforts by volunteer ophthalmologic surgeons to investigate the timescale and developmental factors in visual recovery in adults. The fundamental question asked was if a condition, such as a childhood cataract, is corrected by surgery later in life, how much function recovers immediately and how much emerges only with further experience after the visual input has been recovered?

One of these studies, Project Prakash, *was* associated with an outreach effort in India that followed 11 individuals after they received surgery to correct bilateral childhood cataracts.[174] For the individuals in this study, cataract onset generally occurred before age 1, and surgery between the ages of 8 and 16. Before surgery, most patients had minimal acuity (as measured by tests of hand motion or finger counting at a distance of 1 meter). Within one week after cataract surgery, the patients showed improved but still somewhat deficient contrast-sensitivity functions, especially at high spatial frequencies. By 26 months, contrast sensitivity had further improved for about half the individuals, presumably reflecting learning from participation in normal daily functions (see figure 11.7 for data from some sample subjects). A related study carried out in Ethiopia examined shape recognition following surgery for dense early-onset bilateral cataracts.[175] Visual functions were assessed using visual search for a unique object defined by some feature. The ability to distinguish an object by a low-level feature (such as color, size, or shape) achieved a level of performance nearly comparable to that of controls, while patients continued to exhibit deficits in tasks relying on mid-level visual cues (such as occlusion, shading, three-dimensional shape, or illusory contours). These mid-level deficits had changed little even up to two years after surgery.

The pattern of recovery in these interventions was similar to that of Mike May, perhaps the most studied individual case of adult recovery of vision.[176, 177] As described in chapter 1, May suffered an accident that blinded him at age three but received a full corneal transplant 43 years later. Following his surgery, his ability in low-level tasks, including aspects of simple form, color, and motion perception, compared reasonably with that of normal observers once optical blur was taken into account. He nevertheless remained significantly deficient in more complex tasks of form, object and face recognition, especially for those involving three-dimensional processing.[176] A decade after the intervention, he still had experienced little or no improvement in mid-level and high-level functions.[178] It can be inferred from this and other cases that recovery of more complex feature recognition will be limited by the deprivation experienced by the patient during critical periods of early visual development. Recovery is more likely to be successful when early visual development associated with particular visual functions was normal.

Turning to interventions that are more routine, several training studies examined the adjustments to lenses commonly implanted as part of cataract surgery later in life. One such study examined visual performance following surgical insertion of multifocal intraocular lenses (ReSTOR from Alcon Laboratories or Tecnis ZM900 from Advanced Medical Optics) in both eyes in patients near age 70.[179] Multifocal lenses often require an extensive period of adjustment. These patients practiced threshold orientation-discrimination tasks (angular-difference thresholds) in the nondominant eye, leading to 82% improved judgments near the trained orientations without affecting those in the untrained dominant eye. (One interesting aspect of the data is that these improvements seemed to be larger in individuals with the worst initial performance.) Distance visual acuity and several other measures were also slightly better six months after the training intervention. This suggests that an effective form of explicit training might yield significant functional improvements in individuals experiencing issues in adjusting to new lenses.

On the other hand, advances in neuroscience and ideas about plasticity have increasingly led to technological research that seeks to use prosthetic replacements or other forms of

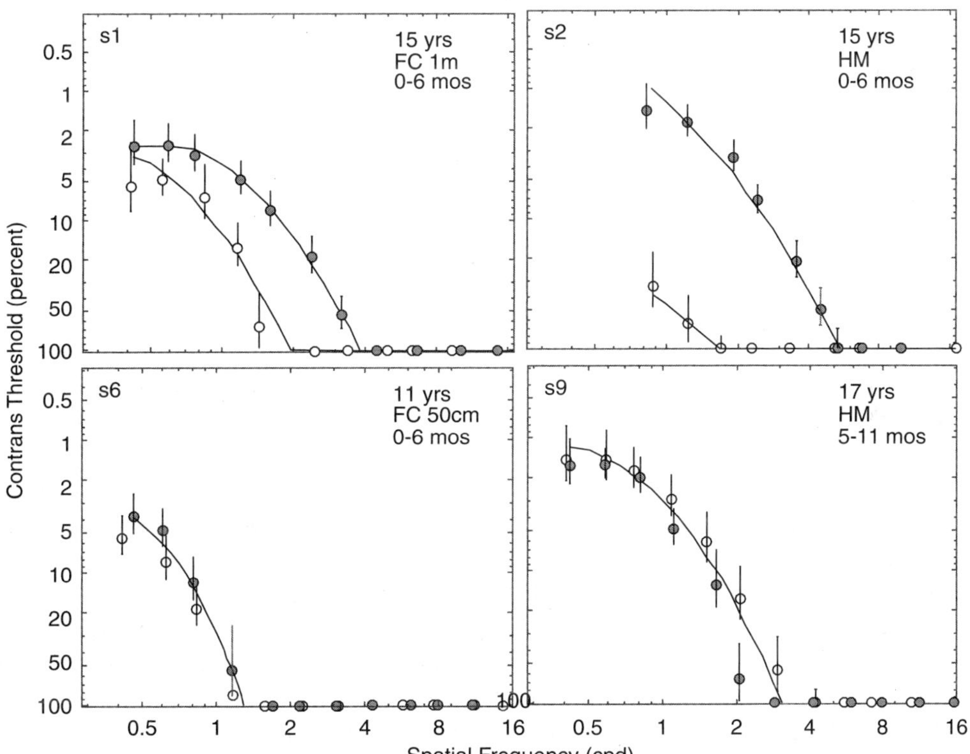

Figure 11.7
Contrast sensitivity immediately (light circles) and 26 months (dark circles) after surgery for visual deprivation caused by dense bilateral childhood cataracts, with samples of individuals showing improvement (top) and not showing improvement (bottom) for different spatial frequencies. Selected from Kaliaet al.,[174] figure 1. Copyright 2013 National Academy of Sciences.

stimulation to replace lost limbs or lost senses. In vision, a number of companies are using prosthetic implants to replace lost sight. A variety of studies have examined more unusual visual prosthetic interventions, including retinal implants,[180, 181] electrical stimulation of the retina of vision-impaired or blind individuals,[182–184] and stimulation of the visual cortex.[185, 186]

To investigate the potential benefits of using prosthetic delivery codes, researchers evaluated the role of learning and the final level of achievable performance in a simulated case of artificial vision in normally sighted individuals.[187] Messages coded as degraded 300-pixel images of pixilated common four-letter (French) words were presented in the periphery. Although these initially led to poor reading, the displays became increasingly usable during the course of about 70 hours of practice over several months—a very reasonable time investment for individuals with serious visual challenges adjusting to a prosthetic device. This suggested that visual prosthetics might ultimately parallel studies evaluating adjustments to cochlear implants.[188] In the case of cochlear implants, the initial level and subsequent improvements after implanting have been shown to vary considerably, depending on the length of the deprivation period in children deafened prelingually[189–191] and the parameters of the device.[192, 193] Adjustments to alterations in signal processing programmed in the device have also been shown to be challenging. Longtime users of cochlear implants who are then exposed to a shifted tonotopy, for example, improved in some measures over a three-month adjustment period yet never achieved the performance levels with the original clinically determined setting,[194] though adjustments were reportedly better for more gradual shifts.[188] Postsurgical or postadjustment protocols have primarily relied on natural exposure to produce adaptation to the device, although active computer-based auditory training protocols have been shown to produce improvements in auditory recognition and in performance with a simulated cochlear implant.[195, 196]

Alongside the early investigations of plasticity in auditory and tactile senses, researchers also developed an interest in the replacement of a damaged sense with another, or sensory substitution (see the brief treatment in subsection 10.3.4).[197–199] Some of these studies involve blind or nearly blind individuals with either sudden or degenerative damage to the visual system, and it must be remembered that such damage may set the stage for unusual compensatory plasticity in areas of the brain that ordinarily represent these inputs, although some believe that these plastic changes in adults may be less than originally believed.[178, 200]

As described earlier, a number of visual-to-auditory substitution devices have been tried.[201] Other examples focused on sensory substitution into two-dimensional tactile stimulation arrays, to replace either visual or vestibular signals.[197, 198, 202] Some of these examined the performance with substitution devices in unaffected individuals, such as tactile displays of visual inputs in blindfolded sighted subjects. More recently, several companies have pursued implant devices to replace visual inputs in blind individuals, typically using an electrode grid on which relatively coarse visual patterns are projected. At present, these replace some sense of the visual input either on the retina or as input to the cortex. The information they convey currently falls far short of restored vision—although any improvement may be useful for everyday function.[203] At the same time, these initial interventions may form the basic research basis for the development of future implant devices.[180, 181, 183, 184, 204] As part of this broader research project, the extraction of useful information seemed to depend on periods of adjustment and training. Some recent work has sought to characterize the percepts of an observer with a retinal implant device (the Argus, from Second Sight Medical Products, Inc.),[205] apparently with the ultimate goal of inserting a reverse-engineered transformation of the stimulus image to improve the effectiveness and consistency with the normally perceived image.[182, 203, 206] A similar process was used in the design of cochlear implant devices that improved the coding schemes as well as relying on plasticity in adjustment of the person to the device.

To summarize, the course of functional visual recovery through surgical interventions, special lenses, and implants all depend to varying degrees on experience or training to achieve the best postintervention recovery. Surgical interventions in individuals deprived of normal visual development seem to enable reasonable but somewhat incomplete recovery of basic or lower-level visual functions, such as luminance, color, size, and motion. Training can improve contrast sensitivity for some individuals, but the recovery of mid-level or high-level visual functions has often been more limited, even with time and experience. One exception to this rule applied to individuals who experienced normal early visual development, a subsequent loss of vision, and then surgical correction, who seem to be more likely to recover full function. Such results parallel similar findings in patients with cochlear implants. In this domain, partial function has resulted from sensory substitution devices. Likewise, useful information has been reported for implanted replacements for retinal inputs in the blind.

Often, improvement and adjustment following medical intervention is left to occur naturally, yet we also know that more systematic forms of training are likely to improve or expedite this natural learning curve. Exactly what protocols would work best for any given intervention, condition, or patient population remain open questions. One promising first step would be to try the most successful training protocols from related conditions (such as in amblyopia with cutoff-frequency training), though the case-by-case successes of these transpositions are still to be measured.

11.5.6 Cortical Blindness

Visual training has also been used to investigate the nature of particular visual deficits, including those from brain injuries resulting in cortical blindness. Cortical blindness occurs following damage to V1 and/or its afferent connections to other visual areas because of a stroke, accident, or tumor. The eye and other parts of the cortex remain intact, yet damage to V1, which represents basic visual features and sends this information upstream to the extrastriate visual cortex, is especially consequential.[207, 208] The result is a loss of conscious vision in the hemifield contralateral to the damage, with significant impact on daily functions.[209, 210] Although in some cases partial sensitivity to motion, form, or color is retained,[211] it is generally one to two log units worse than that in the intact hemifield.[212]

This residual sensitivity can lead to above-chance classification of stimuli in the affected field without conscious awareness, giving rise to the term "blindsight."[211, 213] As with other stroke or brain injury victims, partial recovery of function often occurs over the first three to six months (although retrograde degeneration of the corresponding LGN may also occur). Some researchers, however, find that vision in the blind field may sometimes be further improved with explicit training. This has been the topic of a recent review.[210]

Although aggressive programs of rehabilitation are routinely used following strokes or damage to the motor system, researchers report that a standard protocol for rehabilitation in the case of cortical blindness has yet to be established.[214] For most patients, improvements following damage often involve the use of compensatory eye movements, redirecting fixation to cover the visual field. An illustrative study trained one group to detect a single peripheral light while fixating while another group was trained to detect a square of four lights

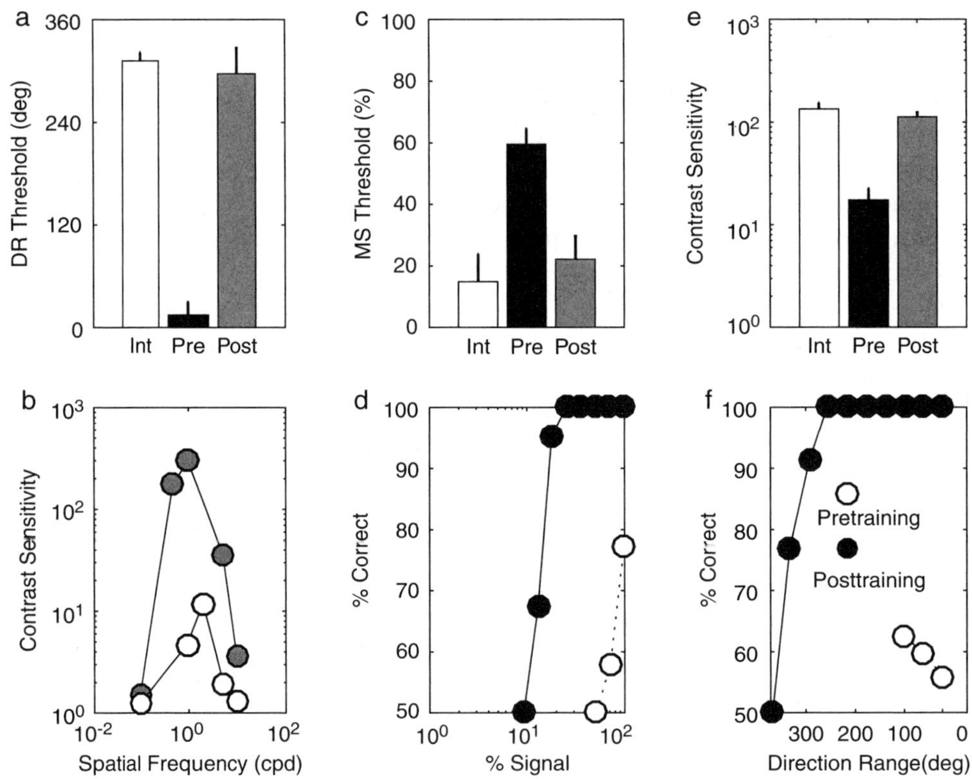

Figure 11.8
Global-motion training in the blind hemifield of patients with cortical blindness leads to retinotopically specific improvements in direction-range thresholds (a, b), left-right direction judgments (c, d), and contrast sensitivity for drifting-grating directions (e, f). After Huxlin et al.,[219] figure 2.

while scanning a board of lights. After four weeks of training, detection accuracy and reaction time had improved when eye movements were permitted but not when fixation was required. Training with eye movements improved daily living skills without changing the blind region, and these learning effects were still present after eight months. A different protocol, visual restitution therapy (VRT), developed by NovaVision, trained patients to detect bright lights at many points along the boundary between the blind and sighted fields.[215] Although initial reports claimed improvements in detection, enlargements of the sighted field, and improvements in rated daily visual function,[216] subsequent trials concluded that the protocol had actually trained small, rapid eye movements toward targets.[217, 218]

Yet another rehabilitative protocol was predicated on exploiting residual function, especially residual sensitivity to transient or moving stimuli. In one study, patients trained for several months at home to detect a pattern patch in the blind field designed to stimulate near peak spatial and temporal frequency sensitivity (one cycle per degree and 10 Hz flicker).[212] Improvements partially generalized to an untrained location in the blind field, and the training also reduced the size of the blind region, as measured with clinical visual field perimetry. A related protocol following this logic trained global motion, which is of particular interest because some visual motion pathways survive V1 damage.[210] Motion-direction thresholds in the blind hemifield were restored essentially to normal at the trained retinal locations, and this partially generalized to other stimuli and tasks (contrast sensitivity for detection of drifting gratings, thresholds for motion in external noise, and detection of luminance increments).[219] The researchers concluded that the training enhanced (reweighted) the pathways connecting residual function to higher visual areas, essentially rerouting around the damaged V1.

Further protocols continue to be developed, some of which have used external-noise methods. In one example, a variant of an external-noise paradigm and the perceptual template model (PTM; see chapter 4) was used to further assess the limiting nature of the damage in the blind hemifield by measuring threshold differences in global-motion directions at different levels of external noise (in this case, implemented as different ranges of direction of randomly moving dots) to yield threshold-versus-noise functions.[220] The thresholds of blind and intact fields were shown to differ primarily in low external noise, indicating that the blind field suffered from high levels of internal noise, with small or no changes in multiplicative internal-noise or external-noise processing.

It has been established across all these studies that cortical blindness related to V1 lesions can benefit from rehabilitative training, at least to some degree. In the same spirit as rehabilitative training after stroke-related motor system damage, it would be appropriate to pursue such training even if its ultimate benefits fall short of full function. At present, there seem to be two promising approaches to training. One focuses on training compensatory eye movement strategies, while the other trains residual visual functions to reweight (or reroute) connections to higher-level visual areas. It has been suggested that training might depend on several physiological mechanisms: learning to weight intact islands of V1; plasticity of intact regions of V1 adjacent to the lesion; and recovery of damaged regions of

V1. Other improvements seem to reflect plastic strengthening of secondary pathways to decision.[207, 221] As discussed previously (subsection 11.5.5), damage to sensory or brain systems is thought to trigger compensatory plasticity in the cortical representations that previously served the damaged input functions, and perhaps also in the development of alternative pathways. Future research should be carried out to develop successful training protocols while taking care to understand when and why any given protocol might be best suited to the needs of the individual patient.

11.6 Summary

Practical applications of perceptual learning are in their early stages but can already be found in a range of fields, most notably in educational pedagogy and medical rehabilitation, in a range of visual conditions. In a number of applications, perceptual analysis may be only one of several aspects of processing that contribute to the target behavior. In these cases, the idea is that some form of perceptual analysis might be a significant *limiting factor* in the overall performance. If this is true, perceptual training might benefit a specific subcomponent, such that visual training might then contribute to overall improvements in more complex behaviors. This idea is illustrated in figure 11.9.

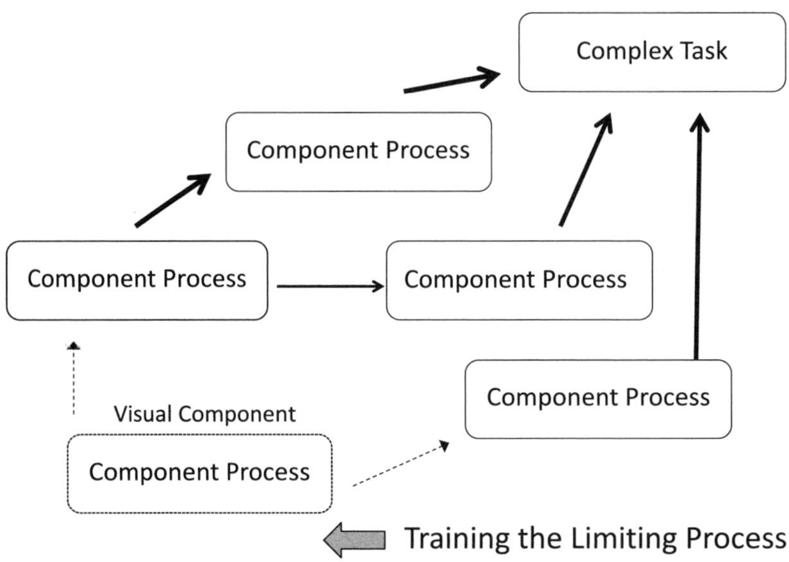

Figure 11.9
Training a weak visual component that feeds other components of a complex task may be an effective way to improve overall performance by training the limiting factor.

One of the most cited, hallmark findings of perceptual learning is its apparent specificity to particular stimuli or tasks (chapter 3). A pattern of observed specificity may have useful theoretical implications, especially when it helps to pinpoint the locus and mechanisms of plasticity. That said, specificity is rarely a virtue for translational applications, where it is almost always desirable for training to work across a broad range of related tasks. Even so, a particular training protocol or app may still be useful when specific to the trained task, so long as that task is of central importance to the learner. Trained improvements in reading, even if specific to a trained font, for example, could still yield functionally significant benefits for a low-vision individual.

Generalizability is certainly an aspiration for many forms of rehabilitative training. This is especially the case for populations with visual deficits for whom one form of training ideally would contribute to a wider range of everyday function.[222] For example, one of the reasons training studies of working memory have been of such interest is that they have been claimed to improve any task relying on fluid intelligence.[36, 37] On the other hand, wide generalization is rarely expected in most domains, and the specificity of certain kinds of learning need not always be taken as a criticism. In mathematics, for example, if a perceptual training module related to linear functions (figure 11.1) improved the solving of linear problems, it would not automatically be expected to extend to full competence in solving quadratic equations or those involving complex numbers. Although improved perception and conceptualization for simple functions might prepare the learner for problems that are more sophisticated, direct training of those more complex functions would almost surely still be required.[223, 224] A number of factors will therefore play a role in determining whether a given learning approach will be of practical value: Does training produce substantial improvement in the individual? What is the practicality or cost of that training? How useful is the trained function itself? How significantly does the training generalize to other relevant tasks?

11.7 Translating Perceptual Learning

Once a training protocol has been studied and shown to be successful, or at least potentially successful, then what? Whether in education, medical remediation, or other areas, the road from the laboratory to the marketplace necessarily involves its own array of factors. Some of these are contingent and seemingly superficial, though others can be substantive. Many revolve around a regulatory landscape. The specific road taken for any given product will have to pass through some combination of these concerns. The development of a commercial training application will likely depend not only on a successful protocol but also on the availability of venture capital; likewise, a medical application may first need to pass through a regulatory process with the Food and Drug Administration (FDA) that may include registered clinical trials. In what follows, we outline a few examples as well as the relevant issues that may arise on the road to commercialization.

11.7.1 Commercialization

This chapter has so far reviewed the successes and challenges of using specialized training protocols in three main areas: (1) the development of domain expertise; (2) educational applications; and (3) in special populations with vision challenges. Many projects have documented training-related improvements significant enough to warrant translation into practical use. Somewhat independent of this academic research, a for-profit sector has sprung up: a wild west of new commercial platforms, apps, and devices. Searching the iTunes store with terms such as "vision training," "visual training," "eye training," or "auditory training" will lead to pages of available apps at a range of prices, several of which require a monthly service charge. Other commercial products use special-purpose computer programs or even special physical devices. (Though most of these apps, programs, and devices were developed independently of academic research, a few of them have ties to academia. Indeed, one of us [Lu] has a commercial interest in a company engaged in developing algorithms and devices for vision testing.)

It is one thing to market training apps or devices primarily for enjoyment or enrichment but another to claim that this training will have significant clinical benefits. A high-end bridge-playing game may engage brain circuits that are important to keep active while aging, but the primary motivation for playing the game is enjoyment, with any cognitive benefits being purely secondary. For the many cognitive and perceptual applications, the relationship is reversed: cognitive or perceptual improvement is primary, with any enjoyment either secondary or instrumental. Some of these apps, programs, and devices are aimed at specific conditions. RevitalVision, Amblyotech, and Vivid Vision are aimed at training in either amblyopia or strabismus; NeuroVision and GlassesOff are meant for myopia and presbyopia, respectively; ULTIMEYES promises broad improvements in vision; and NovaVision aims to improve vision after traumatic brain injury.

Other applications have been marketed within educational domains, aiming to help students overcome specific conditions. Fast ForWord aims to use perceptual training to improve language functions and reading in individuals with poor reading scores. The InsightLT system incorporates perceptual learning into training for mathematics, medicine, and geography and is aimed at broad populations. On a broader scale, several large commercial enterprises have made marketing claims that they help to "train your brain" (and some have been fined by the Federal Trade Commission for making exaggerated claims).

The explosion of digital health monitoring and training has led to large-scale investments in what seems slated to be an enormous industry. Some sources claimed that the projected market value would be $6 billion annually by 2020, with market sectors in biometrics, testing, and training (http://sharpbrains.com/executive-summary/). What is clear is that the various attempts to translate perceptual learning into easily accessible apps and systems, whether successful or not, reflect a burgeoning interest on the part of both investors and consumers.

11.7.2 Challenges

Beyond the vagaries of the marketplace, there are substantive and unavoidable challenges intrinsic to the transition from laboratory research to application. In many ways, these challenges relate to the currently high-profile issue of replicability in the experimental and computational sciences, as well as to practical considerations regarding human clinical trials. The core issue behind these concerns is whether laboratory investigations, which usually use smaller subject samples and more controlled procedures, can be successfully extended to the more varied populations and less controlled training environments of the clinic and/or marketplace. In many cases, another challenge presents itself: how to expand the generalizability of the protocol to a broad enhancement of related functions.

Translation of laboratory findings into real-world applications must involve a more complete testing procedure than is currently typical in basic research. There are several methods or approaches that might be useful in addressing these challenges. Some of these are related to the approaches used in the transition toward carrying out a clinical trial (see table 11.1).

A prudent first step would involve carrying out one or more independent replication studies or a meta-analysis of multiple studies in the literature. Similar to the review of video game training discussed previously, this would need to include the appropriate statistical corrections. Another possibility could involve simply using a larger subject sample than in the original study.[225, 226] The purpose of any of these methods would be to acquire more data. Doing so should help to overcome system-level or subject variability intrinsic to smaller studies and thereby increase the likelihood that the effort and expense of developing a commercial product would prove worthwhile.

Laboratory studies of perceptual learning are often also relatively simple. They tend to examine one or two simple training tasks and, similarly, either do not assess transfer or use at most one or two transfer tests. Both these limitations could be overcome by using testing protocols that are more advanced and efficient. Perhaps the most important principle here would be to include appropriate control groups. Individuals in these groups should be assigned and tested in the same way as the experimental training group and differ only in the omission of the training intervention or the use of selected contrast interventions. Such designs are typical of the randomized controlled trials in clinical research,[227] yet many standard designs in the perceptual learning literature do not use independent control groups, most likely because this would

Table 11.1
Experimental factors important in translation from the laboratory

Replications, larger subject samples, meta-analyses
Control group(s)
Randomized assignment to groups
Pre- and post-assessment batteries
Classification of subject subtypes

require larger sample sizes (see alternative protocols in subsection 3.8.4). Any of the methods discussed would make a sizable contribution to assuaging concerns about replicability.

After this first step, clinical trials in medical applications tend to prize double-blind designs. The analogous situation in training protocols faces its own difficulties. Truly double-blind designs are difficult to achieve in training protocols because, as in exercise or meditation interventions, the person will almost surely find out something about the treatment as it is being administered. Nevertheless, it is still possible to do blind pre- and post-assessment testing and data analysis by research team members who are blind to the group assignment. Other forms of control training interventions might similarly engage the subject in ways that are not expected to be as efficacious (although the ethical issues of withholding optimal treatment should also be considered for each case).

An experimental design feature that would further inform laboratory studies, even if purely for research purposes, would be to include more extensive pre- and post-assessment batteries and tests of transfer in a range of tasks (see figure 11.10).[222] Such designs are by definition more demanding, but they would help to support better classification of population subtypes, while aiding a better understanding of transfer. The subtype classifications of specific populations (different forms of amblyopia, for example, or different forms of dyslexia) would support more nuanced conclusions if the intervention were more successful with one subtype than another. The assessment with a battery of transfer tasks would furthermore provide a better understanding of the range of benefits of the training

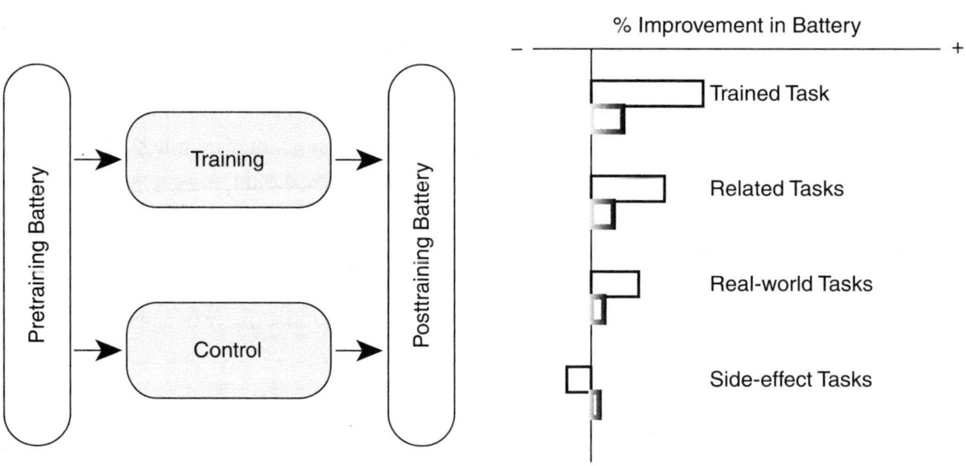

Figure 11.10
Suggested structure of experiments designed to support translation to real applications, illustrating small changes for the control group and larger changes in performance for the training group. The features of these designs are analogous to some features of clinical trials, including control groups, and more comprehensive pre- and posttraining batteries that evaluate benefits for the trained task, related tasks, and other real-world tasks, and potential side effects. After Lu, Lin, and Dosher,[222] figure 1.

intervention, helping to answer a range of questions: Does the intervention improve performance on the trained task or on related tasks? Does it improve outcomes in real-world tasks or activities of daily living? Does the intervention have any unintended side effects in general or specific subpopulations? More complete information of this kind would surely guide future forays in translational product design.

11.7.3 The Regulatory Environment

Any translational perceptual learning product will face a number of hurdles on its way to the marketplace. These hurdles will be especially meaningful if the product is targeted at the remediation of specific conditions or intended for use in clinical populations. In either case, the product must negotiate a given regulatory environment, so the choices behind commercialization have significant consequences. A training system or app, for example, is regulated as a *medical device* by the FDA when its intended use, as conveyed by "labeling claims, advertising materials, or oral or written statements by manufacturers or their representatives" purports to aid in "the diagnosis of disease or other conditions, or the cure, mitigation, treatment, or prevention of disease, or is intended to affect the structure or any function of the body of man, " and the "level of regulatory control necessary to assure safety and effectiveness varies based upon the risk the device presents to public health."[228] Vision-training protocols that are translated into a commercial product making specific health claims—rather than being marketed simply as entertainment or enrichment activities—will require processing through the FDA. Early consultation with FDA experts can save time by guiding specific aspects of product development to pursue a less risky route to commercialization.

If a training system is to be promoted as part of a remedial medical procedure (analogous to the role of physical and occupational therapy following a stroke), then it becomes a *treatment protocol* and is very likely to require successful clinical trials prior to broad application. Whether a particular training product is more suited to testing as a clinical trial or as a medical device, the procedures of either regulatory route raise important questions about potential side effects. In basic science, researchers using human subjects must inform them of the potential risks and benefits of research protocols and assure them that the relevant institutional review boards have approved the protocols. For the clinical application of vision training, the needs and vulnerabilities of the intended population will be central in defining how to evaluate potential risks and benefits. Visual experience plays a unique role in developing visual function during childhood,[229] and this in turn implies a higher standard in assessing and developing training protocols that are intended for children. Alternatively, blindsight patients coping with existing losses in daily function may vary in their willingness to experiment with different forms of rehabilitation or retraining. Fundamentally, the decision whether to develop and market an application as a medical device or as a training protocol must be made on a case-by-case basis, and likewise for appropriate standards in assessment.

Another central issue in the evaluation of medical or commercial applications is the opportunity cost of pursuing the training system as compared to other possible interventions or activities. These costs may be literal if the commercial product is expensive or involves a prescription, but they may also be more diffuse, involving time and effort that could be directed elsewhere. For visual training, which often involves multiple days or sessions, the time demands can be especially important, because they might reduce compliance with the protocol and also because they could displace other, potentially more valuable, forms of remediation. For this reason, vision-training systems should be optimized for both efficacy and efficiency. In addition, both clinical testing and regulation involve their own costs. As a result, once a trial is started, restrictions are put in place to limit the incorporation of new protocols and/or procedures.

Those eager to move training innovations into the marketplace have raised the following questions: What is the most appropriate approach to regulating and overseeing vision-training products? Should the regulatory environment be as strict as it is for the development of drugs? Some commentators have suggested that the FDA should treat therapeutic gaming products similarly to other noninvasive systems for health or fitness. The approach to regulating medical smartphone apps has largely been restricted to the issuance of guidelines for developers and consumers. As the field advances and grows, the regulatory environment might usefully provide a menu of different routes from basic research to application.

Although the process of progressing through the regulatory systems to get an approved protocol or an approved device may be more challenging than developing an enrichment tool, there are significant benefits as well. The consumer benefits from the assurance that the protocol, device, or app has been certified effective and has been screened for potential side effects. A commercial developer benefits from official approval and recognition of the product, effectively legitimizing it for use in certain treatment contexts and lending it valuable promotion. For all these reasons, the regulatory environment is likely to be changing in concert with the field itself.

11.8 Conclusions

Beyond the intrinsic benefits of pure research, there is intense interest in translating the principles and results of perceptual learning research into practical applications. This chapter considered many examples of potential applications inspired by both empirical results and theoretical principles. These included training-based development of expertise in perceptual domains, the use of perceptual training in education, applications of laboratory training principles to clinical populations in vision, and the role of digital games in future developments. Despite this range of domains, researchers have only begun to scratch the surface. There are many more domains, skills, and functions still to be investigated.

The promise of perceptual learning is matched by the challenges of applying it. Perhaps the most serious call for future research involves large-sample experiments that use carefully selected control interventions and random assignment of subjects to conditions—a move toward an investigative model inspired by the clinical trial system in biomedical research. Given the many observations of specificity in the laboratory literature, and the importance of generalizability in most real-world applications, we also advocate the use of broader pre- and posttraining assessment batteries and improved demographic characterization of the subjects. All these features of the ideal experiment, together with the implementation of extensive perceptual training itself, create practical demands on time and resources. This suggests the need for the careful refinement of protocols in smaller-scale laboratory interventions, along with the systematic use of technical models in optimizing training protocols, prior to engaging in large-scale experiments.

Indeed, many or most of the studies considered here began with smaller-scale laboratory investigations. In developing methods and studies, it is important to start with what is already known. From a practical point of view, this often means that assessments are well-known laboratory tests and that the training protocols are the same or similar to ones developed in the laboratory. Existing assessments and training protocols provide an excellent starting point. Principles and models of perceptual learning that have already been tested may further inform possible selections for application. The next step would then be to see whether there are ways to improve the efficiency of learning and the generalization of these improvements to a targeted set of stimuli and tasks. Ultimately, good theories along with good computational models will be central to the enterprise. Chapter 12 examines a formal structure to achieve this optimization.

References

1. Chi MT, Glaser R, Farr MJ. *The nature of expertise*. Psychology Press;2014.
2. De Groot AD, de Groot AD. *Thought and choice in chess*. Vol. 4. Walter de Gruyter;1978.
3. De Groot AD, Gobet F, Jongman RW. *Perception and memory in chess: Studies in the heuristics of the professional eye*. Van Gorcum;1996.
4. Gobet F, Simon HA. Recall of random and distorted chess positions: Implications for the theory of expertise. *Memory & Cognition* 1996;24(4):493–503.
5. Reingold EM, Charness N, Pomplun M, Stampe DM. Visual span in expert chess players: Evidence from eye movements. *Psychological Science* 2001;12(1):48–55.
6. Hoffman RR, Fiore SM. Perceptual (re)learning: A leverage point for human-centered computing. *IEEE Intelligent Systems* 2007;22(3):79–83.
7. Klein GA. *A recognition-primed decision (RPD) model of rapid decision making*. Ablex;1993.
8. Kellman PJ, Massey CM, Son JY. Perceptual learning modules in mathematics: Enhancing students' pattern recognition, structure extraction, and fluency. *Topics in Cognitive Science* 2010;2(2):285–305.
9. Logan GD. Toward an instance theory of automatization. *Psychological Review* 1988;95(4):492.
10. Chase WG, Simon HA. Perception in chess. *Cognitive Psychology* 1973;4(1):55–81.

11. Ericsson KA, Krampe RT, Tesch-Römer C. The role of deliberate practice in the acquisition of expert performance. *Psychological Review* 1993;100(3):363–406.

12. Gladwell M. The picture problem. *The New Yorker* 2004: Dec. 13;74–81.

13. Hambrick DZ, Oswald FL, Altmann EM, Meinz EJ, Gobet F, Campitelli G. Deliberate practice: Is that all it takes to become an expert? *Intelligence* 2014;45:34–45.

14. Mosing MA, Madison G, Pedersen NL, Kuja-Halkola R, Ullén F. Practice does not make perfect: No causal effect of music practice on music ability. *Psychological Science* 2014;25(9):1795–1803.

15. Biederman I, Shiffrar MM. Sexing day-old chicks: A case study and expert systems analysis of a difficult perceptual-learning task. *Journal of Experimental Psychology: Learning, Memory, and Cognition* 1987;13(4):640–645.

16. Tanaka JW, Curran T, Sheinberg DL. The training and transfer of real-world perceptual expertise. *Psychological Science* 2005;16(2):145–151.

17. Peterson MF, Abbey CK, Eckstein MP. The surprisingly high human efficiency at learning to recognize faces. *Vision Research* 2009;49(3):301–314.

18. Gauthier I, Williams P, Tarr MJ, Tanaka J. Training "greeble" experts: A framework for studying expert object recognition processes. *Vision Research* 1998;38(15):2401–2428.

19. Snowden PT, Davies IR, Roling P. Perceptual learning of the detection of features in x-ray images: A functional role for improvements in adults' visual sensitivity? *Journal of Experimental Psychology: Human Perception and Performance* 2000;26(1):379–390.

20. Krasne S, Hillman JD, Kellman PJ, Drake TA. Applying perceptual and adaptive learning techniques for teaching introductory histopathology. *Journal of Pathology Informatics* 2013;4(1):34–41.

21. Guerlain S, Green KB, LaFollette M, Mersch TC, Mitchell BA, Poole GR, Calland JF, Lv J, Chekan EG. Improving surgical pattern recognition through repetitive viewing of video clips. *IEEE Transactions on Systems, Man, and Cybernetics—Part A: Systems and Humans* 2004;34(6):699–707.

22. Cioffi J. Expanding the scope of decision-making research for nursing and midwifery practice. *International Journal of Nursing Studies* 2012;49(4):481–489.

23. DeLoss DJ, Bian Z, Watanabe T, Andersen GJ. Behavioral training to improve collision detection. *Journal of Vision* 2015;15(10):2;1–7.

24. Deveau J, Ozer DJ, Seitz AR. Improved vision and on-field performance in baseball through perceptual learning. *Current Biology* 2014;24(4):R146–R147.

25. Appelbaum LG, Schroeder JE, Cain MS, Mitroff SR. Improved visual cognition through stroboscopic training. *Frontiers in Psychology* 2011;2:276;1–13.

26. Schroeder JE, Appelbaum LG, Cain MS, Mitroff SR. Examining the effects of stroboscopic vision. *Journal of Vision* 2011;11(11):1015

27. Dosher BA, Lu Z-L. Perceptual learning in clear displays optimizes perceptual expertise: Learning the limiting process. *Proceedings of the National Academy of Sciences* 2005;102(14):5286–5290.

28. Gabrieli JD. Dyslexia: A new synergy between education and cognitive neuroscience. *Science* 2009;325(5938):280–283.

29. Mason L. Bridging neuroscience and education: A two-way path is possible. *Cortex* 2009;45(4):548–549.

30. Petitto L-A, Dunbar K. New findings from educational neuroscience on bilingual brains, scientific brains, and the educated mind. Paper presented at Conference on Building Usable Knowledge in Mind, Brain, and Education: 2004.

31. Fischer KW. Mind, brain, and education: Building a scientific groundwork for learning and teaching. *Mind, Brain, and Education* 2009;3(1):3–16.

32. Carew TJ, Magsamen SH. Neuroscience and education: An ideal partnership for producing evidence-based solutions to guide 21st century learning. *Neuron* 2010;67(5):685–688.

33. Ansari D, De Smedt B, Grabner RH. Neuroeducation—a critical overview of an emerging field. *Neuroethics* 2012;5(2):105–117.

34. Kellman PJ, Massey CM. Perceptual learning, cognition, and expertise. *Psychology of Learning and Motivation* 2013;58:117–165.

35. Merzenich MM, Jenkins WM, Johnston P, Schreiner C. Temporal processing deficits of language-learning impaired children ameliorated by training. *Science* 1996;271(5245):77–81.

36. Au J, Sheehan E, Tsai N, Duncan GJ, Buschkuehl M, Jaeggi SM. Improving fluid intelligence with training on working memory: A meta-analysis. *Psychonomic Bulletin & Review* 2015;22(2):366–377.

37. Jaeggi SM, Buschkuehl M, Jonides J, Perrig WJ. Improving fluid intelligence with training on working memory. *Proceedings of the National Academy of Sciences* 2008;105(19):6829–6833.

38. Tallal P, Miller S, Bedi G, e Byrna G, Wang, X, Nagarajan S, Schreiner C, Jenkins W, Merzenich MM. Fast element enhanced speech improves language comprehension in language-learning impaired children. *Science* 1996;271(5245):81–84.

39. Tallal P. Language and reading: Some perceptual prerequisites. *Bulletin of the Orton Society* 1980;30(1):170–178.

40. Strong GK, Torgerson CJ, Torgerson D, Hulme C. A systematic meta-analytic review of evidence for the effectiveness of the "Fast ForWord" language intervention program. *Journal of Child Psychology and Psychiatry* 2011;52(3):224–235.

41. Studdert-Kennedy M, Mody M. Auditory temporal perception deficits in the reading-impaired: A critical review of the evidence. *Psychonomic Bulletin & Review* 1995;2(4):508–514.

42. Borman GD, Benson JG, Overman L. A randomized field trial of the Fast ForWord language computer-based training program. *Educational Evaluation and Policy Analysis* 2009;31(1):82–106.

43. Banai K, Ahissar M. Perceptual learning as a tool for boosting working memory among individuals with reading and learning disability. *Learning & Perception* 2009;1(1):115–134.

44. Simon H, Chase W. Skill in chess. In Levy D, ed., *Computer chess compendium.* Springer; 1988:175–188.

45. Reitman JS. Skilled perception in Go: Deducing memory structures from inter-response times. *Cognitive Psychology* 1976;8(3):336–356.

46. Chi MT. Two approaches to the study of experts' characteristics. In Ericsson KA, Charness N, Hoffman RR, Feltovich RJ, eds., *The Cambridge handbook of expertise and expert performance.* Cambridge University Press;2006:21–30.

47. Jung WH, Kim SN, Lee TY, Jang JH, Choi C-H, Kang D-H, Kwon JS. Exploring the brains of Baduk (Go) experts: Gray matter morphometry, resting-state functional connectivity, and graph theoretical analysis. *Frontiers in Human Neuroscience* 2013;7(633):1–17.

48. Kellman PJ, Massey C, Roth Z, Burke T, Zucker J, Saw A, Aguero KE, Wise JA. Perceptual learning and the technology of expertise: Studies in fraction learning and algebra. *Pragmatics & Cognition* 2008;16(2):356–405.

49. Silva AB, Kellman P. Perceptual learning in mathematics: The algebra-geometry connection. Paper presented at the Twenty-First Annual Conference of the Cognitive Science Society;1999.

50. Boot WR, Blakely DP, Simons DJ. Do action video games improve perception and cognition? *Frontiers in Psychology* 2011;2:226:1–8.

51. Green CS, Bavelier D. Action video game training for cognitive enhancement. *Current Opinion in Behavioral Sciences* 2015;4:103–108.

52. Kristjánsson Á. The case for causal influences of action videogame play upon vision and attention. *Attention, Perception, & Psychophysics* 2013;75(4):667–672.

53. Appelbaum LG, Cain MS, Darling EF, Mitroff SR. Action video game playing is associated with improved visual sensitivity, but not alterations in visual sensory memory. *Attention, Perception, & Psychophysics* 2013;75(6):1161–1167.

54. Li R, Polat U, Makous W, Bavelier D. Enhancing the contrast sensitivity function through action video game training. *Nature Neuroscience* 2009;12(5):549–551.

55. Buckley D, Codina C, Bhardwaj P, Pascalis O. Action video game players and deaf observers have larger Goldmann visual fields. *Vision Research* 2010;50(5):548–556.

56. Caplovitz GP, Kastner S. Carrot sticks or joysticks: Video games improve vision. *Nature Neuroscience* 2009;12(5):527–528.

57. Dye MW, Green CS, Bavelier D. Increasing speed of processing with action video games. *Current Directions in Psychological Science* 2009;18(6):321–326.

58. Dye MW, Green CS, Bavelier D. The development of attention skills in action video game players. *Neuropsychologia* 2009;47(8):1780–1789.

59. Donohue SE, Woldorff MG, Mitroff SR. Video game players show more precise multisensory temporal processing abilities. *Attention, Perception, & Psychophysics* 2010;72(4):1120–1129.

60. Dye MW, Bavelier D. Attentional enhancements and deficits in deaf populations: An integrative review. *Restorative Neurology and Neuroscience* 2010;28(2):181–192.

61. Green CS, Bavelier D. Action video game modifies visual selective attention. *Nature* 2003;423(6939):534–537.

62. Green CS, Bavelier D. Action-video-game experience alters the spatial resolution of vision. *Psychological Science* 2007;18(1):88–94.

63. Feng J, Spence I, Pratt J. Playing an action video game reduces gender differences in spatial cognition. *Psychological Science* 2007;18(10):850–855.

64. Wu S, Spence I. Playing shooter and driving videogames improves top-down guidance in visual search. *Attention, Perception, & Psychophysics* 2013;75(4):673–686.

65. Castel AD, Pratt J, Drummond E. The effects of action video game experience on the time course of inhibition of return and the efficiency of visual search. *Acta Psychologica* 2005;119(2):217–230.

66. Green CS, Bavelier D. Effect of action video games on the spatial distribution of visuospatial attention. *Journal of Experimental Psychology: Human Perception and Performance* 2006;32(6):1465–1478.

67. Trick LM, Jaspers-Fayer F, Sethi N. Multiple-object tracking in children: The "Catch the Spies" task. *Cognitive Development* 2005;20(3):373–387.

68. Green CS, Pouget A, Bavelier D. Improved probabilistic inference as a general learning mechanism with action video games. *Current Biology* 2010;20(17):1573–1579.

69. Sungur H, Boduroglu A. Action video game players form more detailed representation of objects. *Acta Psychologica* 2012;139(2):327–334.

70. Blacker KJ, Curby KM. Enhanced visual short-term memory in action video game players. *Attention, Perception, & Psychophysics* 2013;75(6):1128–1136.

71. Colzato LS, van den Wildenberg WP, Zmigrod S, Hommel B. Action video gaming and cognitive control: Playing first person shooter games is associated with improvement in working memory but not action inhibition. *Psychological Research* 2013;77(2):234–239.

72. Colzato LS, van den Wildenberg WP, Hommel B. Cognitive control and the COMT Val158Met polymorphism: Genetic modulation of videogame training and transfer to task-switching efficiency. *Psychological Research* 2014;78(5):670–678.

73. Karle JW, Watter S, Shedden JM. Task switching in video game players: Benefits of selective attention but not resistance to proactive interference. *Acta Psychologica* 2010;134(1):70–78.

74. Colzato LS, van Leeuwen PJ, van den Wildenberg WPM, Hommel B. DOOM'd to switch: Superior cognitive flexibility in players of first person shooter games. *Frontiers in Psychology* 2010;1:8;1–5.

75. Green CS, Sugarman MA, Medford K, Klobusicky E, Bavelier D. The effect of action video game experience on task-switching. *Computers in Human Behavior* 2012;28(3):984–994.

76. Strobach T, Frensch PA, Schubert T. Video game practice optimizes executive control skills in dual-task and task switching situations. *Acta Psychologica* 2012;140(1):13–24.

77. Cain MS, Landau AN, Shimamura AP. Action video game experience reduces the cost of switching tasks. *Attention, Perception, & Psychophysics* 2012;74(4):641–647.

78. Chiappe D, Conger M, Liao J, Caldwell JL, Vu K-PL. Improving multi-tasking ability through action videogames. *Applied Ergonomics* 2013;44(2):278–284.

79. Boot WR, Kramer AF, Simons DJ, Fabiani M, Gratton G. The effects of video game playing on attention, memory, and executive control. *Acta Psychologica* 2008;129(3):387–398.

80. Gaspar JG, Neider MB, Crowell JA, Lutz A, Kaczmarski H, Kramer AF. Are gamers better crossers? An examination of action video game experience and dual task effects in a simulated street crossing task. *Human Factors: The Journal of the Human Factors and Ergonomics Society* 2014;56(3):443–452.

81. van Ravenzwaaij D, Boekel W, Forstmann BU, Ratcliff R, Wagenmakers E-J. Action video games do not improve the speed of information processing in simple perceptual tasks. *Journal of Experimental Psychology: General* 2014;143(5):1794–1805.

82. Murphy K, Spencer A. Playing video games does not make for better visual attention skills. *Journal of Articles in Support of the Null Hypothesis* 2009;6(1):1–20.

83. Li RW, Ngo C, Nguyen J, Levi DM. Video-game play induces plasticity in the visual system of adults with amblyopia. *PLoS Biology* 2011;9(8):e1001135.

84. Franceschini S, Gori S, Ruffino M, Viola S, Molteni M, Facoetti A. Action video games make dyslexic children read better. *Current Biology* 2013;23(6):462–466.

85. Achtman RL, Green CS, Bavelier D. Video games as a tool to train visual skills. *Restorative Neurology and Neuroscience* 2008;26(4–5):435–446.

86. Bavelier D, Green CS, Pouget A, Schrater P. Brain plasticity through the life span: Learning to learn and action video games. *Annual Review of Neuroscience* 2012;35:391–416.

87. Kiorpes L. Visual processing in amblyopia: Animal studies. *Strabismus* 2006;14(1):3–10.

88. Levi DM. Visual processing in amblyopia: Human studies. *Strabismus* 2006;14(1):11–19.

89. Flynn JT, Schiffman J, Feuer W, Corona A. The therapy of amblyopia: An analysis of the results of amblyopia therapy utilizing the pooled data of published studies. *Transactions of the American Ophthalmological Society* 1998;96:431–453.

90. Holmes JM, Repka MX, Kraker RT, Clarke MP. The treatment of amblyopia. *Strabismus* 2009;14(1);37–42.

91. PEDI Group. A randomized trial of prescribed patching regimens for treatment of severe amblyopia in children. *Ophthalmology* 2003;110(11):2075–2087.

92. PEDI Group. Randomized trial of treatment of amblyopia in children aged 7 to 17 years. *Archives of Ophthalmology* 2005;123(4):437–447.

93. Bradley A, Freeman R. Contrast sensitivity in anisometropic amblyopia. *Investigative Ophthalmology & Visual Science* 1981;21(3):467–476.

94. Hess R, Howell E. The threshold contrast sensitivity function in strabismic amblyopia: Evidence for a two type classification. *Vision Research* 1977;17(9):1049–1055.

95. Levi D, Harwerth R. Contrast evoked potentials in strabismic and anisometropic amblyopia. *Investigative Ophthalmology & Visual Science* 1978;17(6):571–575.

96. Kelly SL, Buckingham TJ. Movement hyperacuity in childhood amblyopia. *British Journal of Ophthalmology* 1998;82(9):991–995.

97. Levi DM, Klein S. Hyperacuity and amblyopia. *Nature* 1982;298(5871):268–270.

98. Simmers AJ, Ledgeway T, Hess RF, McGraw PV. Deficits to global motion processing in human amblyopia. *Vision Research* 2003;43(6):729–738.

99. Hess RF, McIlhagga W, Field DJ. Contour integration in strabismic amblyopia: The sufficiency of an explanation based on positional uncertainty. *Vision Research* 1997;37(22):3145–3161.

100. Bonneh YS, Sagi D, Polat U. Local and non-local deficits in amblyopia: Acuity and spatial interactions. *Vision Research* 2004;44(27):3099–3110.

101. Bonneh YS, Sagi D, Polat U. Spatial and temporal crowding in amblyopia. *Vision Research* 2007;47(14):1950–1962.

102. Levi DM. Crowding—an essential bottleneck for object recognition: A mini-review. *Vision Research* 2008;48(5):635–654.

103. Walraven J, Janzen P. TNO stereopsis test as an aid to the prevention of amblyopia. *Ophthalmic and Physiological Optics* 1993;13(4):350–356.

104. Webber AL, Wood JM, Gole GA, Brown B. Effect of amblyopia on self-esteem in children. *Optometry & Vision Science* 2008;85(11):1074–1081.

105. Huang C-B, Zhou J, Lu Z-L, Feng L, Zhou Y. Binocular combination in anisometropic amblyopia. *Journal of Vision* 2009;9(3):17;1–16.

106. Campbell FW, Hess RF, Watson PG, Banks R. Preliminary results of a physiologically based treatment of amblyopia. *British Journal of Ophthalmology* 1978;62(11):748–755.

107. Levi DM, Li RW. Perceptual learning as a potential treatment for amblyopia: A mini-review. *Vision Research* 2009;49(21):2535–2549.

108. Tsirlin I, Colpa L, Goltz HC, Wong AM. Behavioral training as new treatment for adult amblyopia: A meta-analysis and systematic review. *Investigative Ophthalmology & Visual Science* 2015;56(6):4061–4075.

109. Tytla ME, Labow-Daily LS. Evaluation of the CAM treatment for amblyopia: A controlled study. *Investigative Ophthalmology & Visual Science* 1981;20(3):400–406.

110. Ciuffreda K, Goldner K, Connelly R. Lack of positive results of a physiologically based treatment of amblyopia. *British Journal of Ophthalmology* 1980;64(8):607–612.

111. Nyman K, Singh G, Rydberg A, Fornander M. Controlled study comparing CAM treatment with occlusion therapy. *British Journal of Ophthalmology* 1983;67(3):178–180.

112. Schor C, Wick B. Rotating grating treatment of amblyopia with and without eccentric fixation. *Journal of the American Optometric Association* 1983;54(6):545–549.

113. Levi DM, Polat U. Neural plasticity in adults with amblyopia. *Proceedings of the National Academy of Sciences* 1996;93(13):6830–6834.

114. Levi DM, Polat U, Hu Y-S. Improvement in Vernier acuity in adults with amblyopia: Practice makes better. *Investigative Ophthalmology & Visual Science* 1997;38(8):1493–1510.

115. Li RW, Levi DM. Characterizing the mechanisms of improvement for position discrimination in adult amblyopia. *Journal of Vision* 2004;4(6):7;476–487.

116. Li RW, Provost A, Levi DM. Extended perceptual learning results in substantial recovery of positional acuity and visual acuity in juvenile amblyopia. *Investigative Ophthalmology & Visual Science* 2007;48(11):5046–5051.

117. Polat U, Ma-Naim T, Belkin M, Sagi D. Improving vision in adult amblyopia by perceptual learning. *Proceedings of the National Academy of Sciences* 2004;101(17):6692–6697.

118. Huang C-B, Lu Z-L, Zhou Y. Mechanisms underlying perceptual learning of contrast detection in adults with anisometropic amblyopia. *Journal of Vision* 2009;9(11):24;1–14.

119. Levi DM, Li RW, Klein SA. "Phase capture" in amblyopia: The influence function for sampled shape. *Vision Research* 2005;45(14):1793–1805.

120. Zhou Y, Huang C, Xu P, Tao L, Qiu Z, Li X, Lu Z-L. Perceptual learning improves contrast sensitivity and visual acuity in adults with anisometropic amblyopia. *Vision Research* 2006;46(5):739–750.

121. Huang C-B, Zhou Y, Lu Z-L. Broad bandwidth of perceptual learning in the visual system of adults with anisometropic amblyopia. *Proceedings of the National Academy of Sciences* 2008;105(10):4068–4073.

122. Hou F, Huang C-B, Lesmes L, Feng L-X, Tao L, Zhou Y-F, Lu Z-L. qCSF in clinical application: Efficient characterization and classification of contrast sensitivity functions in amblyopia. *Investigative Ophthalmology & Visual Science* 2010;51(10):5365–5377.

123. Jones RK, Lee DN. Why two eyes are better than one: The two views of binocular vision. *Journal of Experimental Psychology: Human Perception and Performance* 1981;7(1):30–40.

124. Sheedy JE, Bailey IL, Buri M, Bass E. Binocular vs. monocular task performance. *Optometry & Vision Science* 1986;63(10):839–846.

125. Webber AL, Wood JM, Gole GA, Brown B. The effect of amblyopia on fine motor skills in children. *Investigative Ophthalmology & Visual Science* 2008;49(2):594–603.

126. Hess R, Mansouri B, Thompson B. A new binocular approach to the treatment of amblyopia in adults well beyond the critical period of visual development. *Restorative Neurology and Neuroscience* 2010;28(6):793–802.

127. Li J, Thompson B, Deng D, Chan LY, Yu M, Hess RF. Dichoptic training enables the adult amblyopic brain to learn. *Current Biology* 2013;23(8):R308–R309.

128. Vedamurthy I, Nahum M, Huang SJ, Zheng F, Bayliss J, Bavelier D, Levi DM. A dichoptic custom-made action video game as a treatment for adult amblyopia. *Vision Research* 2015;114:173–187.

129. Ooi TL, Su YR, Natale DM, He ZJ. A push-pull treatment for strengthening the "lazy eye" in amblyopia. *Current Biology* 2013;23(8):R309–R310.

130. Xi J, Jia W-L, Feng L-X, Lu Z-L, Huang C-B. Perceptual learning improves stereoacuity in amblyopia. *Investigative Ophthalmology & Visual Science* 2014;55(4):2384–2391.

131. Walsh V, Ashbridge E, Cowey A. Cortical plasticity in perceptual learning demonstrated by transcranial magnetic stimulation. *Neuropsychologia* 1998;36(4):363–367.

132. Seitz AR, Dinse HR. A common framework for perceptual learning. *Current Opinion in Neurobiology* 2007;17(2):148–153.

133. Fagiolini M, Fritschy J-M, Löw K, Möhler H, Rudolph U, Hensch TK. Specific GABAA circuits for visual cortical plasticity. *Science* 2004;303(5664):1681–1683.

134. Hensch TK. Critical period plasticity in local cortical circuits. *Nature Reviews Neuroscience* 2005;6(11):877–888.

135. Hensch TK, Fagiolini M, Mataga N, Stryker MP, Baekkeskov S, Kash SF. Local GABA circuit control of experience-dependent plasticity in developing visual cortex. *Science* 1998;282(5393):1504–1508.

136. Kiorpes L, Movshon JA. Neural limitations on visual development in primates. *Visual Neurosciences* 2003;1:159–173.

137. Foster P, Jiang Y. Epidemiology of myopia. *Eye* 2014;28(2):202–208.

138. Rose KA, Morgan IG, Ip J, Kifley A, Huynh S, Smith W, Mitchell P. Outdoor activity reduces the prevalence of myopia in children. *Ophthalmology* 2008;115(8):1279–1285.

139. Durrie D, McMinn PS. Computer-based primary visual cortex training for treatment of low myopia and early presbyopia. *Transactions of the American Ophthalmological Society* 2007;105:132–138.

140. Tan DT, Fong A. Efficacy of neural vision therapy to enhance contrast sensitivity function and visual acuity in low myopia. *Journal of Cataract & Refractive Surgery* 2008;34(4):570–577.

141. Camilleri R, Pavan A, Ghin F, Battaglini L, Campana G. Improvement of uncorrected visual acuity and contrast sensitivity with perceptual learning and transcranial random noise stimulation in individuals with mild myopia. *Frontiers in Psychology* 2014;5(1234):1–6.

142. Yan F-F, Zhou J, Zhao W, Li M, Xi J, Lu ZL, Huang CB. Perceptual learning improves neural processing in myopic vision. *Journal of Vision* 2015;15(10):12,1–14.

143. Richards OW. Effects of luminance and contrast on visual acuity, ages 16 to 90 years. *Optometry & Vision Science* 1977;54(3):178–184.

144. Chapanis A. Relationships between age, visual acuity and color vision. *Human Biology* 1950;22(1):1–33.

145. Kahn HA, Leibowitz HM, Ganley JP, Kini MM, Colton T, Nickerson RS, Dawber TR. The Framingham eye study. I: Outline and major prevalence findings. *American Journal of Epidemiology* 1977;106(1):17–32.

146. Sekuler R, Owsley C, Hutman L. Assessing spatial vision of older people. *Optometry & Vision Science* 1982;59(12):961–968.

147. Andersen GJ, Atchley P. Age-related differences in the detection of three-dimensional surfaces from optic flow. *Psychology and Aging* 1995;10(4):650–658.

148. Andersen GJ, Ni R, Bower JD, Watanabe T. Perceptual learning, aging, and improved visual performance in early stages of visual processing. *Journal of Vision* 2010;10(13):4, 1–13.

149. Owsley C, Ball K, McGwin G Jr, Sloan ME, Roenker DL, White MF, Overley ET. Visual processing impairment and risk of motor vehicle crash among older adults. *Journal of the American Medical Association* 1998;279(14):1083–1088.

150. Owsley C, Stalvey B, Wells J, Sloane ME. Older drivers and cataract: Driving habits and crash risk. *Journal of Gerontology Series A: Biological Sciences and Medical Sciences* 1999;54(4):M203–M211.

151. Spear PD. Neural bases of visual deficits during aging. *Vision Research* 1993;33(18):2589–2609.

152. Ratcliff R, Thapar A, McKoon G. Aging, practice, and perceptual tasks: A diffusion model analysis. *Psychology and Aging* 2006;21(2):353–371.

153. Mayhew SD, Li S, Storrar JK, Tsvetanov KA, Kourtzi Z. Learning shapes the representation of visual categories in the aging human brain. *Journal of Cognitive Neuroscience* 2010;22(12):2899–2912.

154. Richards E, Bennett PJ, Sekuler AB. Age related differences in learning with the useful field of view. *Vision Research* 2006;46(25):4217–4231.

155. DeLoss DJ, Watanabe T, Andersen GJ. Improving vision among older adults: Behavioral training to improve sight. *Psychological Science* 2015;26(4):456–466.

156. Polat U, Schor C, Tong J-L, Zomet A, Lev M, Yehezkel O, Sterkin A, Levi DM. Training the brain to overcome the effect of aging on the human eye. *Scientific Reports* 2012;2(1):1–6.

157. Elliott DB, Trukolo-Ilic M, Strong JG, Pace R, Plotkin A, Bevers P. Demographic characteristics of the vision-disabled elderly. *Investigative Ophthalmology & Visual Science* 1997;38(12):2566–2575.

158. Friedman DS, O'Colmain BJ, Munoz B, Tomany SC, McCarty C, de Jong PTVM, Nemesure B, Mitchell P, Kempen J, Congdon N. Prevalence of age-related macular degeneration in the United States. *Archives of Ophthalmology* 2004;122(4):564–572.

159. Chung ST. Improving reading speed for people with central vision loss through perceptual learning. *Investigative Ophthalmology & Visual Science* 2011;52(2):1164–1170.

160. Quigley HA. Number of people with glaucoma worldwide. *British Journal of Ophthalmology* 1996;80(5):389–393.

161. Hamel C. Retinitis pigmentosa. *Orphanet Journal of Rare Diseases* 2006;1(1):1–12.

162. Liu L, Kuyk T, Fuhr P. Visual search training in subjects with severe to profound low vision. *Vision Research* 2007;47(20):2627–2636.

163. Somberg BL, Salthouse TA. Divided attention abilities in young and old adults. *Journal of Experimental Psychology: Human Perception and Performance* 1982;8(5):651–663.

164. Kramer AF, Martin-Emerson R, Larish JF, Andersen GJ. Aging and filtering by movement in visual search. *Journal of Gerontology Series B: Psychological Sciences and Social Sciences* 1996;51(4):P201–P216.

165. Anandam BT, Scialfa CT. Aging and the development of automaticity in feature search. *Aging, Neuropsychology, and Cognition* 1999;6(2):117–140.

166. Scialfa CT, Jenkins L, Hamaluk E, Skaloud P. Aging and the development of automaticity in conjunction search. *Journal of Gerontology Series B: Psychological Sciences and Social Sciences* 2000;55(1):P27–P46.

167. Yu D, Cheung S-H, Legge GE, Chung ST. Reading speed in the peripheral visual field of older adults: Does it benefit from perceptual learning? *Vision Research* 2010;50(9):860–869.

168. Chung ST, Legge GE, Cheung S-H. Letter-recognition and reading speed in peripheral vision benefit from perceptual learning. *Vision Research* 2004;44(7):695–709.

169. Kwon M, Nandy AS, Tjan BS. Rapid and persistent adaptability of human oculomotor control in response to simulated central vision loss. *Current Biology* 2013;23(17):1663–1669.

170. Nandy AS, Tjan BS. The nature of letter crowding as revealed by first- and second-order classification images. *Journal of Vision* 2007;7(2):5,1–26.

171. Pelli DG, Tillman KA, Freeman J, Su M, Berger TD, Majaj NJ. Crowding and eccentricity determine reading rate. *Journal of Vision* 2007;7(2):20,1–36.

172. Tillman K, Pelli D, Freeman J, Su M, Berger T, Majaj N. Reading is crowded. *Journal of Vision* 2007;7(9):341.

173. Hussain Z, Webb BS, Astle AT, McGraw PV. Perceptual learning reduces crowding in amblyopia and in the normal periphery. *Journal of Neuroscience* 2012;32(2):474–480.

174. Kalia A, Lesmes LA, Dorr M, Gandhi T, Ghatterjee G, Ganesh S, Bex PJ, Sinha P. Development of pattern vision following early and extended blindness. *Proceedings of the National Academy of Sciences* 2014;111(5):2035–2039.

175. McKyton A, Ben-Zion I, Doron R, Zohary E. The limits of shape recognition following late emergence from blindness. *Current Biology* 2015;25(18):2373–2378.

176. Fine I, Wade AR, Brewer AA, May MG, Goodman DF, Boynton GM, Wandell BA, MacLeod DI. Long-term deprivation affects visual perception and cortex. *Nature Neuroscience* 2003;6(9):915–916.

177. Saenz M, Lewis LB, Huth AG, Fine I, Koch C. Visual motion area MT+/V5 responds to auditory motion in human sight-recovery subjects. *Journal of Neuroscience* 2008;28(20):5141–5148.

178. Huber E, Webster JM, Brewer AA, MacLeod DIA, Wandell BA, Boynton M, Wade AR, Fine I. A lack of experience-dependent plasticity after more than a decade of recovered sight. *Psychological Science* 2015;26(4):393–401.

179. Kaymak H, Fahle M, Mester U, Ott G. Intraindividual comparison of the effect of training on visual performance with ReSTOR and Tecnis diffractive multifocal IOLs. *Journal of Refractive Surgery* 2008;24(3):287–293.

180. Humayun MS, de Juan E, Dagnelie G, Greenberg RJ, Propst RH, Phillips DH. Visual perception elicited by electrical stimulation of retina in blind humans. *Archives of Ophthalmology* 1996;114(1):40–46.

181. Weiland JD, Liu W, Humayun MS. Retinal prosthesis. *Annual Review of Biomedical Engineering* 2005;7:361–401.

182. Beyeler M, Rokem A, Boynton GM, Fine I. Learning to see again: Biological constraints on cortical plasticity and the implications for sight restoration technologies. *Journal of Neural Engineering* 2017;14(5):051003.

183. Luo YH-L, Da Cruz L. The Argus® II retinal prosthesis system. *Progress in Retinal and Eye Research* 2016;50:89–107.

184. Ahuja AK, Dorn J, Caspi A, McMahon MJ, Dagnelie G, daCruz, L, Stanga P, Humayun MS, Greenberg RJ, Argus II Study Group. Blind subjects implanted with the Argus II retinal prosthesis are able to improve performance in a spatial-motor task. *British Journal of Ophthalmology* 2011;95(4):539–543.

185. Dobelle WH, Mladejovsky M, Girvin J. Artificial vision for the blind: Electrical stimulation of visual cortex offers hope for a functional prosthesis. *Science* 1974;183(4123):440–444.

186. Normann RA, Maynard EM, Rousche PJ, Warren DJ. A neural interface for a cortical vision prosthesis. *Vision Research* 1999;39(15):2577–2587.

187. Sommerhalder J, Oueghlani E, Bagnoud M, Leonards U, Safran AB, Pelizzone M. Simulation of artificial vision. I: Eccentric reading of isolated words, and perceptual learning. *Vision Research* 2003;43(3):269–283.

188. Fu Q-J, Galvin JJ. Perceptual learning and auditory training in cochlear implant recipients. *Trends in Amplification* 2007;11(3):193–205.

189. Fryauf-Bertschy H, Tyler RS, Kelsay DM, Gantz BJ, Woodworth GG. Cochlear implant use by prelingually deafened children: The influences of age at implant and length of device use. *Journal of Speech, Language, and Hearing Research* 1997;40(1):183–199.

190. Tyler R, Fryauf-Bertschy H, Gantz B, Kelsay D, Woodworth G. Speech perception in prelingually implanted children after four years. In Honjo I, Takahashi H, eds., *Cochlear implant and related sciences update.* Vol. 52. Karger;1997:187–192.

191. Svirsky MA, Teoh S-W, Neuburger H. Development of language and speech perception in congenitally, profoundly deaf children as a function of age at cochlear implantation. *Audiology and Neurotology* 2004;9(4):224–233.

192. Dorman MF, Loizou PC, Rainey D. Speech intelligibility as a function of the number of channels of stimulation for signal processors using sine-wave and noise-band outputs. *Journal of the Acoustical Society of America* 1997;102(4):2403–2411.

193. Wilson BS, Finley CC, Lawson DT, Wolford RD, Eddington DK, Rabinowitz WM. Better speech recognition with cochlear implants. *Nature* 1991;352(6332):236–238.

194. Fu Q-J, Shannon RV, Galvin JJ III. Perceptual learning following changes in the frequency-to-electrode assignment with the Nucleus-22 cochlear implant. *Journal of the Acoustical Society of America* 2002;112(4):1664–1674.

195. Fu Q-J, Galvin J, Wang X, Nogaki G. Moderate auditory training can improve speech performance of adult cochlear implant patients. *Acoustics Research Letters Online* 2005;6(3):106–111.

196. Eisner F, McGettigan C, Faulkner A, Rosen S, Scott SK. Inferior frontal gyrus activation predicts individual differences in perceptual learning of cochlear-implant simulations. *Journal of Neuroscience* 2010;30(21):7179–7186.

197. Proulx MJ, Brown DJ, Pasqualotto A, Meijer P. Multisensory perceptual learning and sensory substitution. *Neuroscience & Biobehavioral Reviews* 2014;41:16–25.

198. Bach-y-Rita P, Collins CC, Saunders FA, White B, Scadden L. Vision substitution by tactile image projection. *Nature*, 1969;221(5184):963–964.

199. Bach-y-Rita P, Tyler ME, Kaczmarek KA. Seeing with the brain. *International Journal of Human-Computer Interaction* 2003;15(2):285–295.

200. Wandell BA, Smirnakis SM. Plasticity and stability of visual field maps in adult primary visual cortex. *Nature Reviews Neuroscience* 2009;10(12):873–884.

201. Kim J-K, Zatorre RJ. Generalized learning of visual-to-auditory substitution in sighted individuals. *Brain Research* 2008;1242:263–275.

202. Tyler M, Danilov Y, Bach-y-Rita P. Closing an open-loop control system: Vestibular substitution through the tongue. *Journal of Integrative Neuroscience* 2003;2(2):159–164.

203. Fine I, Boynton GM. Pulse trains to percepts: The challenge of creating a perceptually intelligible world with sight recovery technologies. *Philosophical Transactions of the Royal Society B: Biological Sciences* 2015;370(1677):20140208.

204. Da Cruz L, Coley BF, Dorn J, Merlini F, Filley E, Christopher P, Chen FK, Wuyyuru V, Sahel J, Stanga P, Humayun M, Greenberg RJ, Dagnelie G, Argus II Study Group. The Argus II epiretinal prosthesis system allows letter and word reading and long-term function in patients with profound vision loss. *British Journal of Ophthalmology* 2013;97(5):632–636.

205. Nanduri D, Fine I, Horsager A, Boynton GM, Humayun MS, Greenberg RJ, Weiland JD. Frequency and amplitude modulation have different effects on the percepts elicited by retinal stimulation. *Investigative Ophthalmology & Visual Science* 2012;53(1):205–214.

206. Horsager A, Greenwald SH, Weiland JD, Humayun MS, Greenberg RJ, McMahon MJ, Boynton GM, Fine I. Predicting visual sensitivity in retinal prosthesis patients. *Investigative Ophthalmology & Visual Science* 2009;50(4):1483–1491.

207. Cowey A, Stoerig P. The neurobiology of blindsight. *Trends in Neurosciences* 1991;14(4):140–145.

208. Stoerig P, Cowey A. Visual perception and phenomenal consciousness. *Behavioural Brain Research* 1995;71(1):147–156.

209. Brandt T, Thie A, Caplan L, Hacke W. Infarcts in the brain areas supplied by the posterior cerebral artery: Clinical aspects, pathogenesis and prognosis. *Der Nervenarzt* 1995;66(4):267–274.

210. Das A, Huxlin KR. New approaches to visual rehabilitation for cortical blindness: Outcomes and Putative Mechanisms. *The Neuroscientist* 2010;16(4):374–387.

211. Weiskrantz L. *Blindsight.* Oxford University Press;1986.

212. Sahraie A, Trevethan CT, MacLeod MJ, Murray AD, Olson JA, Weiskrantz L. Increased sensitivity after repeated stimulation of residual spatial channels in blindsight. *Proceedings of the National Academy of Sciences* 2006;103(40):14971–14976.

213. Weiskrantz L, Warrington EK, Sanders M, Marshall J. Visual capacity in the hemianopic field following a restricted occipital ablation. *Brain* 1974;97(1):709–728.

214. Ellison A, Walsh V. Perceptual learning in visual search: Some evidence of specificities. *Vision Research* 1998;38(3):333–345.

215. Kasten E, Sabel BA. Visual field enlargement after computer training in brain-damaged patients with homonymous deficits: An open pilot trial. *Restorative Neurology and Neuroscience* 1995;8(3):113–127.

216. Kasten E, Wüst S, Behrens-Baumann W, Sabel BA. Computer-based training for the treatment of partial blindness. *Nature Medicine* 1998;4(9):1083–1087.

217. Balliet R, Blood KM, Bach-y-Rita P. Visual field rehabilitation in the cortically blind? *Journal of Neurology, Neurosurgery & Psychiatry* 1985;48(11):1113–1124.

218. Reinhard J, Schreiber A, Schiefer U, Kasten E, Sabel BA, Kenkel S, Vonthein R, Trauzettel-Klosinski S. Does visual restitution training change absolute homonymous visual field defects? A fundus controlled study. *British Journal of Ophthalmology* 2005;89(1):30–35.

219. Huxlin KR, Martin T, Kelly K, Riley M, Friedman DI, Burgin WS, Hayhoe M. Perceptual relearning of complex visual motion after V1 damage in humans. *Journal of Neuroscience* 2009;29(13):3981–3991.

220. Cavanaugh MR, Zhang R, Melnick MD, Das A, Roberts M, Tadin D, Carrasco M, Huxlin KR. Visual recovery in cortical blindness is limited by high internal noise. *Journal of Vision* 2015;15(10):9,1–18.

221. Weiskrantz L. Roots of blindsight. *Progress in Brain Research* 2004;144:227–241.

222. Lu Z-L, Lin Z, Dosher BA. Translating perceptual learning from the laboratory to applications. *Trends in Cognitive Sciences* 2016;20(8):561–563.

223. Craig SD, Hu X, Graesser AC, Bargagliotti AE, Sterbinsky A, Cheney KR, Okwumabua T. The impact of a technology-based mathematics after-school program using ALEKS on student's knowledge and behaviors. *Computers & Education* 2013;68:495–504.

224. Falmagne J-C, Albert D, Doble C, Eppstein D, Hu X. *Knowledge spaces: Applications in education.* Springer;2013.

225. Begley CG, Ioannidis JP. Reproducibility in science: Improving the standard for basic and preclinical research. *Circulation Research* 2015;116(1):116–126.

226. Lindsay DS. *Replication in psychological science*. Sage Publications;2015.

227. Sibbald B, Roland M. Understanding controlled trials: Why are randomised controlled trials important? *BMJ: British Medical Journal* 1998;316(7126):201.

228. Food and Drug Administration (FDA). Mobile medical applications: Guidancc for industry and Food and Drug Administration staff. FDA;2015.

229. Braddick O, Atkinson J. Development of human visual function. *Vision Research* 2011;51(13):1588–1609.

12

Optimization

The existence of strong computational models of perceptual learning opens the door to new approaches to optimization of training. As part of this new paradigm, optimization would seek to find training protocols that maximize useful qualities, such as the magnitude of learning, the efficiency of training, transfer or generalization to related tasks, and/or better retention. Many problems can now be first explored using computational frameworks in which optimization goals are defined, the domain of potential training manipulations is specified, and a good generating model already makes predictions about outcomes. A search routine would evaluate all possible protocols in relation to the desired goals, and the best protocol(s) could then be experimentally validated. The existing literature already suggests a number of manipulations that are likely candidates to improve learning and generalization, while new techniques in artificial intelligence and machine learning also promise to accelerate research in new directions.

12.1 Harnessing Visual Perceptual Learning

The science of perceptual learning has expanded dramatically over the last few decades. Beginning with the reported observations of early visual plasticity in the 1990s, the field entered a period of intense activity and innovation. New discoveries were further accelerated by the rise of computers, advanced modeling methods, and brain imaging technologies. All these factors pushed research in directions unimagined even as late as the 1980s.

One of the strengths of the field has been the lively theoretical debate about the underlying principles of learning in the visual domain. Some of these focused on phenomenology, such as the dependence of learning on the level of the visual task, while others considered the balance of specificity and transfer. At the center of the debate, however, was the question of functionality and plasticity. Did learning occur primarily through changes in the signal-to-noise ratio? If so, was this the product of changes to encoding (retuning) or decoding (readout or reweighting)? And how were these distinct processes related to the balance between plasticity and stability?

This book was meant to be an assessment of the field on a number of levels: phenomenal, experimental, and theoretical. We have surveyed the experimental territory and the major phenomena of learning, examined the historical and current models, and considered the constraints on these models from known physiology. As we have gone along, we have indicated possible directions for new experiments or models that might further develop research or answer open questions. We have also tried to strike an informed balance between core theoretical dipoles that have structured discussion and research: stability versus plasticity, signal versus noise, and encoding (retuning) versus decoding (readout or reweighting).

As we conclude the book, it is only natural to think about the future of the field as a whole. It seems to us that one of the most exciting avenues for future research involves the integration of theory and practice in ways that accelerate progress in both. One promising way to put new theoretical developments to use in practical applications is to pursue the study of learning within the context of *optimization*.

Humans are by nature goal oriented, and much of our cognitive and perceptual machinery has been optimized by evolution to carry out useful functions. Perceptual learning is such a significant faculty because perception itself is so significant, forming an intrinsic part of more complex and necessary tasks. As researchers work toward an understanding of perceptual learning and how it works, it is increasingly natural to ask, what is the best way to get better? This is where the methods of optimization have a potentially transformative role to play in the future of perceptual learning research.

12.2 An Optimization Framework

Originally developed in mathematics and economics, the theory of optimization and its related methods have found their way into an impressive range of fields from computer science to operations research and even baseball. At its most fundamental, optimization refers to the selection of the best alternative (as defined by some criteria) among a set of alternatives.

The mathematical theory of optimization has precursors dating back to the 1930s and the work of Soviet economist and mathematician Leonid Kantorovich.[1–3] The fundamental idea Kantorovich put forward was that *simulation*, rather than actual empirical testing of every conceivable option, allowed a much more efficient search for the best possible action in order to achieve a desired goal. Computation saves valuable time and effort.

Whether in economics, chemistry, or cognitive science, the science of optimization requires that researchers specify the goal(s) to be optimized. The first significant question to ask in perceptual learning, then, is how to choose the "best" training protocol. When optimizing for training, it seems to us that at least five goals should play a part: magnitude, robustness, generalization, learning to learn, and retention (see table 12.1). For some of these, such as the magnitude of learning or some aspects of generalization, there already exists substantial literature to guide the selection of possible design factors. For others, such as robustness and retention, or even stability, there is considerably less direct evidence to guide protocol selection (although related findings in allied domains can be drawn on).

The theoretical developments in the field of perceptual learning combined with those of mathematical optimization are likely to define one significant frontier of theory and

Table 12.1
Potential goals of optimization of perceptual learning

- *Magnitude* Maximizing the amount of learning with a relatively efficient training protocol. This focuses on training that has a high rate of learning.
- *Robustness* Improve performance of the trained task in contexts beyond the immediate training context. This focuses on training in the laboratory, simulator, or clinic that extends performance in a given task to a wide range of situations.
- *Generalization* Extend benefits of training to similar or related stimuli and tasks. This values training that transfers to new tasks and situations.
- *Learning to learn* Enable training of one task to improve the ability to learn subsequent tasks.
- *Retention* Training that maximizes benefits over longer periods, including conditions in which attention may be focused on other tasks.
- *Stability* Learning in an initial task that survives subsequent training.

practice. It is generally impossible, and surely inefficient, to first investigate all possible training protocols in order to identify the best one. Simply noting that one condition is better or worse than another, while potentially useful, does not constitute optimization. Using an optimization framework is so important because it allows protocols to be evaluated *virtually* by way of computational methods before moving to more costly and time-intensive testing.

The pursuit of optimization also has another virtue in that it provides a strong theoretical context within which to test existing models and develop better ones. The optimization framework may suggest new protocols to test empirically, including those that incorporate approaches to training that are more complex. Selected examples could then inspire empirical tests, and, if necessary, the models could in turn be modified to make their predictions more accurate.

Building on the theoretical insights developed throughout this book, this chapter begins by considering how mathematical optimization could be adapted to visual perceptual learning. Next, we turn to the empirical literature to suggest possible manipulations or factors that can be seen as driving learning and generalizability, thus informing the search domains of potential protocols. We then go on to discuss the implications of different learning rules, the role and requirements for generative models in optimization, and implications of the current discussions for reproducible science. Finally, we discuss the potential effects recent developments in machine learning might have on the next generation of generative models.

12.3 Stages of Optimization

In the context of visual perceptual learning, mathematical optimization invokes a set of procedures that aims to select the best protocol, as defined by some criteria, from among a set of possible protocols. The framework has several key parts: a criterion function, called an *objective function*; the *domain*, or set of possible training protocols; a *predictive engine*, usually a quantitative model, which generates predictions for any protocol; a *search algorithm*, which provides a method of searching the domain; and *validation tests* to measure the predictive accuracy.

The five steps (table 12.2) form a single pass of optimization, which may be repeated, as shown in figure 12.1. If empirical evaluation and validation suggest changes in the generative theory, this in turn could lead to another cycle of optimization. A similar schematic, more specifically focused on optimizing a perceptual learning protocol, is shown in figure 12.2.

One of the first important steps in any optimization problem is to select the goal(s) to be maximized. This process begins with the researcher (or product developer) defining an *objective function*. A simple objective function might have a single goal, such as maximizing the rate of learning, whereas a more complex function might combine several objectives, such as the rate of learning and the amount of transfer to another task. Objective functions specify metrics associated with each goal as well as weights for their multiple outcomes (defining the trade-off between the multiple objectives).[4–6] On the level of nomenclature, the optimization of a complex objective function is called Pareto optimization, the alternative yielding the global optimum is known as the Pareto optimum, and the set of alternatives for which the objectives trade off (which might occur for only a subset of all alternatives) is called the Pareto set.[5] In serendipitous situations, optimizing one goal may also optimize another. In other cases, of course, optimizing one goal damages the other. (Whether that is the case depends on the goals. For example, procedures that optimize the learning rate might do the opposite for retention.)

A *predictive engine* or *generative model* is the central ingredient in the optimization framework. In perceptual learning, it is a computational model that takes a specification

Table 12.2
The five stages of optimizing perceptual learning

- *Objective function* Defining the goal(s) of the optimization. This includes choosing the target task(s) to be learned and specifying a scoring system—some function of the behavior to be improved or maximized.
- *Predictive engine (generative model)* A generative model that takes each input configuration, possibly with specified model parameters, and predicts the performance outcome(s). This could be an equation, a function, or a set of code. In perceptual learning, it is a quantitative model of learning and performance.
- *Domain* An identified set of manipulations (and their ranges) defines the search domain. In perceptual learning, this could include changes in contrast, training accuracy, training schedule, or task formats, among others. These determine the set of possible alternative training protocols from which the optimum alternative will be chosen.
- *Search algorithm* An algorithm or method to search for an optimal training protocol (and possibly model parameters). The search is limited by the size of the search space if basic model parameter values are known; the search process will likely require adaptive procedures to simultaneously find the parameters and the best protocol if model parameter values need to be estimated.
- *Validation* Testing predictions generated by the optimization framework in confirmatory experiments. Failures of predictions may suggest modifications of the generative model, or typical parameters, which will then be used to improve another cycle of optimization.

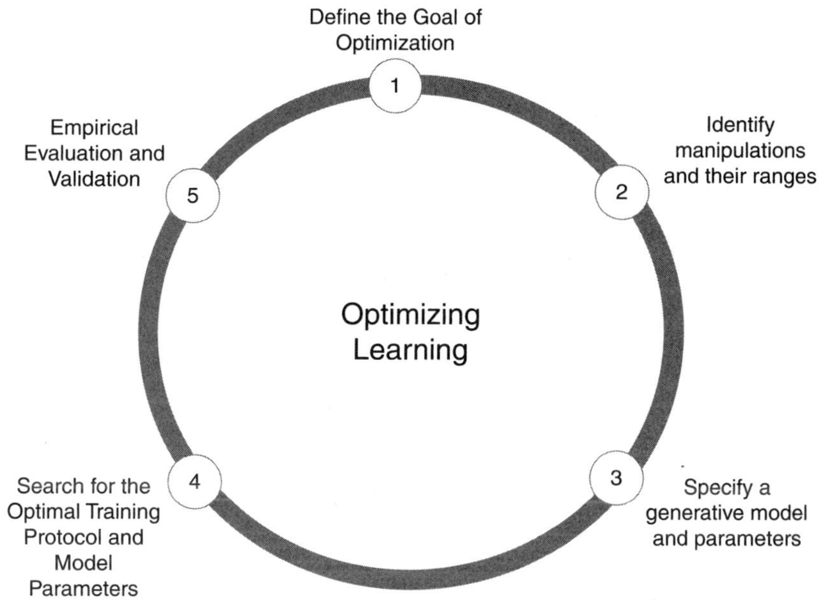

Figure 12.1
A schematic describes the process for optimizing learning by searching through possible manipulations using a generative model and its parameters to make predictions.

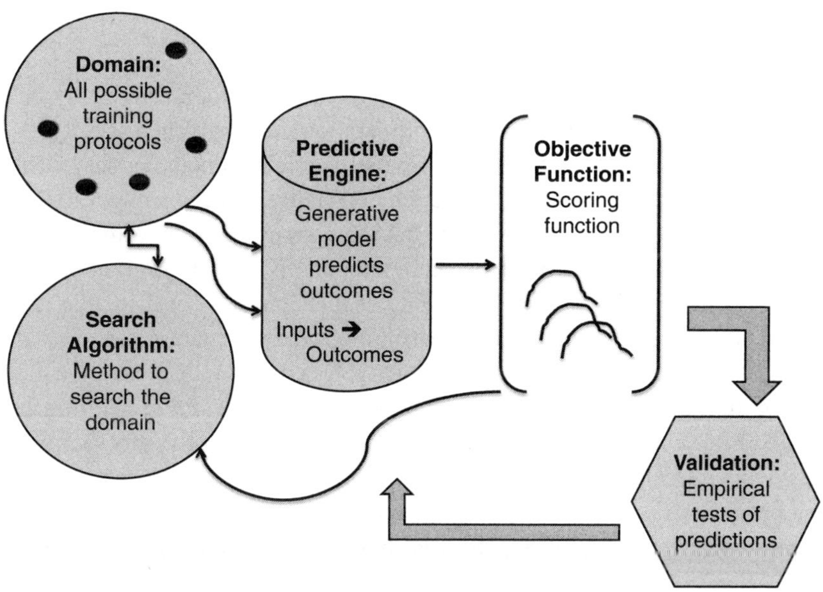

Figure 12.2
A generative model is a key component in optimizing perceptual learning.

of a training protocol and training stimuli and predicts human performance. Such a model can be used to compute the predicted results for many different levels of each factor that might be manipulated and for all their combinations. It uses computations to replace more time-consuming and resource-intensive empirical experimentation, thus saving tests of the human observer for the more promising training schemes. This is especially important in perceptual learning, where multiple training protocols used in the same observer surely interact with one another (so an observer can only be assessed once in relation to a single training protocol).

Developing and selecting a strong generative model is also central to the basic research enterprise. A good model needs to be specific enough to predict outcomes for the various training protocols in a given task domain. For certain visual tasks, we already have strong candidate models; the IRT or the AHRM, or their competitors, for example, might be a good starting point in the domain of pattern judgments.[7–9] In the future, however, it will likely be necessary to create entirely new models, or extend existing ones, to more fully account for performance in different task domains.

A strong generative model is ideal for optimization, but even without one, optimization with alternate methods may still prove fruitful. Although a fully generative model (e.g., one that makes predictions on a trial-by-trial basis for each specific training protocol) would be preferable, an approximate model—derived from empirical relationships, heuristic results, or approximating formulas based on empirical findings—could still save time. Working in this way would help the researcher or designer think through the space of possible manipulations, suggesting promising places to focus empirical efforts or model development.

Even in imperfect situations, optimization can be a useful tool. There may be certain rules of thumb that can still help guide the optimization process. When specifying almost any generative model, it is important to choose reasonable parameter values, such that the model actually generates reasonable predictions. A choice of values might rely on available information derived from fitting the model to previous datasets. A more general and technically sophisticated (and potentially computationally complicated) approach would use Bayesian methods to characterize parameter distributions. In principle, this could involve hierarchical methods that specify distributions of hyperparameters over groups of observers that in turn characterize the parameter values used in the generative model.[10–14] On the level of the optimization space, the set of potential *manipulations* defining the search domain could be selected from the existing literature. Alternatively, this set might incorporate newly devised manipulations inspired by principle or intuition. Several such ideas have been put forward, including the use of easy trials in training (the Eureka effect) or improving transfer by training video games, each with some empirical support.[15, 16] Observing the positive effects of a manipulation in a particular task is only one step toward optimization, however. Since manipulations are often studied in isolation, we tend not to know how general the effects of any given manipulation may be or how multiple design factors might interact. This is where the generative model becomes such a valuable part of the optimization process.

Optimization also requires finding efficient methods to *search through* many possible alternatives. If the number of alternatives is small, then searching should be comparatively easy with any method, including an exhaustive search of all possible options. With more factors, levels, and combinations, however, the search space rapidly becomes too large for an exhaustive approach. In this case, identifying a plausible search method would be an important component of the optimization process, especially for irregular optimization spaces. (A regular space is one in which local variations produce similar predicted results; in such spaces, it may be possible to use traditional search methods, such as gradient descent, designed for differentiable objective functions.)[17] If the search space is huge, it may become necessary to use sampling methods in which information from previous samples helps to focus the search in the more promising regions (e.g., certain genetic algorithms having been developed to handle multiobjective problems).[6] The search process here acquires another level of complexity, although, as mentioned previously, parameters might be estimated from experimental evidence.

In fact, there are at least two levels at which protocols could be optimized. The first, which is the focus of our present discussion, involves computing the results for all possible factor combinations that define a protocol (or at least sampling from among them). In this case, the choices would define a stable protocol throughout training (e.g., if the domain included possible training accuracies such as 55% or 60% to define a staircase target performance level, and the presence or absence of feedback, then one element of the domain might be 60% training accuracy without feedback throughout). A second, higher level of optimization, however, might involve selecting the training for the next trial from all these possibilities (e.g., 75% without feedback on one trial, 95% accuracy with feedback on the next, etc.). This higher-level optimization is guaranteed to find protocols that are at least as good as any simpler one (all of which are special subcases)—but it can also create a combinatoric explosion of possible training sequences with perhaps billions of protocols in the domain (which almost surely would require modern methods such as dynamic programming). Though optimization science can quickly become complicated, it is important to emphasize that even relatively simple forms of optimization could lead to significant advances in our understanding of perceptual learning. The more adaptive trial-by-trial assessment and training methods can be integrated into the field over time as it progresses, thus gradually approaching the complexity of higher-level optimizations.[18, 19]

Once the search space has begun to yield strong candidate training protocols, *validation experiments* should be used to test the predictions of the optimization to make sure that the researcher is on the right track. This is an experiment designed to test predictions of the generative model for several protocols. Often, the most useful validation experiments are those that include several conditions that are predicted to differ in a systematic way. Validation experiments are especially important in determining whether the generative model needs to be improved when possible protocols involve combinations of factors that have previously been tested only separately. Another kind of validation, known as *cross-validation*, examines whether optimization tested on one set of observers or with one set of parameters can be repeated in a new set.[20] Yet another validation level might involve checking the consistency of the experimentally observed data throughout the validation training protocol in relation to the predictions of the generating model.

So far, we have talked about the five steps of optimization as if they were distinct from each other; in some cases, it will be more efficient to merge two or three. Search and empirical evaluation, for example, could occur together; in this case, a generative model and search algorithm would select the next training protocol to examine. Alternatively, parameter values of the model might be updated based on ongoing empirical tests, recomputing the model predictions throughout the course of the optimization exercise. Such an adaptive optimization would be analogous to procedures used in adaptive testing methods, such as for threshold versus contrast functions,[21] contrast-sensitivity functions,[22] or yes-no discrimination.[23] Given that, in most practical applications, several outcomes contribute jointly to robust and useful perceptual learning, a complex criterion set can be used as a compass. In certain cases, however, it may prove more tactically useful to consider each goal separately, analyzing when the goals might be compatible, mutually independent, or conflicting.

With this optimization framework in hand, the research community is poised to explore the vast space of possible training protocols, armed with a tool far more powerful than simple intuition. At the same time, optimizing a given perceptual learning problem might be a computationally and experimentally demanding project. Although barely under way, optimization approaches, in our view, are poised to yield both theoretical and practical benefits. Technical advancements in model generation and innovations in search methods promise to accelerate research.

Although the existing literature has implicitly, and often only qualitatively, focused on the dual objectives of the amount of learning and degree of transfer, both can be examined from distinct perspectives. In what follows, we consider these objectives using first the predictive properties of an expanded Hebbian learning rule and then the potential factors in training and stimulus selection that emerged from the experimental literature. Taken together, these considerations help to sketch a number of possible manipulations that might influence perceptual learning.

12.4 Optimizing the Magnitude of Learning

One of the dominant objectives driving optimization is to maximize learning by using the least amount of training. A number of factors subject to experimental manipulation are relevant here, as suggested by the learning rule: signal strength, task precision, feedback, attention, and rewards (all discussed in prior chapters). The reweighting models, in turn, generate predictions about other aspects of training protocols, such as scheduling, mixtures of task types, or the inclusion of high performance ("easy") stimuli. The factors listed in table 12.3 point to possible variations, with brief descriptions of several factors listed in the following subsections. For the interested reader, further references are also listed.

12.4.1 Analysis of a Learning Rule

A generative model is one that can make relevant predictions for any training protocol. It must simulate exact experimental protocols and generate trial-by-trial predictions. In principle, a single model can make predictions about many different training protocols, although in some cases it may be necessary to develop new representation modules for specific task domains (e.g., for color or motion). For the sake of concreteness, in this subsection we analyze the learning rules of the integrated reweighting theory (IRT) to reveal the possible manipulations that may influence the rate of learning.[7–9, 24]

As we saw in chapter 9, important top-down factors, such as task, attention, and reward, can influence behaviorally observed learning rates. To better understand possible ways that these factors might affect learning, we extended the learning rule of the IRT to incorporate these top-down factors (figure 12.3). The extended augmented Hebbian learning rule is

$$\delta_i = \eta^{A,R,T} a_i^{A,R,T} (o^T - \bar{o}^T)(r^T - \hat{r}^T) \tag{12.1}$$

In the Hebbian equation for weight change, the δ_i for each representation unit i is a product of the model learning rate η, the activity of representation a_i, and the output activation of the decision unit o or $\delta_i = \eta a_i o$, and in the new version, top-down factors can alter each of these parts of the learning equation. Consequently, attention, reward, and task might all influence the learning rate. Such top-down influences are more than hypothetical. As explored in chapter 9, the presence of such factors is undeniable, even though their precise functional influence on learning remains to be fully specified by the empirical literature.

Although not included as an explicit part of the learning rule, the presence of internal noise (and external noise) can also powerfully affect learning. High levels of internal noise inevitably obscure the relevant signals to be learned. Any empirical manipulation that reduces the internal (or external) noise would likely increase the empirical learning rate. This is because better signals would be integrated into the weight changes on each trial (though learning in any given experiment would obviously also depend on the nature of the protocol). Each manipulation of the stimuli, the task, or the scheduling during training could in principle change the observed rate of learning.

In addition to these top-down factors, other potential factors in learning could be considered. Generative models could be developed to incorporate other potentially important mechanisms corresponding with adaptation, consolidation, forgetting, and perhaps other processes of learning. Such extensions will be required in order to make predictions about manipulations such as sleep[25–32] or sensory adaptation[25, 33–35] and their interaction with learning. Revised generative models could also be based on experimentally observed

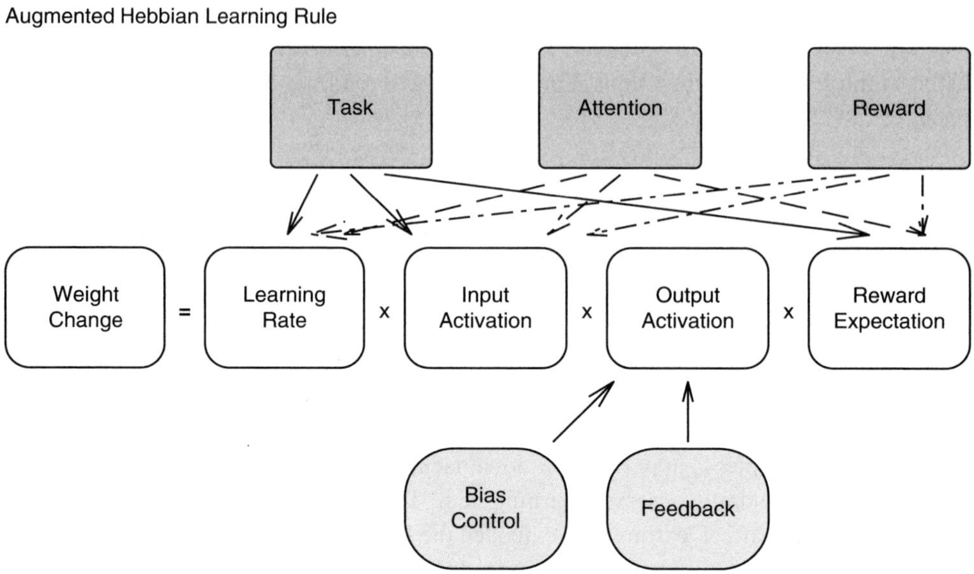

Figure 12.3
The terms of an augmented Hebbian learning rule, and the top-down factors that may affect them (indicated by arrows). Bias and feedback augment Hebbian learning in the IRT by shifting the output activation before learning. See the text for a discussion.

influences as they emerge. Developments of this kind could significantly push the reach of current models.

12.4.2 Manipulations to Increase Learning

In this section, we briefly discuss several obvious manipulations suggested by the literature, listed under the categories of stimulus factors, response factors, task factors, feedback, reward and attention, scheduling, and pharmaceutical or other physiological stimulation. The manipulations listed in table 12.3 correspond with the possible variations in training protocols, while the brief descriptions point to relevant articles for the interested reader.

Stimulus factors Two easily manipulated stimulus factors, the signal *contrast* and the presence of *external noise*, are known to influence learning substantially. Increasing the contrast of the target stimulus should increase the rate of learning, as suggested not only by an analysis of the learning rule but also by previous experimental results. Higher contrasts increase the activity in the input units a_i, which, all else being equal, will increase the response accuracy. If the goal is to optimize learning for low-contrast stimuli, however, what mix of contrasts will be optimal to use during training? Experimental studies that varied training accuracy and feedback[36] or included high-contrast stimuli have

Table 12.3
Examples of potential factors affecting the magnitude of learning

Stimulus
Stimulus contrast
External noise
Judgment precision
Response
Required response accuracy (adaptive staircase levels)
Confidence
Response type (e.g., verbal, button press, joystick, covert biological)
Task
Perceptual judgment (e.g., detection, discrimination, *n*-alternative)
Type of performance measure (i.e., difference threshold, contrast threshold, percentage correct)
Feedback
Modes of trial-by-trial feedback (e.g., consistent, partial, etc.)
Information in feedback (e.g., accuracy, target response)
Block feedback
Exaggerated feedback
Biofeedback (e.g., visualization of EEG, fMRI)
Reward and attention
Exogenous (external) reward
Endogenous (internally generated) reward
Reward magnitude
Reward frequency
Attention manipulation
Scheduling
Trial or session scheduling (e.g., number of trials, number of sessions)
Intermixture of trial judgment difficulty or tasks
Types of adaptive procedure (e.g., long versus short staircases)
Pharmaceutical and stimulation
Pharmaceutical or nutraceutical interventions
Brain modulations (e.g., magnetic stimulation, tDCS)

demonstrated that including some high-contrast stimuli[7, 8] improved performance even in the absence of feedback. The potential consequences of a number of these manipulations have already been predicted using the IRT/AHRM as a generative model and tested empirically. These include simulations of training with a fixed set of contrasts,[37] with multiple short staircases beginning with high-contrast stimuli, and with a single, longer staircase.[38] As part of a more general optimization effort, such predictions should be experimentally validated in the same task and situation to bolster claims regarding the relative effectiveness of the protocols.

External noise is another factor that has been shown to affect learning (and transfer), as well as having an immediate effect on visual judgments. Furthermore, such manipulations occur in natural viewing situations that have visual crowding or camouflage. They also occur in alternative sensor environments (such as night vision or radar) and in medical imaging displays.[39–41] Larger amounts of external noise correspondingly increase internal noisiness, inducing higher internal multiplicative noise. External noise has been found to slow learning even when performance accuracy during training is controlled by adaptive staircases. Such effects are a natural consequence of pushing weights in different directions from trial to trial in response to the external noise. Reweighting models generally predict that training in the presence of external noise should be less efficient for most tasks—although obviously there may be situations in which learning the external noise *is* the task.[42, 43]

Another potentially powerful factor involves the choice of *judgment precision* (which might also be seen as a task training manipulation). Such choices determine the set of stimuli experienced by the observer during training. Even if the goal is to optimize high-precision judgments, an optimization calculation could determine the usefulness of first training in a lower-precision task.[44] This might be especially true if training first in the high-precision task shows no benefits early in training. This means that the stimuli encountered during training may be directly consequential to the rate of learning. By definition, the stimulus set will depend on paradigm selections (such as the selection of threshold difference).[45, 46] It is almost impossible to intuit how all these various factors will trade off and interact, so the availability of a computational model becomes an even greater asset.

Response factors *Response factors* such as the choice of performance measure or the nature of the response will also have consequences for learning. Setting a performance level in an adaptive protocol that maintains, say, 85% accuracy during training will influence both the learning rate and sensitivity to other factors, such as feedback. (It is important to note that while these manipulations are response factors, because they are set based on the observer's responses, they also change the stimulus; e.g., by changing the contrast or the stimulus set.) Generally, training at higher accuracies (e.g., higher stimulus contrasts) has been shown to produce more robust learning, although the value for learning of making some response errors is still not well understood quantitatively.[37] The general principle is that including some high-accuracy trials enhances learning, and this in turn improves performance in lower-accuracy conditions of the same basic task.[15, 37, 47]

Task factors Other *task factors*, such as the nature of the judgment, can influence not only the rate of learning but also what is learned. One such example involves comparisons between detection and discrimination tasks. In detection, performance is thought to benefit from pooling more broadly across stimulus evidence, while in discrimination tasks it is necessary to focus more narrowly on distinguishing evidence in the stimuli. The conclusion that the two kinds of tasks lead to learning different things is one explanation for the findings that training detection had limited transfer to discrimination and vice versa, even when using similar stimuli.[48–51] Another task factor that can impact learning, as discussed in chapter 8, is increasing the number of response categories in a task (e.g.,

n-alternative choice) which leads to lower guessing rates, so each trial will carry more information to drive learning (as well as creating new possibilities for feedback).[52, 53]

Feedback *Feedback* is a powerful factor that has been shown to be important to learning, especially in challenging tasks. Many feedback manipulations have been studied in two-alternative tasks,[24, 54–56] though, as described in chapter 7, the effects of feedback can be complex. Accurate feedback may be critical to learning in some circumstances, while in others it may be relatively unimportant.[36, 54] The pattern of results (at least so far) has been consistent with the predictions of the IRT/AHRM, which permit learning either with or without feedback.[30, 43, 55, 56] The n-alternative forced-choice paradigms also create opportunities for a direct comparison of learning with different levels of supervision (information provided by a "teacher"). In this general framework, so-called response feedback specifies the desired response on both correct and error trials (fully supervised learning), accuracy feedback specifies the desired response only for trials with accurate responses (partial or semisupervised learning), and the absence of feedback provides no information (unsupervised learning). A recently developed n-alternative IRT modeled the consequences of different kinds of supervision, This candidate generative model could provide the broader framework within which to balance the costs of arranging for feedback supervision during learning with the benefits on the rate of learning.

Although in its infancy in the context of visual learning, biometric feedback (e.g., EEG, fMRI, heart rate) is already beginning to be investigated in relation to training. In this field, two forms of feedback have been distinguished: direct and indirect. In the former, feedback about some target aspect of the behavior is provided for the observer (user) to control.[57] In one provocative study, biometric feedback related to V1 pixel activities measured in fMRI was used to train a visual task.[58] In indirect biometric feedback, the observer enhances some biometric indicator that the experimenter believes is related to performance.[57–59] Some researchers, for example, have speculated that gamma-frequency brain waves are related to visual perception, leading to the tentative idea that increasing gamma-frequency brain activity, known to correlate with certain perceptual tasks, might influence the rate of learning.

Reward, attention, and brain stimulation *Reward* and *attention*, along with certain methods of brain stimulation (such as direct current stimulation), are increasingly seen as being relevant to perceptual learning, especially during the early phase of training (see chapter 9 for a review). Though the literature is fragmentary, it has become clear (perhaps unsurprisingly) that exogenous rewards can replace informational feedback in guiding perceptual learning.[60–62] Reward itself, or the differential distribution of rewards, has also been shown to support more rapid learning.[63] The precise effects of the timing, magnitude, and/or distribution of exogenous rewards remain to be investigated, and the elaborated learning rules suggest that they may be significant (see subsection12.4.1 and figure 12.3). A model-based exploration would help to guide these investigations.

The reported effects of pharmaceutical agents such as acetylcholine on learning point to another possible paradigm for training interventions (albeit one in which side effects and other costs need to be carefully weighed).[64, 65] Other modes of physical intervention, such as transcranial magnetic stimulation (TMS), have also begun to be actively investigated.[66–71] Generative models that quantitatively predict the influences of pharmaceutical agents or magnetic stimulation have not yet been developed, although heuristic estimates might be used as first approximations.

Task scheduling One of the most important and obvious manipulations in learning involves the scheduling of training. This is also one of the easiest to manipulate. Scheduling choices are furthermore inevitable, and include such basic factors as the number of sessions, the number of trials per session, and the number of total training days. Beyond this, however, they also include decisions about intermixing different stimuli or tasks, such as mixing easier trials with difficult ones, starting with easier trials, whether to use adaptive staircases and, if so, what kind, and others. One recent investigation sought to identify the smallest number of training trials that still produced learning, and the consequences of distributing the same total number of trials over more sessions.[72–74] In certain contexts, such as texture-discrimination tasks, researchers have suggested that scheduling more trials within a session will lead to more adaptation and less learning—in other words, that less is more.[34] Indeed, achieving approximately the same amount of learning with fewer trials will have obvious practical advantages.

Any scheduling factors to be considered for optimization obviously exist within a large and potentially unspecified set. They will almost surely be motivated by experimental considerations and constrained by practicality. A learner in a high school, for example, might be allotted only three one-hour sessions overall within which to train, whereas the scheduling demands of a learner in a military academy may be more rigorous. The availability of training applications, whether in terms of scheduling, portable display capabilities, or other considerations, will necessarily constrain what is possible. In all these cases, optimization should still be able to contribute significantly to the selection of a promising protocol within the practical constraints.

12.4.3 Summary

Mathematical optimization provides a theoretical framework for efficiently identifying one or more training protocol(s) in order to achieve a set of objectives. The researcher determines the task(s) of interest and selects the set of manipulations within the constraints of practicality—this defines the domain of all protocols over which performance will be computationally optimized. The researcher can then use computer methods to simulate virtual experiments using the generative model, effectively searching through a vast variable space.

The search might be carried out to optimize a given paradigm applied across all trials of training or to find a higher-level optimization wherein different manipulations define the training used for each trial. Regardless of the search method, a number of factors to be manipulated will have to be specified by the researcher. In this section, we examined several such categories, briefly pointing to the ways each might change learning. (Our focus was on the degree of learning, though this is only one possible objective among many, albeit a very common one.) As a rule of thumb, generative models that are theoretically grounded and quantitatively precise will be the most useful in optimization. In cases where such models do not yet exist, however, a model based on some experimentally determined heuristic or on approximate computational rules will still prove tremendously useful.

12.5 Optimizing Robustness, Generalization, and/or Transfer

A second goal of optimization might be to maximize generalization. This could involve generalizing trained improvements to the same task performed in different contexts, to related tasks, or to overall performance in a more complex task.

One intuitive analogy here is tennis. Let's say that our protocol seeks to train a tennis swing by using a machine that spits out balls. The first kind of generalization would occur if training with a machine improves hitting the ball (the same skill) even when the ball is delivered by a human opponent or in different viewing conditions. The second form of generalization can be said to occur if the training also improves the ability to hit a baseball, say, or a squash ball. The third form of generalization occurs when the training serves the broad goal of improving overall tennis performance, of which hitting balls may be only one part. An example might be general physical conditioning, which might affect many aspects of task performance by improving general strength or aerobic capacity.

In visual perceptual learning, the robustness of the trained performance to new and different contexts is an understudied but significant question, discussed more in the applied literature than in the experimental literature. An example in applied psychophysics proposes to use a laboratory training protocol of visual search to improve the search for weapons in x-ray images by airport security.[75, 76] Another example is the use of flight simulators in aviation (of varying degrees of verisimilitude) to improve flying of real planes.[77, 78] The question here involves how much of the real-world task environment must be reinstated in the laboratory in order to promote generalizable learning.[79] Related to this, how much does training in one task extend broadly to other tasks? All these questions are understudied because when transfer is examined, it often occurs only in a single transfer task.

Consider some concrete examples: Does training motion direction in one primary direction improve performance in only close directions, in all directions, or in some graded function between the two? Does training orientation discrimination for patterns of one spatial frequency extend to other spatial frequencies? Does training letter identification using one font generalize to other fonts? Answering these questions would require a battery of transfer tasks.

Another rarely studied question concerns what to train to improve performance in a more complicated composite task. Is training a subcomponent one efficient way to improve overall performance? Or is there a better way to sequence training in two or more subcomponent skills? Or, perhaps, is it better to simply train observers in the more complex task? If, for example, you were training contrast-sensitivity functions in the hopes of improving performance in a wide range of visual tasks in special visual populations (as described in chapter 11), what would be the best protocol to pursue? All these questions largely remain to be answered.

Furthermore, generalization is almost surely intertwined with the amount of learning acquired in the training task. More generalization is almost surely more likely if the amount of learning is substantial to begin with: for example, less proportional generalization in a task with very large training effects may actually lead to better performance than more generalization following modest initial improvements. Evidence for transfer proportional to the original amount of learning has recently been reported in a study comparing different reward conditions.[63] For these reasons, it will almost surely be desirable to jointly optimize the magnitude of learning *and* generalization.

12.5.1 Sketch of a Model for Transfer

In order to optimize for learning *and* generalizability, the objective function would need to include measures of both. The performance measures for the training and transfer task(s) must be identified together with the relative weight (value) of each measure in order to define the objective function. The next goal would then be to identify a generative model that can make predictions about transfer as well as learning. Such a model would need to predict the performance in all the relevant measures for potential training protocols specified in the objective function. Having done this, the predicted performance measures would then be combined to produce a *fitness score* for each protocol in the search space. Then, an appropriate search algorithm would need to be developed.

A challenge in this case will be identifying (or perhaps creating) a strong generating model to successfully predict transfer. Though such a model may exist only on the horizon, a few things can still be said about it in advance. If both learning and generalization are objectives of training, then the model should make predictions about both. At this point, there are only a few such models, so the ability to predict transfer in different conditions is fairly limited. Some predictions about transfer over stimuli emerge naturally from even a relatively simple learning model. For example, the AHRM predicts the extent of transfer between different stimuli in the same task and retinal location. Other forms of transfer, such as location transfer, have recently been developed in similar quantitative form using the IRT.[9, 80] For any generative model based on reweighting, however, no matter the details, the quality of transfer will directly reflect whether weights learned in one task also improve performance in a transfer task. This follows directly from the premises of learning through reweighting.

Another, quite different way of producing a form of generalization might involve changing the *state* of the perceptual system. An intervention that has the effect of reducing the internal noise in the system, for example, could have broad effects, precisely because internal noise limits performance in *any* task and the presence of internal noise slows learning. However, it is not clear at this point what kind of manipulation might reduce internal noise. Some proposals have pointed to training with reduced time to the mask in texture-discrimination tasks as a route to improve temporal response of the visual system, potentially affecting many tasks requiring rapid stimulus analysis.[81]

Such training techniques are (or would be) aimed at generally conditioning the system. Training certain cognitive capacities has also been suggested as a basis of broad generalization. Working-memory capacity, for example, might limit performance in tasks requiring the comparison of multiple stimuli, such as *n*-interval tasks. Many researchers have investigated the training of working memory in the hopes that any improvements could in turn improve an array of functions, including any visual task that relies on memory.[82, 83] Another potentially relevant cognitive capacity that might be trained is the ability to switch from one task to the next. Again, the principle remains the same: training might improve task switching,[84–86] which might in turn improve either performance or learning in other tasks involving switching. Though current model frameworks for perceptual learning do not address these general capacities or how they might be trained, a new generative model might seek to incorporate them.

Future generative models will need to create or extend the model framework to account for the effects of training involving attention, reward, task switching, or working memory, or in multiple stimulus domains (e.g., motion and color). In the meantime, existing empirical evidence as well as several model-based predictions can together suggest several likely manipulations to improve generalization, which we consider next.

12.5.2 Manipulations for Generalization

Almost all existing studies of transfer or generalization examine learning in the trained task and then measure possible effects on immediate performance (or sometimes on subsequent learning) in a closely related transfer task. As indicated previously, stronger versions of these experiments would include several transfer tasks, intermix practice on the training and the transfer tasks, and/or train more basic aspects of vision that would apply broadly to many tasks. If, on the other hand, the most important goal really is to optimize performance on a *specific* transfer task, then direct practice on that task should be used as a benchmark, though it rarely is. A list of factors that could plausibly affect generalization can be found in table 12.4. Since transfer almost surely requires significant learning as a precondition, many of the factors that influence learning are also included here. In what follows, we focus on new manipulations with generalizability in mind.

Stimulus factors One significant issue future research should address is the specificity of perceptual learning to the *performance context*. Should training be carried out in different luminance, lighting, glare, or external-noise contexts? Although potentially critical to

Table 12.4
Potential factors affecting transfer and generalization

Factor	
Stimulus	
	Training and performance context
	Judgment precision and difficulty
	Variability in the training set
Task	
	Compatibility of training and transfer tasks
	Task types (i.e., difference threshold, contrast threshold, percentage correct)
Reward	
Scheduling	
	Intermixture of training different tasks
	Presence of visual adaptation
	Sleep or consolidation
General system factors	
	Train early visual functions
	Training of temporal processing (i.e., time to a mask)
	Training attention deployment
	Training on decision (i.e., to reduce bias)
	Training to task switch or to multitask

predicting how best to train for everyday vision, these basic questions have yet to be systematically investigated. If it were found, for example, that learning in the relatively dark-adapted states typical of the laboratory failed to transfer to bright daylight situations, such a discovery would mandate development of high-illumination training protocols. If so, current measures of light adaptation[87–89] and learning would then need to be integrated into the generative models. Another unanswered question concerns the role of explicit variation in training contexts and ultimate generalization to the intended real-world performance context. Would training in multiple contexts improve generalization, or is training in one or two sufficient? Alternatively, does training in certain special contexts improve generalization? One possible example concerns the special status of training in clear (zero external noise) displays, which often seems to transfer to various external-noise conditions.[47]

Another stimulus factor affecting generalizability is the *judgment precision* of the training and/or transfer task. The literature indicated that there will be less transfer (more specificity) when transferring to high-precision tasks, and furthermore that including easier stimulus variants may support learning of a high-precision task that is so challenging that it cannot be learned on its own.[15, 36, 37] What is not known, however, is whether introducing variations in precision (and therefore variations in stimuli) would improve generalizability. Another largely unexplored question is whether training with variation in accidental features of the stimulus (e.g., varying spatial frequency in orientation judgments or varying orientation in color judgments) improves generalizability. Some of these manipulations fall easily within the current capability of generating models such as the IRT/AHRM, though others will require future model development.

Task factors A number of *task factors* previously identified as manipulations to increase learning may also influence generalization (see table 12.2), though the relationship of these influences is far from clear. A researcher might choose a particular task type for training, without realizing how this selection may impact generalization (see chapter 2).

In particular, using different ways to measure performance could influence generalizability, since the different measures tend to involve different stimulus mixtures.[38] This relates directly back to the discussion in chapter 2 regarding task types: Type I tasks, which measure thresholds along the dimension of discrimination (e.g., orientation-difference thresholds for high-contrast patterns), track increasingly smaller stimulus differences throughout training; Type II tasks, which measure visibility thresholds (e.g., contrast thresholds in orientation discrimination), train stimuli that remain basically unchanged throughout the learning protocol around the visibility threshold; and Type III tasks, which measure performance improvements for identical stimuli, involve no stimulus variability. This task classification should be considered when considering generalization.

Another compelling hypothesis that deserves further evaluation concerns the implications of training with a more variable set of stimuli on generalizability. Finally, yet another alternative hypothesis is that generalizability may in fact reflect the level of judgment required by a task.[90–91] All these hypotheses need to be examined further.

Scheduling The scheduling of training trials, evidently a very open-ended set of manipulations, could also be productively modified to enhance generalization. There are many relevant choices here, including the number and kind of tasks to be trained, the number of secondary task assessments to perform, and how such tasks are to be interleaved in an experiment. It is likely that different choices might trade off against one another, making an optimization exercise all the more useful.

Again, the variable space of possible scheduling factors is vast, complex, and possibly contradictory. *More training*, for example, is required to produce more learning, yet more training can also produce more specificity to the stimuli and context of the original training task.[92] *Intermixed training*, on the other hand, may be successful in some cases, so long as the tasks are sufficiently different, though here, too, intermixture (or roving) can interfere with learning if the tasks are too similar (e.g., varying base contrasts in a contrast-increment task).[93–95] Further scheduling options are also available. In *double training*, transfer to other retinal locations has been shown to improve by training a different task promoter.[96–99] Indeed, strong claims have been made about the role of double training in releasing generalization. Task scheduling might also manipulate *adaptation*[25, 100] or overnight *sleep*.[25, 29]

Taken collectively, the choices regarding length of training sessions, the mixture composition, the presence of overnight sleep, nap, and rest, and so on all yield a seemingly endless range of options for training protocols. It will be impossible, not to say impractical, to evaluate all these empirically, by trial and error. Optimization based on a strong generative model (one extended to include adaptation or consolidation) would replace expensive testing with computation to identify plausible candidates for further empirical investigation.

Conditioning basic functions One approach to broadly improving visual performance used in certain applied settings has been to train the limits on basic functions. In training programs whose overall goal is to improve vision across a wide range of ordinary visual tasks, training will almost surely include several levels of visual representation, different kinds of decisions, and different interactions with motor execution. One idea has been to identify certain basic early processes and then focus training there. One example of this can be found in commercial programs that seek to improve reading by training early visual responses.[101–105] Another involves training detection near the high spatial-frequency cutoff of the contrast-sensitivity function in special populations (e.g., in amblyopes, for whom such training improves measures of both acuity and motion perception).[106, 107] Yet another example involves training short displays in a classic texture-discrimination task, a protocol that was claimed to benefit other tasks with rapid displays.[81]

Training decision or attention[108] might also yield broad improvements either through reductions in internal noise or through improved external-noise filtering. Many popular claims have been put forward in this domain. Training in attention in video games,[109] working memory,[110] and multitasking[111] have all been said to widely improve both performance and learning itself. This is typified in the slogan "learning to learn,"[112, 113] a mantra most associated with reports that individuals first trained in action video games may learn subsequent tasks faster.[16, 109, 113–118]

12.5.3 Summary

Cracking the "generalization problem" is one of the most pressing and exciting challenges facing the field of perceptual learning. Optimizing generalization will necessarily involve maximizing a multiple-objective function that includes performance on multiple tasks as well as learning—as generalization is unlikely to be important if little is learned to begin with. In this section, we considered several factors that might influence the degree of generalization over stimuli and/or tasks. One recurring point was the enormous variable space of possible manipulations and how little we currently know about it. To pursue the study of generalization by trial-and-error experimentation will yield little more than local insights, so methods that are more rigorous will be required.

Future research should model potential factors relevant to learning and generalization, and the effects of their *combinations*. The number of such combinations may be so large that specialized search methods may be required (e.g., dynamic programming). We believe that some existing models could provide the initial foundations for a full generative model to be used in an optimization procedure. In the future, extensions for other stimulus input domains may be developed in addition to extensions that incorporate other potential factors, such as the role of sleep, consolidation, attention, and working memory, for example. One important reason for thoroughly studying a few selected optimization problems that include generalization would be to build a theoretical toolkit or template, and a sense of which kinds of manipulations might be most promising. In this sense, the "generalization problem" is at once a metaphor and model for the research enterprise as well.

12.6 New Generative Models

The science of modeling is changing rapidly, and the implications for the field of perceptual learning may be significant. Generative models for optimizing learning fall on a spectrum from the simple and approximate to the complex and biologically grounded. On the simple and approximate end, a model might consist of empirically grounded but still quantitative approximating functions. On the other end, one might seek to create a model of the whole brain with its many regions and connections.[119] Among those of intermediate complexity, models such as the AHRM[7, 8] and the IRT[9] aim to strike a balance. These models incorporate key biologically inspired computations (as in the representation front end that mimics the early visual cortex), yet they remain relatively simplified while still providing a complete computational account from stimulus input to response output. Several other variants of the IRT are also examples.[120, 121] Adequately parameterized, these are generative models that can predict human performance on a trial-by-trial basis for individual observers.

As the field moves forward, newer and/or more complicated models of perceptual learning will undoubtedly emerge. However, the six fundamental goals of modeling, discussed in chapter 6.1, still remain the same. In the context of optimization, generative models must satisfy the additional requirement of being able to generate predictions for individual trials and individual observers from experimental protocols (i.e., visual stimuli, training procedure, etc).

Recently, the growing interest in multilevel deep neural networks (DNNs), or convolutional deep-learning networks has led some researchers to apply them to model perceptual learning. (Schematic illustrations of several of these network models appear in figure 12.4.).[122, 123] The DNNs include a larger number of hidden layers in a feed forward network. In the visual applications, they begin with pixel representations and use massive training with labeled object images to set the weights in early layers. They may then be trained again to produce perceptual learning in a particular task, often with training sequences decoupled from the actual protocols. The two recent applications of deep learning to visual perceptual learning[122, 123] sought to capitalize or extend previous claims[124, 125] of an isomorphism between the first few layers of a network and the early visual cortical responses. One of the claims made on behalf of deep learning is that its networks mimic functional physiology. It has been suggested that once trained to identify a large set of object images, the responses of the first few layers will bear a resemblance to the responses of neurons in early visual cortical areas.[122, 126] The DNNs can yield impressive applications

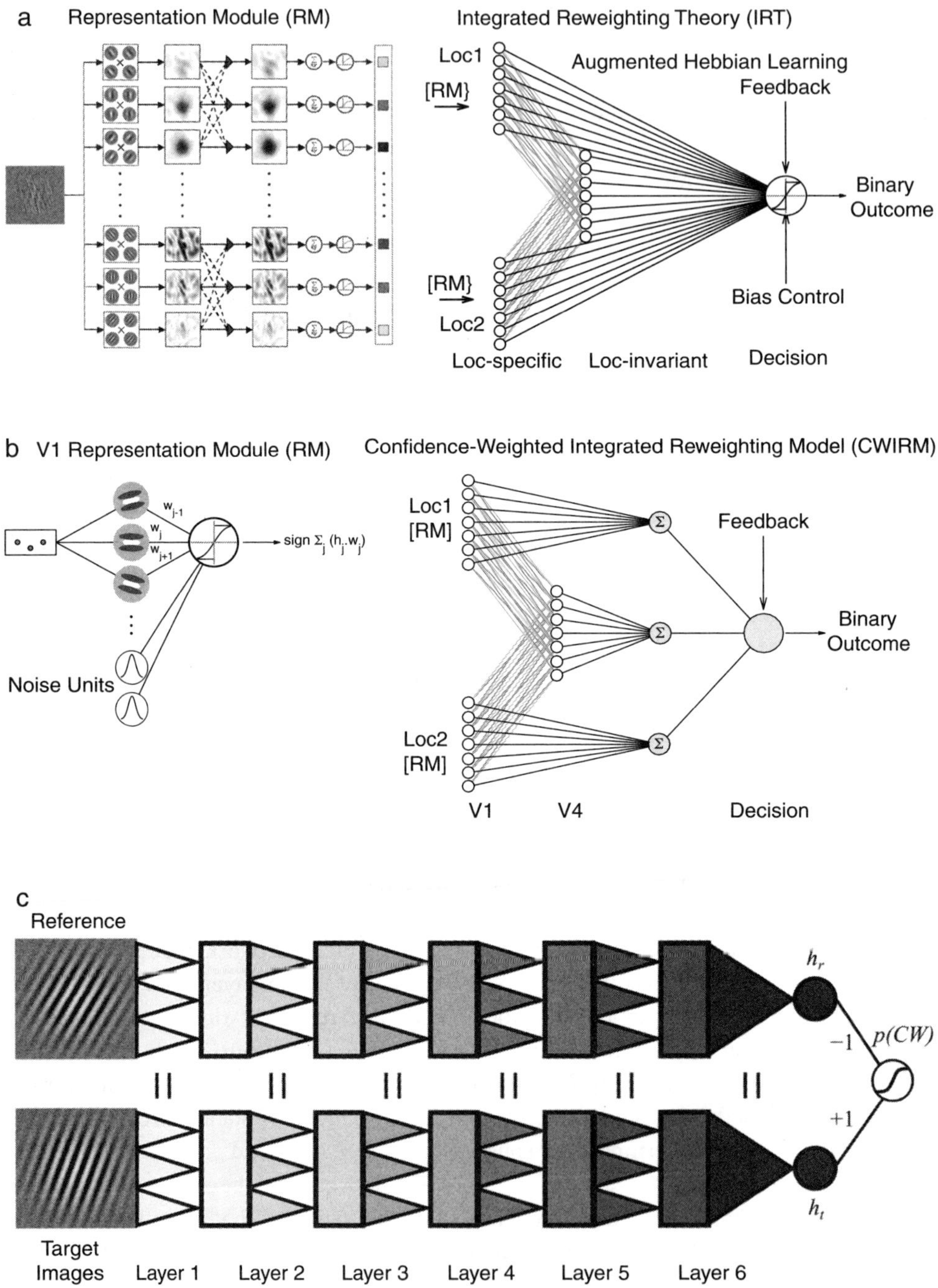
a
Representation Module (RM)
Integrated Reweighting Theory (IRT)
Loc1
[RM}
Augmented Hebbian Learning
Feedback
Binary Outcome
[RM}
Loc2
Bias Control
Loc-specific
Loc-invariant
Decision
b
V1 Representation Module (RM)
Confidence-Weighted Integrated Reweighting Model (CWIRM)
w j-1
w j
w j+1
sign Σj (hj.wj)
Noise Units
Loc1 [RM]
Feedback
Binary Outcome
Loc2 [RM]
V1
V4
Decision
c
Reference
h r
−1
p(CW)
+1
h t
Target Images
Layer 1
Layer 2
Layer 3
Layer 4
Layer 5
Layer 6

to artificial intelligence, but they are designed to solve classification problems, not mimic human behavior. The internal functions often remain opaque.[127–130] There has been limited attempt to account for the empirical data and wide range of phenomena in perceptual learning. So far, the applications to perceptual learning have identified similarities rather than fit specific data.[122–124] The case for the usefulness of these abstract deep-learning approaches within the framework of cognitive science, where understanding the human is paramount, faces a number of challenges.[131] On the other hand, DNNs also face many of the theoretical issues we encounter in perceptual learning, such as specificity of learning and the trade-off between plasticity and stability. The principles we learned from perceptual learning and cognitive science may help improve the architecture of DNN and other multilevel networks. We hope that such cross fertilization could lead to better multilevel generative models of perceptual learning.

Beyond deep learning, there are other computational approaches that seem particularly relevant to perceptual learning, which is often specific to a particular task and only sometimes affects performance in other tasks. As in learning, models of natural language processing (e.g., hierarchical adaptive networks) implement readouts of patterns that are dependent on the context (e.g., following the inherently sequential context of language). Such a property, in which the readout resulting from the same inputs depends critically on the context, is especially apt in the context of perception, where the multiplexed learning of different tasks seems to occur by assigning a specific weight structure to each task. In language, this reflects the fact that the same small sequence of words can serve different functions, depending on their position in the ongoing stream of language. In perception, different learned visual task structures may be relevant in different functional visual contexts. There may be yet other principles and algorithms that will prove useful in next-generation generative models of perceptual learning in the future.

Throughout our research program, we have used reweighting models to generate detailed predictions for many perceptual learning experiments. These reweighting models are relatively simple when compared to DNNs in that they use only a few network layers for learning the task. On the other hand, they are relatively complex in that they include somewhat intricate, albeit fixed, representation modules with multiple layers of biologically inspired computations incorporating nonlinearity and internal noise. Other scientists have extended our models by adding another layer of decision to provide greater flexibility, along with

Figure 12.4 (opposite)
Simplified schematic illustrations of two integrated reweighting models and one deep learning model of perceptual learning. (a) The representation module and the network structure of the integrated reweighting theory (IRT).[9] (b) The representation module and network of the confidence-weighted integrated reweighting model (CW-IRM).[121] (c) The structure of a deep neural network (DNN),[122] where the early layers stand in for a representation module and the task (for two-interval discrimination). Panel (a), with permission of the authors, based on Dosher et al.,[9] figure 1; panel (b, left), redrawn from Sotiropolous, Seitz, and Seriès,[120] figure 1, with permission; panel (b, right), redrawn based on Talluri et al.,[121] figure 1; panel (c) after Wenliang and Seitz,[122] figure 1A (open access).

computations of the reliability of different representations that may drive their relative weighting,[120, 121] yielding more complex and functional predictions from only a limited number of modules. Given the current state of research, there may be a number of advantages to continuing to use simpler reweighting models, at least for now, especially in the context of optimization. These models exactly simulate the sequence of trials in the training protocols; they include internal noise to model the stochastic nature of visual processing and decision; they produce simulated behavior directly comparable to the behavioral data; and they have parameters that can be adjusted to account for differences in performance between tasks or individuals. All these points are fundamental and help to define the distinct research objectives of artificial intelligence and cognitive science.

We suggest that the best—in the sense of the most computationally and practically useful—generative model for an optimization enterprise will be the *simplest* model that is capable of making reliable predictions over the domain of learning protocols of interest. This model may not be the one that best reflects brain structure or function, mimics specific neural responses, or even yields the most powerful simulations (at least not yet, given current limits on computability). The best candidate model will be the one that is *just complicated enough.* To repeat George Box's well-known saying quoted at the beginning of our book, "All models are wrong, but some are useful."

Whatever the particular model, or family of models, selected by a researcher, the field of perceptual learning has much to learn from the broader practice of optimization. Even initial forays armed with this new methodology promise to push protocol design far beyond its current rather rudimentary state. Experience in the development of objective functions could lead to novel real-world applications for training and rehabilitation, while developments in computational search algorithms could make the relevant processing problems much more accurate and efficient. As secular trends toward big data continue, researchers may furthermore be able to mine vast datasets of naturalistic behavior or physiological biometrics to reveal de facto learning protocols in different practical contexts. A formal optimization framework has a critical role to play as part of all these trends, promising to accelerate both theoretical development and practical application.

12.7 The Future of Theories

Throughout this book, we have sought to develop a scientific understanding of visual perceptual learning based on a systematic consideration of the field. Our discussion included surveys of the core phenomena of learning alongside analyses of the experimental data. While being sensitive to the powerful pull of intuition and hunches, we have tried to prioritize deep structural principles and quantitative predictive models. We believe that without these models, our knowledge of learning will not progress as quickly.

Given the rate of progress in neuroscience and computational techniques, the quantitative models of visual learning will almost surely be extended, elaborated, or outright replaced in the coming years. They will be applied to new kinds of evidence, as we learn more about a whole panoply of relevant mechanisms and phenomena—attention, reward, consolidation, adaptation, and sleep—as well as the underlying physiology. As these factors are modeled and technology grows more powerful, researchers in cognitive science will have to work to strike a balance between computational power and an understanding of the limitations of the human system. In this sense, our work, though of course occurring within the broader context of the "revolution" in the methods of artificial intelligence and machine learning seen in society at large, has a distinct ultimate objective of understanding human cognitive processes.

The same tool can be used differently for different purposes. As we move into an age of big data, powered by the ubiquitous availability of enormous information sets and the algorithmic ability to extract patterns from them, machine learning has already transformed areas such as natural language processing and automatic image recognition. Such research is being pursued to create a range of applications, from self-driving vehicles to predictive health systems, under the assumption that such systems will provide performance not only equal to but ultimately above and beyond human capabilities. Indeed, this has already occurred in certain domains for specific well-structured tasks.

As cognitive scientists, our goals are slightly different. We are not trying to build machines that can surpass humans at a given task but rather are trying to understand what exactly defines human performance for that task. And while the opportunistic extraction of patterns in large datasets by powerful learning algorithms may lead to predictions or proposed actions, the reasons for those outcomes may not be at all transparent to human users. The stimulus features driving such decisions are often hard to understand, and in some cases replicate unknown and unwanted biases.[128–130] There are still many circumstances in which the flexibility and inventiveness of human judgments, though intrinsically limited in other ways, are valuable for achieving good outcomes. As cognitive scientists, that characteristic human balance is our object of study.

While this book has been focused on visual learning, we hope that the principles and the themes developed will also inform research into the other sensory modalities. Though our specific models will likely be superseded in 5 to 10 years, we have tried to develop a dialogue between phenomena, modeling, and theory that is based on scientific principles that are more timeless, such as balancing the stability and adaptability of any system, the fact that humans are amazingly robust but never perfect, and the value of dialogue across disciplines. Even as we look forward to coming advances in the understanding of learning systems, these systems will need to model the key phenomenological properties of human performance, which is simultaneously impressive and imperfect.

References

1. Kantorovich L. The mathematical method of production planning and organization. *Management Science* 1939;6(4):363–422.

2. Intriligator MD. *Mathematical optimization and economic theory.* Prentice-Hall;1971.

3. Yang X-S. *Introduction to mathematical optimization: From linear programming to metaheuristics.* Cambridge International Science Publishing;2008.

4. Coello CAC. A comprehensive survey of evolutionary-based multiobjective optimization techniques. *Knowledge and Information Systems* 1999;1(3):269–308.

5. Miettinen K, Ruiz F, Wierzbicki AP. Introduction to multiobjective optimization: Interactive approaches. In Branke J, Deb K, Miettinen K, Słowiński R, eds., *Multiobjective optimization.* Springer;2008:27–57.

6. Tan KC, Lee TH, Khor EF. Evolutionary algorithms for multi-objectiveoptimization: Performance assessments and comparisons. *Artificial Intelligence Review* 2002;17(4):253–290.

7. Petrov AA, Dosher BA, Lu Z-L. The dynamics of perceptual learning: An incremental reweighting model. *Psychological Review* 2005;112(4):715–743.

8. Petrov AA, Dosher BA, Lu Z-L. Perceptual learning without feedback in non-stationary contexts: Data and model. *Vision Research* 2006;46(19):3177–3197.

9. Dosher BA, Jeter P, Liu J, Lu Z-L. An integrated reweighting theory of perceptual learning. *Proceedings of the National Academy of Sciences* 2013;110(33):13678–13683.

10. Gaudet CE. *Review of doing Bayesian data analysis: A tutorial with R, JAGS, and Stan.* Taylor & Francis;2017.

11. Gelman A, Lee D, Guo J. Stan: A probabilistic programming language for Bayesian inference and optimization. *Journal of Educational and Behavioral Statistics* 2015;40(5):530–543.

12. Jackman S. *Bayesian analysis for the social sciences*, Wiley;2009.

13. Kruschke J. *Doing Bayesian data analysis: A tutorial with R, JAGS, and Stan.* Academic Press;2014.

14. Lee MD, Wagenmakers E-J. *Bayesian cognitive modeling: A practical course.* Cambridge University Press;2014.

15. Rubin N, Nakayama K, Shapley R. Abrupt learning and retinal size specificity in illusory-contour perception. *Current Biology* 1997;7(7):461–467.

16. Green CS, Bavelier D. Action video game training for cognitive enhancement. *Current Opinion in Behavioral Sciences* 2015;4:103–108.

17. Liang X-B, Wang J. A recurrent neural network for nonlinear optimization with a continuously differentiable objective function and bound constraints. *IEEE Transactions on Neural Networks* 2000;11(6):1251–1262.

18. Gu H, Kim W, Hou F, Lesmes LA, Pitt MA, Lu Z-L, Myung JI. A hierarchical Bayesian approach to adaptive vision testing: A case study with the contrast sensitivity function. *Journal of Vision* 2016;16(6):15,1–17.

19. Kim W, Pitt MA, Lu Z-L, Steyvers M, Myung JI. A hierarchical adaptive approach to optimal experimental design. *Neural Computation* 2014;26(11):2465–2492.

20. Kohavi R. A study of cross-validation and bootstrap for accuracy estimation and model selection. Paper presented at International Joint Conference on Artificial Intelligence; 1995.

21. Lesmes LA, Jeon S-T, Lu Z-L, Dosher BA. Bayesian adaptive estimation of threshold versus contrast external noise functions: The quick TvC method. *Vision Research* 2006;46(19):3160–3176.

22. Lesmes LA, Lu Z-L, Baek J, Albright TD. Bayesian adaptive estimation of the contrast sensitivity function: The quick CSF method. *Journal of Vision* 2010;10(3):17,1–22.

23. Lesmes LA, Lu Z-L, Baek J, Tran N, Dosher BA, Albright TD. Developing Bayesian adaptive methods for estimating sensitivity thresholds (d′) in yes-no and forced-choice tasks. *Frontiers in Psychology* 2015;6,1070:1–24.

24. Dosher BA, Lu Z-L. Hebbian reweighting on stable representations in perceptual learning. *Learning & Perception* 2009;1(1):37–58.

25. Censor N, Karni A, Sagi D. A link between perceptual learning, adaptation and sleep. *Vision Research* 2006;46(23):4071–4074.

26. Karni A, Sagi D. Where practice makes perfect in texture discrimination: Evidence for primary visual cortex plasticity. *Proceedings of the National Academy of Sciences* 1991;88(11):4966–4970.

27. Karni A, Sagi D. The time course of learning a visual skill. *Nature* 1993;365:250–252.

28. Mednick S, Nakayama K, Stickgold R. Sleep-dependent learning: A nap is as good as a night. *Nature Neuroscience* 2003;6(7):697–698.

29. Mednick SC, Nakayama K, Cantero JL, Atienza M, Levin AA, Pathak N, Stickgold R. The restorative effect of naps on perceptual deterioration. *Nature Neuroscience* 2002;5(7):677–681.

30. Aberg KC, Tartaglia EM, Herzog MH. Perceptual learning with Chevrons requires a minimal number of trials, transfers to untrained directions, but does not require sleep. *Vision Research* 2009;49(16):2087–2094.

31. Stickgold R, Whidbee D, Schirmer B, Patel V, Hobson JA. Visual discrimination task improvement: A multi-step process occurring during sleep. *Journal of Cognitive Neuroscience* 2000;12(2):246–254.

32. Yotsumoto Y, Sasaki Y, Chan P, Vasios CE, Bonmassar G, Ito N, Náñez JE Sr, Shimojo S, Watanabe T. Location-specific cortical activation changes during sleep after training for perceptual learning. *Current Biology* 2009;19(15):1278–1282.

33. Ashley S, Pearson J. When more equals less: Overtraining inhibits perceptual learning owing to lack of wakeful consolidation. *Proceedings of the Royal Society B*; 2012;247:4143–4147.

34. Censor N, Sagi D. Benefits of efficient consolidation: Short training enables long-term resistance to perceptual adaptation induced by intensive testing. *Vision Research* 2008;48(7):970–977.

35. Harris H, Gliksberg M, Sagi D. Generalized perceptual learning in the absence of sensory adaptation. *Current Biology* 2012;22(19):1813–1817.

36. Liu J, Lu Z-L, Dosher BA. Augmented Hebbian reweighting: Interactions between feedback and training accuracy in perceptual learning. *Journal of Vision* 2010;10(10):29,1–14.

37. Liu J, Lu Z-L, Dosher BA. Mixed training at high and low accuracy levels leads to perceptual learning without feedback. *Vision Research* 2012;61:15–24.

38. Hung S-C, Seitz AR. Prolonged training at threshold promotes robust retinotopic specificity in perceptual learning. *Journal of Neuroscience* 2014;34(25):8423–8431.

39. Burgess A, Shaw R, Lubin J. Noise in imaging systems and human vision. *Journal of the Optical Society of America A* 1999;16(3):618.

40. Parkes L, Lund J, Angelucci A, Solomon JA, Morgan M. Compulsory averaging of crowded orientation signals in human vision. *Nature Neuroscience* 2001;4(7):739–744.

41. Snowden PT, Davies IR, Roling P. Perceptual learning of the detection of features in X-ray images: A functional role for improvements in adults' visual sensitivity? *Journal of Experimental Psychology: Human Perception and Performance* 2000;26(1):379–390.

42. Dosher BA, Lu Z-L. Perceptual learning in clear displays optimizes perceptual expertise: Learning the limiting process. *Proceedings of the National Academy of Sciences* 2005;102(14):5286–5290.

43. Liu J, Lu Z, Dosher B. Augmented Hebbian learning accounts for the complex pattern of effects of feedback in perceptual learning. *Journal of Vision* 2010;10(7):1115.

44. Jeter PE, Dosher BA, Petrov A, Lu Z-L. Task precision at transfer determines specificity of perceptual learning. *Journal of Vision* 2009;9(3):1,1–13.

45. Law C-T, Gold JI. Neural correlates of perceptual learning in a sensory-motor, but not a sensory, cortical area. *Nature Neuroscience* 2008;11(4):505–513.

46. Schoups AA, Vogels R, Orban GA. Human perceptual learning in identifying the oblique orientation: Retinotopy, orientation specificity and monocularity. *Journal of Physiology* 1995;483(3):797–810.

47. Lu Z-L, Liu J, Dosher BA. Modeling mechanisms of perceptual learning with augmented Hebbian reweighting. *Vision Research* 2010;50(4):375–390.

48. Sowden PT, Rose D, Davies IR. Perceptual learning of luminance contrast detection: Specific for spatial frequency and retinal location but not orientation. *Vision Research* 2002;42(10):1249–1258.

49. Lu Z-L, Hua T, Huang C-B, Zhou Y, Dosher BA. Visual perceptual learning. *Neurobiology of Learning and Memory* 2011;95(2):145–151.

50. Weiss Y, Edelman S, Fahle M. Models of perceptual learning in Vernier hyperacuity. *Neural Computation* 1993;5(5):695–718.

51. Fahle M. Specificity of learning curvature, orientation, and Vernier discriminations. *Vision Research* 1997;37(14):1885–1895.

52. Gold J, Bennett P, Sekuler A. Signal but not noise changes with perceptual learning. *Nature* 1999;402(6758):176–178.

53. Gauthier I, Williams P, Tarr MJ, Tanaka J. Training "greeble" experts: A framework for studying expert object recognition processes. *Vision Research* 1998;38(15):2401–2428.

54. Herzog MH, Fahle M. The role of feedback in learning a Vernier discrimination task. *Vision Research* 1997;37(15):2133–2141.

55. Shibata K, Yamagishi N, Ishii S, Kawato M. Boosting perceptual learning by fake feedback. *Vision Research* 2009;49(21):2574–2585.

56. Liu J, Dosher B, Lu Z-L. Modeling trial by trial and block feedback in perceptual learning. *Vision Research* 2014;99:46–56.

57. Contestabile M, Recupero S, Palladino D, De Stefanis M, Abdoirahimzadeh S, Suppressa F, Gabrieli CB. A new method of biofeedback in the management of low vision. *Eye* 2002;16(4):472–480.

58. Shibata K, Watanabe T, Sasaki Y, Kawato M. Perceptual learning incepted by decoded fMRI neurofeedback without stimulus presentation. *Science* 2011;334(6061):1413–1415.

59. Scharnowski F, Hutton C, Josephs O, Weiskopf N, Rees G. Improving visual perception through neurofeedback. *Journal of Neuroscience* 2012;32(49):17830–17841.

60. Başar-Eroglu C, Strüber D, Schürmann M, Stadler M, Başar E. Gamma-band responses in the brain: A short review of psychophysiological correlates and functional significance. *International Journal of Psychophysiology* 1996;24(1–2):101–112.

61. Seitz AR, Kim D, Watanabe T. Rewards evoke learning of unconsciously processed visual stimuli in adult humans. *Neuron* 2009;61(5):700–707.

62. Frankó E, Seitz AR, Vogels R. Dissociable neural effects of long-term stimulus–reward pairing in macaque visual cortex. *Journal of Cognitive Neuroscience* 2010;22(7):1425–1439.

63. Zhang P, Hou F, Yan F-F, Xi X, Lin B-R, Zhao J, Yang J, Hhen G, Zhang M-Y, He Q, Dosher BA, Lu Z-L, Huang C-B. High reward enhances perceptual learning. *Journal of Vision* 2018;18(8):11,1–21.

64. Rokem A, Silver MA. Cholinergic enhancement augments magnitude and specificity of visual perceptual learning in healthy humans. *Current Biology* 2010;20(19):1723–1728.

65. Rokem A, Silver MA. The benefits of cholinergic enhancement during perceptual learning are long-lasting. 2013;7,66:1–7.

66. Filmer HL, Mattingley JB, Dux PE. Improved multitasking following prefrontal tDCS. *Cortex* 2013; 49(10):2845–2852.

67. Fertonani A, Pirulli C, Miniussi C. Random noise stimulation improves neuroplasticity in perceptual learning. *Journal of Neuroscience* 2011;31(43):15416–15423.

68. Peters MA, Thompson B, Merabet LB, Wu AD, Shams L. Anodal tDCS to V1 blocks visual perceptual learning consolidation. *Neuropsychologia* 2013;51(7):1234–1239.

69. Clark VP, Coffman BA, Mayer AR, Weisend MP, Lane TDR, Calhoun VD, Raybourn EM, Garcia CM, Wassermann EM. TDCS guided using fMRI significantly accelerates learning to identify concealed objects. *Neuroimage* 2012;59(1):117–128.

70. Pirulli C, Fertonani A, Miniussi C. The role of timing in the induction of neuromodulation in perceptual learning by transcranial electric stimulation. *Brain Stimulation* 2013;6(4):683–689.

71. Antal A, Nitsche MA, Kincses TZ, Kruse W, Hoffmann KP, Paulus W. Facilitation of visuo-motor learning by transcranial direct current stimulation of the motor and extrastriate visual areas in humans. *European Journal of Neuroscience* 2004;19(10):2888–2892.

72. Hussain Z, Sekuler AB, Bennett PJ. How much practice is needed to produce perceptual learning? *Vision Research* 2009;49(21):2624–2634.

73. Hussain Z, Bennett PJ, Sekuler AB. Versatile perceptual learning of textures after variable exposures. *Vision Research* 2012;61:89–94.

74. Molloy K, Moore DR, Sohoglu E, Amitay S. Less is more: Latent learning is maximized by shorter training sessions in auditory perceptual learning. *PloS One* 2012;7(5):e36929.

75. Van Wert MJ, Horowitz TS, Wolfe JM. Even in correctable search, some types of rare targets are frequently missed. *Attention, Perception, & Psychophysics* 2009;71(3):541–553.

76. Wolfe JM, Brunelli DN, Rubinstein J, Horowitz TS. Prevalence effects in newly trained airport checkpoint screeners: Trained observers miss rare targets, too. *Journal of Vision* 2013;13(3):33,1–9.

77. Bell HH, Waag WL. Evaluating the effectiveness of flight simulators for training combat skills: A review. *International Journal of Aviation Psychology* 1998;8(3):223–242.

78. Hays RT, Jacobs JW, Prince C, Salas E. Flight simulator training effectiveness: A meta-analysis. *Military Psychology* 1992;4(2):63–74.

79. Hays RT, Singer MJ. *Simulation fidelity in training system design: Bridging the gap between reality and training.* Springer;2012.

80. Dosher B, Lu Z-L. Perceptual learning and models. *Annual Review of Vision Science* 2017;3:343–363.

81. Wang R, Cong L-J, Yu C. The classical TDT perceptual learning is mostly temporal learning. *Journal of Vision* 2013;13(5):9,1–9.

82. Au J, Sheehan E, Tsai N, Duncan GJ, Buschkuehl M, Jaeggi SM. Improving fluid intelligence with training on working memory: A meta-analysis. *Psychonomic Bulletin & Review* 2015;22(2):366–377.

83. Jaeggi SM, Buschkuehl M, Jonides J, Perrig WJ. Improving fluid intelligence with training on working memory. *Proceedings of the National Academy of Sciences* 2008;105(19):6829–6833.

84. Zinke K, Einert M, Pfennig L, Kliegel M. Plasticity of executive control through task switching training in adolescents. *Frontiers in Human Neuroscience* 2012;6:41, 1–15.

85. Strobach T, Frensch PA, Schubert T. Video game practice optimizes executive control skills in dual-task and task switching situations. *Acta Psychologica* 2012;140(1):13–24.

86. Strobach T, Frensch PA, Soutschek A, Schubert T. Investigation on the improvement and transfer of dual-task coordination skills. *Psychological Research* 2012;76(6):794–811.

87. Dao DY, Lu Z-L, Dosher BA. Adaptation to sine-wave gratings selectively reduces the contrast gain of the adapted stimuli. *Journal of Vision* 2006;6(7):6,739–759.

88. Larsson J, Landy MS, Heeger DJ. Orientation-selective adaptation to first- and second-order patterns in human visual cortex. *Journal of Neurophysiology* 2006;95(2):862–881.

89. Montaser-Kouhsari L, Landy MS, Heeger DJ, Larsson J. Orientation-selective adaptation to illusory contours in human visual cortex. *Journal of Neuroscience* 2007;27(9):2186–2195.

90. Green CS, Kattner F, Siegel MH, Kersten D, Schrater PR. Differences in perceptual learning transfer as a function of training task. *Journal of Vision* 2015;15(10):5,1–14.

91. Stewart N, Chater N. The effect of category variability in perceptual categorization. *Journal of Experimental Psychology: Learning, Memory, and Cognition* 2002;28(5):893–907.

92. Jeter PE, Dosher BA, Liu S-H, Lu Z-L. Specificity of perceptual learning increases with increased training. *Vision Research* 2010;50(19):1928–1940.

93. Herzog MH, Aberg KC, Frémaux N, Gerstner W, Sprekeler H. Perceptual learning, roving and the unsupervised bias. *Vision Research* 2012;61:95–99.

94. Zhang J-Y, Kuai S-G, Xiao L-Q, Klein SA, Levi DM, Yu C. Stimulus coding rules for perceptual learning. *PLoS Biology* 2008;6(8):e197.

95. Tartaglia EM, Aberg KC, Herzog MH. Perceptual learning and roving: Stimulus types and overlapping neural populations. *Vision Research* 2009;49(11):1420–1427.

96. Xiao L-Q, Zhang J-Y, Wang R, Klein SA, Levi DM, Yu C. Complete transfer of perceptual learning across retinal locations enabled by double training. *Current Biology* 2008;18(24):1922–1926.

97. Wang R, Zhang J-Y, Klein SA, Levi DM, Yu C. Task relevancy and demand modulate double-training enabled transfer of perceptual learning. *Vision Research* 2012;61:33–38.

98. Wang R, Zhang J-Y, Klein SA, Levi DM, Yu C. Vernier perceptual learning transfers to completely untrained retinal locations after double training: A "piggybacking" effect. *Journal of Vision* 2014;14(13):12,1–10.

99. Dosher BA, Lu Z-L. Mechanisms of perceptual learning. *Vision Research* 1999;39(19):3197–3221.

100. Sagi D. Perceptual learning in vision research. *Vision Research* 2011;51(13):1552–1566.

101. Mahncke HW, Connor BB, Appelman J, Ahhsanuddin ON, Hardy JL, Wood RA, Joyce NM, Boniske T, Atkins SM, Merzenich MM. Memory enhancement in healthy older adults using a brain plasticity-based

training program: A randomized, controlled study. *Proceedings of the National Academy of Sciences* 2006;103(33):12523–12528.

102. Merzenich MM, Jenkins WM, Johnston P, Schreiner C. Temporal processing deficits of language-learning impaired children ameliorated by training. *Science* 1996;271(5245):77–81.

103. Tallal P. Language and reading: Some perceptual prerequisites. *Bulletin of the Orton Society* 1980;30(1):170–178.

104. Tallal P, Merzenich MM, Miller S, Jenkins W. Language learning impairments: Integrating basic science, technology, and remediation. *Experimental Brain Research* 1998;123(1–2):210–219.

105. Tallal P, Miller S, Bedi G, Byrne G, Wang X, Nagarajan S, Schreiner C, Jenkins WM, Merzenich MM. Language comprehension in language-learning impaired children improved with acoustically modified speech. *Science* 1996;271(5245):81–84.

106. Huang C-B, Lu Z-L, Zhou Y. Mechanisms underlying perceptual learning of contrast detection in adults with anisometropic amblyopia. *Journal of Vision* 2009;9(11):24.1–14.

107. Huang C-B, Zhou Y, Lu Z-L. Broad bandwidth of perceptual learning in the visual system of adults with anisometropic amblyopia. *Proceedings of the National Academy of Sciences* 2008;105(10):4068–4073.

108. Dosher BA, Han S, Lu Z-L. Perceptual learning and attention: Reduction of object attention limitations with practice. *Vision Research* 2010;50(4):402–415.

109. Green CS, Bavelier D. Effect of action video games on the spatial distribution of visuospatial attention. *Journal of Experimental Psychology: Human Perception and Performance* 2006;32(6):1465–1478.

110. Banai K, Ahissar M. Perceptual learning as a tool for boosting working memory among individuals with reading and learning disability. *Learning & Perception* 2009;1(1):115–134.

111. Dux PE, Tombu MN, Harrison S, Rogers BP, Tong F, Marois R. Training improves multitasking performance by increasing the speed of information processing in human prefrontal cortex. *Neuron* 2009;63(1):127–138.

112. Bavelier D, Green CS, Pouget A, Schrater P. Brain plasticity through the life span: Learning to learn and action video games. *Annual Review of Neuroscience* 2012;35:391–416.

113. Green CS, Bavelier D. Exercising your brain: A review of human brain plasticity and training-induced learning. *Psychology and Aging* 2008;23(4):692–701.

114. Green CS, Bavelier D. Action video game modifies visual selective attention. *Nature* 2003;423(6939):534–537.

115. Green CS, Bavelier D. Action-video-game experience alters the spatial resolution of vision. *Psychological Science* 2007;18(1):88–94.

116. Green CS, Bavelier D. Enumeration versus multiple object tracking: The case of action video game players. *Cognition* 2006;101(1):217–245.

117. Green CS, Pouget A, Bavelier D. Improved probabilistic inference as a general learning mechanism with action video games. *Current Biology* 2010;20(17):1573–1579.

118. Green CS, Sugarman MA, Medford K, Klobusicky E, Bavelier D. The effect of action video game experience on task-switching. *Computers in Human Behavior* 2012;28(3):984–994.

119. Maniglia M, Seitz AR. Towards a whole brain model of perceptual learning. *Current Opinion in Behavioral Sciences* 2018;20:47–55.

120. Sotiropoulos G, Seitz AR, Seriès P. Performance-monitoring integrated reweighting model of perceptual learning. *Vision Research* 2018;152:17–39.

121. Talluri BC, Hung S-C, Seitz AR, Seriès P. Confidence-based integrated reweighting model of task-difficulty explains location-based specificity in perceptual learning. *Journal of Vision* 2015;15(10):17,1–12.

122. Wenliang LK, Seitz AR. Deep neural networks for modeling visual perceptual learning. *Journal of Neuroscience* 2018;38(27):6028–6044.

123. Cohen G, Weinshall D. Hidden layers in perceptual learning. *Proceedings of the IEEE Conference on Computer Vision & Pattern Recognition (CVPR)* 2017:4554–4562.

124. Cadieu CF, Hong H, Yamins D, Pinto N, Majaj NJ, DiCarlo JJ. The neural representation benchmark and its evaluation on brain and machine. *arXiv preprint 13013530.* 2013.

125. Cadieu CF, Hong H, Yamins DL, Pinto N, Ardila D, Solomon EA, Majaj NJ, DiCarlo JJ. Deep neural networks rival the representation of primate IT cortex for core visual object recognition. *PLoS Computational Biology* 2014;10(12):e1003963.

126. Bakhtiari S. Can deep learning model perceptual learning? *Journal of Neuroscience* 2019;39(2):194–196.

127. Došilović FK, Brčić M, Hlupić N. Explainable artificial intelligence: A survey. Paper presented at 41st International Convention on Information and Communication Technology, Electronics and Microelectronics (MIPRO);2018.

128. Gunning D. Explainable Artificial Intelligence (XAI). *Defense Advanced Research Projects Agency (DARPA), Web* 2017;2.

129. Adadi A, Berrada M. Peeking inside the black-box: A survey on Explainable Artificial Intelligence (XAI). *IEEE Access* 2018;6:52138–52160.

130. Samek W, Wiegand T, Müller K-R. Towards explainable artificial intelligence. In Samek W, Montavon G, Vedaldi A, Hansen L, Muller KR, eds., *Understanding, visualizing and interpreting deep learning.* Springer;2019:5–22..

131. Cichy RM, Kaiser D. Deep neural networks as scientific models. *Trends in Cognitive Sciences* 2019;23(4):305–317.

Index